Second Supplements to the 2nd Edition of

RODD'S CHEMISTRY OF CARBON COMPOUNDS

ELSEVIER SCIENCE B.V.
Sara Burgerhartstraat 25
P.O. Box 211, 1000 AE Amsterdam, The Netherlands

Erratum

Sainsbury (Ed.), Second Supplements to the 2nd Edition of Rodd's Chemistry of Carbon Compounds

Volume III BCD (partial), which appeared in 1995, should have been numbered Volume III BC. Chapter 12 is the last chapter of Part C, but it was erroneously stated on the cover and title page to be a chapter in Volume III, Part D.

ISBN: 0-444-82552-5

This book is printed on acid-free paper.

Printed in The Netherlands.

Second Supplements to the 2nd Edition of

RODD'S CHEMISTRY OF CARBON COMPOUNDS

VOLUME I

ALIPHATIC COMPOUNDS

★

VOLUME II

ALICYCLIC COMPOUNDS

★

VOLUME III

AROMATIC COMPOUNDS

★

VOLUME IV

HETEROCYCLIC COMPOUNDS

★

VOLUME V

MISCELLANEOUS

GENERAL INDEX

★

Second Supplements to the 2nd Edition of

RODD'S CHEMISTRY OF CARBON COMPOUNDS

A modern comprehensive treatise

Edited by
MALCOLM SAINSBURY
School of Chemistry, The University of Bath,
Claverton Down, Bath BA2 7AY, England

Second Supplement to

VOLUME III AROMATIC COMPOUNDS

Part D: Monobenzenoid and Phenolic Aralkyl Compounds, their Derivatives and Oxidation Products; Depsides, Tannins, Lignans, Lignin and Humic Acid

Part E: Monobenzenoid Compounds with Unsaturated or Polyhydroxylated Side-chains, their Derivatives and Oxidation Products

Part F: Polybenzenoid Compounds; Hydrocarbon Ring Assemblies; Polyphenyl-substituted Aliphatic Compounds (Partial: Chapters 20-22 in this Volume)

1996
ELSEVIER
Amsterdam – Lausanne – New York – Oxford – Shannon – Tokyo

Contributors to this Volume

ALAN ARMSTRONG
Department of Chemistry, University of Nottingham, University Park, Nottingham NG7 2RD, England

JOHN CHARALAMBOUS
Department of Applied Chemistry and Life Sciences, University of North London, Holloway, London N7 8DB, England

HAK-FUN CHOW
Department of Chemistry, The Chinese University of Hong Kong, Shatin, New Territories, Hong Kong

E.G. EVAGOROU
The School of Pharmacy, University of London, Centre for Polymer Therapeutics, 29–39 Brunswick Square, London WC1N 1AX, England

S.G.R. GUINOT
Department of Chemistry, University of Central Lancashire, Preston PR1 2HE, Lancashire, England

K. HALE
Department of Chemistry, University College London, 20 Gordon Street, London WC1H 0AJ, England

JOHN D. HEPWORTH
Department of Chemistry, University of Central Lancashire, Preston PR1 2HE, Lancashire, England

J.A. HUGH MACBRIDE
Department of Chemistry, Science Laboratories, University of Durham, South Road, Durham DH1 3LE, England

WILLIAM S. MURPHY
Department of Chemistry, University College, Cork, College Road, Cork, Ireland

B. SARSAM
Department of Chemistry, University College, Cork,
College Road, Cork, Ireland

NEVILLE H.P. SMITH
Department of Chemistry, University of Manchester Institute of
Science and Technology, Sackville Street, Manchester M60 1QD, England

Y. SOUMA
Osaka National Research Institute, 1-8-31 Midorigaoka, Ikeda,
Osaka 563, Japan

R.A. STOCKMAN
School of Chemistry, University of Bath,
Claverton Down, Bath BA2 7AY, England

MUTSUO TANAKA
Osaka National Research Institute, 1-8-31 Midorigaoka, Ikeda,
Osaka 563, Japan

Preface

The structure and content of the chapters in this volume follow those in the 2nd Edition of Rodd's Chemistry of Carbon Compounds, although this was completed 25 years ago! Since then, of course, organic chemistry has moved apace and to account for this progress the relative space allocated to each chapter varies widely.

Whenever possible it is my policy to allow individual contributors to determine the depth of coverage necessary to reflect the importance of modern trends within each chapter. Only in the case of chapter 23 was supplementation unnecessary. The main advances in the subject matter of this chapter are fully disclosed elsewhere in Volume 3.

I would like to thank all the authors of this Volume for their patience and perseverance during the time that it has taken to bring the work to completion and to congratulate them on a job well done.

Malcolm Sainsbury

Bath
June 1996

Contents
Volume IIID/E/F (partial)

Aromatic Compounds; Monobenzenoid Hydrocarbon Derivatives with Functional Groups in an Acyclic Side Chain; Aralkylamines and their Derivatives, Aralkanols and their Oxidation Products, Monobenzenoid Carbaldehydes and Monoketones; Phenolic Aralkanols, their Oxidation Products and Derivatives including Carboxylic Acids; Depsides, Hydrolysable Tannins, Lignin and Humic Acid; Monobenzenoid Hydrocarbon Derivatives with Functional Groups in Separate Side Chains; Dihydric and Polyhydric Aralkanols and their Oxidation Products; Monobenzenoid Hydrocarbons with Unsaturated Side Chains and their Derivatives; Polybenzenoid Hydrocarbons and their Derivatives; Hydrocarbon Ring Assemblies, Polyphenyl-substituted Aliphatic Hydrocarbons and their Derivatives (Partial).

Chapter 15. Phenolic Aralkylamines, Monohydric Alcohols, Monocarbaldehydes, Monoketones and Monocarboxylic Acids
by W.S. MURPHY and B. SARSAM

Chapter 16. Depsides, Hydrolysable Tannins, Lignans, Lignin and Humic Acids
by W.S. MURPHY and B. SARSAM

Chapter 17. Dihydric and Polyhydric Aralkanols and their Oxidation Products with Functional Groups in Separate Side Chains

by J. CHARALAMBOUS and E.G. EVAGOROU

Chapter 18. Dihydric and Polyhydric Aralkanols and their Oxidation Products
by H.-F. CHOW

Chapter 19. Benzene Derivatives with One or More Unsaturated Side Chains (January 1980–January 1995)
by K.J. HALE

Chapter 20. Aryl Benzenes and their Derivatives

by J.A.H. MACBRIDE

Chapter 21. Di-, Tri- and Tetra-phenylmethanes, their Derivatives and Oxidation Products
by J.D. HEPWORTH and S.G.R. GUINOT

Chapter 22. Diphenyl-alkanes, -alkenes, and -alkynes, their Derivatives and Oxidation Products
by A. ARMSTRONG and R.A. STOCKMAN

List of Common Abbreviations and Symbols Used

A	acid
Å	Ångström units
Ac	acetyl
a	axial
as, asymm.	asymmetrical
at.	atmosphere
B	base
Bu	butyl
b.p.	boiling point
c, C	concentration
CD	circular dichroism
conc.	concentrated
crit.	critical
D	Debye unit, 1×10^{-18} e.s.u.
D	dissociation energy
D	dextro-rotatory; dextro configuration
d	density
dec., decomp	with decomposition
deriv.	derivative
E	energy; extinction; electromeric effect
E	entgegen (opposite) configuration
*E*1, *E*2	uni- and bi-molecular elimination mechanisms
E1cB	unimolecular elimination in conjugate base
ESR	electron spin resonance
Et	ethyl
e	nuclear charge; equatorial
f.p.	freezing point
G	free energy
GLC	gas liquid chromatography
g	spectroscopic splitting factor, 2.0023
H	applied magnetic field; heat content
h	Planck's constant
Hz	hertz
I	spin quantum number; intensity; inductive effect
IR	infrared
J	coupling constant in NMR spectra
J	Joule
K	dissociation constant
k	Boltzmann constant; velocity constant
kcal	kilocalories
L	laevorotatory, laevo configuration
M	molecular weight; molar; mesomeric effect
Me	methyl

m	mass; mole; molecule; *meta*-
m.p.	melting point
Ms	mesyl (methanesulphonyl)
[M]	molecular rotation
N	Avogadro number; normal
NMR	nuclear magnetic resonance
NOE	Nuclear Overhauser Effect
n	normal; refractive index; principal quantum number
o	*ortho*-
ORD	optical rotatory dispersion
P	polarisation; probability; orbital state
Pr	propyl
Ph	phenyl
p	*para*-; orbital
PMR	proton magnetic resonance
R	clockwise configuration
S	counterclockwise configuration; entropy; net spin of incompleted electronic shells; orbital state
S_N1, S_N2	uni- and bi-molecular nucleophilic substitution mechanism
S_Ni	internal nucleophilic substitution mechanism
s	symmetrical; orbital
sec	secondary
soln.	solution
symm.	symmetrical
T	absolute temperature
Tosyl	*p*-toluenesulphonyl
Trityl	triphenylmethyl
t	time
temp.	temperature (in degrees centrigrade)
tert	tertiary
UV	ultraviolet
v	velocity
Z	zusammen (together) configuration
α	optical rotation (in water unless otherwise stated)
$[\alpha]$	specific optical rotation
ϵ	dielectric constant; extinction coefficient
μ	dipole moment; magnetic moment
μ_B	Bohr magneton
μg	microgram
μm	micrometer
λ	wavelength
ν	frequency; wave number
χ, χ_d, χ_μ	magnetic; diamagnetic and paramagnetic susceptibilities
~	about
(+)	dextrorotatory
(–)	laevorotatory
–	negative charge
+	positive charge

Second Supplements to the 2nd Edition of Rodd's Chemistry of Carbon Compounds, Vol.III D/E/F (Partial), edited by M. Sainsbury

Chapter 13

ARALKANOLS AND ARALKYLAMINES

N.H.P. SMITH

1. Aralkanols

(a) Preparation

Benzyl alcohol has been produced by the oxidation of toluene using *Pseudomonas oleovorans* mono-oxygenase (H. Fu *et al., J. Amer. Chem. Soc.,* **113,** 5878).

New variants of well-established reactions have been developed for the preparation of benzyl and substituted benzyl alcohols.

(i) Replacement of halogen by -OH

Benzylic (and other primary) halides are readily hydrolysed to the corresponding alcohols by heating them in benzene or THF with an ion exchange resin in the carbonate form. The yields of alcohol are consistently in the range 85-95% (G. Cardillo *et al., Synthesis* 1981, 793)

The indirect replacement of a benzylic halogen atom has been achieved via the benzeneselenenate (1) (T. Oida *et al., Bull., Chem. Soc.* Japan, 1991, **64,** 702):

$$\underset{(1)}{PhSeOCH_2CH_2SiMe_3} \xrightarrow{a.} \underset{100\%}{PhCH_2OH}$$

a. $PhCH_2Br/Bu_4NF/THF$, room temperature

It has been postulated that the reaction proceeds by O-benzylation of the selenenate and the cleavage of the selenium-trimethylsilylethyl bond. The products then undergoes hydrolysis to a benzyl alcohol and phenylselenenic acid. It is noteworthy that $PhCH{=}CHCH_2Br$ under the same conditions gives $PhCH{=}CHCH_2OH$ (65%) but that saturated halides (chloride, bromide and iodide) give yields of alcohols not exceeding 49% together with (several) other products.

(ii) Reduction methods

(1) Reduction of aldehydes and ketones.

(a) 4-Substituted benzaldehydes are reduced by $Zn/ZnCl_2$ in aqueous THF to mixtures of the benzyl alcohol (2) and a 1,2-diol (3) (K. Tanaka *et al.*, *J. Org. Chem.*, 1990, **55** 2981):

$$4\text{-}XC_6H_4CHO \rightarrow \underset{(2)}{4\text{-}XC_6H_4CH_2OH} + \underset{(3)}{[4\text{-}XC_6H_4CH(OH)]_2}$$

The yields, however, are variable (Table 1).

TABLE 1

X =	H	Me	Cl	Br	CN	Ph	
(2)	39	81	82	72	0	49	%
(3)	11	7	16	27	90	38	%

(b) Consistently high yields (ca. 98%) of benzylic alcohols are obtained from benzaldehyde and its 4-methyl, -methoxy and -chloro and 3,4-methylenedioxy derivatives using $SbCl_3$-Al (or Zn) in aqueous DMF at room temperature.

When the reaction is performed in the presence of allyl bromide and sodium iodide allyl(benzyl) alcohols are formed in good (61-99%) yield (W.B. Wang *et al.*, *Tetrahedron,* 1990, **46,** 3315):

$$ArCHO + CH_2{=}CHCH_2Br + NaI \xrightarrow{a.} ArCH(OH)CH_2CH{=}CH_2$$

a. $SbCl_3$-Al(or Fe) : Ar=Ph; 4-Cl, -MeO or $-HOC_6H_4$; $2-HOC_6H_4$

(c) Alkylthiobenzaldehydes are reduced in high (70-94%) yield to benzyl alcohol by a reagent made *in situ* from nickel(II) chloride hexahydrate and sodium borohydride (superiority to Raney nickel for de-sulphurizing thio-compounds is claimed for this reagent) (M.R. Euerby and R.D. Waigh, *Synth. Comm.* 1986, **16,** 779).

(d) Benzaldehyde gives a 90% yield of benzyl alcohol when it is treated with sodium N,N-dimethylaminoborohydride [$Na(Me_2N)BH_3$]. This reagent, however, offers no advantage over the conventional borohydride reagents for the reduction of arylaldehydes. Its value lies in the reduction of esters and tertiary amides to primary alcohols (see Section 2.2). (R.O. Hutchins *et al., J. Org. Chem.,* 1984, **49,** 2438).

(e) Synthetic analogues of NADPH reduce arylaldehydes (4*a*) and (selectively) chloroalkyl aryl ketones (4*b*) to the corresponding benzylic alcohols (5) (V. Levacher *et al., Tetrahedron,* 1991, **47,** 429):

ArCOR (4) ArCH(OH)R (5)

(*a*) Ar=Ph; 4-$CH_3C_6H_4$
R=H

(*b*) Ar=Ph
R=CH_2Cl; $(CH_2)_2Cl$

(2) Reduction of esters and tertiary amides.
Borane-dimethyl sulphide ($H_3B{:}SMe_2$) is very effective for reducing a variety of substituted ethyl benzoates to the corresponding benzyl alcohols (Table 2)(H.C. Brown *et al., J. Org. Chem.,* 1982, **47,** 3153).

TABLE 2
YIELDS OF SUBSTITUTED BENZYL ALCOHOLS

$XC_6H_4CO_2Et \rightarrow XC_6H_4CH_2OH$

X:	H	4-Cl	3-Br	4-NO_2	3-MeO	4-MeO
Yield (%):	90	96	94	97	91	94

Similarly, the hydride-delivering reagents $Na(RR'NBH_3)[R=R'=Me$ and $R'=H, R={}^tBu]$ convert methyl (and benzyl) esters of benzoic acid and N,N-dialkylbenzamides into benzyl alcohol (R.O. Hutchins *et al.*, *J. Org. Chem.*, 1984, **49**, 2438).

(3) O-Benzylhydroxylamine is reduced to benzyl alcohol at room temperature and pH 9.1 by dithiols in the presence of Fe(II). Dithiols are effective only when Fe(II) is present; in the absence of Fe(II) or when Co(II), Mn(II), Cu(II), Zn(II) or Mo(VI) replace it, no reaction occurs. Effective dithiols include dihydrolipoic acid (6), its amide (7) and 3-oxapentane-1,5-dithiol (8); in these cases the yield of benzyl alcohol is 95-100%. 2,3-Dihydroxybutane-1,4-dithiol (9) is less effective and monothiols - e.g. ethanethiol and thiophenol - are ineffective.

$HS(CH_2)_2CH(SH)CH_2(CH_2)_3COX$
(6; X=OH)
(7; $X=NH_2$)

$HS(CH_2)_2O(CH_2)_2SH$
(8)

$HSCH_2CH(OH)CH(OH)CH_2SH$
(9)

(*iii*) Benzylamine, dissolved in anhydrous ether at -75°C, reacts with NOCl to form benzyldiazonium chloride. This solution when treated with ethanoic acid or 70% perchloric acid gives benzyl alcohol in yields varying from 10-64%, depending upon the conditions.

(*iv*) A reaction of limited scope involves a Wittig rearrangement of the aryl ether (10). Overall it effects the insertion of a methylene unit between the OH group and aryl ring of a phenol:

$$\underset{(10)}{ArOCH_2SiMe_3} \xrightarrow[b.]{a.} \underset{(11)}{ArCH(OH)SiMe_3} \xrightarrow{c.} ArCH_2OH$$

a. RLi b. H_2O c. KOH/EtOH

With R=s-Bu and Ar=Ph, a high (85%) yield of (11), and subsequently benzyl alcohol, is obtained; but when Ar=4-MeO- or -$NO_2C_6H_4$, the incursion of competing reactions results in the failure of the formation of this intermediate (J.J. Eisch *et al., J. Org. Chem.*, 1982, **47,** 5051).

(b) *Physical properties*

V. Gold and his collaborators (*J. Chem. Soc., Perkin II,* 1986, 1279) have measured the rates of proton exchange in CF_2Br_2 of several primary alcohols and shown that the basicity of the oxygen atom is in the order:

ethanol 〉 allyl alcohol 〉 benzyl alcohol

(c) *Chemical behaviour*

1. Oxidation

Benzyl alcohol is oxidized to benzaldehyde in high (81%) yield by sodium bromate, catalysed by Ru(III), in a buffered (pH 10) two-phase (aqueous dichloromethane) medium (S. Kanemoto *et al., Bull. Chem. Soc. Japan,* 1986, **59,** 105).

Sodium benzylates are converted by dioxygen in basic DMSO into the enol ether (12) of an arylethanal from which the corresponding aldehyde (13) is obtained by hydrolysis (H. Langhals *et al., Liebig's Annalen* 1982, 2216):

$$\text{4-RC}_6\text{H}_4\text{CH}_2\text{O}^-\text{Na}^+ \xrightarrow{\text{a.}} \text{4-RC}_6\text{H}_4\text{CH=CHOCH}_2\text{C}_6\text{H}_4\text{(4-R)} \quad (12)$$

$$(12) \xrightarrow{\text{b.}} \text{4-RC}_6\text{H}_4\text{CH}_2\text{CHO} \quad (13)$$

a. O_2 ; DMSO/Et_3N/room temperature or DMSO/KOH/70 °C:
b. aq. $HClO_4$

	R:	H	MeO	Me	Cl
Yield of (12) (%)		92	80	66	56

Thus, this reaction has potential as the oxidative step in the homologation of arylaldehydes:
ArCHO → $ArCH_2OH$ → $ArCH=CHOCH_2Ar$ → $ArCH_2CHO$

An indirect method for oxidizing primary and secondary aralkanols involves treating their corresponding bromides with bis(4-methoxyphenyl) selenoxide (14); with a weak base ($NaHCO_3$) the resulting alkoxyselenonium salt (15) affords a substituted benzaldehyde (16, R=H) in high (83-93%) yield. Substituents such as methyl, bromo or nitro are unaffected by the reaction conditions, as is also the olefinic group in vinylogous alcohols:

$$\text{(4-MeOC}_6\text{H}_4\text{)}_2\text{SeO (14)} \xrightarrow{\text{a.}} \text{(4-MeOC}_6\text{H}_4\text{)}_2\text{SeOCHRArBr (15)}$$

$$(15) \xrightarrow{\text{b.}} \text{ArC(R)=O (16)} + \text{(4-MeOC}_6\text{H}_4\text{)}_2\text{Se}$$

$$(14) \xrightarrow{\text{c.}} \text{(4-MeOC}_6\text{H}_4\text{)}_2\text{SeOCH}_2\text{CH=CHPhBr} \xrightarrow{\text{b.}} \text{PhCH=CHCHO}$$

a. $XC_6H_4CH(R)Br$; MeCN/75 °C or THF/reflux [X=H; 2- or 3-Me; 3- or 4- Br, $-NO_2$]; b. $NaHCO_3$; c. $PhCH=CHCH_2Br$; other conditions as a.

The value of this method lies in the mildness of the conditions employed and the lack of over-oxidation of the aldehyde formed (K. Ariyoshi *et al., Chem. Letters,* 1984, 1984).

Methods for the electrochemical oxidation of the hydroxymethyl group have been developed. Depending upon the conditions employed - mediator used, pH - substituted benzaldehydes or benzoic acid are formed almost quantitatively. Thus, with triphenylamine derivatives as mediator and sodium carbonate or 2,6-dimethylpyridine as base, substituted benzyl alcohols give the corresponding benzaldehydes; substituents such as methoxy and halogens are unaffected (K. G. Brinkhall *et al., Acta Chem. Scand.,* 1983, **B37**, 499). Similarly, with mediation by RuO_4 and RuO_2 in a heterogeneous medium (saturated NaCl-CCl_4) and control of pH (7), benzyl alcohol was oxidized to benzoic acid (S. Torii *et al., J. Org. Chem.,* 1986, **51,** 155).

2. Reduction

When treated with the hydride-delivering reagents di-isobutylaluminium hydride (DIBAH) or lithium 9-borabicyclononane (Li 9-BBN), benzyl alcohol liberates hydrogen gas rapidly and quantitatively, but reduction is not observed and the alcohol is restored upon acidification of the reaction mixture (H.C. Brown *et al., J. Org. Chem.,* 1984, **49,** 3091; N.M. Yoon and Y.S. Gyoung, *J. Org. Chem.,* 1985, **50**, 2443). Potassium tris-isopropoxyborohydride [K(^{i}PrO$_3$BH] in THF at 0°C, is similarly unreactive towards benzyl alcohol and other primary, secondary and tertiary alcohols (H.C. Brown *et al., J. Org. Chem.,* 1984, 49, 885). However, primary, secondary and tertiary aralkanols are smoothly reduced under mild conditions by P_2I_4 (H. Suzuki *et al., Chem. Letters,* 1983, 247):

$$\mathrm{ArCRR'(OH)} \xrightarrow{\text{a.}} \mathrm{ArCHRR'} \quad 59\text{-}98\%$$

a. P_2I_4/C_6H_6, reflux; N_2:
R=R´=H; R=H; R´=Me, ${}^tBu$ or Ph: R=R´=Me or Ph:
Ar=Ph; 4-XC_6H_4 (X=Br, NO_2, CN, CO_2Et); 2-,4-,6-$Me_3C_6H_2$

Advantages are: mild conditions which are compatible with a wide range of nuclear substituents and the absence of isomerization of the side chain in the case of hindered alcohols (such as occurs when, for example, hydriodic acid and red phosphorus are used).

3. O-Acylation

A large number of substituted benzyl carbamates (17) has been prepared by reacting aralkanols with N,N-dimethylaminoformyl chloride in the presence of sodium hydride (C. Yamagami *et al., Chem. Pharm. Bull.,* 1982, **30,** 4175):

$$\mathrm{RC_6H_4CH_2OH} \xrightarrow{\text{a.}} \mathrm{RC_6H_4\,CH_2OCONMe_2} \quad (17)$$

a. Me_2NCOCl/NaH/pyridine or C_6H_6

The acid chloride (18) reacts with benzyl alcohols to give the ester (19) from which the alcohol can be regenerated by anodic oxidation (R.W. Johnson *et al., Tetrahedron Lett.*, 1981, **22,** 3719):

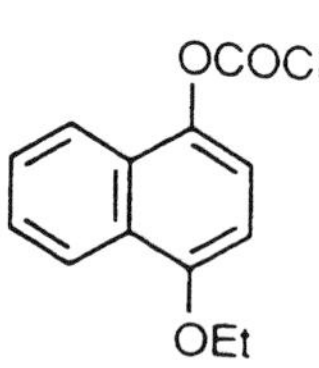

(18)

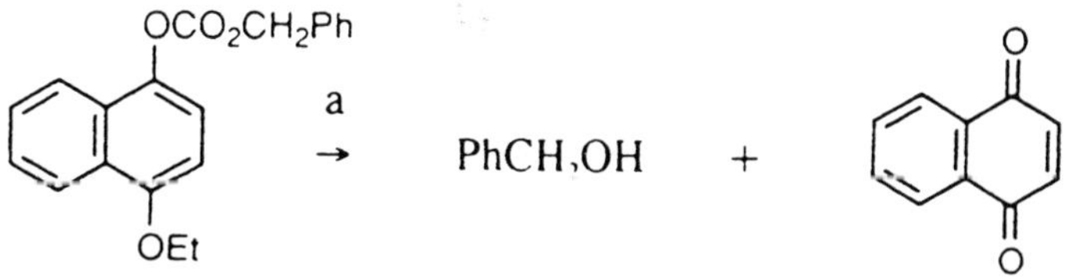

(19)

a. Electrochemical oxidation/aq. Me_2CO; room temperature or -20°C.

This provides a valuable means of protecting the hydroxyl group of those benzyl alcohols (and, incidentally, the amino group of benzylamines) containing substituents that are not susceptible to electrochemical oxidation.

The mixed anhydride (20) (readily prepared from sodium formate and 4-methoxybenzoyl chloride) selectively formylates benzyl alcohol under mild conditions (room temperature, dichloromethane solvent) (W.K. Fife and Z. Zhang, *J. Org. Chem.*, 1986, **51**, 3744).

$$4\text{-}MeOC_6H_4COOCHO$$

(20)

An almost quantitative yield of benzyl 2-aminobenzoate (21) is obtained from benzyl alcohol and isatoic anhydride in the presence of 4-N,N-dimethylaminopyridine (M.C. Venuti, *Synthesis,* 1982, 266):

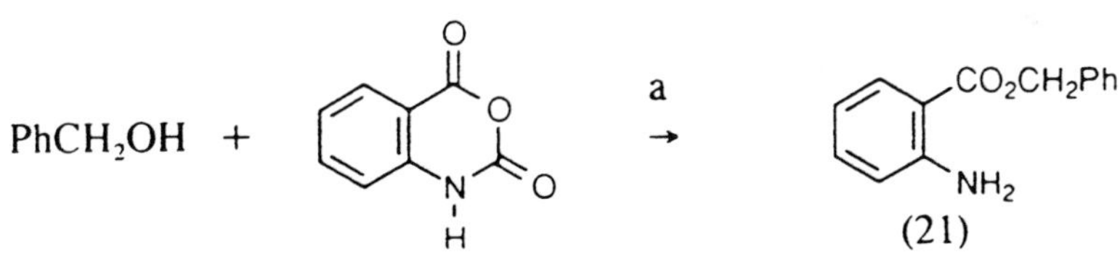

(21)

a. 4-Me_2N-pyridine /DMF; room temperature -60°C.

The reaction between benzyl alcohol and dimethylketene at 25°C is catalysed by the Lewis acid, $SbCl_3$. A study of the kinetics of the reaction has shown that the rate-determining step involves a complex formed from the alcohol and the Lewis acid (N.P. Poon and D.P.N. Satchell, *J. Chem. Soc., Perkin II,* 1985, 1551):

$$PhCH_2OH + SbCl_3 \rightarrow PhCH_2OH.SbCl_3$$
$$PhCH_2OH.SbCl_3 + Me_2C{=}C{=}O \rightarrow Me_2CHOOPh + SbCl_3$$

Benzyl alcohol is esterified by a number of carboxylic acids at room temperature, modern activators include:

Activating agent	*Carboxylic acid*	*Ref.*
2-Chloro-3,5-dinitropyridine	$RC_6H_4CO_2H$ (R=H; 4-NO_2; 3,5-$(NO_2)_2$; 2,4,6-Me_3)	1
N-Acyl-N-benzyl-imidazolium halide	$MeCO_2H$ $PhCO_2H$	2
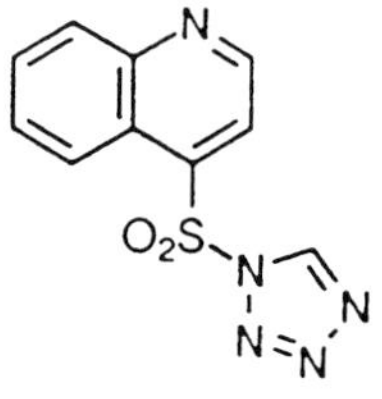	RCO_2H (R=Me, ^{i}Pr, ^{t}Bu; Ph; PhCH=$CHCH_2$)	3

The yields of benzyl esters are reported to be 75%-98%.

References.

1. S.Takimoto *et al., Bull. Chem. Soc. Japan,* 1983, **56,** 639.
2. T. Kamijo *et al., Chem. Pharm. Bull.,* 1982, **30,** 4242.
3. H. Takaku *et al., Chem. Pharm. Bull.,* 1982, **30,** 2633.

4. O-Alkylation (-arylation)

Epichlorohydrin and substituted benzyl alcohols react readily at room temperature in the presence of a phase-transfer catalyst to give high (87-99%) yields of glycidyl ethers. It is claimed that this method of preparation avoids the disadvantages of others (G. Mouzin *et al.*, *Synthesis,* 1983, 117).

$$RC_6H_4CH_2OH + ClCH_2\underset{\diagdown O \diagup}{CH-CH_2} \xrightarrow{a.} RC_6H_4CH_2OCH_2\underset{\diagdown O \diagup}{CH-CH_2}$$

a. $NaOHaq/Bu_4NHSO_4$
R=H; 3-Me; 2-OMe, -OEt; 4-OMe.

The chloro-thioether (22) reacts similarly giving the mixed ethers (23) which on reaction with sulphuryl chloride form the chloromethyl ethers (24) (T. Benneche *et al., Synthesis,* 1983, 762):

$$\underset{(22)}{RC_6H_4CH_2OH + ClCH_2SMe} \xrightarrow{a.} \underset{(23)}{RC_6H_4CH_2OCH_2SMe}$$

$$(23) \xrightarrow{b.} \underset{(24)}{RC_6H_4CH_2OCH_2Cl}$$

a. NaH/NaI, DME: b. SO_2Cl_2/CH_2Cl_2; room temperature
R=H; 2-OMe; 3-CF_3; 4-Cl, -OMe

Here again, it is claimed that this synthesis of the compounds (24) overcomes the inconvenience and limitations of earlier methods. Bromoalkanes (RBr, R= Et, ^{n}Pr, ^{n}Bu, cyclo C_6H_{11} and ^{t}Bu have been reported to react at room temperature with benzyl alcohol in the presence of HgO and HBF_4. The resulting alkyl benzyl ethers were obtained in yields ranging from 63 to 80% (J. Barlengua *et al., Synthesis,* 1983, 53).

Benzyl and other primary alcohols are converted into their ethers by means of a potassium Y-type zeolite. Because of the neutral conditions involved, this procedure is recommended for protection of primary alcohols (M. Onaka *et al., Chem. Lett.*, 1983, 1101).
Assisted by phase-transfer catalysts, benzyl alcohol readily forms aryl benzyl ethers from both nitro-aryl and heterocyclic chloro-compounds:

$$ArCl + PhCH_2OH \rightarrow ArOCH_2Ph$$

Substrate	*Phase transfer catalyst*	*Ref*
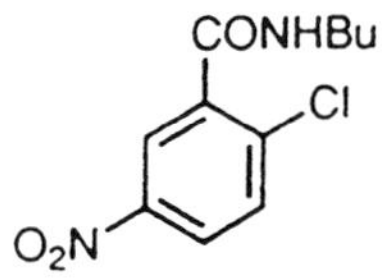 (25)	Bu_4NCl or Me_3NCH_2PhCl	1
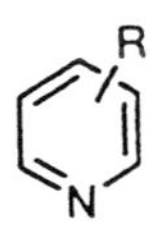 R=2-Cl; 4-Cl; 2-Cl, 3-NO_2	tris(3,6-dioxaheptyl)amine [TDA-1]	2

References.

1. D. Nisato *et al., Synthesis,* 1982, 1081.
2. P. Ballesteros *et al., Tetrahedron,* 1987, 43 2557.

In the case of compound (25) it appears that the N-H bond provides anchimeric assistance in the displacement of the chlorine atom, for the corresponding Me_2N- analogue is unreactive.

The benzyl imino-ether (26) is formed in modest yield from benzyl alcohol and the imino-chloride derived from ϵ-caprolactam (J. Jurczak *et al., Synthesis,* 1983, 382):

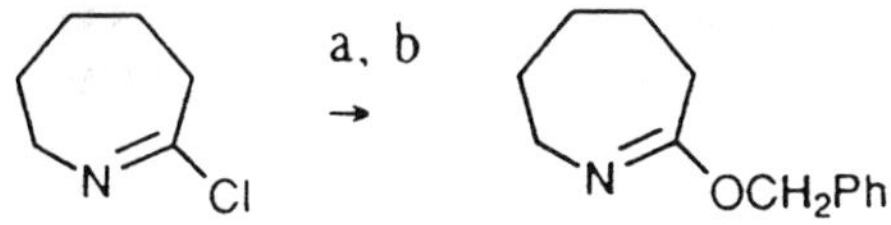

(26)

a. Ph_2CHOH, -10°C: b. NaOHaq., <0°C

A number of silicon containing compounds react with benzyl alcohol to form benzyloxysilane derivatives. Thus, chloro- and cyano-trimethylsilane afford benzyloxytrimethylsilane. The reagent (27) and benzyl alcohol yield bis(benzyloxy)dimethylsilane (28); 2-hydroxybenzyl alcohol forms the cyclic analogue (29). In all cases, the reaction conditions are mild and the yields high:

Me_3SiX + $PhCH_2OH$ → $PhCH_2OSiMe_3$
X=Cl, CN

$Me_2Si[OC(OMe)=CMe_3]_2$ + $PhCH_2OH$ → $Me_2Si(OCH_2Ph)_2$
(27) (28)

(27) + $2\text{-}HOC_6H_4CH_2OH$ →

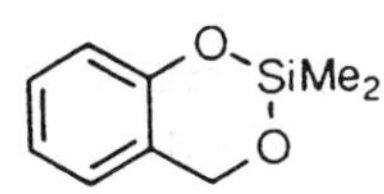

(29)

(M.Lissel and J. Weiffen, *Synth.Comm.,* 1981, **11,** 545. K.Mai and G.Partil, *J.Org.Chem.,* 1986, **51,** 3545. Y.Kita *et al, Tetrahedron Lett.,* 1983, **24,** 1273).

The benzyloxy group has been introduced into crown ethers (e.g. 15-crown-5) as a 'lariat'-like side-chain, but the hoped-for improvement in capacity to bind potassium or sodium ions was not realized (D.M.Dishong *et al., J.Amer.Chem.Soc.,* 1983, **105,** 586).

5. Replacement of OH

(a) by halogen

Several procedures, invariably involving mild conditions, have been developed for the preparation of benzyl chloride, bromide and iodide, and they are summarized in Table 3.

TABLE 3

Benzyl halide	*Procedure*	*Reference*
chloride	A, B	1, 2
bromide	A, B, C, D, E	1,3,4,5,6,7
iodide	C, D, F, G	3,4,6,8,9

A. Polymer-supported PCl_5 or PBr_3: B. Polymer-supported PPh_3/CCl_4 or CBr_4:

C. BF_3 etherate and (i) Et_3N Br (or I) or (ii) NaBr (I):

D. 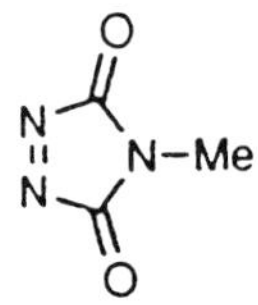 $/PBr_3/MeBr$ or MeI

E. $PPh_3/EtO_2CN{=}NCO_2Et/LiBr$: F. Me_3SiCl_3/Na:

G. $PhCH_2OCOOCH(Cl)Me/NaI$.

References

1. G. Cainelli *et al., Synthesis,* 1983, 306.
2. C. R. Harrison *et al., J.Org.Chem.,* 1983, **48,** 3721.
3. A. K. Mandal and S. W. Mahajan, *Tetrahedron Lett.,* 1985, **26,** 3863.
4. Y. D. Vankar and C. T. Rao, *Tetrahedron Lett.,* 1985, **26,** 2717.
5. P. Hodge and E. Khoshdel, *J.Chem.Soc., Perkin I,* 1984, 195.
6. T. Oshikawa and M. Yamashita, *Bull.Chem.Soc. Japan,* 1984, **57,** 2675.

7. S. Manna *et al., Synth.Comm.* 1985, **15,** 663.
8. G. A. Olah *et al., J.Org.Chem.,* 1983, **48,** 3667.
9. J. J. Brunet *et al., Tetrahedron Lett.,* 1985, **26,** 5445.

In addition, benzyl bromide has been indirectly obtained from the alcohol *via* its ethers or esters by treatment with Me_3SiBr and IBr (E. C. Friedrich and G. DeLucca, *J.Org.Chem.,* 1983, **48,** 1678):

$$PhCH_2OX \xrightarrow{a.} PhCH_2Br \qquad (50\text{-}82\%)$$

a. Me_3SiBr/IBr; room temperature:
X = Et, CH_2Ph, cycloC_6H_{11}, $SiMe_3$; COMe, COPh.

This has potential as a method for de-protecting benzylic alcohols, and at the same time activating the benzylic methylene group for further reaction.

(b) by thioethers and thioesters.
Mixed thioethers are obtained when sodium benzylate reacts with the salt derived from a primary alkyl halide and N,N-tetramethylthiourea, in a one-pot procedure (S. Fujisaki *et al., Bull. Chem.Soc. Japan.,* 1985, **58,** 2429):

$$PhCH_2ONa + [(Me_2N)_2\,CSR]\,X \xrightarrow{a.} PhCH_2SR$$

a. DMF or MeCN/room temperature: R = Me, Et, $PhCH_2$, allyl:
X = Br, I.
The advantage claimed for this procedure is that malodorous thiols are not used as reactants. The reaction is, however, confined to primary alkyl halides; and thiourea cannot be substituted as a starting material because of the formation of stable imines, $HN{=}C(NH_2)SR$.
Substituted benzyl phenyl sulphides are readily formed from the benzyl alcohol and thiophenol under catalysis by zinc iodide (Y. Guindon *et*

al., J.Org.Chem., 1983, **48,** 1357):

$$4\text{-}XC_6H_4CH(OH)R \xrightarrow{a.} 4\text{-}XC_6H_4CH(SPh)R$$

a. $PhSH/ZnI_2$; room temperature: X = H, MeO, CO_2Me: R = H, Me, ^nBu.

The most effective solvents for this reaction are dichloromethane or 1,2-dichloroethane; acetonitrile, diethyl ether, N,N-dimethylformamide, tetrahydrofuran and ethyl ethanoate are unsatisfactory because of their tendency to coordinate with the Lewis acid.

Thiol esters (and hence, indirectly, thiols) result when a substituted benzyl alcohol and a thiocarboxylic acid are treated with either zinc chloride (J. Y. Gauthier *et al., Tetrahedron Lett.*, 1986, **27,** 15) or triphenylphosphine and di-isopropyl azodicarboxylate (R. P. Volante, *Tetrahedron Lett.*, 1981, **22**, 3119):

$$4\text{-}XC_6H_4CH(OH)R \xrightarrow{a.\text{or } b.} 4\text{-}XC_6H_4(SCOY)R \xrightarrow{c.} 4\text{-}XC_6H_4CH(SH)R$$

a. $YCOSH/ZnCl_2/CH_2Cl_2$: b. $YCOSH/Ph_3P$, $^iPrO_2CN{=}NCO_2{}^iPr$; THF, O °C: c. saponification or reduction:
X = H, MeO, CO_2Me: R = H, Me, ^nBu: Y = Me or Ph

(c) by selenyl ether

N-Phenylselenophthalimide, benzyl alcohol and tributylphosphine react readily to give benzyl phenyl selenide in excellent (95%) yield (K. C. Nicolaou *et al., Tetrahedron,* 1985, **41,** 4835):

$PhCH_2OH +$ 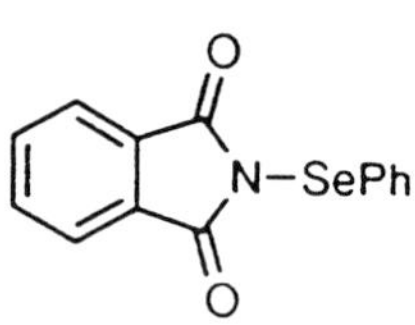 $\xrightarrow{a.}$ $PhCH_2SePh$

a. $^nBu_3P/CH_2Cl_2$ (or THF); -20 °C.

It is claimed that this method is superior to that using PhSeCN, a reagent that is difficult to prepare, malodorous and unstable.

(d) by nitrogenous groups

The ruthenium complex $RuCl_2(PPh_3)_3$ catalyses the reaction between an aniline and benzyl alcohol, leading to N-benzylation of the former:

				Ref.
$4\text{-}XC_6H_4NH_2$	+ $PhCH_2OH$	$\xrightarrow{a.}$	$4\text{-}XC_6H_4NHCH_2Ph$	1.
cf. $PhCONH_2$	+ $PhCH_2OH$	$\xrightarrow{b.}$	$PhCONHCH_2Ph$	2.

a. $RuCl_2(PPh_3)_3$/150°C: b. $RuCl_2(PPh_3)_3$/180°C:X=H, Me, MeO, Cl

References

1. Y. Watanabe *et al., J. Org. Chem.,* 1984, **49,** 3356.
2. Y. Watanabe *et al., Bull. Chem. Soc. Japan,* 1983, **56,** 2647.

For these N-benzylations the ruthenium complex illustrated was the most effective of a number that was tried. With this catalyst, the ratio of mono- to di-benzylation was dependent upon the mole ratio, benzyl alcohol/amine. By contrast, however, only mono-N-benzylation of benzamide occurred even when an excess of benzyl alcohol was used.

Triphenylphosphine and diethyl azodicarboxylate bring about the reaction between benzyl alcohol and either primary or secondary amines, but although the yields of secondary and tertiary amines, respectively, are reported to be not spectacular, nevertheless the reaction has been used as part of a synthesis (P.G. Sammes and S. Smith, *J. Chem. Soc., Perkin I.,* 1984, 2415):

$$PhCH_2OH + PhNH_2 \xrightarrow{a.} PhCH_2NHPh$$

$PhCH_2OH$ + 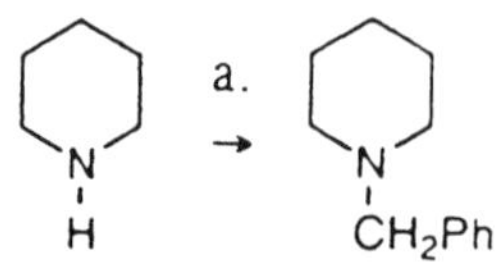

a. $PPh_3/EtO_2CN{=}NCO_2Et$; 0 °C - room temperature/DMF or THF.

Benzylic alcohols are converted into the corresponding azides either directly (using HN_3, and $TiCl_4$ as catalyst) or indirectly as part of a one-pot synthesis of benzylamines. In the latter case the azide is not isolated and only primary alcohols react.

	Ref.
$PhCH(OH)R \xrightarrow{a.} PhCH(N_3)R$	1

a. $HN_3/TiCl_4$: R = H, Me

$4\text{-}XC_6H_4CH_2OH \xrightarrow{b,c.} [4\text{-}XC_6H_4CH_2N_3] \xrightarrow{d,e.} 4\text{-}XC_6H_4CH_2NH_2$ (X = H, MeO)	2

b. $CBrCl_3/PPh_3$: c. $NaN_3/^{n}Bu_4NBr/DMF$: d. $P(OEt)_3$: e. HCl aq.

References

1. A. Hassner *et al., J.Chem.Soc., Perkin I,* 1981, 4237.
2. A. Koziara, *J.Chem.Soc.,* (*Synopses*), 1989, 296.

The replacement by NH_2 of the OH groups in benzylic alcohols is also accomplished in a 'Gabriel'-like manner using $(EtO)_2P(O)NHCO_2{}^tBu$ as the phthalimide equivalent (E. Slusarska and A. Zwierak, *Liebig's Annalen,* 1986, 402). Noteworthy is the mildness of the conditions for each step. When chiral alcohols are used the configuration is completely inverted in the product.

$$PhCH_2OH \xrightarrow{a.} (EtO)_2P(O)N(CH_2Ph)CO_2{}^tBu \xrightarrow{b.} PhCH_2N^+H_3\ Cl^-$$

a. $(EtO)_2P(O)NHCO_2{}^tBu/EtO_2CN{=}NCO_2Et/C_6H_6$, room temperature: b. $HCl(g)/C_6H_6$, room temperature.

(e) Miscellaneous

(i) Benzyl alcohol is readily converted *via* (30) into the monobenzyl ester (31) of phosphorous acid; such monoesters are useful starting materials for the preparation of monoesters of phosphoric acid (D.E.Gibbs and C.Larsen, *Synthesis,* 1984, 410):

$PhCH_2OH + (F_3CCH_2O)_2P(O)H \xrightarrow{a.}$ 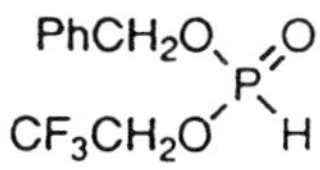

(30)

(30) $\xrightarrow{b}$ 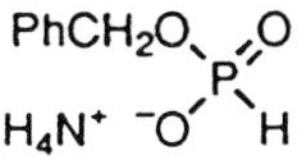

(31)

a. MeCN/70 °C: b. NH_3aq.

G. A. A.Claesen *et al.* (*Rec.Trav.Chim.,* 1985, **104,** 119) have shown that benzyl phosphorodichloridite (32) - a reagent useful in nucleotide synthesis - is formed almost quantitatively by briefly contacting benzyl alcohol with phosphorus trichloride (in acetonitrile) at room temperature.
The corresponding reaction involving Ph_2PCl affords high yields of (33; X=H or MeO) (A. D. Buss and S. Warren, *J.Chem.Soc., Perkin I,* 1985, 2313).

$PhCH_2OPCl_2$	$4\text{-}XC_6H_4P(O)Ph_2$
(32)	(33)

(ii) A reagent made by heating $^{n}Bu_2SnO$ and $(^{n}BuO)_3PO$ at 250 °C catalyses the regioselective cleavage of unsymmetrical epoxides by alcohols (J. Otera *et al., Tetrahedron Lett.,* 1985, **26**, 3219). With benzyl alcohol, compounds (34) are formed exclusively; with other alcohols (e.g. methanol), while analogues of (34) are the main products, increasing amounts of the compounds (35) are formed.

$PhCH_2OH$ + 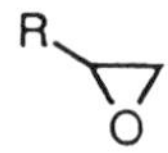 → $RCH(OH)CH_2OCH_2Ph$ (34)

room temperature or reflux: R = ⁿhexyl, CH_2Cl, OAc.

$RCH(OMe)CH_2OH$ (35)

The bis-lithio derivative of benzyl alcohol (36) reacts with oxirane (ethylene oxide) to form (37); the latter when treated with benzylamine gives the tertiary amine (38) (S. Murahashi *et al ., Tetrahedron Lett.,* 1982, **23**, 229):

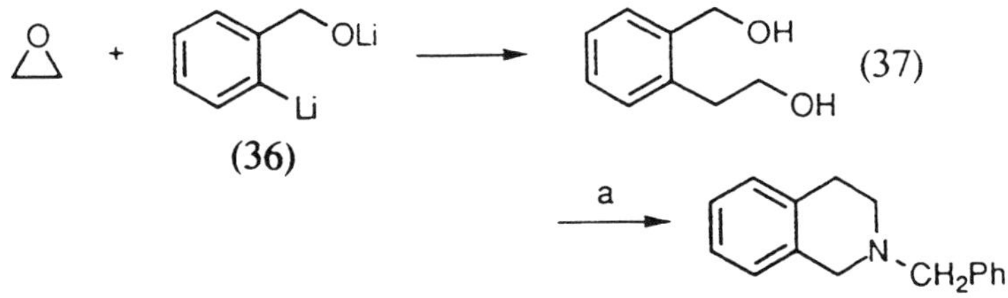

(36) (37) (38)

a. $PhCH_2NH_2$

In a similar manner, aziridines are cleanly cleaved by benzyl alcohol in the presence of boron trifluoride-etherate. This reaction has been developed into a synthesis of L-serine derivatives (K. Nakajima *et al., Bull.Chem.Soc. Japan,* 1982, **55,** 3049).

(iii) A one-pot procedure has been developed for the preparation of O-alkyl S-methylxanthates which uses mild conditions throughout; it is applicable to other primary and secondary alcohols, besides benzyl alcohol (R. Chênevert *et al., Synth.Comm.,* 1981, **11,** 817):

$$CS_2 + PhCH_2OH \xrightarrow{a.} PhCH_2OC(S)^{-}SK^{+} \xrightarrow{b.} PhCH_2OC(S)SMe$$

a. 18-crown-6/^{t}BuOK/CH_2Cl_2; room temperature:
b. MeI/CH_2Cl_2; room temperature.

(iv) Benzylic alcohols have been found to react readily with the reagents $SOCl(ONO_2)$ and $SO(ONO_2)_2$ at 25 °C, yielding the corresponding nitrates, $PhCH(ONO_2)$ R [R = H, Et]. The reagents are without effect upon a wide range of substituents in the aromatic ring (e.g. Me, OMe, NH_2, NO_2, CO_2H). However, the reagents do not discriminate between alcoholic and phenolic hydroxyl groups. The simplicity of the procedure, mildness of the conditions and high yields, etc. combine to make this reaction a valuable addition to the methods for protecting the hydroxyl group of alcohols and phenols (G. I. Hakimelahi *et al., Helv. Chim. Acta,* 1984, **67,** 906).
(v) The 2-position of the tetrahydropyran derivative (39) is susceptible to nucleophilic attack by the oxygen atom of benzyl alcohol affording the mixed acetal (40) (D.S. Brown *et al., Tetrahedron Lett.,* 1988, **29,** 4873; *Tetrahedron,* 1991, **47,** 1329):

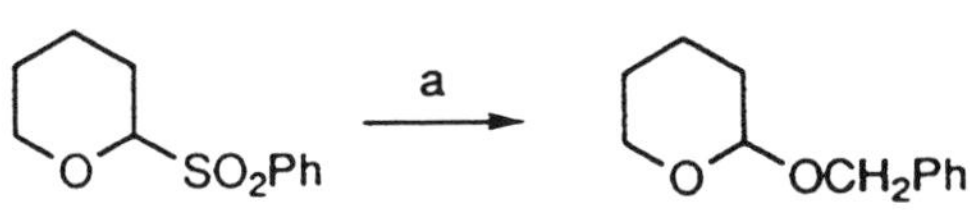

(39) (40)

a. $PhCH_2OH/MgBr_2.Et_2O/NaHCO_3$/THF; room temperature.

This and other examples establish the reaction as being useful for making unsymmetrical acetals under very mild conditions.
J. W. VanCleve (*Carbohydrate Res.,* 1982, **106,** 239) has detailed the procedure for the synthesis of the benzyl α-and ß-D-erythrosides, starting from 4,6-O-ethylidene-D-glucose and the thermolysis of

sucrose in benzyl alcohol has been shown to give mainly benzyl D-fructopyranoside and D-glucose (W. Moody and G. N. Richards, *Carbohydrate Res.*, 1981, **97**, 247).
The substituted aromatic lactones (41) are reduced by di-isobutylaluminium hydride (DIBAH) to the hemiacetals (42) and the ratio of hemiacetal to (hydroxymethyl)benzaldehyde (43) has been measured (Table 4) (G.Wulff and G. Wolf, *Chem. Ber.*, 1986, **119**, 1876):

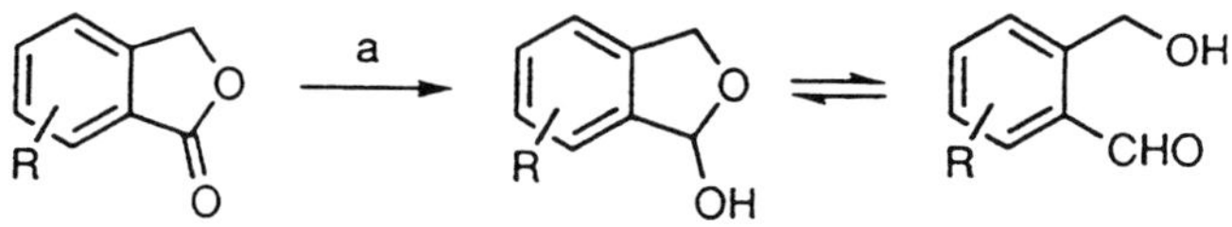

(41) (42) (43)

a. DIBAH/PhMe; - 80 °C: R = H; 7-OMe; 6-NO_2, -Br

TABLE 4 Ratios for the equilibrium (42)⇄(43)

R	Ratio, (42) : (43) Solution	Solid
H	3:1	0:1
7-OMe	9:1	1:0
6-NO_2	9:1	1:0
6-Br	4:1	0:1

(vi) (Trimethylsilyl)diazomethane is a less hazardous substitute for diazomethane in the Arndt-Eistert reaction. When the diazoketones (44) are allowed to decompose in benzyl alcohol good yields of the benzyl esters of the homologated carboxylic acids are obtained (T. Aoyama and T. Shioiri, *Chem. Pharm. Bull.*, 1981, **29**, 3249):

$$\mathrm{Me_3SiCHN_2 + RCOCl \xrightarrow{a.} [RCOC(SiMe_3)N_2] \xrightarrow{b.} RCH_2CO_2CH_2Ph}$$
(44)

R = 4-H, -OMe, -ClC$_6$H$_4$; α-naphthyl: PhCH$_2$CH$_2$; cycloC$_6$H$_{11}$.
a. Et$_3$N: b. PhCH$_2$OH/2,4,6-Me$_3$-pyridine, 180 °C.

(vii) Not only does benzyl alcohol add to the carbonyl group of ketones [see (vi) above], it also participates in conjugate addition reactions. Thus, with ethyl crotonate a product [presumably (45)] is formed which is reduced to the benzyl ether (46) (E. Dimitriadis and R. A. Massy-Westropp, *Austral. J.Chem.*, 1984, **37**, 619):

$$\mathrm{PhCH_2OH + MeCH{=}CHCO_2Et \xrightarrow{a.} MeCH(OCH_2Ph)CH_2CO_2H}$$
(45)

$$\mathrm{(45) \xrightarrow{b.} MeCH(OCH_2Ph)CH_2CH_2OH}$$
(46)

a. KOH/150 °C: b. LiAlH$_4$

Similarly, lithium benzylate combines with the ester (47) in a stereospecific double 'Michael'- type reaction (M. Joucla *et al.*, *Tetrahedron Lett.*,1985, **26**, 1221):

$$\mathrm{PhCH_2OLi + H_2C{=}C(Br)CO_2CH_2Ph \rightarrow}$$
(47)

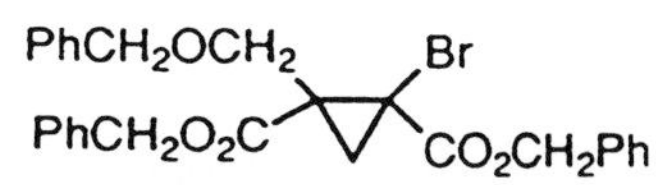

The O-benzyl ether of sinapyl alcohol (49), a constituent of lignin, has been synthesized by the addition of benzyl alcohol to the quinonemethide (48) (A. Zanarotti, *J. Org.Chem.*, 1985, **50**, 941):

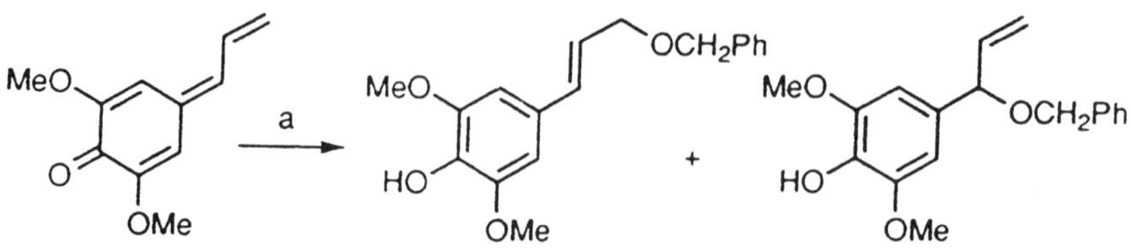

(48) (49)

a. 4-$MeC_6H_4SO_3H$

(viii) Vinyl alkyl ethers react with N-iodophthalimide in the presence of benzyl alcohol and form iodo-diethers (50) in good yield. In turn, these undergo elimination of HI to give high yields of ketene acetals (51) (D. S. Middleton and N. S. Simpkins, *Synth. Comm.* 1989, **19,** 21):

$$ROCH{=}CH_2 \xrightarrow{a.} ROCH(OCH_2Ph)CH_2I \xrightarrow{b.} ROC(OCH_2Ph){=}CH_2$$

(50) (51)

a. 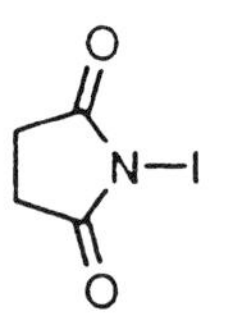 /$PhCH_2OH/CH_2Cl_2$, -30°C: b. ^{t}BuOK/THF:

R=Et, CH_2Ph

(ix) Benzyl alcohol and potassium 2-naphtholate react at 200 °C to form 1-benzyl-2-naphthol. Performing the reaction at lower (100-170 °C) temperatures allows the isolation of a number of precursors which show that under the conditions obtaining the reaction is not a conventional electrophilic substitution (T. Kito *et al., J. Org. Chem.,* 1985, **50,** 4628).

Benzylation of the aromatic ring occurs when 2,6-dimethylphenol and its O-methyl derivative are treated at 30 °C with a solution of benzyl alcohol in diethyl ether containing concentrated sulphuric acid (B. Miller, *et al.*, *J.Org. Chem.*, 1982, **47,** 710):

2,6-$Me_2C_6H_3OR$ →
(R = H, Me)

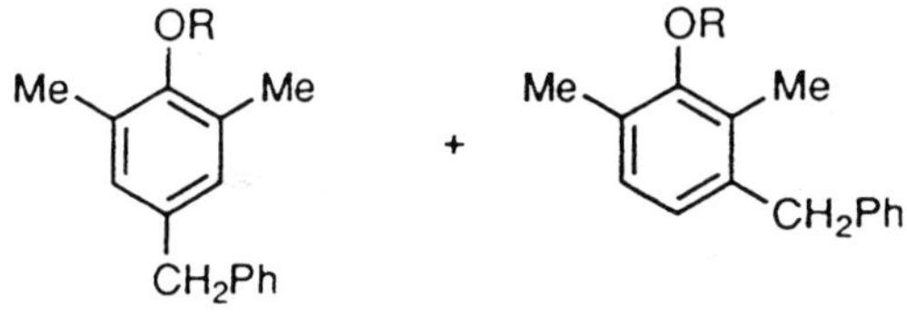

(x) The C=C bond of αß-unsaturated acids and esters (52) is reduced by transfer hydrogenation when treated with a benzylic alcohol in the presence of a chiral ruthenium catalyst (K. Yoshinaga *et al.*, *Bull. Chem. Soc. Japan,* 1983, **56,** 1786).

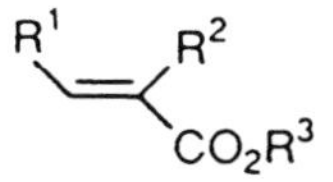

(52)

R^1=H, Me, Ph, CO_2H

R^2=Me, CH_2CO_2H

R^3=H, Me, Et, ^{n}Bu, CH_2Ph

Using benzyl alcohol as hydrogen donor the yields vary from *ca.* 10 to 50%; better yields are obtained from the esters (R^3 = Me, Et etc.) using (±)-1-phenylethanol as the hydrogen source. Irrespective of the alcohol used, the enantiomeric excess of the products is not high (ca. 2-18%).

(xi) In the presence of a cobalt catalyst and either an amine or alcohol, substituted benzyl alcohols undergo carbonylation by carbon monoxide under mild conditions and form N-alkylamides or esters, respectively (T. Imamoto *et al.*, *Bull. Chem. Soc. Japan,* 1982, **55,** 643):

a.

4-$XC_6H_4CH_2OH$ + R_2NH + CO → 4-$XC_6H_4CH_2CONR_2$

a. ethyl polyphosphate/$Co_2(CO)_8$/NaI, room temperature:
X = H, MeO, Cl, Br, NO_2 : R = Et, ^{i}Pr, cycloC_6H_{11}

$$PhCH_2OH + ROH + CO \xrightarrow{b.} PhCH_2CO_2R$$

b. $Co_2(CO)_8$/^{n}Bu_3N/room temperature: R = Me, Et, ^{t}Bu, Ph
The general acceptability of the yields and the mildness of the conditions make for an attractive route to the esters and amides of arylacetic acids.

(xii) Azo pigments bonded to silica gel have been produced by anchoring benzyl groups on to the surface and then introducing aryl azo groups by the usual sequence of reactions (nitration, reduction of NO_2, diazotization of NH_2 and coupling of the diazonium salt with an end-component). The colours of the pigments so obtained correspond closely with those of the PhN=NAr analogues (M. Suzuki *et al. Chem. Letters,* 1981, 1785).

(xiii) The hydrolysis of tetrachlorosilane by mixtures of aqueous benzyl alcohol produces polymers which have properties (e.g. surface tension, specific heat, flash point, viscosity etc.) similar to those of commercial-grade silicones. It is claimed that such 'benzyl silicones' can be used as lubricants, hydraulic fluids, heat transfer media and water-repellent and mould-release agents (V. Krishnasamy and B. Janardhanan, *J. Indian Chem.Soc.,* 1985, **62,** 275).

2. Aralkylamines

(a) *Preparation*

(i) It has been reported (V. F. Anikin *et al., Ukr. Khim.Zh.,* 1982, **48,** 439) that N-alkylated phthalimides are cleaved in high yield by fusion (250-350 °C) with sodium hydroxide; thus, benzylamine is obtained in 72% yield.

(ii) Benzylamine and its ring-substituted analogues have been prepared by the reduction of amides, azides and nitriles using a variety of reducing agents.

Amides

Primary amides have been selectively reduced to amines by lithium borohydride; secondary amide and carboxylate groups do not interfere. (K.Soai *et al., J.Chem.Soc., Chem.Comm.,* 1983, 668).

$$PhCONH_2 \xrightarrow{a.} PhCH_2NH_2 \ (92\%)$$

a. $LiBH_4$/diglyme-MeOH

Sodium N,N-dimethylaminoborohydride similarly reduces benzamide to benzylamine, but when it reacts with tertiary amides (e.g. $PhCONR_2$; R = Me, Et), benzyl alcohol is formed preferentially (R. O. Hutchings *et al., J.Org.Chem.* 1984, **49,** 2438).
An extensive study by H. C. Brown *et al.,* (*J.Org.Chem.,* 1982, **47,** 3153) has shown that benzamide, its N-Me-, N,N-Me_2- and nuclear-substituted derivatives are reduced by $H_3B{:}SMe_2$ to the corresponding benzylamines in high yields. Chlorine atoms and nitro groups in the substrates are unaffected by this reagent.
A reagent made from $NaBH_4$ and ethane-1,2-dithiol readily reduces benzamide to benzylamine; under the same conditions it has no effect on nitriles. Investigation suggests the structure (53) for the active reducing agent (W. C. Guida *et al., J.Org.Chem.,* 1984, **49,** 3024).

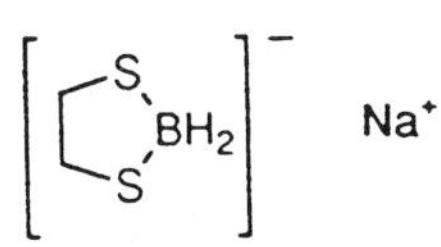

(53)

Nitriles

The reagent $H_3B{:}SMe_2$ also reduces substituted benzonitriles to the corresponding benzylamines in yields in the range 60-85%.
In non-polar solvents (e.g. benzene) lithium aluminium hydride is

solubilized by phase-transfer agents such as 15-crown-5 or benzyltriethylammonium chloride, and such two-phase systems are capable of reducing benzonitrile almost quantitatively to benzylamine (V. Gevorgyan and E. Lukevics, *J.Chem.Soc., Chem.Comm.*, 1985, 1234).

Azides

Benzyl azide has been reduced in high yield to benzylamine by means of $NaBH_4$ in the presence of a phase-transfer catalyst (F. Rolla, *J.Org.Chem.*, 1982, **47**, 4327):

$$PhCH_2N_3 \xrightarrow{a.} PhCH_2NH_2$$

a. $NaBH_4$ aq./PhMe/[$Bu_3P(^nC_{16}H_{31})$]Br/70 °C.

(iii) From benzyl halides. An alternative to the Gabriel procedure has been developed for the preparation of benzylamine and its N-alkyl derivatives from benzyl chloride (P. A. Harland et al., *Synthesis,* 1984, 941):

$$PhCH_2Cl \xrightarrow{a.} \underset{(54)}{PhCH_2NHCOCF_3} \xrightarrow{b.\ or\ c.} PhCH_2NH_2$$

$$(54) \xrightarrow[e.]{d.} PhCH_2NRCOCF_3 \xrightarrow{b.\ or\ c.} PhCH_2NHR$$

a. $F_3CCONHNa$/DMF
b. KOH(R=H or $PhCH_2$)
c. $NaBH_4$ (R=H or Me)
d. NaH
e. MeI or $PhCH_2Cl$
R=Me or $PhCH_2$

The yields reported for the N-acetylated amines (54) and final products are uniformly high: (54), 72%; $PhCH_2NHR$, R=H, 95%; Me, 90%; $PhCH_2$.

The method appears to have potential for exploitation, for the vinylogue (55); X=NH_2) is readily prepared from the corresponding halide (X=Br).

$$\underset{(55)}{PhCH{=}CHCH_2X}$$

(b) Physical properties

The basicities of benzylamine and its substituted derivatives have been measured under different conditions, as have also the equilibrium constants for the reactions of these bases with cations and with 2,4-dinitrophenol. Thus, the basicities of benzylamine and a number of its derivatives have been shown to be significantly lower in aqueous DMSO (DMSO, 80% water, 20%) than in water alone. (G. Velinov and M. Georgieva, *Anal. Chim. Acta,* 1982, **140,** 339).
R. D. Hancock *et al., (Inorg. Chem.,* **22,** 2531) have reported the pKa value for benzylamine and the formation constants for the complexation of it with Ag(I) and Hg(II) in 0.1M nitric acid.
The pKa values for benzylamine and its N-methyl (and -ethyl) and N,N-dimethyl derivatives (in water), and the equilibrium constants (K_b) for the reactions of these compounds with 2,4-dinitrophenol (in benzene) have been determined (V. Freuna *et al., J. Chem. Soc., Perkin II,* 1985, 1865). The linear correlation between logK_b and pKa has been interpreted as evidence that in both solvents similar steric and electronic effects operate to determine the basicities of the amines.

(c) Chemical behaviour

1. N-Alkylation
Amongst the reactions involving the direct nucleophilic attack of benzylamine on a leaving-group bearing carbon atom, the following may be cited:

Ref.

(a) $PhCH_2CH(OTs)CH{=}CH_2$ 1.

a.
$\rightarrow PhCH_2CH(NHCH_2Ph)CH{=}CH_2$

(b) 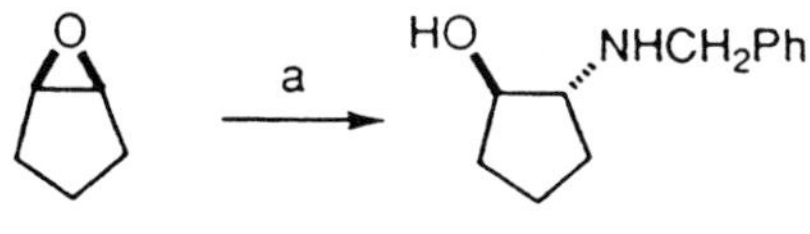 2.

a. $PhCH_2NH_2$

References

1. R. Brettle and I. A. Jafri, *J. Chem. Soc., Perkin I,* 1983, 387.
2. P. M. Barkworth and T. A. Crabb, *Org. Mag. Res.,* 1981, **17**, 260.

Routes leading to the N-alkylation of benzylamine via other intermediates include:
(i) The formation and C-alkylation of $PhCH_2NHCH_2CN$, followed by de-cyanation (L. E. Overman and R. M.Burk, *Tetrahedron Lett.*, 1984, **25,** 1635):

$$PhCH_2NH_2 \xrightarrow{a.} PhCH_2NHCH_2CN \xrightarrow{b.} PhCH_2NHCH_2R \ (58\text{-}81\%)$$

(56)

a. CH_2O/KCN or $ClCH_2CN$: b. RLi(orRMgBr)/THF, -70→0°C.

R= n-, sec- ot tert-Bu; $PhCH_2$; $H_2C{=}C(OEt)$-; 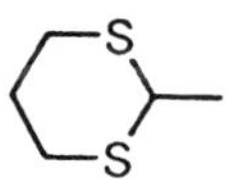

An advantage of this procedure is that by the introduction of vinyl ether or dithiane units, ketonic or aldehydic carbonyl groups, respectively, may subsequently be generated in the products (56).

(ii) N-Ethylation of benzylamine is accomplished in acceptable - high yield (depending upon temperature) by the reaction between benzylamine and diethoxytriphenylphosphorane (DTPP), $(EtO)_2PPh_3$ (J. W. Kelly *et al., J. Org. Chem.,* 1986, **51**, 95). This indicates a particular instance of a more general type of N-C bond-forming process, for when a ß-hydroxyamine is used, the appropriately-substituted aziridine (57) is formed in high yield. With chiral starting materials, there is total retention of chirality in the products.

DTPP

$PhCH_2NH_2$ → $PhCH_2NHEt$ + $PhCH_2NEt_2$

60°C	63%	20%
-30°C	98%	2%

$H_2NCHPh(CH_2OH)$ → 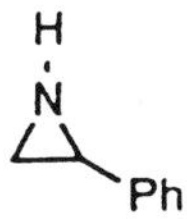 (85%)

(57)

2. N-Acylation

Direct N-formylation of benzylamine (and other primary amines) has been achieved in good (~60%) yield by heating the amine under reflux with anhydrous formic acid in DMF solvent (J. M. Aizpurua and C. Palomo, *Synth. Comm.,* 1983, **13,** 745).

Benzylamine reacts exothermically at 25 °C with the mixed anhydride (58) to form N-formylbenzylamine in 94% yield:

$$PhCH_2NH_2 + {}^tBuCOOCHO \xrightarrow{a.} PhCH_2NHCHO$$

(58)

a. $CHCl_3$/25 °C

Other primary amines, both aliphatic and aromatic, react similarly, giving excellent (91 - 97%) yields of product. The reaction is especially useful for the N-acylation of those amines which are unstable when released from their salts. It is highly regio-selective, no N-pivaloyl derivative being formed (E. J. Vlietstra *et al., Rec. Trav. chim., Pays-Bas,* 1982, **101,** 460).

Other N-acylation reactions involving transfer from other acyl group carriers include:

(i) polymer-bound quinoline esters [*Ref.* 1], (ii) pyrazine esters [*Ref.*2], (iii) S-acyl derivatives [*Ref.*3].

References
1. X. Huang *et al., Synth. Comm.* 1982, **12,** 709.
2. A. Ohta *et al., J. Heterocyclic Chem.,* 1983, **20,** 951.
3. Y. Nambu *et al., Bull. Chem. Soc. Japan,* 1983, **56,** 815.

3. Miscellaneous

(a) P. Brun *et al.,* (*Tetrahedron,* 1985, **41,** 5019) have shown that the imine (59), formed from benzylamine and cinnamaldehyde, is capable of being distilled at 10mm Hg pressure without decomposition.

$$PhCH_2N{=}CHCH{=}CHPh$$
(59)

(b) 4-Pyridyl isothiocyanate (60) reacts at 0 °C almost quantitatively with benzylamine to give the thiourea (61) ([E. T. Hansen and H. J. Petersen, *Synth. Comm.,* 1984, **14,** 537)].

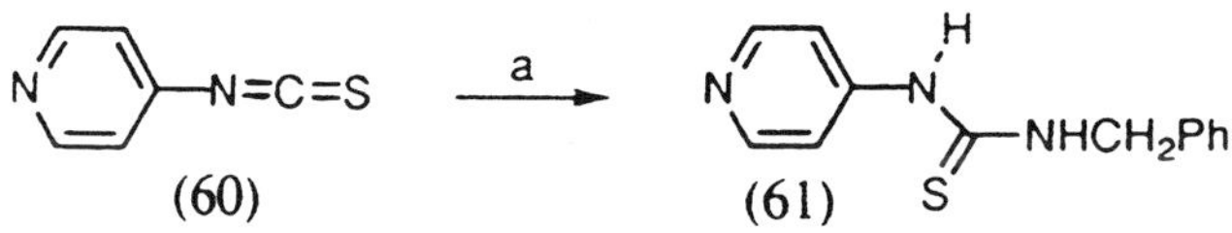

(60) (61)

a. $PhCH_2NH_2$/O °C/Et_2O-Pr_2O

(c) The adducts (63) are formed at room temperature by combining benzylamine and (62); pyrolysis of these adducts in xylene affords the substituted butadienes (64) in excellent yield (M. Adamczyck and M. Mokrosz, *Synthesis,* 1981, 802).

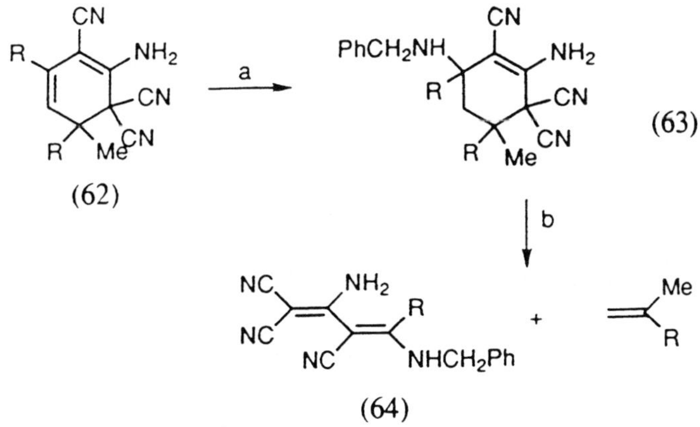

a. $PhCH_2NH_2$/r.t./dioxan or DMF: b. ▲/xylene:R=Me, Et

Second Supplements to the 2nd Edition of Rodd's Chemistry of Carbon Compounds, Vol.III D/E/F (Partial), edited by M. Sainsbury

Chapter 14

MONOCARBALDEHYDES AND MONOKETONES OF THE BENZENE SERIES

M. TANAKA AND Y. SOUMA

Aryl aldehydes and ketones are very important because of the reactivity of the carbonyl group. This leads to the formation of numerous derivatives. In this chapter, recent approaches to the synthesis of these compounds are highlighted, and the methods used are classified according to the starting material used.

1. Synthesis of Aromatic Monocarbaldehydes

(a) From Aryl Alcohols

(i) Chromium Oxidants

Chromium(VI) reagents are classical and versatile oxidants, which can be used to oxidize benzyl alcohols to benzaldehydes. However, the reactions are not always selective, and the desired products can be over-oxidized to benzoic acids. Despite this pyridinium chlorochromate can be used under mild conditions, as are other chromium bearing reagents, frequently in combination with certain amines (F. A. Luzzio and F. S. Guziec, Jr., *Org. Prep. Proc. Int.*, 1988, **20**, 533; G. Piancatelli *et al., Synthesis,* 1982, 245). These include $HCrO_3F{\bullet}C_5H_5N$ (M. N. Bhattacharjee *et al., Synthesis,* 1982, 588), $HCrO_3Cl{\bullet}C_5H_5N$ (M. Fourmigué *et al., J. Am. Chem. Soc.*, 1993, **115**, 3752; A. Osuka *et al., J. Org. Chem.*, 1993, **58**, 3582), $HCrO_3Cl{\bullet}(C_5H_4N)_2$ (F. S. Guziec, Jr. and F. A. Luzzio, *Synthesis,* 1980, 691), $H_2Cr_2O_7{\bullet}(C_5H_5N)_2$ (S. Kanemoto *et al., Bull. Chem. Soc. Jpn.*, 1988, **61**, 3607), and $H_2Cr_2O_7{\bullet}(3\text{-}C_5H_4NCO_2H)_2$ (F. P. Cossio *et al., Tetrahedron,* 1987, **43**, 3963). Generally, oxidation capacity decreases with the basicity of the amine and this provides a measure of control in those cases where over-oxidation is troublesome. Other chromium bearing reagents such as esters, $(t\text{-}BuO)_2CrO_2$ and $CH_2(CMe_2O)_2CrO_2$, are also effective oxidants (F. A. Luzzio and W. J. Moore, *J. Org. Chem.*, 1993, **58**, 512). A good illustration of the method is the oxidation of 2-chlorobenzyl alcohol by $HCrO_3F{\bullet}C_5H_5N$ in chloroform. This gives the corresponding benzaldehyde in 95% yield (N. Narayanan and

T. R. Balasubramanian, *Ind. J. Chem.*, 1986, **25B**, 228).

(ii) Manganese Oxidants

Manganese reagents are the most versatile and vigorous of the commonly used oxidants (A. J. Fatiadi, *Synthesis,* 1987, 85) and they have similar applications to chromium reagents. Manganese dioxide has been used for the oxidation of alcohols under mild conditions (S. Tsuboi *et al., Bull. Chem. Soc. Jpn.,* 1990, **63**, 1888; L. A. Martinez *et al., Tetrahedron Lett.*, 1993, **34**, 5293), but its activity is dependent upon the method used to make it. On the other hand, the combination of potassium permanganate and triethylamine can be used to oxidize benzyl alcohol to benzaldehyde in almost quantitative yield (W. Li and L. K. Liu, *Synthesis,* 1989, 293).

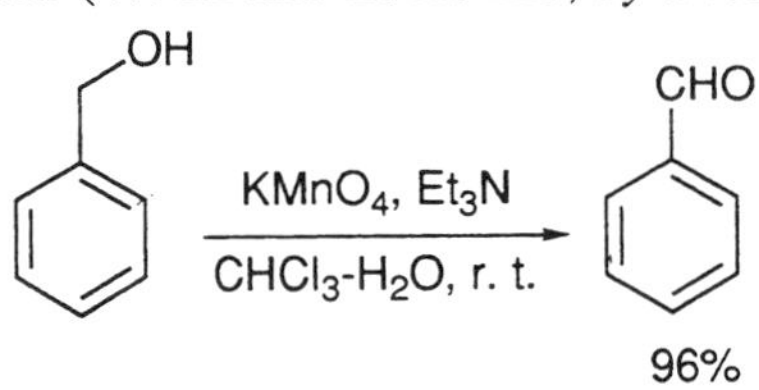

(iii) Ruthenium Oxidants

Ruthenium oxides generated *in situ* can be used to oxidize primary alcohols to the corresponding aldehydes and this reagent is recommended if acid-sensitive groups are present in the substrate. The reactions can be made catalytic with respect to ruthenium if a co-oxidant is added (P. Müller and J. Godoy, *Tetrahedron Lett.,* 1981, **22**, 2361; W. P. Griffith *et al., J. Chem. Soc., Chem. Commun.,* 1987, 1625; A. C. Dengel *et al., J. Chem. Soc. Dalton Trans.,* 1990, 737; M. M. T. Khan *et al., J. Mol. Catal.,* 1991, **67**, 309; M. Navarro *et al., Synth. Commun.,* 1990, **20**, 399; S. Kanemoto *et al., Bull. Chem. Soc. Jpn.,* 1988, **61**, 3607). For example, 4-methoxybenzyl alcohol affords anisaldehyde in 93% yield when it is treated with ruthenium trichloride and sodium bromate plus a suitable quaternary ammonium salt (A. J. Bailey *et al., Inorg. Chem.,* 1993, **32**, 268).

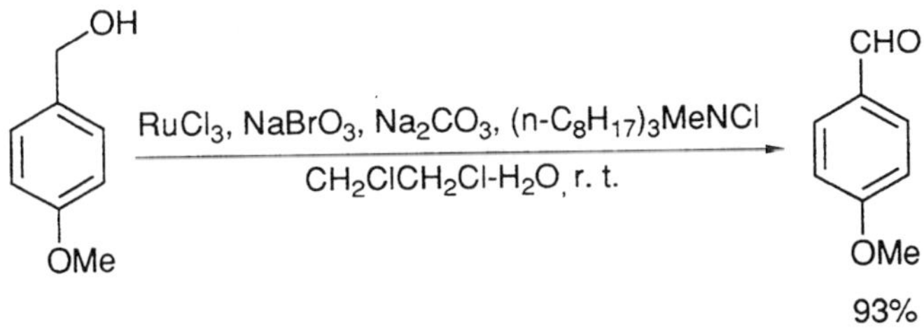

(iv) Oxygen and Peroxides

Oxygen and peroxides are activated by transition metal complexes so that they may be used to oxidize aromatic alcohols to aldehydes. Although synthetic applications are few in number, such reactions are studied extensively because they mimic those which occur in biological systems (P. Müller and H. Idmoumaz, *J. Organomet. Chem.*, 1988, **345**, 187; H. -C. Tung and D. T. Sawyer, *J. Am. Chem. Soc.*, 1990, **112**, 8214; H. Sugimoto *et al.*, *J. Am. Chem. Soc.*, 1988, **110**, 2465; A. M. Sakharov and I. P. Skibida, *J. Mol. Catal.*, 1988, **48**, 157; W. -Y. Sun *et al.*, *Tetrahedron,* 1993, **49**, 1357). In a typical procedure, the substrate and a Co-complex, prepared *in situ*, are intimately mixed with oxygen. In this way, benzyl alcohol affords benzaldehyde in 87% yield (A. Sobkowiak and D. T. Sawyer, *J. Am. Chem. Soc.*, 1991, **113**, 9520).

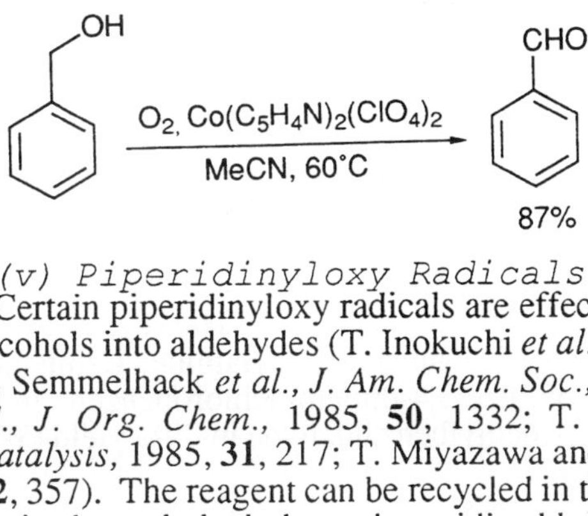

(v) Piperidinyloxy Radicals

Certain piperidinyloxy radicals are effective reagents for the conversion of alcohols into aldehydes (T. Inokuchi *et al., J. Org. Chem.*, 1990, **55**, 462; M. F. Semmelhack *et al., J. Am. Chem. Soc.*, 1984, **106**, 3374; T. Miyazawa *et al., J. Org. Chem.*, 1985, **50**, 1332; T. Miyazawa and T. Endo, *J. Mol. Catalysis,* 1985, **31**, 217; T. Miyazawa and T. Endo, *J. Mol. Catalysis,* 1985, **32**, 357). The reagent can be recycled in the presence of a co-oxidant. Thus, 3-nitrobenzyl alcohol can be oxidized by a catalytic amount of 4-methoxy-2,2,6,6-tetramethylpiperidine N-oxyl in the presence of sodium hypochlorite to give 3-nitrobenzaldehyde in quantitative yield (P. L. Anelli *et al., J. Org. Chem.*, 1987, **52**, 2559).

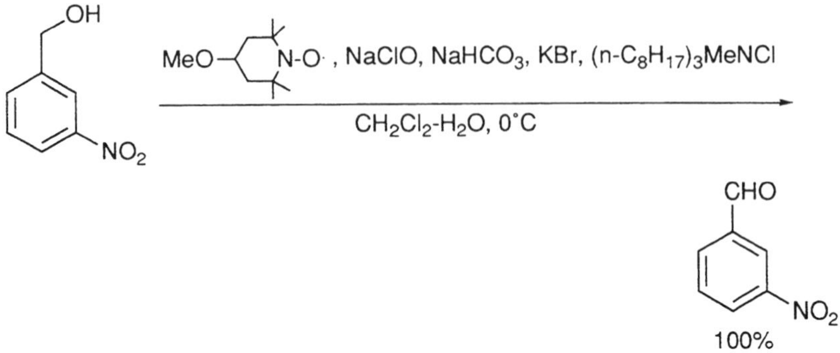

(vi) Other Oxidants

2,3-Dichloro-5,6-dicyanobenzoquinone (H. Becker *et al., J. Org. Chem.*, 1980, **45**, 1596) and trichloromelamine (S. Kondo *et al., J. Org. Chem.*, 1993, **58**, 5003) are also recommended as effective oxidants for primary alcohols. In addition, it is also possible to use aryl halides as oxidants in the presence of a palladium catalyst and a base (Y. Tamaru *et al., J. Org. Chem.*, 1983, **48**, 1286). Yet another reagent is lead tetraacetate, which can be employed to convert α,β-epoxyalcohols into aldehydes through oxidative cleavage (T. J. Brocksom *et al., J. Chem. Res. Synop.*, 1981, 334). Similarly the electro-oxidative bond scission of α,β-diols also affords aldehydes (S. Torii *et al., J. Org. Chem.*, 1984, **49**, 1830).

(b) From Arylalkanes

An activated methyl group, such as that of certain toluenes, is easily converted into a formyl unit by a variety of oxidants such as cerium(IV) methanesulfonate (R. P. Kreh *et al., J. Org. Chem.*, 1989, **54**, 1526), cerium(IV) sulfate (M. Morita *et al., Bull. Chem. Soc. Jpn.*, 1993, **66**, 2747), 2,3-dichloro-5,6-dicyanobenzoquinone (H. Lee and R. G. Harvey, *J. Org. Chem.*, 1983, **48**, 749) and potassium permanganate/triethylamine. Using the last reagent system, it is possible to oxidize 3,4-dimethylchlorobenzene in a selective manner to afford 4-chloro-2-methylbenzaldehyde in 89% yield (W. Li and L. K. Liu, *Synthesis*, 1989, 293).

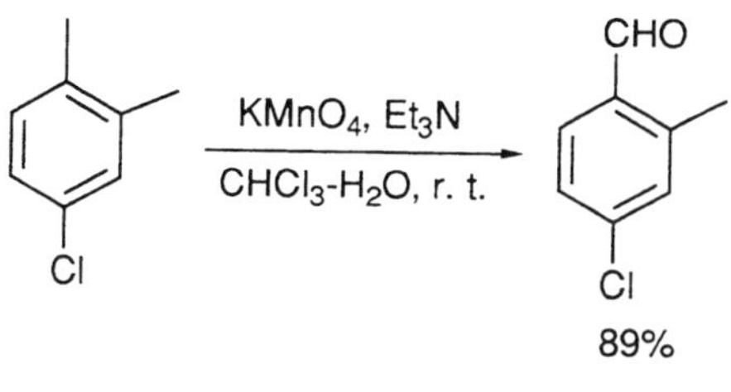

The direct anodic oxidation of para-substituted toluenes also gives aldehydes (I. Nishiguchi and T. Hirashima, *J. Org. Chem.*, 1985, **50**, 539).

(c) From Arylalkenes

Alkenes undergo oxidative cleavage to aldehydes by using either chromium trioxide/trimethylsilyl chloride (J. G. Lee and D. S. Ha, *Tetrahedron Lett.*, 1989, **30**, 193), solid-supported potassium permanganate (S, Baskaran *et al.*, *J. Org. Chem.*, 1989, **54**, 5182; D. G. Lee *et al.*, *J. Org. Chem.*, 1993, **58**, 2918), triethylsilyl hydrotrioxide (G. H. Posner *et al.*, *J. Am. Chem. Soc.*, 1987, **109**, 278), polymer-supported osmium tetraoxide (G. Cainelli *et al.*, *Synthesis,* 1989, 47), or ozone (T. Sugimoto *et al.*, *J. Org. Chem.*, 1993, **58**, 135) as the reagents. For example, ozonized oxygen when bubbled through a solution of 3,4-dimethoxyphenylpropene in methanol produces 3,4-dimethoxybenzaldehyde in 88% yield (D. Gupta *et al.*, *Tetrahedron,* 1982, **38**, 3013).

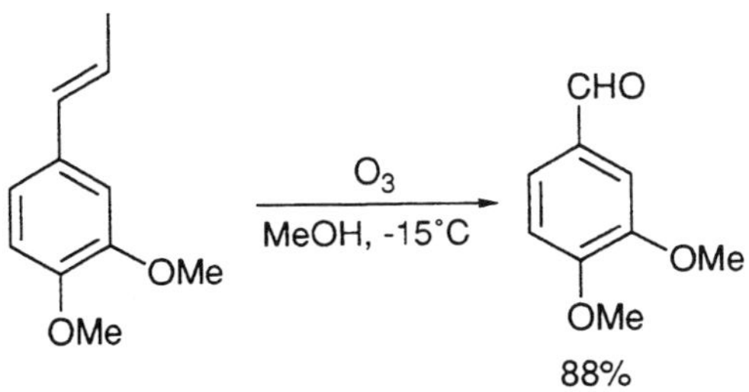

(d) From Aryl Halides

(i) Organometallic Reagents with Formylating Reagents

The reaction of an organometallic reagent with a formylating reagent provides a very versatile method with which to obtain aldehydes from aryl halides. Various formylating reagents, such as N,N-dimethylformamide (G. Voss and H. Gerlach, *Chem. Ber.*, 1989, 1199; G. A. Olah *et al.*, *Synthesis,* 1984, 228), N-formylpiperidine (G. A. Olah and M. Arvanaghi, *Angew. Chem. Int. Ed. Engl.*, 1981, **20**, 878), lithium formate (M. Bogavac *et al.*, *Tetrahedron Lett.*, 1984, **25**, 1843), carbon dioxide (G. Zadel and E. Breitmaier, *Angew. Chem. Int. Ed. Engl.*, 1992, **31**, 1035), and N-formylmorpholine can be used. As an illustration, 1-naphthylmagnesium bromide reacts with N-formylmorpholine to give 1-formylnaphthalene in 92% yield (G. A. Olah *et al.*, *J. Org. Chem.*, 1984, **49**, 3856).

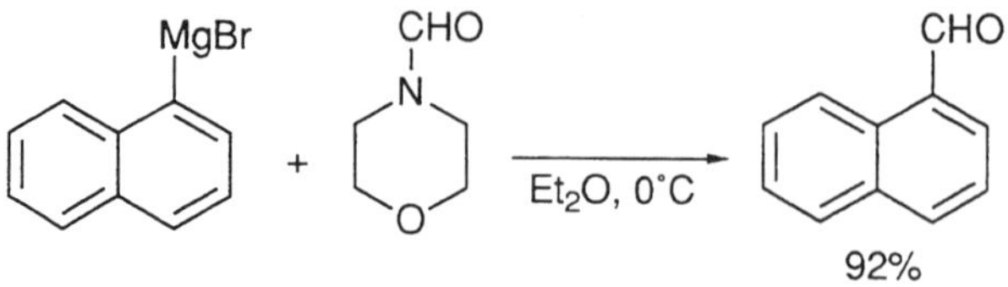

(ii) Formylation with Carbon Monoxide

Aryl halides, especially aryl iodides, are converted into aldehydes by reaction with carbon monoxide and hydrogen donors in the presence of a palladium catalyst (J. K. Stille, *Angew. Chem. Int. Ed. Engl.*, 1986, **25**, 508; R. Mutin *et al.*, *J. Chem. Soc., Chem. Commun.*, 1988, 896; I. Pri-Bar and O. Buchman, *J. Org. Chem.*, 1984, **49**, 4009; Y. Ben-David *et al.*, *J. Chem. Soc., Chem. Commun.*, 1989, 1816). A number of functional groups are tolerated under these conditions. In a typical procedure, a pressure vessel is used, and methyl 4-iodobenzoate is reacted with carbon monoxide at 45 psig in contact with tetrakis(triphenylphosphine)palladium(0) and tributyltin hydride to afford methyl 4-formylbenzoate in 90% yield (V. P. Baillargeon and J. K. Stille, *J. Am. Chem. Soc.*, 1986, **108**, 452).

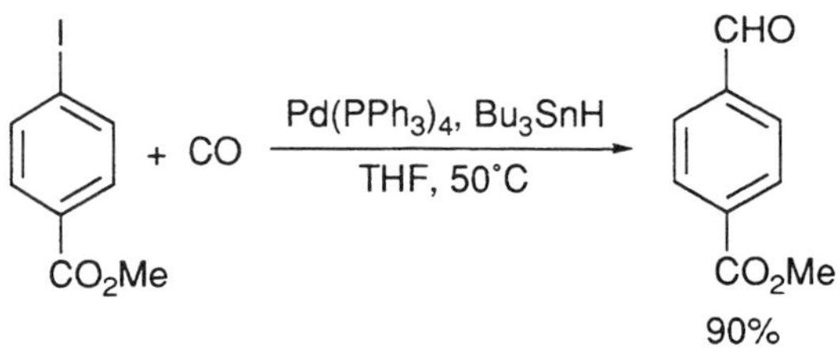

A palladium catalyst is not always necessary, and the free-radical carbonylation of aromatic iodides with carbon monoxide, tributyltin hydride, and 2,2'-azobis(isobutyronitrile) can give reasonably good results. Thus 4-methoxyiodobenzene gives anisaldehyde in 78% yield (I. Ryu *et al.*, *Tetrahedron Lett.*, 1990, **31**, 6887; *J. Am. Chem. Soc.*, 1990, **112**, 1295).

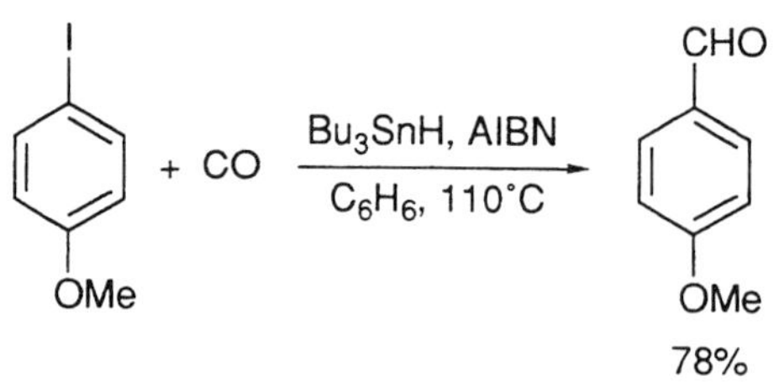

(e) From Carbonyl Compounds

(i) Oxidation of Aryl Acetic Acid Derivatives

Oxidative decarboxylations of phenylacetic acids with sodium hypochlorite (H. J. Carlsen, *Acta Chem. Scand.*, 1984, **B38**, 343), calcium hypochlorite (S. O. Nwaukwa and P. M. Keehn, *Tetrahedron Lett.*, 1982, **23**, 3135), N-iodosuccinimide (T. R. Beebe *et al., J. Org. Chem.*, 1982, **47**, 3006), or tetra(n-butyl)ammonium periodate (E. Santaniello *et al., Synthesis,* 1980, 563) all afford benzaldehydes. The last reagent, for example, degrades 4-chlorophenylacetic acid into 4-chlorobenzaldehyde in 65% yield (E. Santaniello *et al., Tetrahedron Lett.*, 1980, **21**, 2655).

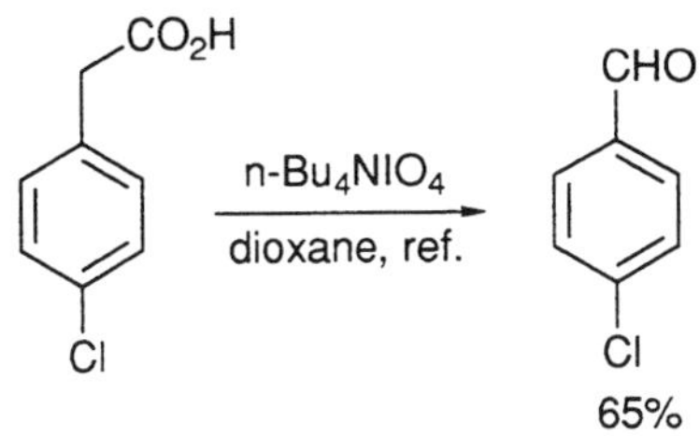

(ii) Reduction of Aryl Acid Derivatives

In general, the reduction of carbonyl compounds is difficult to control and alcohols are produced. However, by careful control of the reaction conditions and selection of the reductant aldehydes can be obtained. For example, acid halides can be reduced to aldehydes using sodium borohydride (I. D. Entwistle and P. Boehm, *J. Chem. Soc., Perkin Trans. 1,* 1980, 27; J. H. Babler and B. J. Invergo, *Tetrahedron Lett.*, 1981, **22**, 11; J. H. Babler, *Synth. Commun.*, 1982, **12**, 839), bis(cyanotrihydroborato)-tetrakis(triphenylphosphine) dicopper(I) (R. O. Hutchins and M. Markowitz, *Tetrahedron Lett.*, 1980, **21**, 813), polystyryl diphenyl phosphine tetrahydroborate copper(I) (W. Amaratunga and J. M. J. Frécht, *Polym. Prepr. Am. Chem. Soc., Div. Polym. Chem.*, 1981, **22**, 151), and pentacoordinated silicon hydrides (R. J. P. Corriu *et al., Tetrahedron Lett.*, 1988, **29**, 1271). Carboxylic acids can be reduced with thexylchloroborane (H. C. Brown *et al., J. Org. Chem.*, 1987, **52**, 5400), thexylbromoborane (J. S. Cha *et al., J. Org. Chem.*, 1987, **52**, 5030), 9-borabicyclo[3.3.1]nonane (J. S. Cha *et al., Tetrahedron Lett.*, 1987, **28**, 4575; J. S. Cha *et al., Tetrahedron Lett.*, 1987, **28**, 6231), and tri(t-butoxy)lithiumaluminium hydride (T. Fujisawa *et al., Tetrahedron Lett.*, 1983, **24**, 1543). Benzamides too can be reduced, and suitable reagents for this type of reaction are di(i-butyl)aluminium hydride (Y. Nagao *et al., J. Chem. Soc., Perkin Trans. 1,* 1980, 2470) and lithiumaluminium hydride. More than one functional group can be reduced at the same time: thus the diacid chloride of benzene-1,3-dicarboxylic acid can be reduced by treatment with tri(t-butoxy)sodiumaluminium hydride to give 1,3-diformylbenzene in 95% yield

(J. S. Cha and H. C. Brown, *J. Org. Chem.*, 1993, **58**, 4732).

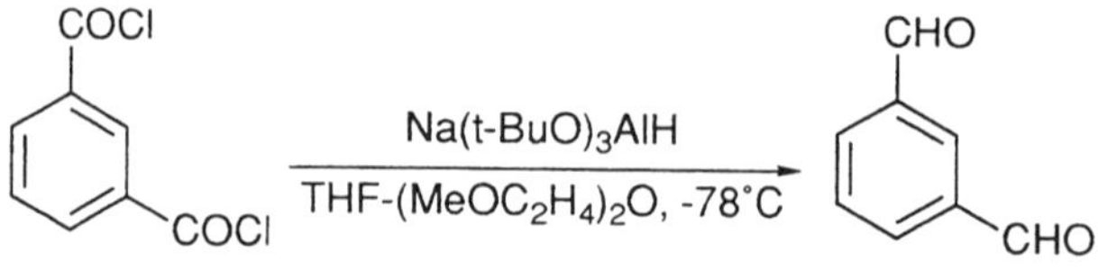

95%

Selectivity is also possible: thus 4-nitrobenzoic acid can be reduced to 4-nitrobenzaldehyde in a high yield by reaction with thexylchloroborane-dimethyl sulfide (H. C. Brown *et al.*, *J. Am. Chem. Soc.*, 1984, **106**, 8001).

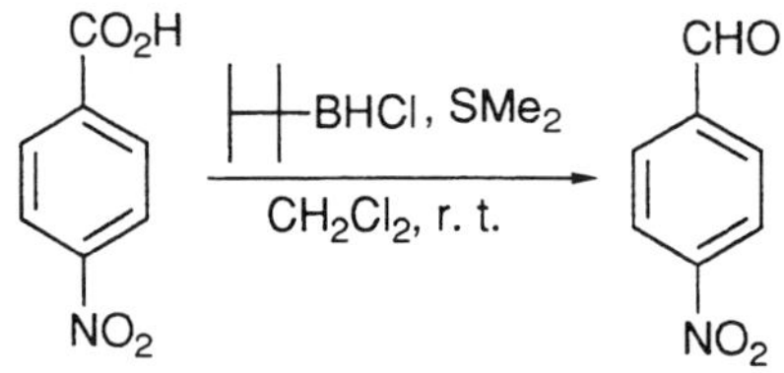

86%

Lithium aluminium hydride can be used to reduce the amide function of the diarylsulfide (1) without effecting C-S bond fission (K. Šindelář *et al.*, *Collect. Czech. Chem. Commun.*, 1982, **47**, 1367).

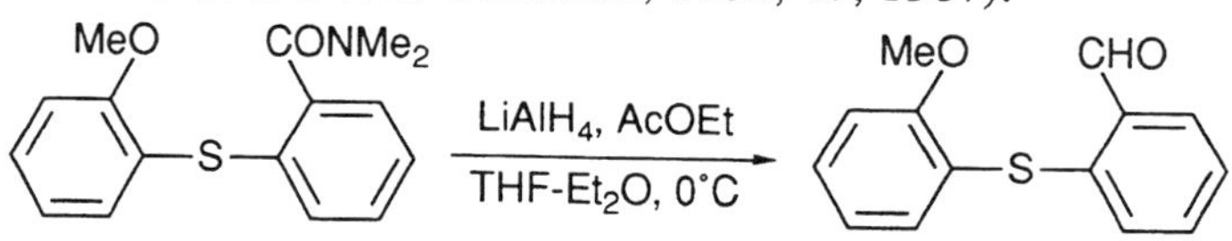

(1) 89%

(f) From Other Compounds

Benzyl halides are oxidized by dimethyl sulfoxide (A. J. Mancuso and D. Swern, *Synthesis,* 1981, 165), bis(*p*-methoxyphenyl)selenoxide (K. Ariyoshi *et al., Chem. Lett.*, 1984, 891), and 4-dimethylaminopyridine N-oxide (S. Mukaiyama *et al., Bull. Chem. Soc. Jpn.*, 1981, **54**, 2221) to aldehydes. The oxidation of nitro paraffin salts using potassium permanganate (N. Kornblum *et al., J. Org. Chem.*, 1982, **47**, 4534), ozone (J. E. McMurry and J. Melton, *Org, Synth.*, 1988, VI, 648), and ceric ammonium nitrate (G. A. Olah and B. G. B. Gupta, *Synthesis,* 1980, 44) gives aldehydes. Sulfones are oxidized in a similar manner using n-butyl lithium and bis(trimethylsilyl)perioxide (J. R. Hwu, *J. Org. Chem.*, 1983, **48**, 4432). The oxidations of benzyltrimethylsilanes by iodosobenzene in the presence of Fe and Mn porphyrins (E. Baciocchi and O. Lanzalunga, *Tetrahedron,* 1993, **49**, 7267), benzylamines by arylsulfonyl peroxides (R. V. Hoffman and A. Kumar, *J. Org. Chem.*, 1984, **49**, 4011; 1984, **49**, 4014), and benzyl ethers by metallic nitrates (T. Nishiguchi, and M. Bougauchi, *J. Org. Chem.*, 1989, **54**, 3001) all give the corresponding aldehydes. The reaction of diazonium salts with carbon monoxide in the presence of a palladium catalyst and a hydrosiloxane (K. Kikukawa *et al., J. Organomet. Chem.*, 1984, **270**, 283) produces aldehydes, and 1,4-cyclohexadienes (B. Raju and G. S. K. Rao, *Synthesis*, 1987, 197; B. Raju and G. S. K. Rao, *Synthesis*, 1985, 779) and cyclohexane-1,3-diones (A. R. Katritzky and C. M. Marson, *J. Org. Chem.*, 1987, **52**, 2726) react with N,N-dimethylformamide in the presence of thionyl chloride, or phosphoryl trichloride, to give aromatic aldehydes.

(g) From Aromatic Compounds

(i) Friedel-Crafts Type Formylation

The formylation of aromatic compounds by electrophilic substitution is a classic and important method for the preparation of aldehydes. The reaction of arenes with carbon monoxide in a super acidic mixture of Brønsted and Lewis acids is named after Gattermann and Koch (B. L. Booth *et al., J. Chem. Soc. Perkin Trans. 1*, 1980, 181; G. A. Olah *et al., J. Org. Chem.*, 1985, **50**, 1483; L. Toniolo and M. Granziani, *J. Organoment. Chem.*, 1980, **194**, 221; M. Tanaka *et al., J. Org. Chem.*, 1992, **57**, 2677; M. Tanaka and Y. Souma, *J. Chem. Soc., Chem. Commun.*, 1991, 1551). An illustration is the formylation of biphenyl with carbon monoxide in contact with hydrogen fluoride and antimony pentafluoride, which effects disubstitution and gives mainly the 4,4'-dialdehyde with a minor amount of the 2,4'-isomer (M. Tanaka *et al., J. Org. Chem.*, 1993, **58**, 3213).

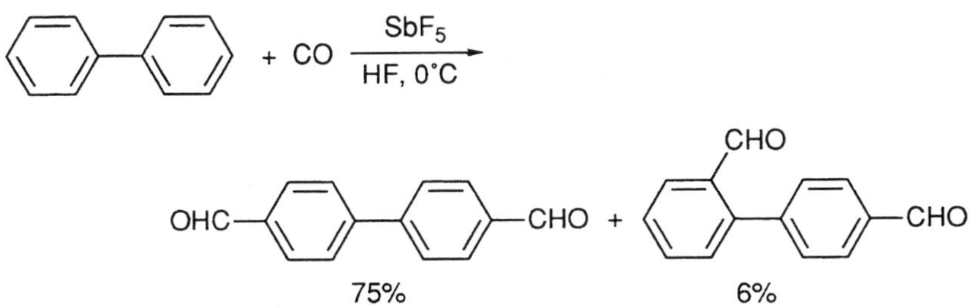

Recently, either hydrogen fluoride, fluorosulfonic acid, or trifluoromethanesulfonic acid have been used as the Brønsted acid in combination with either boron trifluoride, antimony pentafluoride, tantalum pentafluoride, or niobium pentafluoride, as the Lewis acid component. All form extremely acidic media which promote formyl cation formation, even though the formyl cation is very unstable. More stable equivalents of this cation have been sought, and the following reagents are recommended, dichloromethyl methyl ether (P. C. Meltzer *et al., J. Org. Chem.*, 1985, **50**, 3121), cyanide salts (K. Chen *et al., J. Org. Chem.*, 1985, **50**, 3997), hexamethylenetetramine (D. L. Ladd *et al., Synth. Commun.*, 1985, **15**, 61),

N,N-dimethylformamide (I. M. Downie *et al., Tetrahedron,* 1993, **49**, 4015; D. St. C. Black *et al., J. Chem. Soc., Chem. Commun.,* 1993, 819), tris(phenylthio)methane-dimethyl(methylthio)sulfonium tetrafluoroborate, and imidazole-acid anhydride. All these are used in the presence of a strong acid, or a Lewis acid. Even fluoro arenes can be formylated in this way, and 3,4-dimethoxyfluorobenzene when treated with dichloromethyl methyl ether and titanium tetrachloride gives 2-fluoro-4,5-dimethoxybenzaldehyde in 60% yield (D. C. Furlano and K. L. Kirk, *J. Org. Chem.,* 1986, **51**, 4073).

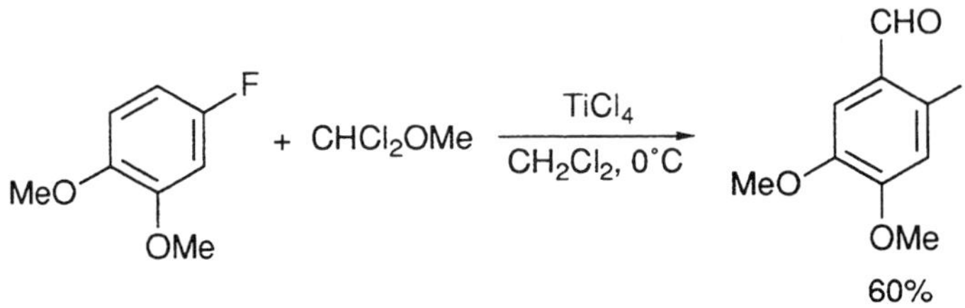

Dibenzo-18-crown-6 reacts with hexamethylenetetramine and trifluoroacetic acid to give di(4-formylbenzo)-18-crown-6 in 90% yield (F. Wada *et al., Bull. Chem. Soc. Jpn.,* 1980, **53**, 1473), whereas more reactive aromatics such as 2,6-dihydroxytoluene can be formylated with zinc cyanide in the presence of hydrogen chloride (J. Barber and J. Staunton, *J. Chem. Soc., Perkin Trans. 1,* 1981, 1685).

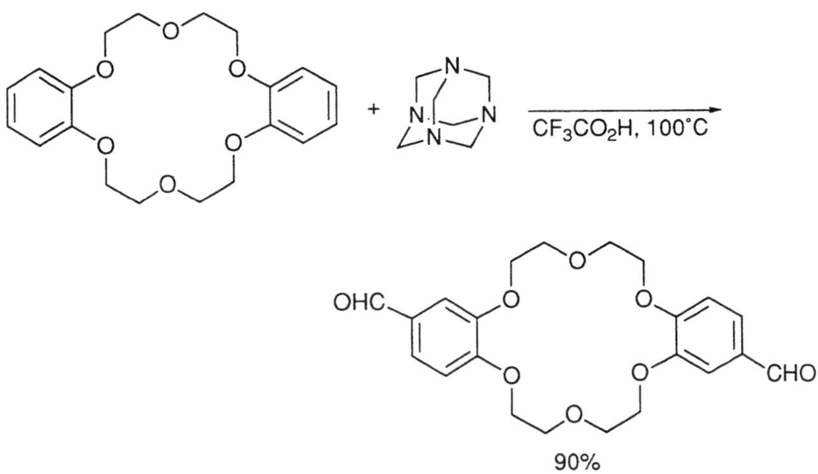

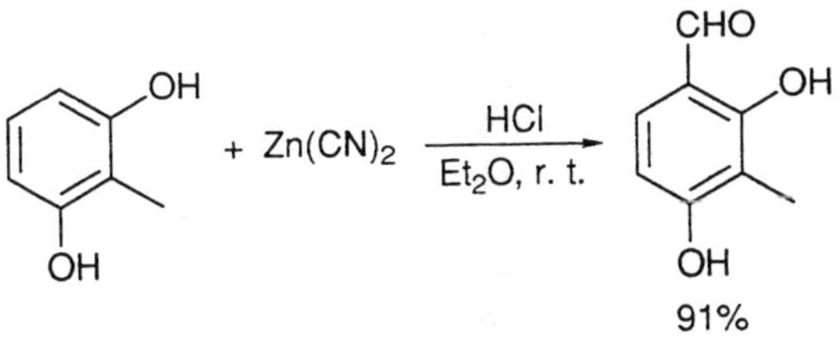

Under conventional Vilsmeir conditions (N,N-dimethylformamide/phosphoryl trichloride), julolidine (2) affords a 70% yield of the corresponding aldehyde (G. Cai *et al.*, *J. Am. Chem. Soc.*, 1993, **115**, 7192).

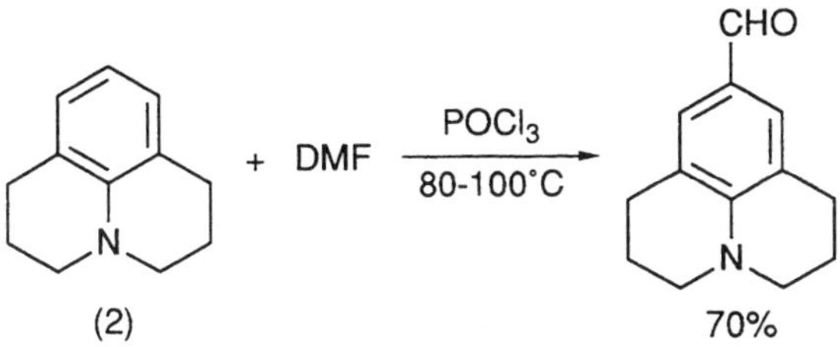

(2)

Selective 1-formylation of 2-naphthol occurs when tris(phenylthio)methane and dimethyl(methylthio)sulfonium tetrafluoroborate are the reagents (R. A. J. Smith and A. R. B. Manas, *Synthesis,* 1984, 166).

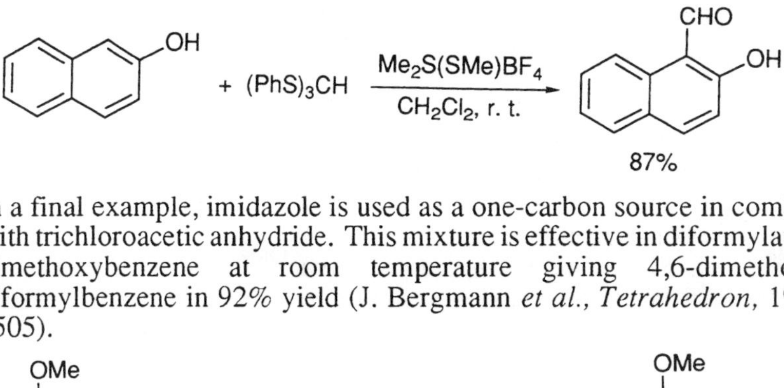

In a final example, imidazole is used as a one-carbon source in combination with trichloroacetic anhydride. This mixture is effective in diformylating 1,3-dimethoxybenzene at room temperature giving 4,6-dimethoxy-1,3-diformylbenzene in 92% yield (J. Bergmann *et al.*, *Tetrahedron,* 1980, **36**, 2505).

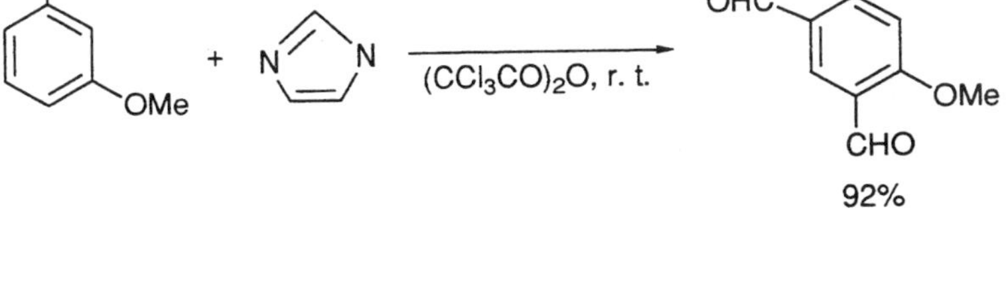

Phenols are converted into hydroxyaldehydes by treatment with chloroform under basic conditions (Reimer-Tiemann reaction) (C. W. Bird *et al.*, *Tetrahedron,* 1985, **41**, 4685; A. Thoer *et al., Synth. Commun.,* 1988, **18**, 2095), however, yields are generally poor. The addition of either α- or β-cyclodextrin (β-CD) offers some improvement (M. Komiyama and H. Hirai, *Bull. Chem. Soc. Jpn.,* 1981, **54**, 2053; M. Komiyama and H. Hirai, *Polym. J.,* 1986, **18**, 375) and in this way, phenol, for example, undergoes regioselective formylation giving 4-hydroxybenzaldehyde in 65% yield (Komiyama and H. Hirai, *J. Am. Chem. Soc.,* 1983, **105**, 2018).

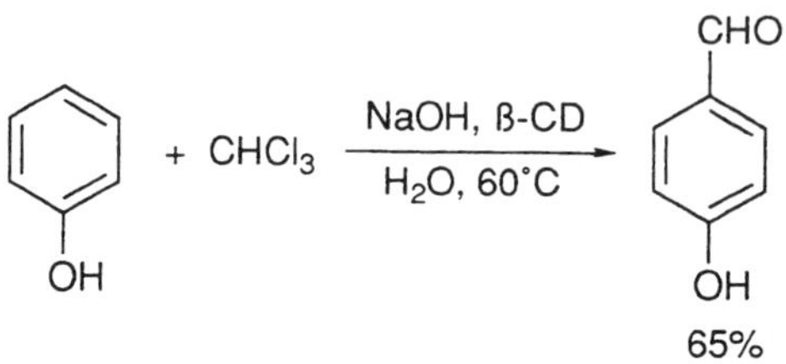

(ii) Lithiation of Aromatic Compounds

Aromatic compounds which have protons *ortho* to substituent groups bearing hetero atoms (N, O, or S) can be lithiated by treatment with alkyl lithium. Subsequent reaction of the lithiated derivatives with N,N-dimethylformamide gives excellent yields of the corresponding benzaldehydes (L. A. Paquette and M. M. Schulze, *Tetrahedron Lett.,* 1993, **34**, 3235; A. Basha and D. W. Brooks, *J. Org. Chem.,* 1993, **58**, 1293). The formylation of 3,4-dimethoxy-N,N-diethylbenzamide is an example of this *ortho*-lithiation formylation procedure giving the amido aldehyde as the only product in 88% yield (S. O. de Silva *et al., Tetrahedron,* 1992, **48**, 4863).

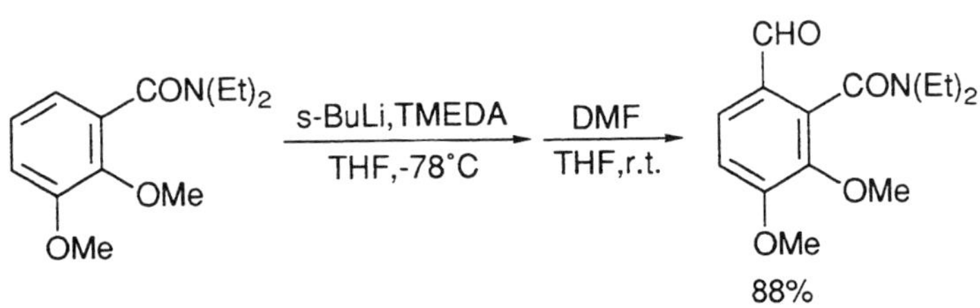

2. Synthesis of Aromatic Monoketones

(a) From s-Arylalkanols

(i) Chromium Oxidants

Over oxidation is now less of a problem, and oxidants containing chromium(VI) are commonly used for the synthesis of ketones from various s-arylalkanols. These reagents are popular because of their ready availability and operational simplicity (F. A. Luzzio and F. S. Guziec, Jr., *Org. Prep. Proc. Int.,* 1988, **20**, 533; G. Piancatelli *et al., Synthesis,* 1982, 245). Oxychromium-amine complexes are recommended, and these include $HCrO_3F{\bullet}C_5H_5N$ (M. N. Bhattacharjee *et al., Synthesis,* 1982, 588; T. Nonaka *et al., Bull. Chem. Soc. Jpn.,* 1984, **54**, 2019), $HCrO_3Cl{\bullet}C_5H_5N$ (J. Herscovici *et al., J. Chem. Soc., Perkin Trans. 1,* 1982, 1967), $HCrO_3Br{\bullet}C_5H_5N$ (N. Narayanan and T. R. Balasubramanian, *Ind. J. Chem.,* 1986, **25B**, 228), $H_2Cr_2O_7{\bullet}(C_5H_5N)_2$ (S. Kanemoto *et al., Bull. Chem. Soc. Jpn.,* 1988, **61**, 3607), and $H_2Cr_2O_7{\bullet}(3\text{-}C_5H_4NCO_2H)_2$ (F. P. Cossio *et al., Tetrahedron,* 1987, **43**, 3963). Yields are good, and for example, 1-(3,4-methylenedioxyphenyl)ethanol gives the corresponding ketone in almost quantitative yield when $HCrO_3Cl{\bullet}(C_5H_4N)_2$ is employed (F. S. Guziec, Jr. and F.A. Luzzio, *Synthesis,* 1980, 691).

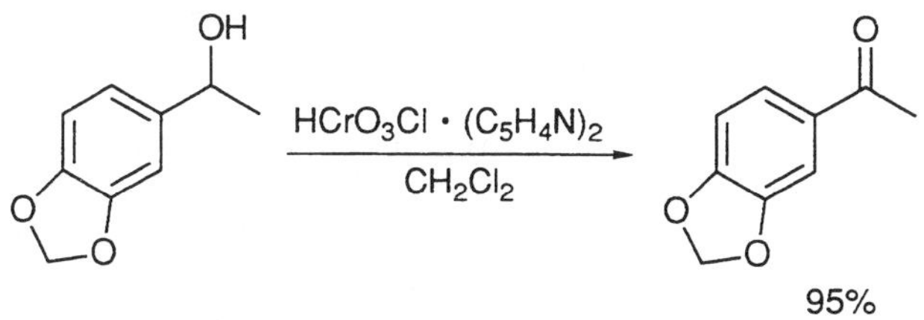

Other chromium bearing reagents such as esters, $(t\text{-}BuO)_2CrO_2$ and $CH_2(CMe_2O)_2CrO_2$, are also suggested as effective oxidants for secondary arylalkanols (F. A. Luzzio and W. J. Moore, *J. Org. Chem.,* 1993, **58**, 512).

(ii) Manganese Oxidants

Manganese reagents have been extensively employed in acid, alkaline, and neutral media for the oxidation of secondary arylalkanols (A. J. Fatiadi, *Synthesis,* 1987, 85; M. Trömel and M. Russ, *Angew. Chem. Int. Ed. Engl.,* 1987, **26**, 1007). Manganese dioxide is especially useful (S. Tsuboi *et al., Bull. Chem. Soc. Jpn.,* 1990, **63**, 1888; T. Kimura *et al., Chem. Lett.,* 1988, 1387; M. Uemura *et al., J. Org. Chem.,* 1993, **58**, 1238) and be applied under mild conditions. However, its activity varies depending upon the method used to prepare it. On the other hand, potassium permanganate in combination with triethylamine is very effective, and this procedure is suitable for diphenylmethanol to benzophenone in 96% yield (W. -S. Li and L. K. Liu, *Synthesis,* 1989, 293).

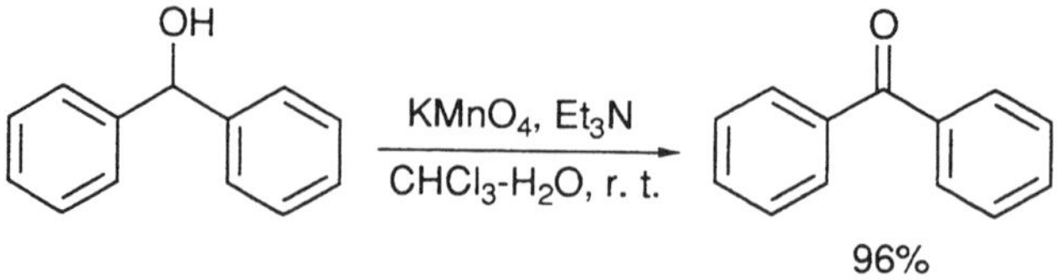

(iii) Ruthenium Oxidants

The value of ruthenium salts as oxidants for primary alcohols has already been mentioned, and this extends to secondary alcohols (P. Müller and J. Godoy, *Tetrahedron Lett.*, 1981, **22**, 2361; M. Navarro *et al., Synth. Commun.*, 1990, **20**, 399; S. Murahashi *et al., J. Org. Chem.*, 1993, **58**, 7318; S. Kanemoto *et al., Bull. Chem. Soc. Jpn.*, 1988, **61**, 3607). Thus another, although less efficient, method for the oxidation of diphenylmethanol is to react it with ruthenium(III) chloride in the presence of a co-oxidant, sodium bromate. Once again, this type of reaction can be made catalytic with respect to ruthenium (A. J. Bailey *et al., Inorg. Chem.*, 1993, **32**, 268).

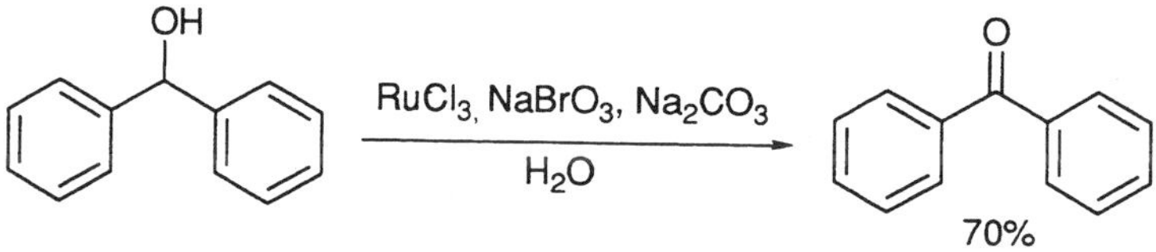

(iv) Oxygen and Peroxides

Biomimetic oxidation using oxygen, or peroxides, in combination with a transition metal complex has been employed to obtain aryl ketones from alcohols (H. -C. Tung and D. T. Sawyer, *J. Am. Chem. Soc.*, 1990, **112**, 8214; C. Sheu *et al., J. Am. Chem. Soc.*, 1990, **112**, 1936; H. Sugimoto *et al., J. Am. Chem. Soc.*, 1988, **110**, 2465; A. Sobkowiak *et al., J. Am. Chem. Soc.*, 1993, **115**, 609; W. -Y. Sun *et al., Tetrahedron,* 1993, **49**, 1357), but the potential of this reaction has not been fully exploited. In one case, however, it has been shown that tris(triphenylphosphine)rhodium chloride in nitrobenzene containing t-butyl hydroperoxide is effective in oxidizing 1-phenylethanol to afford acetophenone in 98% yield (P. Müller and H. Idmoumaz, *J. Organomet. Chem.*, 1988, **345**, 187).

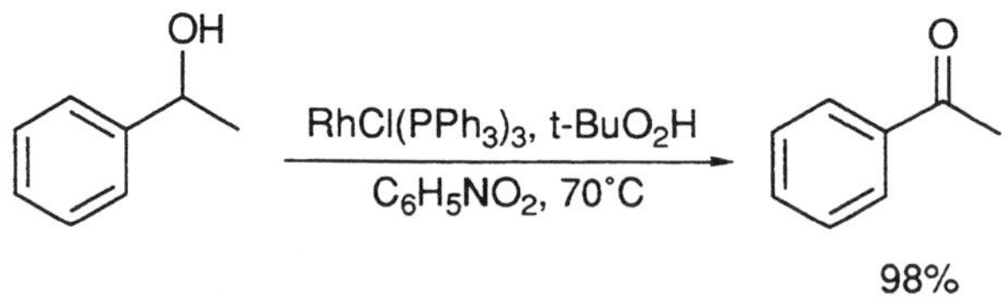

(v) Piperidinyloxy Radicals

Certain piperidinyloxy radicals are also valuable oxidants for both primary and secondary alcohols. Indeed, as mentioned earlier, these reagents can be recycled if a co-oxidant is added (T. Miyazawa *et al., J. Org. Chem.*, 1985, **50**, 1332; Z. Ma *et al., J. Org. Chem.*, 1993, **58**, 4837). For example, in the case of 1-phenylethanol, 4-benzoyloxy-2,2,6,6-tetramethylpiperidine N-oxyl can be used to form acetophenone, and the reagent can be regenerated by the addition of sodium bromite (T. Inokuchi *et al., J. Org. Chem.*, 1990, **55**, 462).

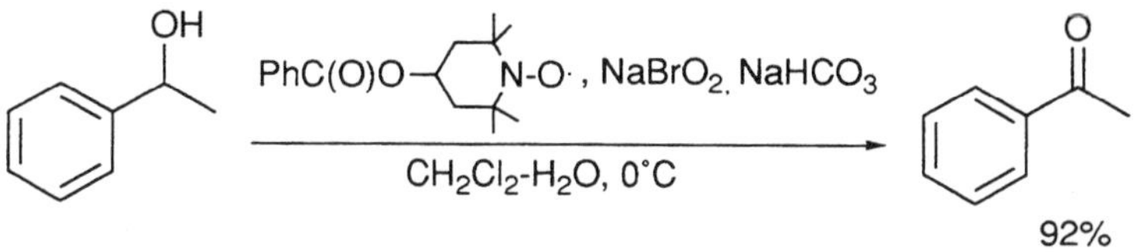

(vi) Other Oxidants

Calcium hypochlorite is a commercially available and cheap oxidant which can be used to obtain ketones from alcohols (S. O. Nwaukwa and P. M. Keehn, *Tetrahedron Lett.*, 1982, **23**, 3135). Other reagents include trichloromelamine (S. Kondo *et al., J. Org. Chem.*, 1993, **58**, 5003), N-bromosuccinimide (P. C. Bulman *et al., Tetrahedron,* 1992, **48**, 7265), 2,3-dichloro-5,6-dicyanobenzoquinone (H. Becker *et al., J. Org. Chem.*, 1980, **45**, 1596), and activated dimethyl sulfoxide (A. J. Mancuso and D. Swern, *Synthesis,* 1981, 165). In addition, cyclic peroxides, dioxiranes, offer synthetic potential as new and powerful oxidants (W. Adam *et al., Acc. Chem. Res.*, 1989, **22**, 205; R. W. Murray, *Chem. Rev.*, 1989, **89**, 1187; W. Adam *et al., Tetrahedron Lett.*, 1993, **34**, 8427; L. D'Accolti *et al., J. Org. Chem.*, 1993, **58**, 3600). Zirconium dioxide is also a known oxidant for arylalkanols when there is a suitable ketone as a hydrogen acceptor (H. Kuno *et al., Bull. Chem. Soc. Jpn.*, 1990, **63**, 1943).

(b) From Arylalkanes

Benzylic methylene groups are easily oxidized to carbonyl units, and $HCrO_3Cl{\bullet}C_5H_5N$ (R. Rathore *et al., Synth. Commun.*, 1986, **16**, 1493; E. J. Parish *et al., Synth. Commun.*, 1986, **16**, 1371; W. J. Lee and C. H. Park, *J. Org. Chem.*, 1993, **58**, 7149), chromium trioxide (Y. Ohba *et al., Bull. Chem. Soc. Jpn.*, 1993, **66**, 828), potassium permanganate (W. Li and L. K. Liu, *Synthesis,* 1989, 293), ruthenium oxides (C. -M. Che *et al., J. Chem. Soc. Dalton Trans.*, 1991, 1259; S. Murahashi *et al., Tetrahedron Lett.*, 1993, **34**, 1299), ceric ammonium nitrate (W. Y. Lee *et al., J. Am. Chem. Soc.*, 1993, **115**, 1184), cerium(IV) methanesulfonate (R. P. Kreh *et al., J. Org. Chem.*, 1989, **54**, 1526), and 2,3-dichloro-5,6-dicyanobenzoquinone (H. Lee and R. G. Harvey, *J. Org. Chem.*, 1983, **48**, 749) have all been used to afford ketones. A good illustration is provided by the conversion of the hydrocarbon (3) into the diketone (4) using $HCrO_3Cl{\bullet}C_5H_5N$ and Celite suspended in benzene (W. Y. Lee *et al., J. Org. Chem.*, 1992, **57**, 4074).

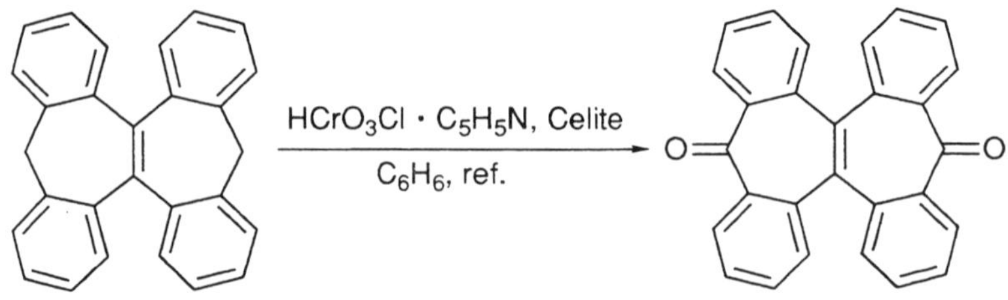

(3) (4)77%

Benzylic oxidation is also promoted during anodic oxidation, and for example, 6-methoxytetralin affords 6-methoxytetralone when it is electrolyzed with tetraethylammonium tosylate as the supporting electrolyte (I. Nishiguchi and T. Hirashima, *J. Org. Chem.*, 1985, **50**, 539).

(c) From Arylalkenes

Arylalkenes can be oxidized and cleaved by ozone to give arylketones (T. Sugimoto *et al.*, *J. Org. Chem.*, 1993, **58**, 135) and similar reactions are observed during oxidations catalyzed by palladium (J. Tsuji, *Synthesis*, 1984, 369; J. Tsuji and M. Minato, *Tetrahedron Lett.*, 1987, **28**, 3683). Other processes of this type include photochemical oxidations using molecular oxygen in the presence of a sensitizer (U. T. Bhalerao and M. Sridhar, *Tetrahedron Lett.*, 1993, **34**, 4341; A. A. Frimer and J. Weiss, *J. Org. Chem.*, 1993, **58**, 3660; M. L. Graziano *et al.*, *Synthesis*, 1983, 125). Thus, for example, 1,1-diphenylethene is oxidatively cleaved to give benzophenone in 71% yield when it is irradiated in contact with oxygen and titanium dioxide (M. A. Fox and C. -C. Chen, *J. Am. Chem. Soc.*, 1981, **103**, 6757).

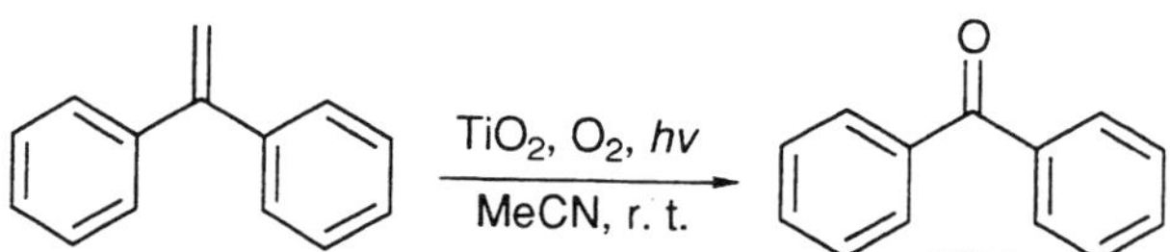

71%

The oxidation of various arylalkenes to the corresponding ketones bearing hydroxyl or thiol groups, as well as halogens, have been documented recently exploiting potassium permanganate (S. Baskaran *et al.*, *J. Org. Chem.*, 1989, **54**, 5182), chromium trioxide-trimethylsilyl chloride (J. G. Lee and D. S. Ha, *Tetrahedron Lett.*, 1989, **30**, 193), sodium bromite (T. Kageyama *et al.*, *Chem. Lett.*, 1983, 1481), ferric chloride (A. Kohda *et al.*, *J. Org. Chem.*, 1981, **46**, 509), and dimethylsulfoxide-dimethyl(methylthio)sulfonium tetrafluoroborate (B. M. Trost *et al.*, *J. Am. Chem. Soc.*, 1982, **104**, 3228) as the reagents.

(d) From Arylalkynes

The oxidations of arylalkynes with either chromium oxidants in combination with iodosobenzene (B. Rihter *et al., J. Am. Chem. Soc.*, 1993, **115**, 3918), ruthenium(III) ion and sodium iodate (R. Zibuck and D. Seeback, *Helv. Chim. Acta,* 1988, **71**, 237), or hydrogen peroxide-mercuric acetate (F. P. Ballistreri *et al., J. Org. Chem.*, 1988, **53**, 830) all give α,β-diketones. For example, diphenylethyne is oxidized by tris(triphenylphosphine)ruthenium dichloride and iodosobenzene to afford benzil in 86% yield (P. Müller and J. Godoy, *Helv. Chim. Acta.*, 1981, **64**, 2531).

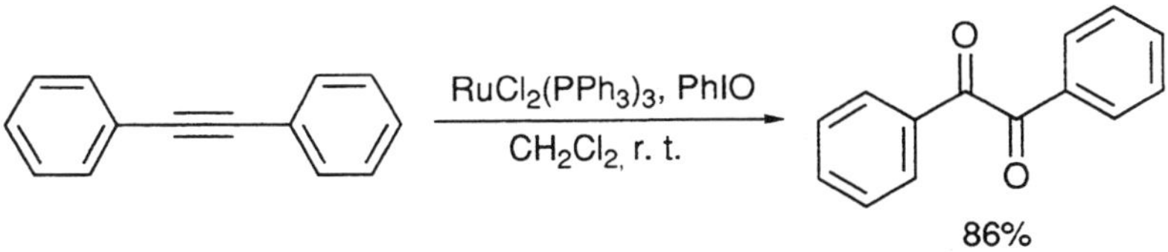

(e) From Aryl Halides

(i) Carbonylation with Carbon Monoxide

Coupling reactions between aryl halides and carbon monoxide and alkenes, alkynes, organometallics, or halides, are catalyzed by transition metals (N. E. Schore, *Chem. Rev.*, 1988, **88**, 1081; K. S. Chan *et al., J. Organomet. Chem.*, 1987, **334**, 9; M. Tanaka, *Synthesis,* 1981, 41), especially palladium (J. K. Stille, *Angew. Chem. Int. Ed. Engl.*, 1986, **25**, 508; N. A. Bumagin *et al., J. Organometal. Chem.*, 1989, **365**, 379; J. M. Tour and E. Negishi, *J. Am. Chem. soc.*, 1985, **107**, 8289; E. Negishi, *Acc. Chem. Res.*, 1987, **20**, 65; S. Torii *et al., Tetrahedron,* 1993, **49**, 6773). For example, 4-cyanoiodobenzene reacts with carbon monoxide and di(i-butyl)phenylaluminium in the presence of $PdCl_2(MeCN)_2$ to give 4-cyanobenzophenone in 94% yield (N. A. Bumagin *et al., Tetrahedron Lett.*, 1985, **26**, 4819).

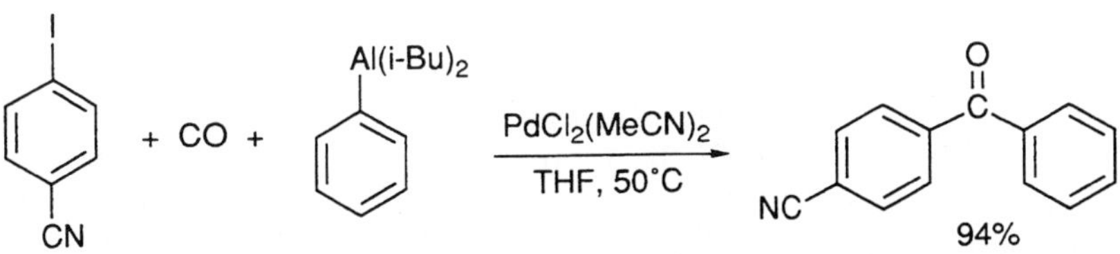

Radical coupling reactions between aryl halides, carbon monoxide, and an allylstannane have become very popular. Considering the multi-component nature of these processes, yields are good, and in an illustration, 4-iodotoluene undergoes a tandem coupling with carbon monoxide, ethyl acrylate, and allyltributyltin to give ethyl 2-(4-methylbenzoylmethyl)pent-4-enoate in 65% yield (I. Ryu *et al., J. Am. Chem. Soc.*, 1993, **115**, 1187).

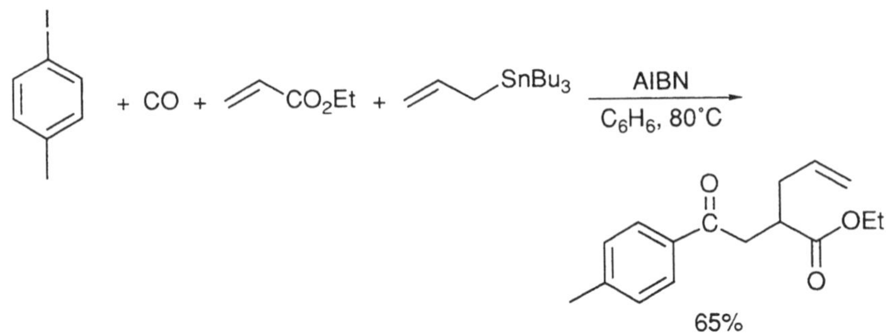

(ii) Other Reactions

s-Benzyl halides are oxidized by ruthenium oxides to yield ketones (A. J. Bailey *et al., Inorg. Chem.,* 1993, **32**, 268), and 4-(N,N-dimethylamino)pyridine N-oxide (S. Mukaiyama *et al., Bull. Chem. Soc. Jpn.,* 1981, **54**, 2221), and also bis(*p*-methoxyphenyl)selenoxide (K. Ariyoshi *et al., Chem. Lett.,* 1984, 891) can be used in similar cases. Aryl ketones are available though the nucleophilic reactions of organoarenes and carbonyl compounds. The problem, of course, is to prevent the formation of t-alcohols, and in this regard, suitable lithium (M. R. Paleo *et al., J. Org. Chem.,* 1993, **58**, 2763), manganese (G. Friour *et al., Synthesis.* 1984, 37), mercury (N. A. Bumagin *et al., J. Organometal. Chem.,* 1989, **365**, 379), and palladium (R. A. Holton and K. J. Natalie, Jr., *Tetrahedron Lett.,* 1981, **22**, 267) reagents have been prepared from aryl halides.

(f) From Aryl Ethers

Aryl epoxides are converted into aryl ketones through ring-opening reactions with samarium hypoiodite (J. Prandi *et al., J. Organomet. Chem.,* 1985, **285**, 449), silver tetrafluoroborate (K. Griesbaum *et al., Chem. Ber.,* 1981, **114**, 1858), cesium acetate-lead tetraacetate (T. Satoh *et al., Bull. Chem. Soc. Jpn.,* 1986, **59**, 946), palladium catalysts (Y. D. Vankar *et al., Synth. Commun.,* 1986, **16**, 1621; M. Suzuki *et al., J. Am. Chem. Soc.,* 1980, **102**, 2095), and by electrocatalytic isomerization (J. Delaunay *et al., J. Chem. Soc., Chem. Commun.,* 1982, 387). In each case, however, there can be problems in controlling the regioselectivity of the ring-opening process, unless groups are present which exert a clear influence on the electronic properties of the epoxide. For instance, the nitroepoxide (5) when treated with tetrakis(triphenylphosphine)palladium(0) and triethylamine gives benzil in 63% yield (Y. D. Vankar and S. P. Singh, *Chem. Lett.,* 1986, 1939).

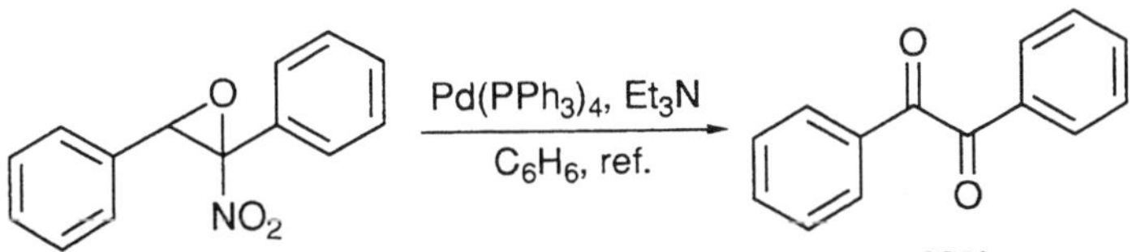

(5) 63%

A more useful reaction is the oxidative dimerization of the arylsiloxycyclopropane (6) in the presence of copper(II) tetrafluoroborate to produce 1,6-diphenylhexane-1,6-dione in 78% yield (I. Ryu *et al., J. Am. Chem. Soc.*, 1993, **115**, 12330).

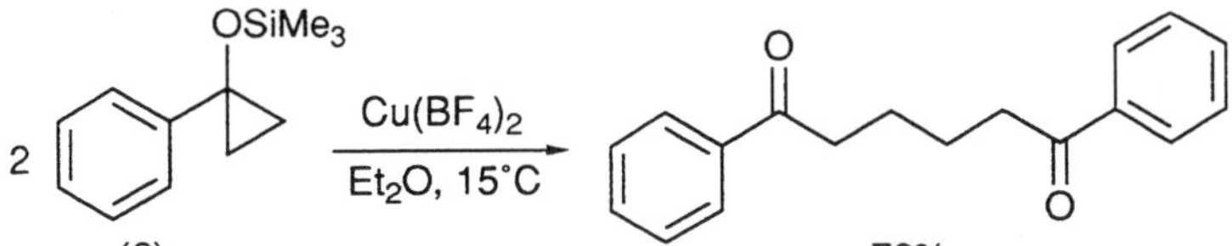

(6) 78%

Other reagents, for example, mercury(II) acetate (I. Ryu *et al., Tetrahedron Lett.*, 1980, **21**, 4283) and ceric ammonium nitrate (A. B. Paolobelli *et al., Tetrahedron Lett.*, 1993, **34**, 6333) can also be employed for the ring-opening reactions of arylsiloxycyclopropanes.

(g) From Carbonyl Compounds

(i) From Aroyl Halides

The reaction of organolithium (A. R. Katritzky *et al., Synthesis,* 1990, 666; W. Hayden *et al., Monatsh. Chem.*, 1987, **118**, 415; M. Sekine *et al., Bull. Chem. Soc. Jpn.*, 1982, **55**, 218) or Grignard (C. Cardellicchio *et al., Tetrahedron Lett.*, 1987, **28**, 2053) reagents with acid halides has provided an important method for ketone synthesis for many decades. There is a problem, however, since the ketones which are produced can react with more of the reagent affording tertiary alcohols. To circumvent this, various organometallic reagents including those of vanadium (T. Hirao *et al., Tetrahedron Lett.*, 1986, **27**, 929), cadmium (E. R. Burkhardt and R. D. Rieke, *J. Org. Chem.*, 1985, **50**, 416), nickel (S. Inaba and R. D. Rieke, *Tetrahedron Lett.*, 1983, **24**, 2451; *J. Org. Chem.*, 1985, **50**, 1373), titanium (A. N. Kasatkin *et al., J. Organomet. Chem.*, 1988, **346**, 23; T. Chou and S. Huang, *Tetrahedron Lett.*, 1983, **24**, 2169), zinc (R. A. Grey, *J. Org. Chem.*, 1984, **49**, 2288; T. Sakamoto *et al., Tetrahedron*, 1993, **49**, 9713; Y. Tamaru *et al., Angew. Chem.*, 1987, **99**, 1193), tin (C. Nativi *et al., Tetrahedron*, 1989, **45**, 1131; N. A. Bumagin *et al., J. Organomet. Chem.*, 1985, **282**, 421; L. S. Liebeskind *et al., J. Org. Chem.*, 1993, **58**, 3543; J. P. Gillet *et al., Synthesis*, 1986, 538), mercury (R. C. Larock and Y. Lu, *Tetrahedron Lett.*, 1988, **29**, 6761; *J. Org. Chem.*, 1993, **58**, 2846), lead (J. Yamada and Y.

Yamamoto, *J. Chem. Soc., Chem. Commun.*, 1987, 1302), and aluminum (K. Wakamatsu *et al., Bull. Chem. Soc. Jpn.*, 1985, **58**, 2425) have all been tried with varying degrees of success. Organozinc, tin, lead, and aluminum reagents are usually used in the presence of a palladium catalyst, and organomercury reagents require catalysis by a Lewis acid. Organocopper species also combine with aroyl halides to give aryl ketones (P. Knochel *et al., J. Org. Chem.*, 1988, **53**, 2390; 1993, **58**, 2694; *J. Am. Chem. Soc.*, 1993, **115**, 7027; J. Wei and R. O. Hutchins, *J. Org. Chem.*, 1993, **58**, 2920; N. Jabri *et al., Tetrahedron,* 1986, **42**, 1369), and recently these types of reactions have been shown to work using highly reactive copper powder (R. D. Rieke *et al., J. Org. Chem.*, 1993, **58**, 2483; 1988, **53**, 4482 ; *Tetrahedron,* 1989, **45**, 443; *J. Am. Chem. Soc.,* 1992, **114**, 5110). These examples illustrate the types of reactions which are possible thus: 4-bromophenyl(N,N-diethylamino)methyl ketone can be obtained from 4-bromobenzoyl chloride and tributyl(N,N-diethylamino)methylstannane (J. Verlhac and J. Quintard, *Tetrahedron Lett.,* 1986, **27**, 2361), 2-cyanobenzophenone is formed in almost 100% yield from benzoyl chloride and 2-cyanophenylzinc bromide (L. Zhu *et al., J. Org. Chem.,* 1991, **56**, 1445), and ethyl 5-(4-methoxyphenyl)pent-5-onoate is produced in 93% yield through the coupling of benzoyl chloride and ethyl 4-bromobutanoate in the presence of copper powder at -35°C in tetrahydrofuran (R. M. Wehmeyer and R. D. Rieke, *Tetrahedron Lett.*, 1988, **29**, 4513).

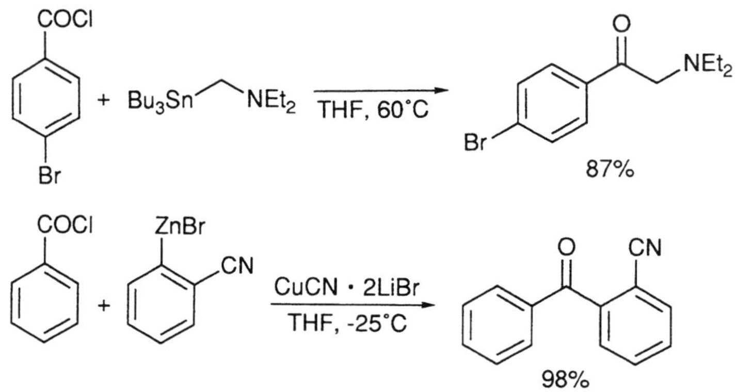

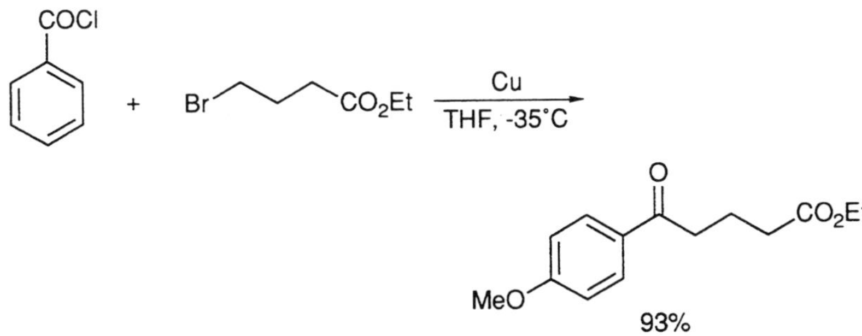

The coupling reaction of acid halides with carbonyl compounds mediated by samarium diiodide gives ketones (P. Girard *et al., Tetrahedron Lett.*, 1981, **22**, 3959; J. Collin *et al., Tetrahedron Lett.*, 1989, **30**, 7407). α,β-Unsaturated ketones are obtained by the reaction of acid halides with ketones (T. Mitsudo *et al., Tetrahedron Lett.*, 1985, **26**, 5143) or with acid halides (M. Kadokura *et al., J. Chem. Soc., Chem. Commun.*, 1986, 252) through decarbonylation in the presence of a palladium catalyst. In an illustration of the samarium diiodide approach, benzoin is obtained by the reductive coupling of benzoyl chloride and benzaldehyde with a suspension of this reagent in tetrahydrofuran (J. Collin *et al., J. Org. Chem.*, 1991, **56**, 3118).

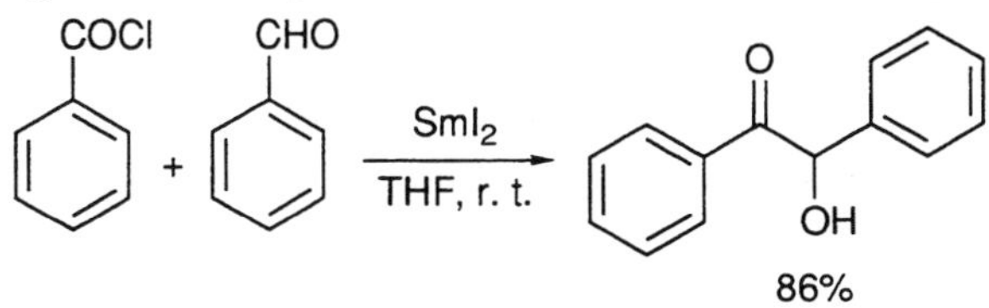

Other reductive coupling reactions are recommended including electrolytic monodecarbonylation of aroyl halides (J. Folest *et al., Tetrahedron Lett.*, 1993, **34**, 7571).

(ii) From Aromatic Aldehydes

Thiazolium salt-catalyzed additions of aromatic aldehydes to enones gives ketones (H. Stetter and H. Bender, *Chem. Ber.*, 1981, **114**, 1226; H. Stetter and J. Nienhaus, *Chem. Ber.*, 1980, **113**, 979; B. Chang and Y. L. Chang, *J. Chinese Chem. Soc.*, 1983, **30**, 55). This has been exploited in the synthesis of 1,4-di(4-chlorophenyl)butane-1,4-dione through the addition of sodium 3-(4-chlorobenzoyl)acrylate to 4-chlorobenzaldehyde (H. Setter and F. Jonas, *Chem. Ber.*, 1981, **114**, 564).

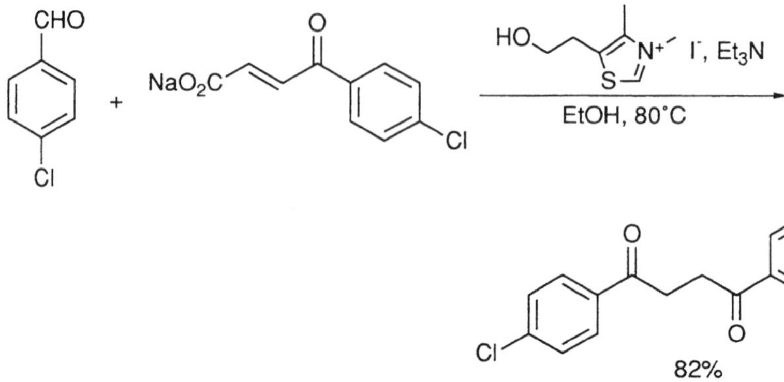

The treatment of aldehydes with vanadium acetylides generated from vanadium trichloride and Grignard, or lithium compounds gives α,β-acetylenic ketones via oxidative nucleophilic addition (T. Hirao *et al.*, *Tetrahedron Lett.*, 1986, **27**, 933). On the other hand, cobalt(II) chloride catalyzes the coupling of aromatic aldehydes with acetic anhydride to diketones through radical formation (S. Ahmad and J. Iqbal, *J. Chem. Soc., Chem. Commun.*, 1987, 692).

(iii) From Other Carbonyl Compounds

Arylacetic acid derivatives can be oxidized to ketones using tetra(n-butyl)ammonium periodate (E. Santaniello *et al., Tetrahedron Lett.*, 1980, **21**, 2655), copper(I) oxide (O. Toussaint *et al., Tetrahedron Lett.,* 1984, **25**, 3819), or N-iodosuccinimide (T. R. Beebe *et al., J. Org. Chem.*, 1982, **47**, 3006). In the case of 2-hydroxy-2-phenylpropionic acid, for example, oxidation with sodium hypochlorite simply gives acetophenone in 93% yield (H. J. Carlson, *Acta Chem. Scand.*, 1984, **B38**, 343).

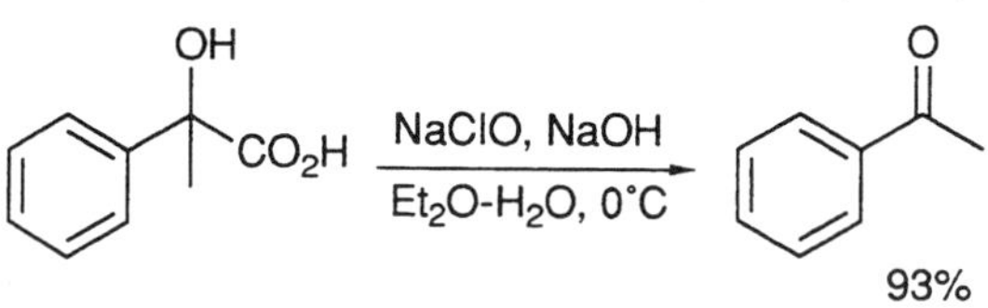

The reaction of organolithium reagents with aryl amides (G. A. Olah *et al., Synthesis,* 1984, 228), esters (C. Alexandre *et al., Synthesis,* 1992, 547), carboxylic acids (T. Fujisawa *et al., Chem. Lett.*, 1983, 1267; 1983, 1791; *Tetrahedron Lett.*, 1982, **23**, 5059; K. B. Wiberg *et al., J. Org. Chem.*, 1993, **58**, 1372), and carbon dioxide (G. Zadel and E. Breitmaier, *Angew. Chem. Int. Ed. Engl.,* 1992, **31**, 1035) gives ketones. For example, 3-hydroxybenzoic acid reacts with methyllithium and trimethylsilyl chloride to afford 3-hydroxyacetophenone in 94% yield (G. M. Rubattom and C. Kim,

J. Org. Chem., 1983, **48**, 1550).

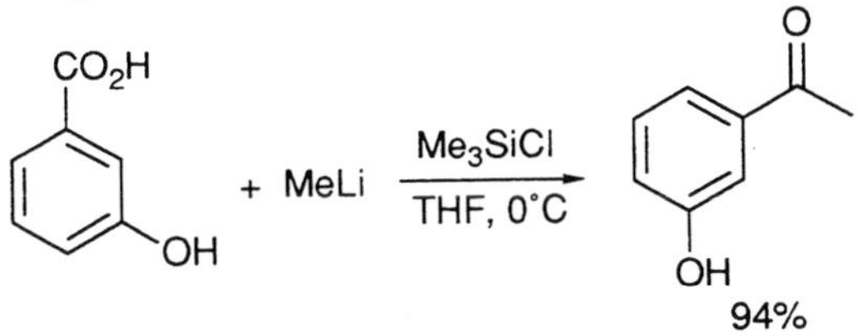

Grignard reagents react with aroic acids (V. Fiandanese *et al., Tetrahedron Lett.*, 1983, **24**, 3677), or carboxamides (M. A. de las Heras *et al., J. Org. Chem.*, 1993, **58**, 5862), in the presence of a nickel catalyst to give ketones. Carboxamides also react with organozinc reagents in a similar way (C. Kashima *et al., J. Org. Chem.*, 1993, **58**, 793). Phenyl selenoesters (D. L. Boger and R. J. Mathvink, *J. Org. Chem.*, 1988, **53**, 3379; 1989, **54**, 1777) and telluroesters (C. Chen and D. Crich, *Tetrahedron Lett.*, 1993, **34**, 1545) can act as precursors of benzoyl radicals, and these may then be reacted with alkenes to give aryl ketones. Alternatively, acyl- or aroyllithiums, generated by lithium-tellurium exchange, react with carbonyl compounds to afford α-hydroxy ketones (T. Hiiro *et al., J. Am. Chem. Soc.*, 1990, **112**, 455). Finally, 1,2-diketones are obtained by the electrochemical reductive-coupling of aromatic esters (M. Heintz *et al., Tetrahedron*, 1993, **49**, 2249).

(h) From Other Compounds

The oxidations of certain nitro compounds with potassium permanganate (N. Kornblum *et al., J. Org. Chem.*, 1982, **47**, 4534), or sulfones with bis(trimethylsilyl)perioxide (J. R. Hwu, *J. Org. Chem.*, 1983, **48**, 4432) are yet other sources of ketones, similarly, aryl triflates (A. M. Echavarren and J. K. Stille, *J. Am. Chem. Soc.*, 1988, **110**, 1557) and arene diazonium salts (K. Kikukawa *et al., Chem. Lett.*, 1982, 35) couple with carbon monoxide and organotin reagents to give ketones. Aryl cyanides react with organolithiums (A. Couture *et al., J. Chem. Soc., Perkin Trans. 1*, 1993, 2463) and Grignards (K. Šindelář *et al., Collect. Czech. Chem. Commun.*, 1982, **47**, 1367) to produce the corresponding ketones, and isoxazolines afford the corresponding ketones when they are treated with either Raney-nickel (D. P. Curran, *J. Am. Chem. Soc.*, 1983, **105**, 5826), palladium catalysts (V. Mancuso and C. Hootelé, *Tetrahedron Lett.*, 1988, **29**, 5917), dimethyl sulfate-sodium nitrite (S. Kwiatkowski, *J. Chem. Soc., Chem. Commun.*, 1987, 1496), or iron pentacarbonyl (M. Nitta and T. Kobayashi, *J. Chem. Soc., Perkin Trans. 1*, 1984, 2103).

(i) From Aromatic Compounds

Friedel-Crafts acylation is a versatile and well tried method for ketone synthesis. Acylation of aromatic compounds with carboxylic acid derivatives in the presence of Lewis acids such as aluminium trichloride (T. F. Buckley III and H. Rapoport, *J. Am. Chem. Soc.*, 1981, **103**, 6157), iron trichloride (F. Effenberger and D. Steegmüller, *Chem. Ber.*, 1988, **121**, 117), and lanthanide trifluoromethanesulfonate (A. Kawada *et al., J. Chem. Soc., Chem. Commun.*, 1993, 1157), or dehydration reagents such as polyphosphoric acid (PPA) (B. S. Verma *et al., Indian J. Chem., Sect. B,* 1993, **32B**, 239; I. M. Downie *et al., Tetrahedron,* 1993, **49**, 4015; X. -M. Wang *et al., J. Org. Chem.*, 1993, **58**, 7498) and methanesulfonic acid (S. E. Clayton *et al., Tetrahedron,* 1993, **49**, 939) are most commonly used. The use of PPA is common place, and this reagent has been used, to modest effect, in the double cyclization of benzene-1,3-di(butan-4-oic acid). This gives 3,4,5,6-tetrahydro-1(2H),8(7H)-anthracenedione in 34% yield (W. C. Christopfel and L. L. Miller, *J. Org. Chem.*, 1984, **49**, 5198).

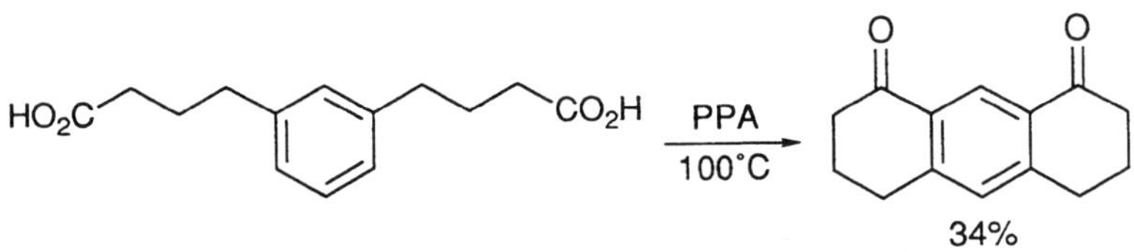

Recently, carboxylic trifluoromethanesulfonic anhydrides (F. Effenberger *et al., Chem. Ber.*, 1983, **116**, 1195; T. Keumi *et al., Bull. Chem. Soc. Jpn.*, 1988, **61**, 455) and carboxylic dihalophosphoric anhydrides have been recommended as strong acylating reagents, which combine with arenes without the need for co-reagents. For example, i-propionic dichlorophosphoric anhydride reacts with anisole at room-temperature in dichloromethane to afford 4-methoxyphenyl i-propyl ketone in 83% yield (F. Effenberger *et al., Chem. Ber.*, 1981, **114**, 926).

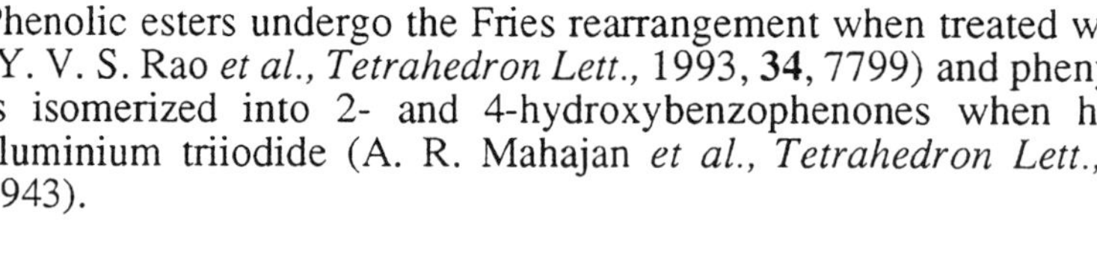

Phenolic esters undergo the Fries rearrangement when treated with zeolites (Y. V. S. Rao *et al., Tetrahedron Lett.*, 1993, **34**, 7799) and phenyl benzoate is isomerized into 2- and 4-hydroxybenzophenones when heated with aluminium triiodide (A. R. Mahajan *et al., Tetrahedron Lett.,* 1990, **31**, 3943).

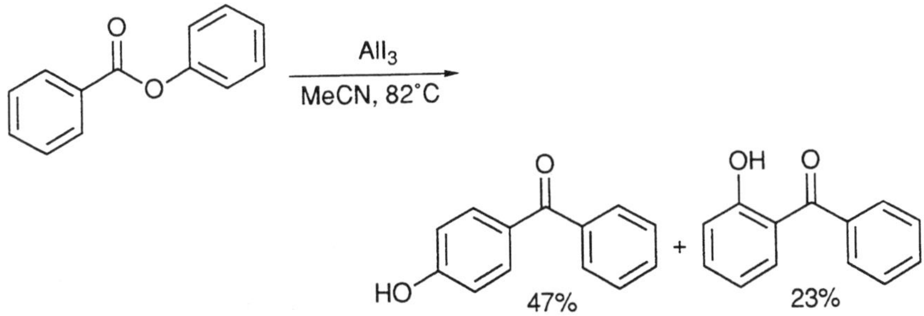

Photochemically induced Fries rearrangements can also be achieved (A. V. Veglia and R. H. de Rossi, *J. Org. Chem.*, 1993, **58**, 4941; D. J. Owen *et al.*, *J. Org. Chem.*, 1993, **58**, 6203) and aryl methyl ketones are available through the electrochemical acetylation of arenes with acetic anhydride (N. Gatti, *Tetrahedron Lett.*, 1990, **31**, 3933).

(j) From Ketones

(i) From Silyl Enol Ethers

Silyl enol ethers have been extensively studied (P. Brownbridge, *Synthesis*, 1983, 85), and it is known that they are versatile starting compounds for the synthesis of α-substituted ketones when reacted with a range of electrophiles in the presence of Lewis acids (V. V. Zhdankin *et al.*, *J. Org. Chem.*, 1989, **54**, 2605; T. K. Hollis *et al.*, *Tetrahedron*, 1993, **49**, 5415; J. Morgan *et al.*, *J. Chem. Soc., Perkin Trans. 1*, 1993, 1677; C. L. Roux *et al.*, *J. Org. Chem.*, 1993, **58**, 1835), transition metal complexes (I. Kuwajima and H. Urabe, *J. Am. Chem. Soc.*, 1982, **104**, 6831), and other activators (I. Kopka and M. W. Rathke, *J. Org. Chem.*, 1981, **46**, 3771; R. V. Hoffman *et al.*, *J. Org. Chem.*, 1985, **50**, 5148; S. Kobayashi and K. Nishio, *J. Org. Chem.*, 1993, **58**, 2647; F. A. Davis and A. C. Sheppard, *J. Org. Chem.*, 1987, **52**, 954). Such reactions can proceed enantioselectively with appropriate chiral auxiliaries or reagents (T. Hayashi *et al.*, *J. Am. Chem, Soc.*, 1988, **110**, 5579; M. Yamaguchi *et al.*, *J. Am. Chem. Soc.*, 1993, **115**, 3362; A. J. M. Caffyn and K. M. Nicholas, *J. Am. Chem. Soc.*, 1993, **115**, 6438; S. Kobayashi and K. Nishio, *J. Org. Chem.*, 1993, **58**, 2647). Some illustrative synthesises are : the reaction of 1-phenyl-1-trimethylsilyloxyethene with methyl N-chloromethyl-N-methyl carbamate in the presence of titanium(IV) chloride to give the N-arylalkanoylcarbamate (7) (S. Danishefsky *et al.*, *Tetrahedron Lett.*, 1980, **21**, 2033), and the reaction of the same enol ether with 1,1-difluoro-2,3,3,3-tetrachloroethane and dichlorotris(triphenylphosphine)ruthenium to form the phenyl propenyl ketone (8) (T. Okano *et al.*, *J. Org. Chem.*, 1993, **58**, 5163), or with sulfuryl chloride fluoride in dichloromethane at -78°C to give phenacyl chloride (G. A. Olah *et al.*, *J. Org. Chem.*, 1984, **49**, 2032).

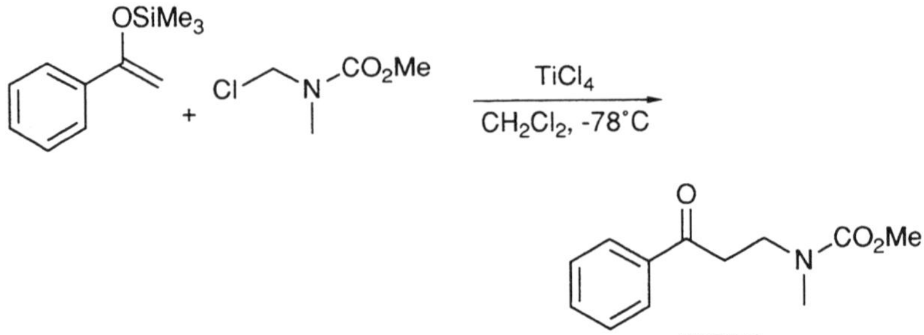

(7)73%

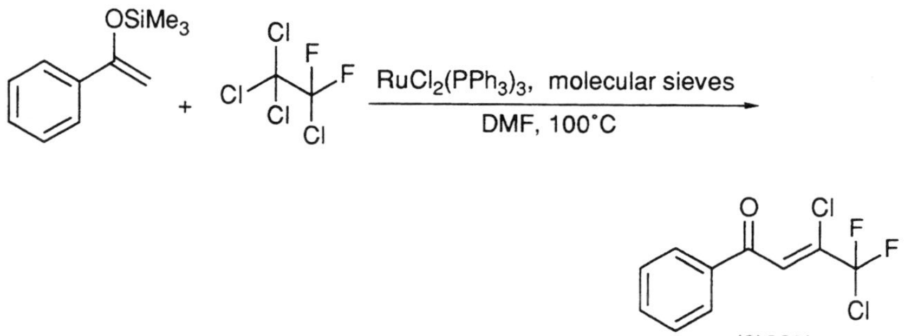

(8)80%

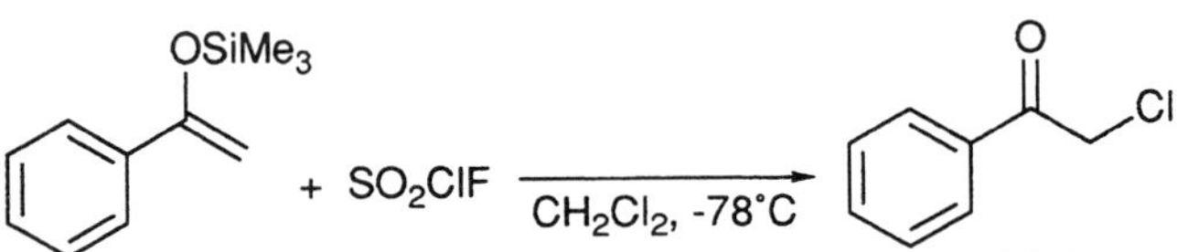

90%

In a further example, a mixture of the isomeric enol ethers (9) and benzaldehyde combine in the presence of a catalytic amount of tris(diethylamino)sulfonium difluorotrimethylsiliconate to give a mixture of the erythro and threo β-siloxyketones (10) (R. Noyori *et al., J. Am. Chem. Soc.*, 1981, **103**, 2106).

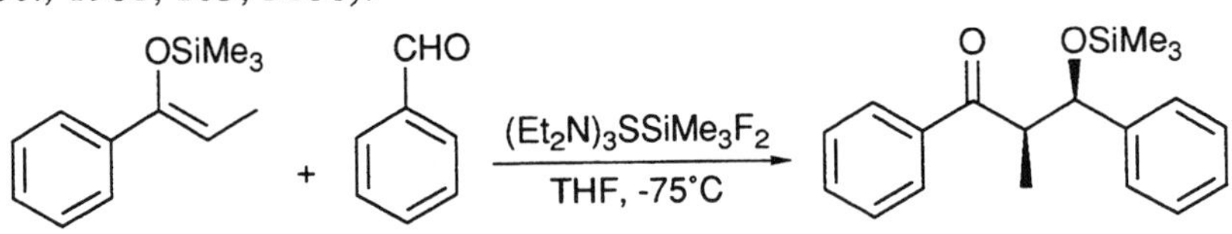

(9)E : Z = 1 : 99

(10)erythro : threo = 95 : 5
75%

(ii) Other Reactions

The condensation of the enolates of aryl alkyl ketones with electrophiles is a classic way of preparing more complicated (polyfunctionalized) derivatives. Lithium reagents (A. D. M. Curtis *et al., Tetrahedron,* 1993, **49**, 187; A. R. Katritzky and P. A. Harris, *Tetrahedron.,* 1990, **46**, 987; D. A. Oare and C. H. Heathcock, *J. Org. Chem.,* 1990, **55**, 157; R. J. Mears and A. Whiting, *Tetrahedron,* 1993, **49**, 177) are most commonly used, although other alkaline metal (F. A. Davis *et al., J. Am. Chem. Soc.,* 1990, **112**, 6679; M. Periasamy *et al., J. Org. Chem.,* 1993, **58**, 4997; E. Austin *et al., Tetrahedron,* 1993, **49**, 4495; C. Dell'Erba *et al., Tetrahedron,* 1992, **48**, 3250) and alkaline earth metal reagents (J. Brocard *et al., Tetrahedron,* 1990, **46**, 6995; G. Sartori *et al., Tetrahedron Lett.,* 1993, **34**, 7339) are also employed. A typical procedure is the reaction of the lithium enolate of acetophenone with trimethylorthoformate in the presence of boron trifluoride to give the keto acetal (11) in 84% yield (M. Suzuki *et al., Tetrahedron Lett.,* 1982, **23**, 5395).

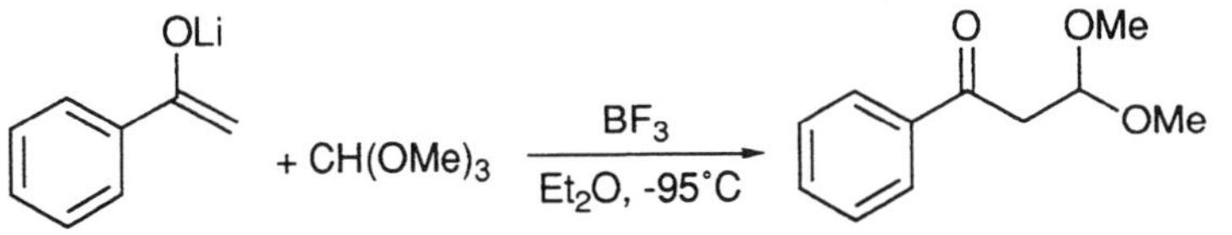

(11)84%

A less usual variant is the arylation of 3-methoxyacetophenone with the azosulfide (12) to produce benzyl 3-methoxyphenyl ketone in contact with potassium t-butoxide (C. Dell'Erba *et al., Tetrahedron,* 1993, **49**, 235).

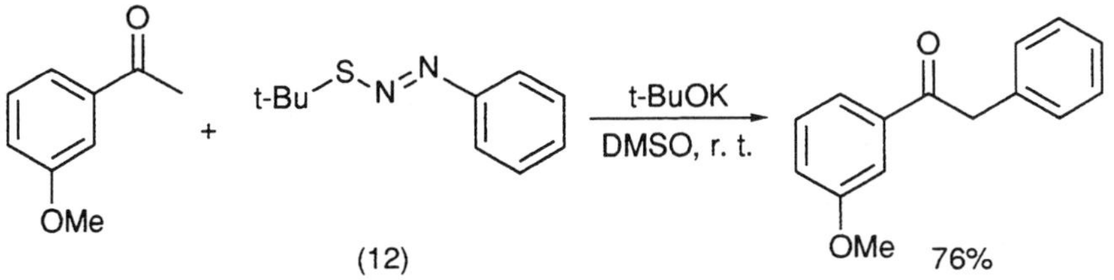

(12) 76%

Aldol reactions, such as that between acetophenone and 4-chlorobenzaldehyde, can be promoted with barium hydroxide. In this particular case, the product is the chalcone (13) (A. Fuentes *et al., Tetrahedron Lett.,* 1987, **28**, 4541).

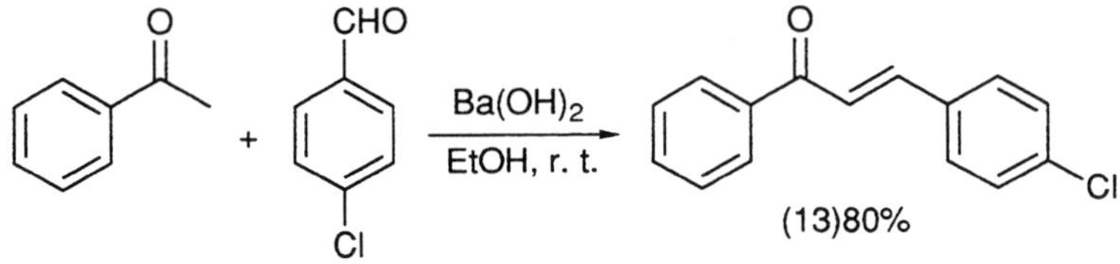

(13)80%

However, under conditions of Lewis acid catalysis aldolization products can be isolated (M. Uemura *et al., Tetrahedron Lett.,* 1989, **30**, 6383; I. Paterson *et al., Tetrahedron,* 1990, **46**, 4663; H. C. Brown *et al., J. Am. Chem. Soc.*, 1989, **111**, 3441; I. Paterson and J. M. Goodman, *Tetrahedron Lett.*, 1989, **30**, 997). For example, ethyl phenyl ketone and benzaldehyde in contact with boron trichloride and di(i-propyl)ethylamine afford mainly the erythro isomer (14) with only minor amounts of the alternative diastereomer (15) (H. -F. Chow and D. Seeback, *Helv. Chim. Acta Pays-Bas,* 1986, **69**, 604).

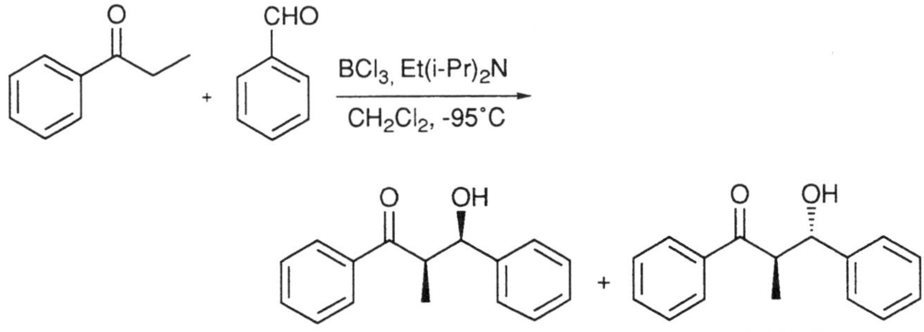

(14)87% (15)8%

Various α-halo ketones and enones can be reacted with electrophiles in the presence of tin reagents (T. Mukaiyama *et al., Chem. Lett.,* 1982, 1601; J. Ichikawa and T. Mukaiyama, *Chem. Lett.,* 1985, 1009; M. Yasuda *et al., J. Chem. Soc., Perkin Trans. 1,* 1993, 859; I. Shibata *et al., Tetrahedron Lett.*, 1933, **34**, 6567). For example, 1,2-dibenzoylethene combines with 3-phenylpropanal in contact with bis(2-pyridinethiolato)tin(II), formed *in situ* from 1,1'-dimethylstannocene and pyridine-2-thiol, to give the two diastereomers (16) and (17) (T. Mukaiyama *et al., Chem. Lett.*, 1985, 1539).

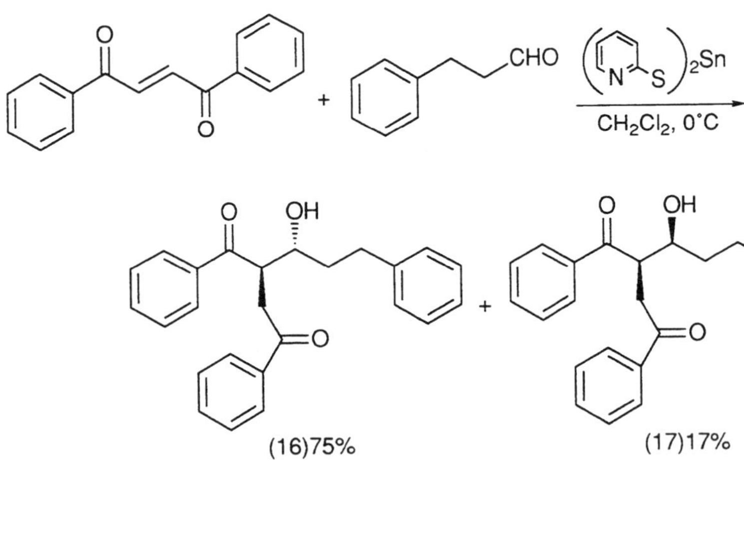

(16)75% (17)17%

α-Diazoketones react with either aldehydes (A. Padwa *et al., J. Org. Chem.*, 1990, **55**, 5297) or alkynes (A. Padwa *et al., J. Am. Chem. Soc.*, 1993, **115**, 2637; P. H. Mueller *et al., Tetrahedron Lett.*, 1993, **34**, 4285) to form the corresponding ketones. In the case of the reaction between diazoacetophenone and tri(n-hexyl)borane, homologation takes place to afford n-heptyl phenyl ketone in 80% yield (H. Kono and J. Hooz, *Org. Synth.*, 1988, **IV**, 919).

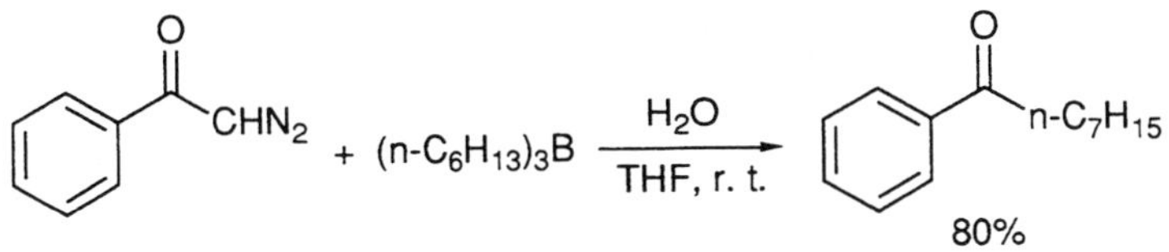

Second Supplements to the 2nd Edition of Rodd's Chemistry of Carbon Compounds, Vol.III D/E/F (Partial), edited by M. Sainsbury

Chapter 15

PHENOLIC ARALKYLAMINES, MONOHYDRIC ALCOHOLS, MONOCARBALDEHYDES, MONOKETONES AND MONOCARBOXYLIC ACIDS

W.S. MURPHY AND B. SARSAM

1. Phenolic aralkylamines

Synthesis of aminoalkylphenols using the Mannich reaction has retained its popularity (H.Heaney in 'Comprehensive Organic Synthesis',Vol. 2, Pergamon Press ,1991). Many factors are involved: the degree of ring substitution, phenol reactivity, pH , the Mannich reagent , the steric size of the amine component and whether it is primary or secondary.

Although polyalkylation of phenols in the Mannich reaction is common, there is a marked preference for *ortho*-substitution. This observation lead to an early proposal that an H-bonded cyclic transition state may be involved (J.K.Burckhalter, J.N.Wells and W.J.Mayer, Tetrahedron Lett., **1964**,1353) with Claisen rearrangement-like characteristics. It has been shown that aminals or aminolethers are the reactive intermediates (J.H.P.Tyman, E.L Short and V.Tychopoulos, J.Chem.Technol. Biotechnol., 1992,**53**, 389). Reaction occurs at the 2-position, the 1-position and the 7-position, respectively, when 1-naphthol, 2-naphthol and 4-methyltropolone are heated with dipiperidylmethane (P.L.Pauson, P.B.Kelly and R.J.Porter, J.Chem.Soc.,(C), **1970**, 1323).

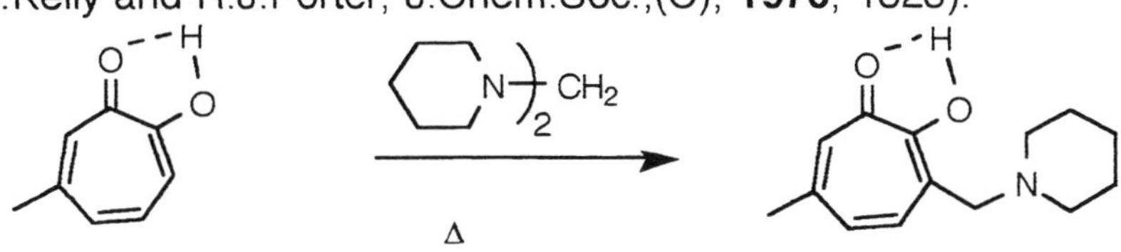

In non-aqueous conditions preformed iminium salts are effective. Also, sulphor dioxide is found to catalyse the reaction. The mechanism is outlined below (R.A.Fairhurst *et al.,* Tetrahedron Lett., 1988, **29**,5801).

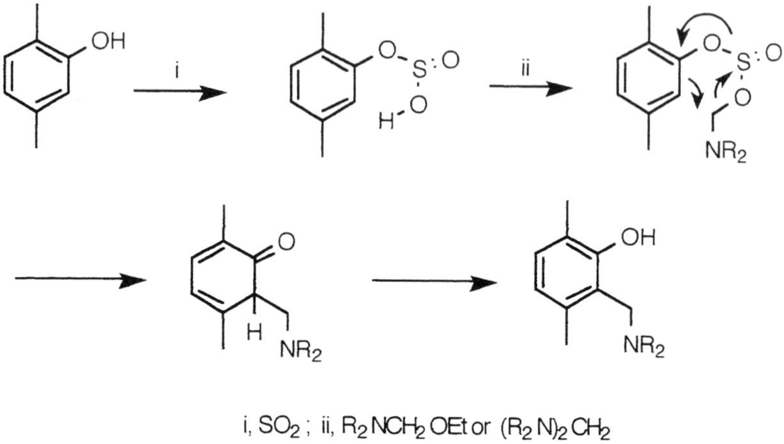

As with secondary amines, phenols react with aldehydes and primary amines principally at the *ortho* position. It is now well established that a benzoxazine is readily formed such as illustrated below with methyl hydroquinone (D.D.Reynolds, J.Heterocycl. Chem., 1970,**7**,1397). However, the heterocyclic ring is readily cleaved reversibly.

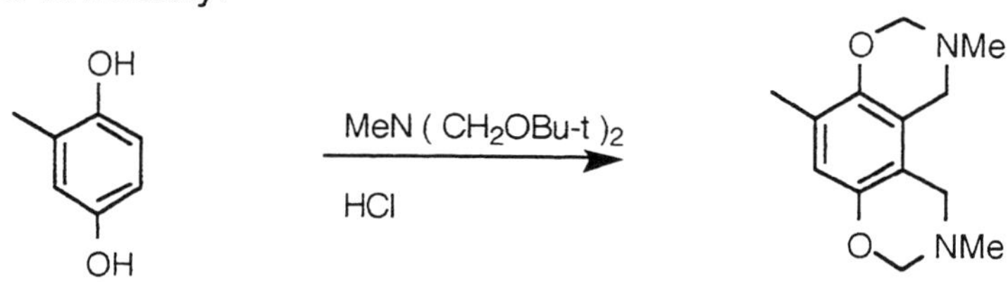

This ring-chain tautomerism has been investigated. The equilibration in the presence of primary or secondary amines involves Mannich bases and ring-opened products(D.Tzschoppe *et al.*, Bull. Soc. Chim.Belg., 1986, **95**, 45). Benzoxazines have been employed as Mannich reagents (W.J. Burke *et al.*, J.Org.Chem., 1965,**30**, 3423) in reactions with phenols, such as that illustrated in the following example.

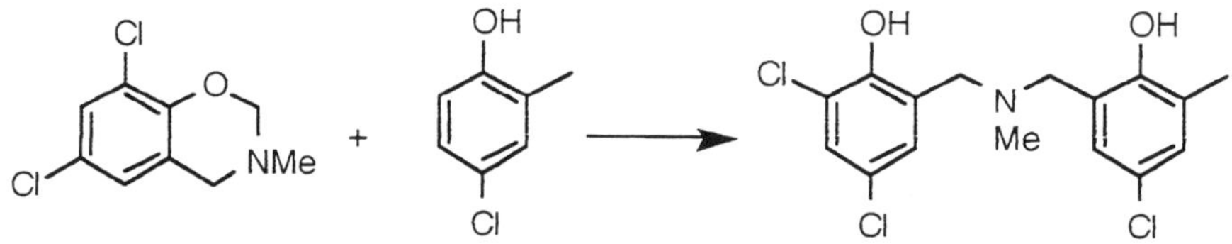

Amidoalkylation is similarly employed and provides an alternative route to the same type of product. For example, a simple method for introducing the 2-hydroxybenzyl group into the tetrazole ring involves the use of paraformaldehyde and succinimide. The reaction involves the intermediate formation of N-(2-hydroxybenzyl) succinimide (A.F.H.Hachiam *et al.*, Syn.Commun., 1994,**24**,665). Amidoalkylation has been reviewed several times (Comprehensive Organic Synthesis Vol 2, Ch4.5; H.E.Zaugg and W.B. Martin, Org React., 1965,**14**,52; E. Zaugg, Synthesis,**1970**.49; *idem*, **1984**, 181; W.N.Speckamp and H.Hiemstra, Tetrahedron, 1985, **41**,4367).

The following are some recently reported reactions of phenolic amines. Electrophilic substitutions of donor - substituted arylamines using tricarbonyliron-complexed cyclohexadienylium cations lead regio-and stereo-selectively to the correponding tricarbonylcyclohexadieneiron complexes. The C-C bond formation occurs regioselectively in the *ortho* -position with respect to the amino group, with arylamines containing a substituent in the *para* -position. The initially formed N-alkylated arylamine is shown to rearrange in an intermolecular process to the C-alkylated arylamine. With activated manganese dioxide, oxidative cyclisation to the carbazole occurs at room temperature (H.J.Knolker *et al.*, Chem. Ber.,1992,**125**,2783).

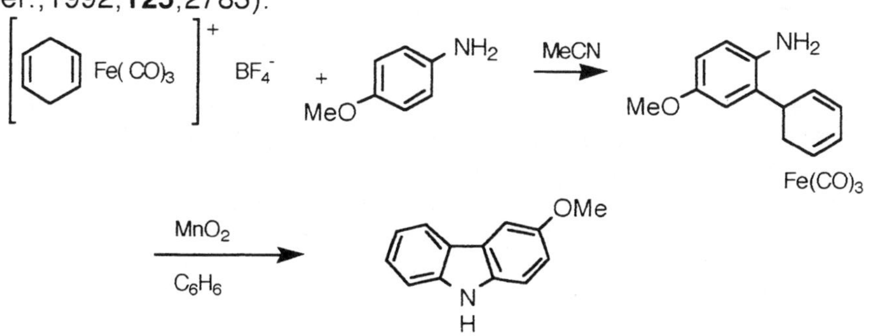

It is found that Fremy's salt promotes the oxidative degradation of *p*-hydroxybenzylamines and *p*-hydroxylbenzamides and provides a novel approach to *p*-quinones (J.M.Saa, A.Llobera and P.M. Deya, Chem.Lett., **1987**,771). This methodology has been extended to the synthesis of prenylated quinones (*idem*, J.Org.Chem.,1993, **58**, 328).

An improved method for the synthesis of N-acyltetrahydro - quinolines from 2-arylethylamines, acid chlorides and aldehydes has been reported (A.P.Venkov and L.K.Lukanov, Syn. Commun., 1992,**22**,3235). o-(2-Arylphenyl)-hydroxylamines undergo interesting transformations to the corresponding azepinone, in acid. The intermediate 2-amino-2-phenyl-3,5-cyclohexadienone readily dimerises and may be trapped with trifluoracetic anhydride (Y.Endo *et al.*, Tetrahedron Lett., 1992,**33**,3339).

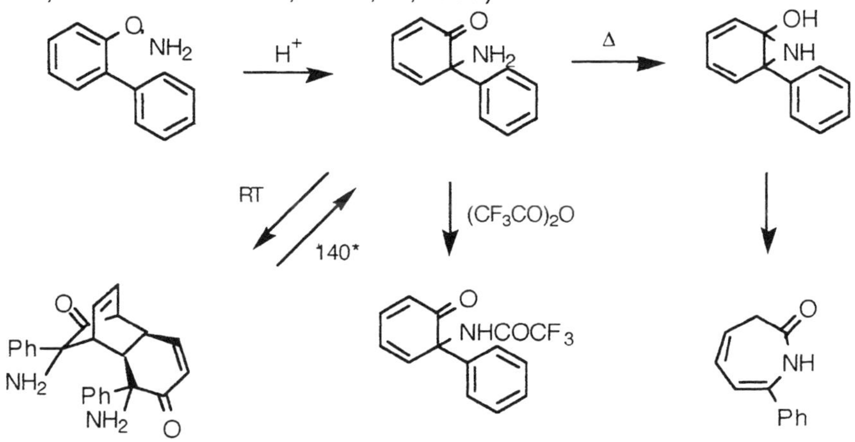

A study of 3,5-dialkoxy-4-hydroxybenzylamine antiooxidants has been undertaken with an emphasis on means of peroxyl radical scavenging (M.J. Yu *et al.*, J.Med Chem.,1993,**36**,1262). N-Arylalkyl -N-phenylhydroxylamines when photolysed under aerobic conditions induce single - strand DNA cleavage. This process functions as the trigger to initiate the new and controllable DNA cleavage process (J.R.Hwu *et al.*, J.Chem. Soc. Chem. Commun., **1994**,1427).

A series of *cis* - and *trans* -2-arylcyclohexylamines were synthesised and resolved. The absolute configurations were determined by single-crystal X-ray analyses. Some of these compounds and especially 2-(3-hydroxyphenyl)-N,Ndipropylcyclohexylamine showed a marked locomotor stimulation in reserpinized rats but displayed only minor effects on dopamine and 5-hydroxytryptamine synthesis rates (B. Andersson *et al.*, Eur.J. Med. Chem.,1993, **28**, 63).

The biosynthesis of *p*-hydroxyphenylacetaldehyde oxime from L- tyrosine has been elucidated. It is catalysed by cytochrome P-450_{TYR}. It is proposed that a key intermediate is N,N-dihydroxytyrosine which after dehydration and tautomerisation yields the oxime (O.Sibbesen *et al.*, J.Biol.Chem., 1995,**270**,3506).

A dihydroxyphenylalanine derivative incorporating the 1,2-dicarba-*closo*-dodecaborane cage has been synthesised from 3,4-dihydroxybenzaldehyde. The hydroxy groups were protected by benzylation , the carborane cage introduced *via* an acetylene substituent and the aminoacid function *via* diethylformamido-malonate (J.K.Prashar *et al.*, Tetrahedron Lett., 1993,**34**,6799). This procedure is claimed to have the potential to become a general method for the synthesis of other boronated hydroxylated aromatic aminoacids, for testing as boron delivery agents in neutron capture therapy.

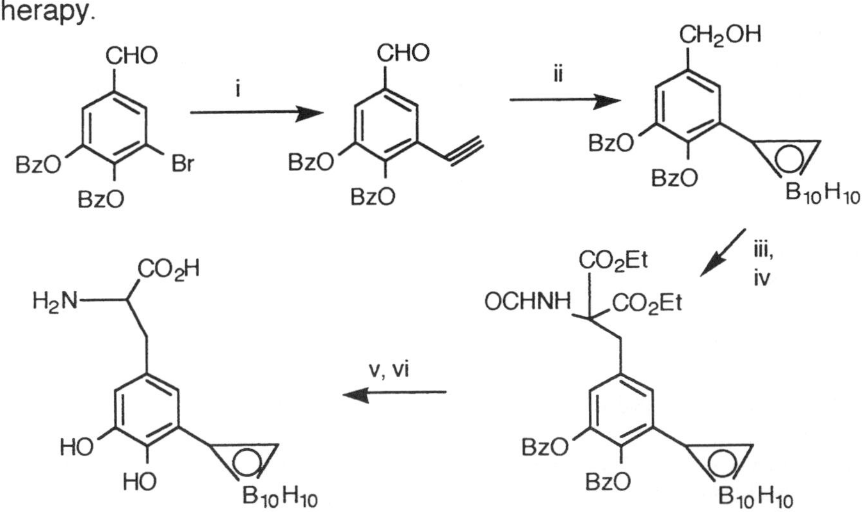

Reagents : i , $HCCSiMe_3$, Et_3N ; ii, $B_{10}H_{10}(CN)_2$;
iii ,PBr_3 ; iv , diethylformamidomalonate, Na,EtOH;
v, H_2 / Pd-C ; vi , HCl

2. Hydroxyaryl alcohols

It was found that the catalyst system Pt / C - $CeCl_3$ - $Bi_2(SO_4)_3$ in aqueous methanol is effective for the selective oxygen oxidation of *m* -hydroxybenzyl alcohol to *m* -hydroxybenzaldehyde (R.Oi and S.Takenaka, Chem.Lett., **1988,** 1115). The ceric ion appears to stabilise the benzaldehyde hydrate and in this way inhibit further oxidation. Formation of the hydrate in CD_3OD/D_2O in the presence of $CeCl_3$.was proved by nmr.

The oxidation of 3,4-dihydroxybenzyl alcohol , one of the sclerotizing precursors for the tanning of the ootheca of cockroach *Periplaneta americana*, is reported for the first time. Mushroom tyrosinase catalyzed the oxidation of 3,4-dihyroxybenzyl alcohol *via* the corresponding quinone which was found to be unstable and readily transformed to produce 3,4-dihydroxybenzaldehyde as the stable product , probably through the intermediate formation of a quinone methide (M.Sugumaran *et al.*, Arch.Insect. Biochem. Physiol.,1991,**16**,31).

Buckminsterfullerene C60 underwent a [4+2] cycloaddition reaction across a 6.6 ring junction with *o* -quinonemethide generated *in situ* from *o* -hydroxybenzyl alcohol in 1,2-dichlorobenzene at 180°C (M.Ohno *et al.*, Chem.Lett., **1993**, 1833).

The major C_6-C_2 phenolic compounds found in the vegetation water of fruits of *Olea europaea*, Dritta and Cipressino cultivars, were 4-hydroxyphenylethanol, 3,4-dihydroxyphenyl - ethanol and 3,4-dihydroxyphenylglycol. The latter, a formal metabolite of norepinephrine, is first reported as a major component of the olive phenolic fraction (G.Bianchi and N.Pozzi, Phytochemistry, 1994, **35,** 1335).

3. Hydroxyarylcarbaldehydes

Considerable effort has been expended on the synthesis of substituted benzaldehydes. The various methods of formylating aromatic rings have been reviewed (G.A.Olah *et al.*, Chem.Rev., 1987,**87**, 671). One of these, the Reimer -Tiemann reaction, has been reviewed (H.Wynberg and E.W. Meijer, Org.Reactions, 1982, **28**,1; *idem*, 'Comprehensive Organic Synthesis', vol.2, 1991) and comparison made with other methods. Other than the Reimer-Teimann reaction, direct introduction of a formyl group into the aromatic nucleus normally involves acidic conditions. In the Gattermann reaction the entering group usually occupies the position *para* - to the hydroxyl group, whereas the Duff reaction fails with polyhydric phenols and with phenols carrying electron-donating substitutions. The Reimer-Tiemann is on occasion the only common method for the direct formylation of phenols.

Although abnormal Reimer- Teimann products - usually resulting in the formation of *ipso* -substitution products - is well known, the formal displacement of a methoxyl group remains a unique reaction. Thus, when the complex phenol (1) is treated (C.W.Bird *et al.*, Tetrahedron, 1985, **41**, 4685) with chloroform in aqueous sodium hydroxide, the salicylaldehyde (2) is formed in 25% yield. The proposed mechanism is as follows:

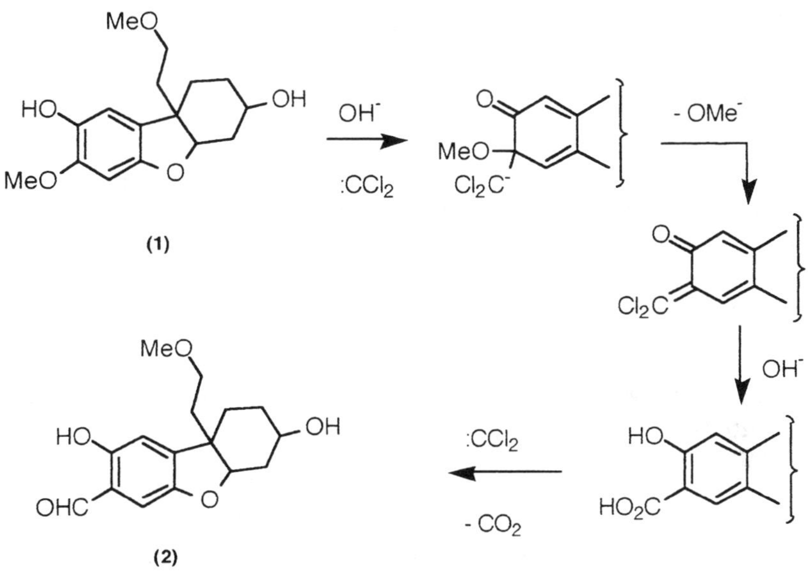

A convenient method for the 6 - substitution of salicylaldehydes from the parent aldehyde involves the dianion of a 1,3-dimethylimidazolidine (3) (M.Gray and P.J.Parsons, Synlett., **1991**,729). The intermediate (3) is readily prepared from the aldehyde and N,N'-dimethylethylenediamine.

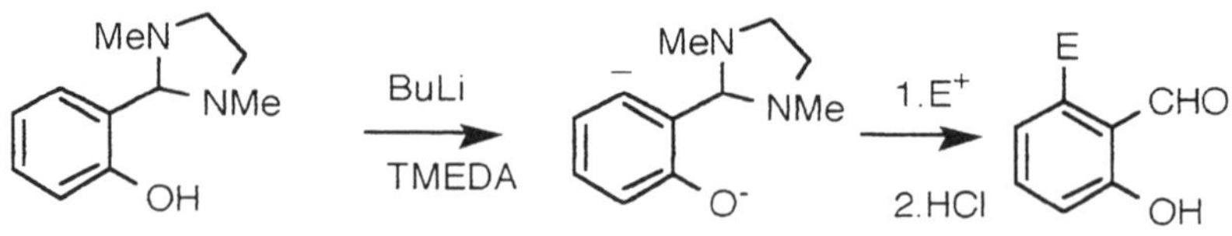

(3)

Condensation of guaiacol with glyoxylic acid results in the formation of 3-methoxy-4-hydroxymandelic acid. When oxidised with copper (II) hydroxide, vanillin is formed in 88% yield (R.G. Kalikar *et al.*, J.Chem. Technol. Biotechnol.,1986,**36**,38). A simple synthesis of vanillin-d_3 (4-hydroxy-3-methoxy d_3-benzaldehyde) has been achieved in 35% yield by stirring 3,4-dihydroxybenzaldehyde with NaOH and iodomethane - d_3 (S.Schneider and C.Rolando, J.Lab.Comp. Radiopharm., 1992, **31**, 489).

Selective oxidation of *para* - cresol derivatives to the corresponding *para* - hydroxybenzaldehydes by oxygen with a cerium catalyst is reported (T. Yoshikuni, J.Mol.Cat.,1992,**72**,29). Related oxidations with dioxygen were observed using copper (II) - amine complex catalysts (K.Takehira *et al.*, Tetrahedron Lett., 1990, **31**, 2607) under very mild conditions.

Microbial routes to the important flavor components *p* - hydroxybenzaldehyde and vanillin were explored (J.Casey and R.Dobb, Enz.Microb.Tech., 1992, **14**, 739). The most successful was *proteus vugaris* . This organism transformed phenylalanine, tyrosine, and methoxytyrosine into the corresponding arylpyruvic acids. These in turn were readily converted with base to benzaldehyde, *p* -hydroxybenzaldehyde and vanillin.

Monomethylation of 3,4-dihydroxybenzaldehyde with dimethylsulphate in aqueous buffer was investigated kinetically. The ratio of vanillin to isovanillin was determined by hplc. *meta* - Methylation was predominant in the presence of Cu(II). The reaction may serve as a chemical model for catechol-O-methyltransferase (S.Sugata *et al.*, Chem. Pharm.Bull.,1989,**37**,1143). Loupy and Majdoub (Org.Prep.Proc. Internat , 1990,**22**,99) investigated the same reaction under solid-liquid PTC conditions in the absence of added organic solvent. A mixture of isovanillin (*p* -methylation) and veratraldehyde(di-O-methylation) were formed uniquely. The reagents were methyl iodide, potassium carbonate, and Aliquat 336.

Salicylaldehydes react with isonitriles in the presence of an ammonium salt with the formation of 2,3-diaminobenzofuran Schiffs bases (R.Bossio *et al.*, Synthesis, **1991**,999). The proposed mechanism is as follows:

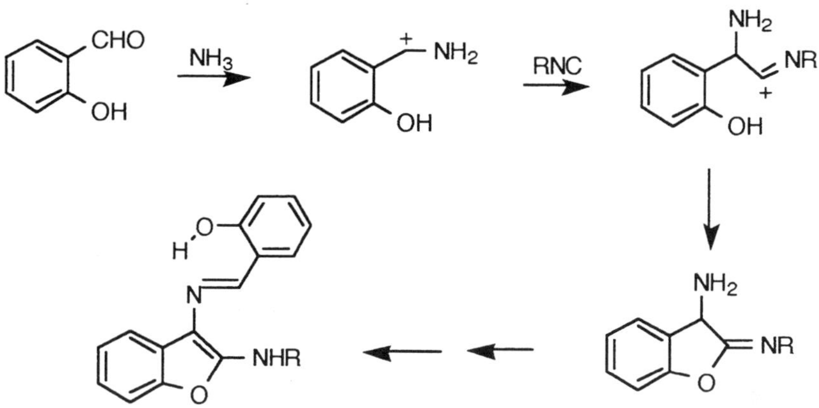

Benzylidenemalononitrile derivatives synthesised from aromatic aldehydes and malononitriles are known as tyrphostins. They display an enhanced ability to inhibit the epidermal growth factor(EGF) receptor tyrosine kinase domain, as the number of hydroxyl groups increases. Thus, the tyrophostin derived from 3,4,5-trihydroxybenzaldehyde is the most effective (B.G. Brunton *et al.*, Anti-Cancer Drug Design, 1994,**9**,291).

The synthesis and analytical properties of 3,4-dihydroxybenzaldehyde guanylhydrazone have been described. The reagent gave coloured complexes with Co(II), Fe(III), Mo(VI), W(VI) and V(V). The method was employed in the determination (a) of Co(II) in the presence of other cations at excess, (b) of Fe(III) in a city drinking water sample without preconcentration, and (c) of Mo(VI) in a standard steel sample (E.Kavlentis, Mikrochimica Acta, 1986,**1**,27).

The interesting rearrangement products (5) and (6), in addition to the normal products, were detected in the *para* -Claisen rearrangement of 2-(allyloxy)- and 2-(crotyloxy)-3-hydroxy -benzaldehyde, (4a) and (4b), respectively (S.N.Kilényi *et al.*, J.Org.Chem.,1991,**56**,2591). The formation of such products was unprecedented.

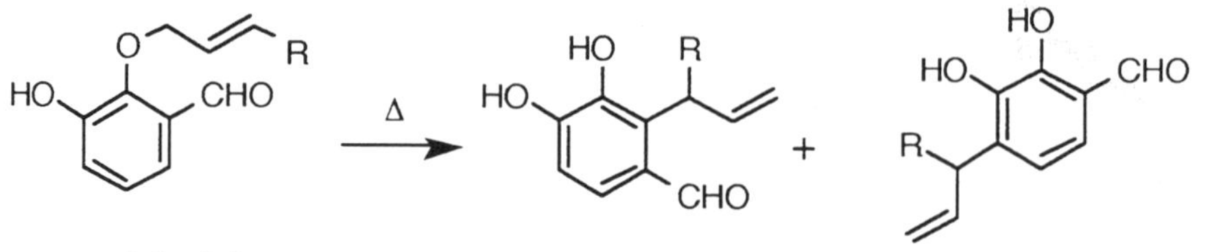

(**4a, b**) (**5a, b**) (**6a, b**)

These results were rationalised acording to the following scheme:

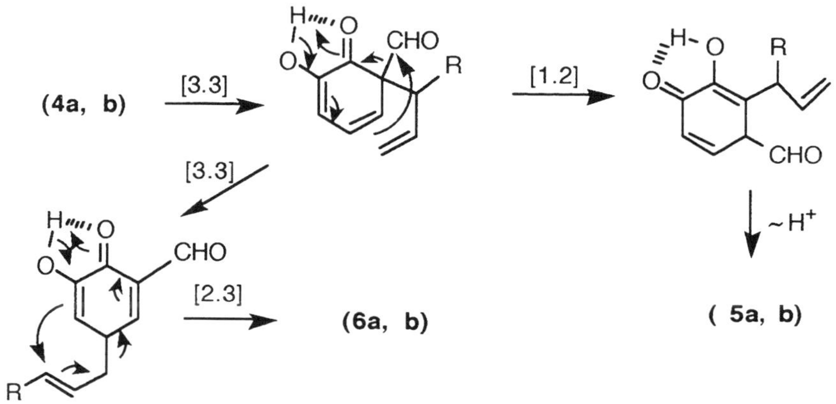

4. Hydroxyphenyl ketones

The antioxidant property of dehydrozingerone and its analogs (E-ArCH:CHCOMe) were investigated in the ferric - ascorbate induced lipid peroxidation model in fat brain homogenate. Those with hydroxy groups at both *meta* - positions were found to be more active than dehydrozingerone. Based on their IC50 values, several of these compounds are more potent than vitamin E and hence may be of potential use in various diseases caused by oxidant stress (D.V.Rajakumar and M.N.A.Rao, Pharmazie, 1994,**4 9**,516).

Some phenolic highly congested benzophenones and benzaldehydes undergo stereoselective photocyclization to 2-arylbenzocyclobutanol methylethers. Hydrogenolysis of either diasteromer provided a 1-methoxy-2-arylbenzocyclobutene mixture enriched in the *cis* -isomer. These are established intermediates for the synthesis of lignans (G.Coll *et al* ., J.Org.Chem.,1992, **5 7**, 6222).

Thus, heat induces stereospecific ring opening to the *ortho*-quinodimethane, which can be trapped *in situ*. The following is an outline of the sequence:

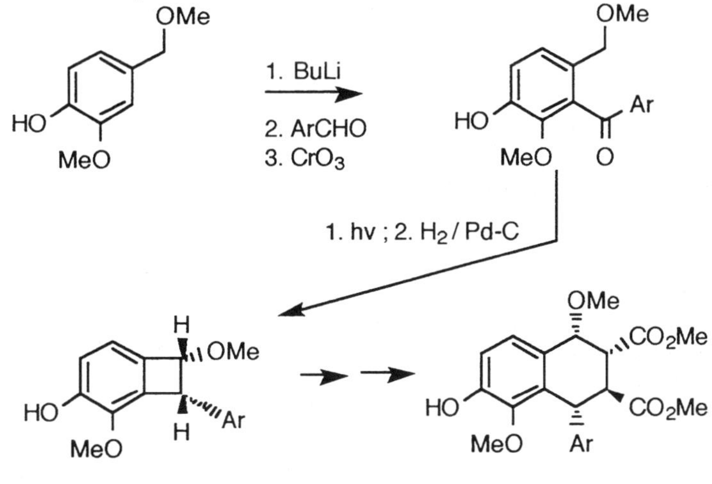

Ar =2,4-$(OMe)_2C_6H_3$

A simple one - pot synthesis of flavonols has been achieved by condensation of a suitable arylaldehyde in alkaline aqueous medium with the α-hydroxyacetophenone, followed by oxidation with H_2O_2 (F.Fringuelli *et al.*, Tetrahedron ,1994,**5 0**, 11499):

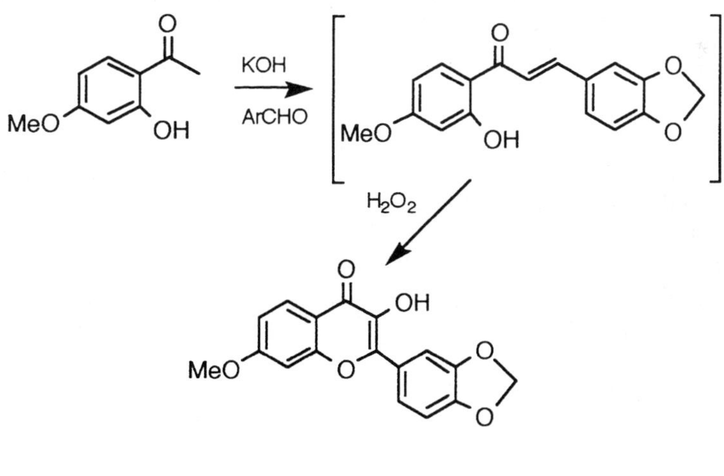

5. Phenolic monocarboxylic acids : the hydroxybenzoic acids

2,6-Di-*tert.*-butyl-4-(2,2-disubstituted vinyl)phenols, such as SF6847 (X=Y=CN) and its analogues, exhibit potent uncoupling activity with

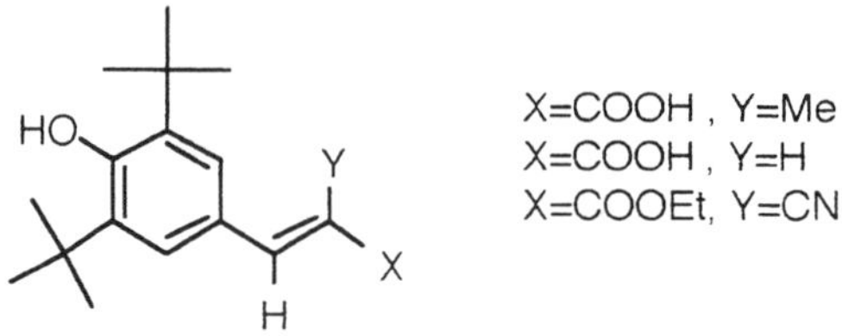

energy-transducing biomembranes, such as rat-liver mitochondria and spinach chloroplasts. The acid dissociation constant (K_a) is one of the important factors which determine the uncoupling potency. H.Miyoshi *et al.* (Bull.Chem.Soc.Japan, 1993,**66**,269) studied the various factors which influence the pK_a values. The inductive and resonance effects of the X-substituents and the inductive effects of the Y-substituents are significant . The stability of the phenolate anion is governed by the electron-withdrawing ability of the vinyl moiety. The planar conformation of the phenolate anion is the most stable one.

The alkaline extract of barley straw cell walls (D.Stewart *et al.*, Biol.Mass Spectrom.,1994,**23**,71) was analysed, using gcms. The components were assigned as the three 5,5'-bisdehydroferulic acids (see below), in part on the basis of the fragmentation patterns. However, since six peaks in the single ion chromatograph at mit 674 were detected, the structures of the three other isomers have not been conclusively identified as yet.

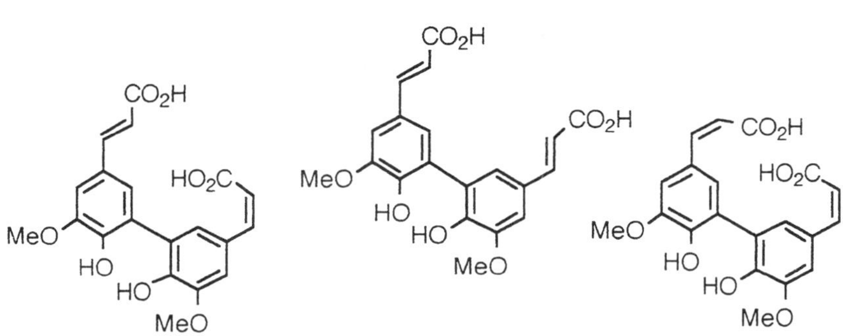

The enzymatic formation of *p*-hydroxybenzoic acid from *p* -coumaric acid has been detected in cell free exracts of *Lithospermum erythrorhizon* cell cultures. In the absence of NAD, *p* -hydroxybenzaldehyde is formed instead of *p* -hydroxybenzoic acid. The reaction is specific to *p* -coumaric acid. The results suggest that the reaction mechanism is non-oxidative (K.Yazaki *et al.*, Phytochemistry, 1991, **30**, 2233). The oxidative decarboxylation of 3,4-dihydroxymandelic acid to 3,4-dihydroxy-benzaldehyde, has been studied by two groups : one using electrochemical methods (T.H.Czapla *et al.*, Biochim.Biophys.Acta, 1991, **1077**, 400) and the other employed tyrosinase - catalysis ((M.Sugumaran *et al.*, Biochem. J.,1992,**281**,353). Both groups independently deduced the same mechanistic pathway, namely, initial orthoquinone formation, followed by rapid decarboxylation to an intermediate *p* -quinone methide. The latter in turn, then rearranges to the 3,4-dihydroxybenzaldehyde.

Tandem reduction - intramolecular hydroxyalkylation of 4-(3-hydroxyphenyl)butanoates has been reported (G.Guanti *et al.*, Tetrahedron Lett., 1992,**33**,3919). The sequence results in the regiospecific formation of 1,8-dihydroxytetralins.

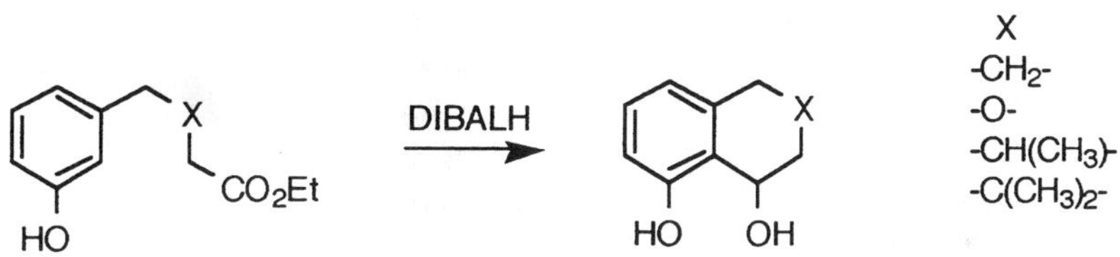

A synthetic carboxylic acid ionophore with eight ether oxygens and five aromatic rings, mediated selective K^+ ion transport through a liquid membrane (H.Kuboniwa *et al.*, J.Chem.Soc.,Chem. Commun., **1985**, 1468). The coiled carboxylate - encapsulated potassium ion is highly lipophilic, even more so than the origional shorter model (H.Kuboniwa *et al.*, Chem.Lett., **1982**, 1937).

Second Supplements to the 2nd Edition of Rodd's Chemistry of Carbon Compounds, Vol.III D/E/F (Partial), edited by M. Sainsbury

Chapter 16

DEPSIDES, HYDROLYSABLE TANNINS, LIGNANS, LIGNIN AND HUMIC ACIDS

W.S.MURPHY AND B.SARSAM

1. Depsides, depsidones and depsones

(a) Isolation and detection

Substitution of methyl *tert.*- butyl ether (MTBE) for diethyl ether in the standardised tlc method of lichen metabolite analysis, has been a simple but effective improvement (C.F.Culberson *et al.*, J.Chromatog., 1982,**238**,483). R_F values remain virtually unchanged but there are no problems due to the high volatility of the diethyl ether. A compilation of such data is now available (J.A.Elix *et al.*, 'A Catalogue of Standardised Thin Layer Chromatoghraphic Data and Biosynthetic Relationships for Lichen Substances', Aust.Natl.Univ., Canberra, 1987). Droplet countercurrent chromatography using non-aqueous solvent systems is useful for separation and isolation of weakly polar depsides (B. Domo *et al.*, J. Chromatog.,1982, **246**, 133). The scope and limitations of *tert.*-butanolysis of depsides has been investigated (S.Huneck, Phytochem., 1984,**23**,2697). Neither 2-O-methylated nor 2-O-acetylated compounds are cleaved by heating with *tert.*- butanol.

More sopisticated methodologies such as specialised ms and 2D-nmr are more frequently employed as need arises. The application of negative FAB and ms/ms to screening of lichen compounds has been reviewed (G.Holtzmann *et al.*, Phytochem., 1990, **29**, 2277). Fast - atom ionization ms of depsides of the orcinol series I (R,R^1= OH, OMe), II (R,R^1= OH, OMe; R^1,R^3= H, Me) and depsidone III were measured. Unlike EI spectra, those of these substrates revealed the corresponding protonated molecular ions.The

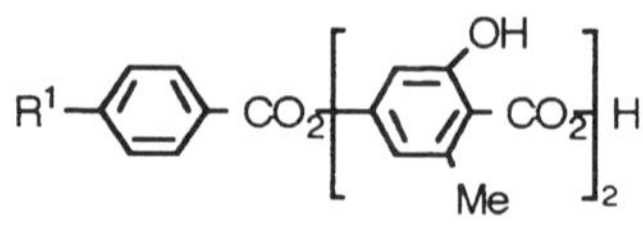

I

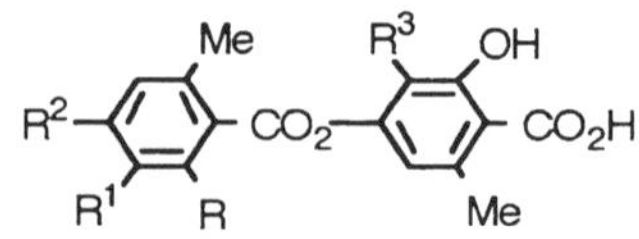

II

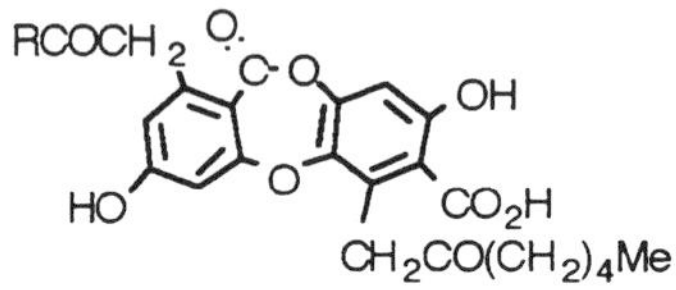

III

^{13}C nmr spectra of depsides, depsidones and depsones of the orcinol series have been correlated and tabulated.(E.G.Sundholm *et al.*, Chem.Scr., 1980,**16**,197). A.I.Scott has reviewed the applications of ^{13}C nmr to the study of biosynthetic mechanisms (J.Nat.Prod., 1985, **48**,689). As for example, Auranticins A (1) and B (2), two new antimicrobial depsides obtained from a mangrove isolate of the fungus *Preussia aurantiaca,* had their structures determined through analysis of selective INEPT, decoupling, COSY, and NOESY experiments (G.K.Poch *et al.*, J.Nat Prod.,1991,**54**, 213).

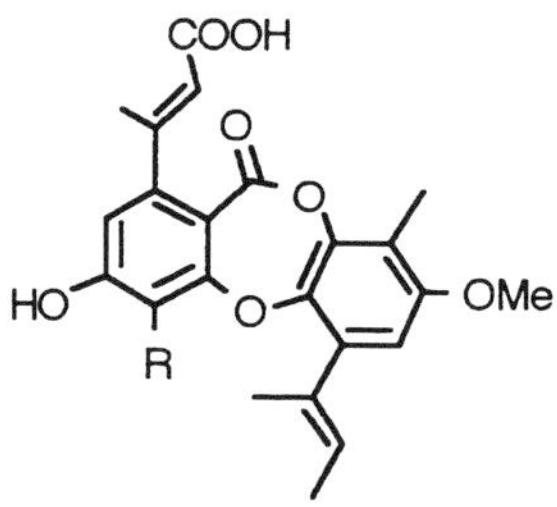

(1): R = CH_2OH

(2): R = CHO

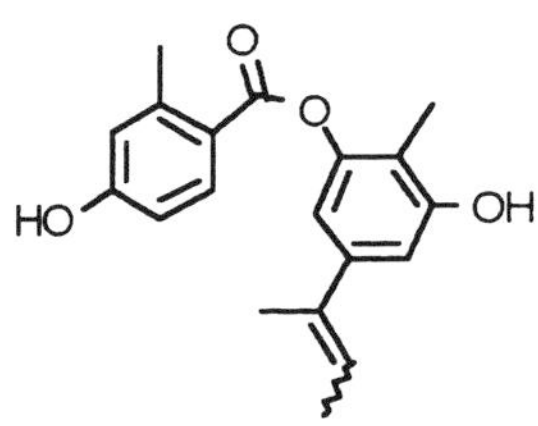

(3)

(b) Biosynthesis

The first non-chlorinated fungal depside (3) was isolated (J. Sierankiewicz *et al.*, Acta Chim. Scand.,1972,**26**,455) from the mycelium of *Aspergillus nidulans* grown in chloride free substrate. If the organism is grown in chloride free medium until a good production of (3) is obtained, addition of KCl causes its disappearance and simultaneous appearance of nidulin. ^{14}C-Labelled (3) is converted into nidulin with high incorporation.

Patterns of labelling were determined (J.P.Colonna, Café, Cacao, Thé, 1986,**30**,83) for the depsides chlorogenic, cryptochlorogenic and neochlorogenic acids in the leaves of the young coffee plants ,*C. robusta* and *C. arabica* after photosynthesis in an $^{14}CO_2$ atmosphere for two hours. In *C.arabica* , 3-feruloylquinic acid was labelled at similar levels to chlorogenic acid.These two depsides may play an equivalent role in coffee, providing phenylpropanoic units in lignin formation. In a recent brief review [M. Zamfir, Rev.Chim.(Bucharest), 1993,**44**,482], it is suggested that depside formation is an unsuccessful evolutionary attempt at building macromolecules!

The biosynthesis of *meta* - depsides had been the subject of speculation and a number of proposals put forward. Now it has been discovered (J.A.Elix *et al.*, Aust.J.Chem.,1986,**39**,613) that *meta*- and *para* - scrobiculins equilibrate very rapidly under mild conditions. This result provides strong evidence for the biosynthesis of *meta*-depsides and why no *para*-depsides have been discovered containing an additional hydroxy group in the B-ring. It appears that C-hydroxylation of the *para*-depside is followed by rapid intramolecular O-acyl rearrangement to the thermodynamically more stable *meta*-depside. The co-occurrence of closely related *para*- and *meta*-depsides, such as divaricatic acid and sekikaic acid, has been observed in a number of *Ramalin* a spp, and provides further circumstantial evidence for this proposal.

There has been a number of proposed biosynthetic pathways leading to depsidones. Sala and Sargent (J.Chem.Soc., Perkin Trans I, **1981**,855) had suggested that depsidones are derived independently of depsides through the phenolic coupling of benzophenones and the intermediacy of spirobenzofuran-3-ones. Elix and co-workers now propose (J.A.Elix *et al* ., Aus.J.Chem., 1987,**40**,1451) the alternative pathway involving, as key step, the Smiles rearrangement of the depside. The biosynthetic interelationships between *para* -depsides and a range of other lichen metabolites is summarised below. Elix subsequently accomplished the synthesis of the depsidones, divaric and stenosporonic acids, using a biomimetic approach based on this chemistry and will be discussed below.

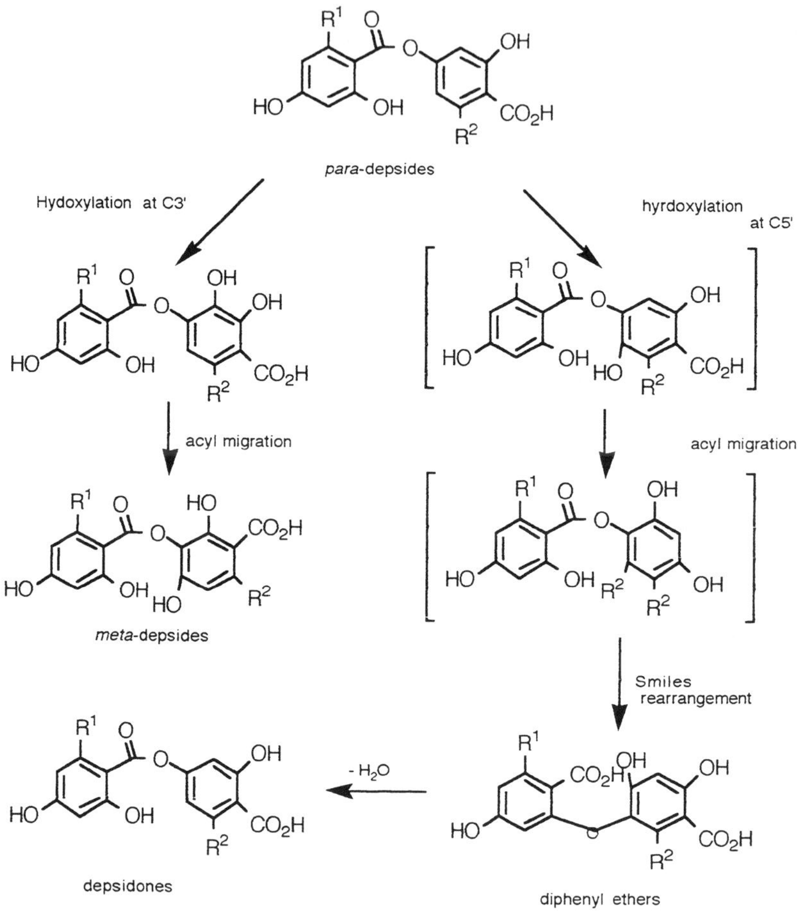

Proposed biosynthetic transformations of *para*-depsides

4-O-Methylisocryptochlorophaeic acid isolated by Archer and Elix(Mycotaxon.,1992,**45**,417) from *Pertusaria paradoxica* is the first natural *meta*-depside known to involve substitution at the 5'-position of the B-ring and supports the biosynthetic theory outlined above(J.A. Elix *et al.*, Aust.J.Chem., 1993,**4 6**,301).

It is now thought that algae are not needed for catalysis in lichen depside and depsidone biosynthesis. Instead , it is suggested (C.F.Culberson *et al.*, Exp. Mycol., 1992,**1 6**,52) that the characteristic secondary metabolism of natural lichens is linked to their aerial habit of growth.

(c) Depsides

(i) Synthesis

The overall strategy is to prepare the suitably substituted benzoic acids and phenols as A- and B-ring precursors, respectively, condense these in the presence of trifluoroacetic anhyride (or N,N'-dicyclohexylcarbodiimide) and subsequently deprotect the intermediate depside ester. This procedure is now well established (J.A.Elix *et al.,* Aust.J.Chem.,1993,**4 6**, 301). Total synthesis is most frequently employed to prove structures unambiguously. Often this is the only alternative to x-ray crystallography. For example, patagonic acid (4), isolated from the lichen *Bunodophoron patagonicum* , was synthesised (J.A.Elix *et al.*, Aust.J.Chem., 1994, **4 7**,1335) along these lines, as outlined below.

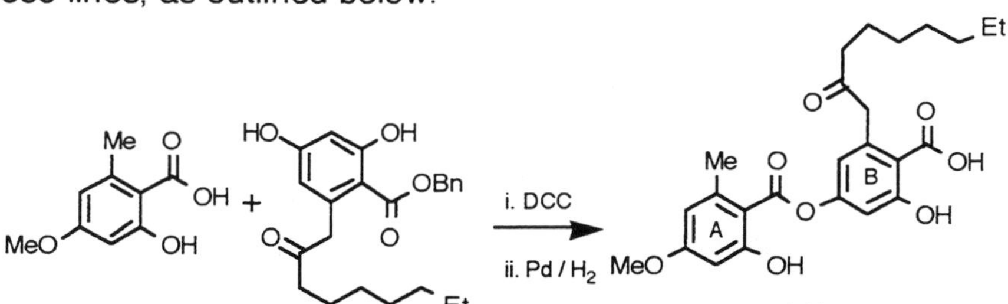

(4)

White and Johnson (J.Org.Chem.,1994,**5 9**, 3347) reported the total aymmetric synthesis of the novel lichen metabolite, the aliphatic depside, (+)-bourgeanic acid (5) in twelve steps. This substance had been isolated some twenty years earlier from several *Ramalina* species. The key intermediate, the homoallylic alcohol (7) was employed as the basis of both the A-and the B-portions of the target molecule.Thus (7) was converted to the β-lactone (8) and reacted with lithio-(7). The resulting ester was ozonolysed and in the final step, oxidised with hydrogen peroxide in acetic acid and deprotected. The sequence confirmed that (5) is a self condensation of acid (6).

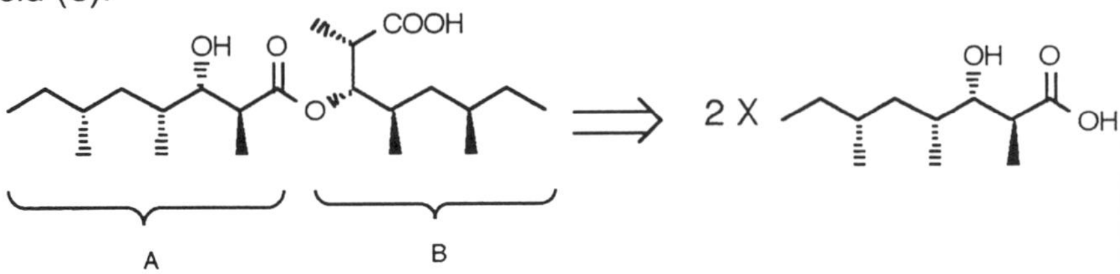

(5) (6)

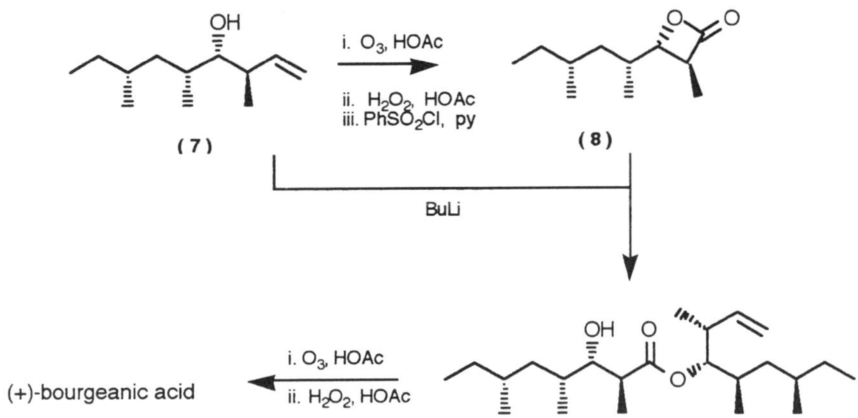

(5)

(ii) Individual depsides

Two new polysubstituted depsides, **brialmontins 1** and **2**, were isolated from *Lecanis brialmontii* (C.Vinet *et al.*, J.Nat.Prod., 1990,**5 3**,5002). Brialmontin 1 (9) was isolated as white prisms mp 102-4°($CHCl_3$) and brialmontin 2 (10) also as white prisms mp106-9°. Their structures were determined by spectroscopic means, particularly ^{13}C nmr.

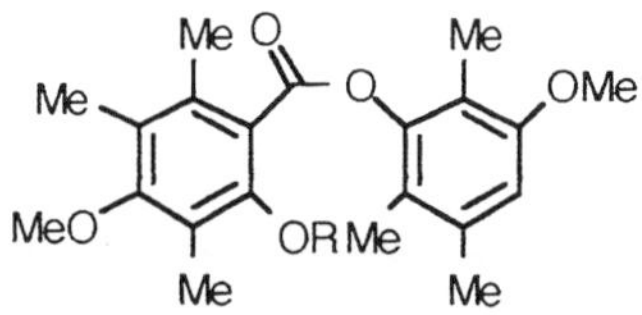

(9) brialmontin 1, R = Me
(10) brialmontin 2, R = H

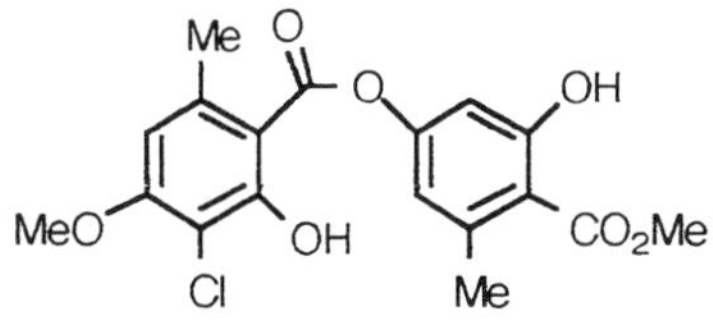

(11)

Wrightiin (11), a new chlorinated depside was isolated from *Erioderma wrightii* had its structure determined by x-ray crystallography. The mass spectrum was anomolous (W.S.Maass and A.W. Hanson, Z.Naturforsch.,Sect.B, 1986,**41**, 1589).

A new *ortho* - depside, **isolecanoric acid** was isolated from *Parmelia tinctorum* Despr. 2-(2,4-Dihydroxy-6-methylbenzoyloxy)-4-hydroxy-6-methylbenzoic acid was the structure assigned to this substance from studies on the hydrolysis products and from analyses of the ^{1}H and ^{13}C nmr spectra. This is the first *ortho* -depside isolated from the genus *Lichen* (A.Sakurai and Y.Goto, Bull.Chem.Soc. Japan, 1987,**60**,1917).

Isosalvianolic acid C, a new, novel depside possessing a dibenzooxepin skeleton (12), was isolated from the aqueous extract of *Salvia chinensis* (T.X.Qian and L.N.Li, Phytochemistry,1992, **31**,1068). A new depside, **salvianolic acid J**, was isolated from the aqueous extract of *Salvia flava* as a trace component, and was assigned structure (13) (*idem,ibid.*.,1994,**37**,907).

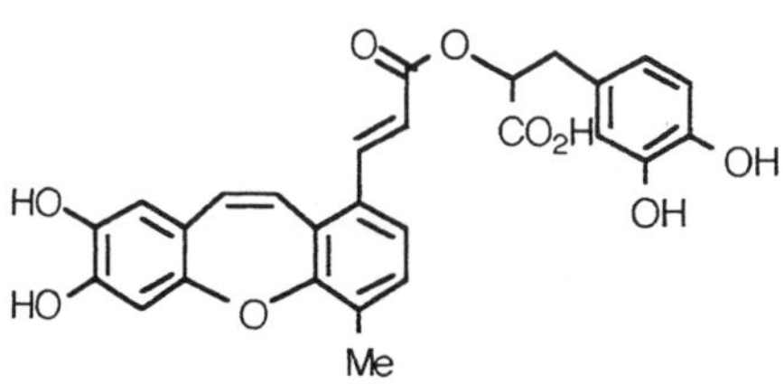

(12)

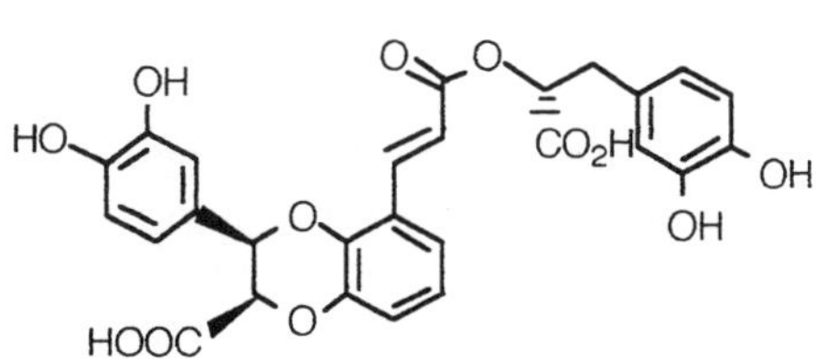

(13)

(d) Depsidones

(i) Synthesis

Synthesis of depsidones presents the problem of devising a method of accomplishing the aryl aryl oxygen linkage. One answer to this problem is to employ a cyclohexenone as ring-A synthetic equivalent (J.A.Elix *et al.*, Aus.J.chem., 1984,**37**,2349).This methodology was later superceded by the employment of the Smiles rearrangement. The necessary prerequisite is that the electophilic component bear an electron withdrawing group, though not necessarily at the respective *ortho* - or *para* - position. In this way Elix *et al.*, synthesised divaronic (14) and stenosporonic (15) acids referred to above (Aust.J.Chem.,1987,**40**,1451):

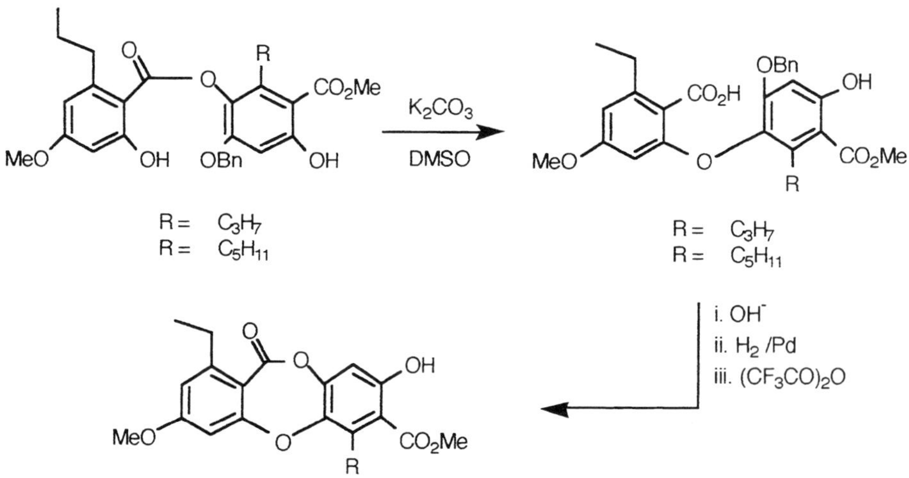

(14) divaronic acid; R = C_3H_7

(15) stenosporonic acid; R = C_5H_{11}

Synthesis of the depsidone (16), a derivative of furfuric acid was achieved by employing a biomimetic route wherein an intermediate grisadienone was thermally converted to the depsidone. In this case since the reaction could not be regiocontrolled, two grisadienediones were formed and required separation (J.Gunzinger and R.Tabacchi , Helv.Chim.Acta, 1985, **68**,1940). Conversion of the grisadienedione to the depsidone(16) proceeded satisfactorally.

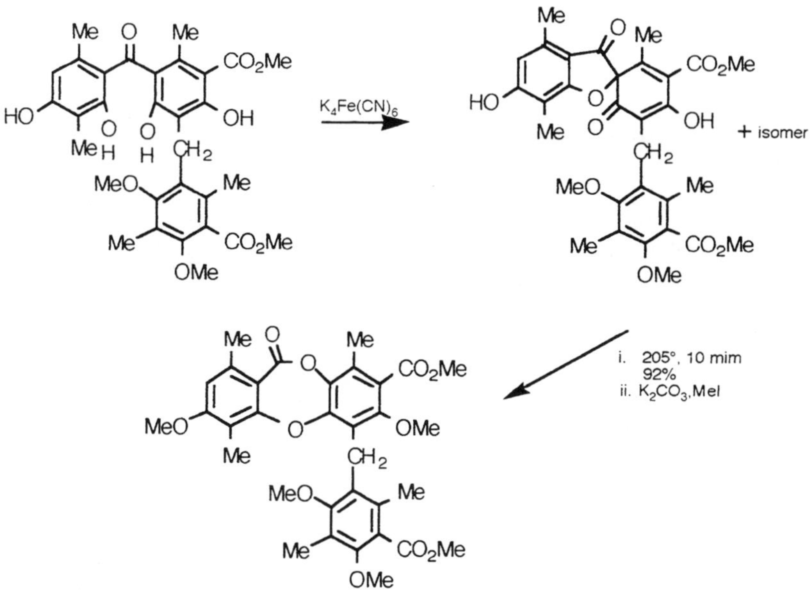

(16)

(ii) Individual depsidones

Pennarin (17) is a natural depsidone the detailed conformation of which has been determined by X-ray crystallography (SHELXS86) (D.Blaser and H.Stoeckli-Evans, Acta Cryst., C, **1991**, 2624). The central dioxepine-11-one is essentially boat shaped.

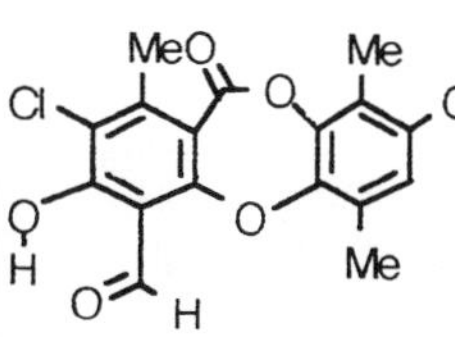

(17)

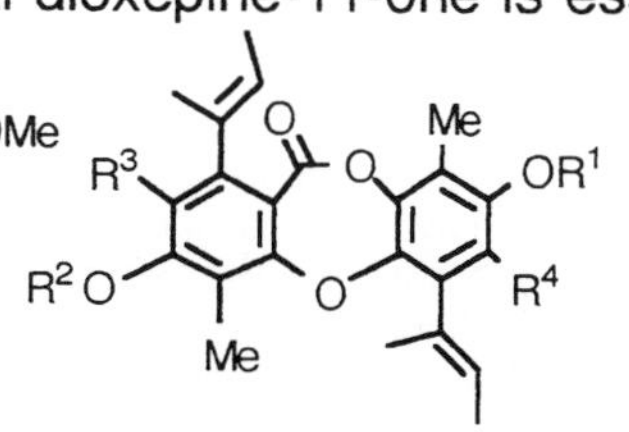

(18) emeguisin- A; $R^1=R^2=R^3=H$, $R^4=Cl$
(19) emeguisin- B; $R^1=R^3=H$, $R^2=Me$, $R^4=Cl$
(20) emeguisin- C; $R^1=Me$, $R^2=H$, $R^3=R^4=Cl$
(21) folipastatin; $R^1=R^2=R^3=R^4=H$,

Auranticin A(1) was isolated as a white crystalline solid and **B**(2) as an orange oil and were referred to earlier. They are members of an increasing number of depsidones from non-lichen sources. Auranticin A is much more potent than B and accounts for most of the antibiotic activity against *Bacillus subtilis* and *Staphylococcus aureaus.*

Emeguisin (A)(18),**(B)**(19) and **(C)**(20) are further examples of fungal depsidones and were isolated from *Emericella unguis.* Their structures were assigned from extensive spectroscopic analysis.The E,E -stereochemistry was assigned (N.Kawahara *et al.*, J.Chem. Soc. Perkin Trans.1, **1988**, 2611) on the basis that the others present in the fungus, 2-chlorounguinol, unguinol, and nidulin, are known to have this stereochemistry. It is of note that these fungal depsidones (1), (2), (18)-(20) and the depside (3) (presumed *cis*) have in common, this unusual *cis* - dimethylvinyl substituent. A more recent isolate, **folipastatin** (21) (K.Hamano *et al.*, J.Antibiotics, 1992, **45**, 1195) from *Aspergillus-unguis* also has these structural features. Folipastatin is an inhibitor of phospholipase A_2.

(e) Depsones

The new depsones **subpicrolichenic acid** (22) and **superpicrolichenic acid** (23) have been isolated from the lichens *Pertusaria amara* and *P.truncata*, respectively (J.A.Elix *et al.*, Aust. J. Chem., 1991,**44**, 1487). The assigned structures were confirmed by unambiguous synthesis. Four more new depsones (24)-(26),were recently isolated from the lichen *Pertusaria-truncata* (J.Elix *et al.*, Aust. J.Chem., 1994,**47**,1345). The structures assigned were confirmed by unambiguous synthesis.

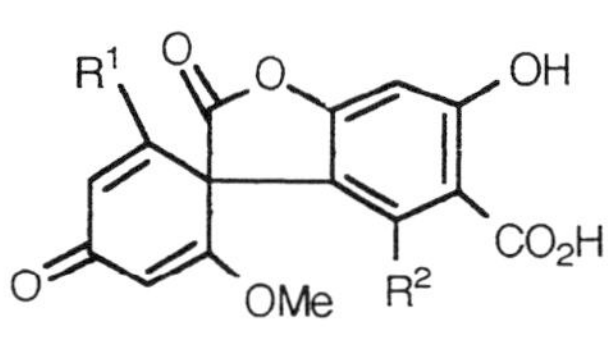

	R^1	R^2
(22) subpicrolichenic acid;	C_3H_7	C_5H_{11}
(23) superpicrolichenic acid;	C_7H_{15}	C_7H_{15}
(24) hyperpicrolichenic acid;	C_5H_{11}	C_7H_{15}
(25) isohyperpicrolichenic acid;	C_7H_{15}	C_5H_{11}
(26) megapicropilichenic acid;	C_7H_{15}	C_9H_{19}
(27) isomegapicrolichenic acid;	C_9H_{19}	C_7H_{15}

2. Hydrolysable Tannins

(a) Introduction

There has been a major revival of interest in the tannins (representative references may be found in the following reviews: E.Haslam 'Plant Polyphenols',Cambridge University Press,1989; *idem* in 'Natural Products of Woody Plants', Vol. I, J.W.Rowe, ed., Springer -Verlag 1989; *idem.* in R.W.Hemingwy and P.E.Laks,eds. 'Plant Polyphenols' Plenum Press,1992; L.J.Porter, Methods Plant Biochem., 1989, **1**, 389). This has been the consequence of major developments in separatory techniques and in structural elucidation methodologies. Investigation of structure - biological activity relationships is now possible (see for example, Y. Kashiwada *et al.*, J.Nat.Prods., 1992, **55**, 1033; T.Okuda *et al.* in 'Plant Polyphenols',1992). The tannins possess a variety of biological activities including DNA-breaking capabilities, anti-mutagenic effects, lipid peroxidation inhibition, in addition to anti-human immunodeficiency virus' properties and anticancer activities. For example, the complex flavonal camelliatannin A (1) inhibited the cytopathetic effects induced by HIV (T.Hatano *et al.*, Yakugaku Zasshi, 1992, **34**, 510) and chebulagic acid (2) *inter alia*, exhibited moderately selective cytotoxicity against melanoma cells.

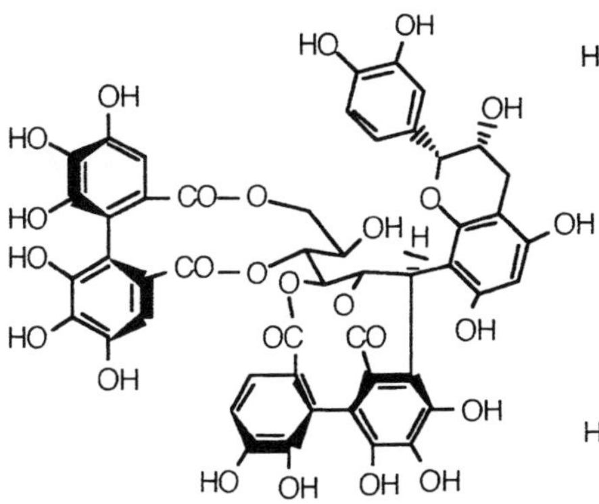

(1)

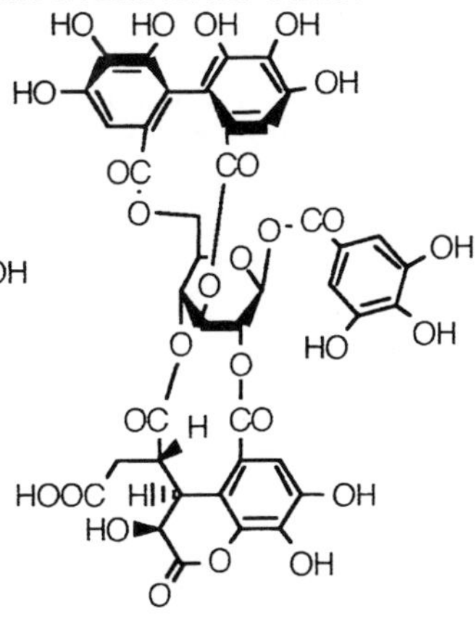

(2)

(b) Classification of tannins

Traditionally the vegetable tannins have been divided into two principal classes, the hydrolysable and the condensed. Only the former will be considered here. These tannins constitute a class of natural products of such great structural diversity that they are now more conveniently considered under a number of headings: gallotannins , ellagitannins, complex tannins and hydrolysable oligotannins. The gallotannins are further subdivided into four groups based on their alcohol core: those with a glucose (3), with a quinic acid (4) ,with a shikimic acid (5), and with a hamamelose residue (6).

Ellagitannins may have a glucose core in the pyranose form (7) or in the open chain form (8), as O- or C- glycosides, with a triterpenoid moiety (9), or with a modified hexahydroxydiphenoyl and / or related group (see below). There are also caffeetannins *e.g.* 3,5-di-O-caffeoylquinic acid, in which the polyphenolic group is the caffeoyl group (T.Okuda *et al.*, Yakugaku Zasshi, 1986,**106**,894).

In addition, the oligomeric hydrolyzable tannins have developed as a new class (review: T.Okuda *et al.*, Heterocycles, 1990, **30**,1195). Between 1982 when the first dimer agrimoniin (10)

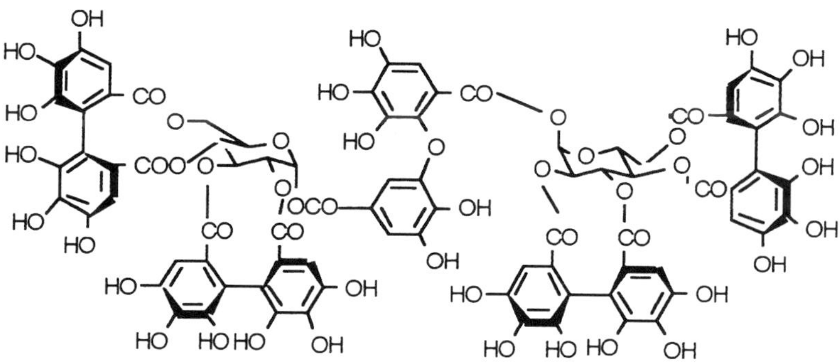

(10)

was identified and 1990, more than sixty others were detected.This rate of discovery has persisted. This class consists of dimers, trimers and tetramers. Higher oligomers have not yet been recorded. Classification of these oligomers is based on the constitution of the monomers and the the linkage between them (T.Okuda *et al.*, Phytochemistry, 1993,**32**,507). Finally, the complex tannins constitute a separate class. These are composed of both a hydrolysable tannin and a condensed tannin residue, exemplified by chevulagic acid (2).

It is thus apparent that classification of tannins is not simple. The system based on biogenetic relationships developed by Haslam ('Plant Polyphenols', 1989) is the most rational and effective. It will be considered briefly in the section on biosynthesis.

(c) Analysis and structure determination

The important biological activities exhibited by tannins are critically dependent on structure. This has highlighted the need for precise analytical methods. Centrifugal partition chromatography combined with gel-partition chromatography, have facilitated preparative separations. New methods of analysing tannins have been reviewed (T.Okuda *et al.*, J. Nat Prods, 1989, **52**,1).

Oligomeric hydrolysable tannins are generally more polar and unstable and generally more sensitive to heat. Extraction must not be undertaken above ambient temperatures. In addition to fabms and high-performance gel permeation chromatography (hp-gpc), normal - phase hplc is particularly useful for the separation of hydrolysable oligotannins. Anomers and hemiacetal tautomers are separable by reverse phase hplc. For example the equilibrium exhibited by geraniin (11) was first detected by hplc (T.Hatano *et al.*, J. Chromatogr., 1988, **435**, 285).

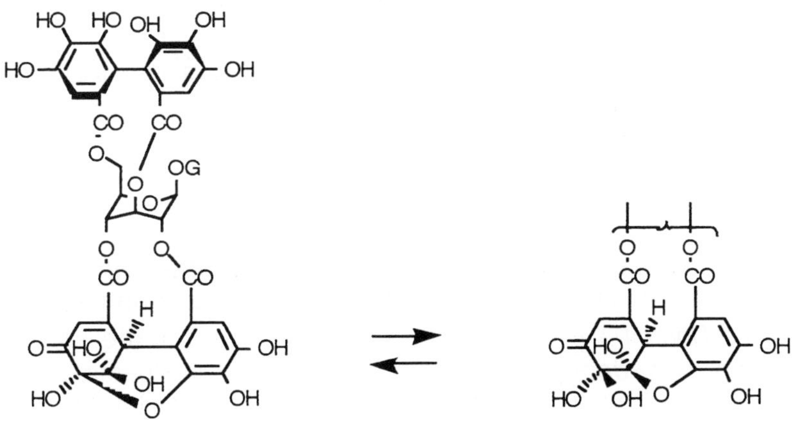

(11)

Fabms does not require derivatisation and is therefore most effective for the molecular weight determination of tannins that are highly polar, thermolabile, and nonvolatile. Both ^1H and ^{13}C nmr spectroscopy have dramatically improved detailed structural assignement. For example, the signals of glucose moieties in tannins may be fully assigned using ^1H -^1H and ^1H -^{13}C selective decoupling and 2D-nmr techniques. The absolute configurations of the chiroptical aromatic substituents are determined directly by means of cd absorption spectroscopy.

(d) Biosynthesis

The current state of knowledge regarding the biosynthesis of hydrolysable tannins has been reviewed (G.G.Gross, Methods Plant Biochem., 1993, **9**, 25). The overall biosynthetic pathway may be summarised under the following headings: (a) the origin of the first specific intermediate, β-glucogallin (12), (b) the metabolic sequence from this monoester to pentagalloylglucose (13) and (c) the secondary transformation of this pivotal intermediate to gallotannins and ellagitannins and ultimately, the oligotannins.

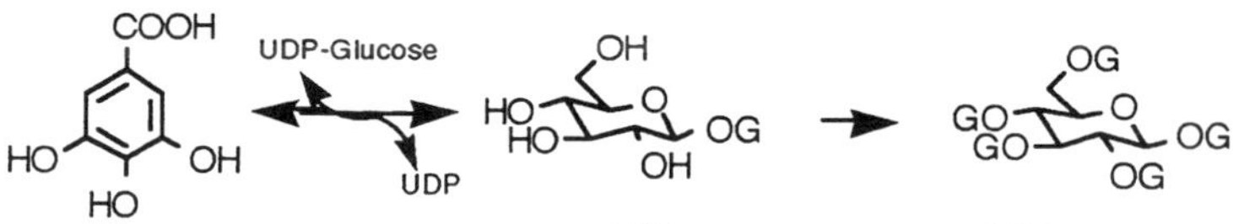

(12) **(13)**

Biosynthesis of β-glucogallin(12) (UDP, uridine -5'-diphosphate) and 1,2,3,4,6-penta-O-galloylglucose

β-Glucogallin (12) is a highly reactive galloyl donor and in a sequence of enzyme catalysed disproportionation steps, it leads specifically to the sequential formation of 1,6-digalloyl- , 1,2,6-trigalloyl- ,1,2,3,6-tetragalloyl- and finally the penta-O-galloyl-glucose (13). Preliminary results regarding the gallotannins are now available (*loc.cit.*.). β-Glucogallin is again involved in the enzyme catalysed *meta*-depside bond formation of 1,2,3,4,6-penta-O-galloylglucose resulting in the biosynthesis of the gallotannins (14), (15), and (16).

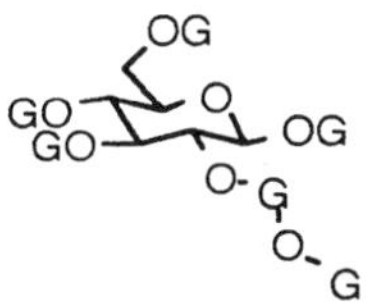

(14)

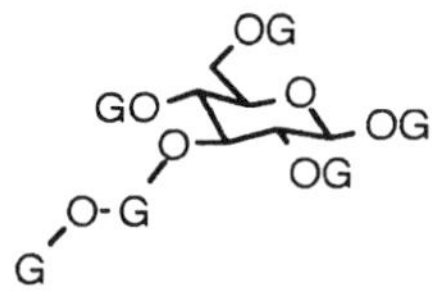

(15)

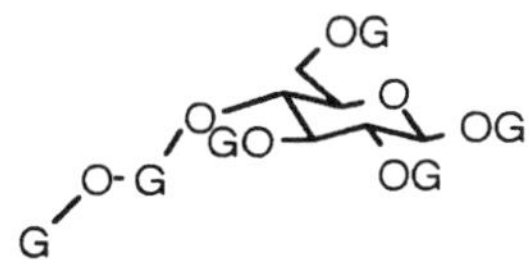

(16)

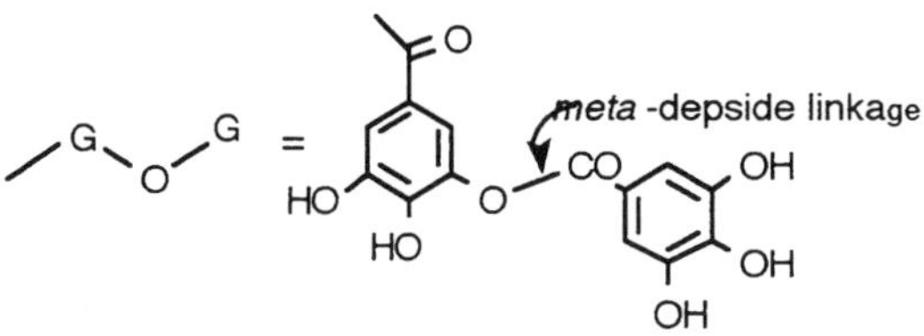

Haslam (*loc.cit.*,1989) has employed such biogenetic relationships to classify tannins, although many of the steps proposed have not been confirmed in detail as yet. The first group is the gallotannins which have a typical *meta*-depside linkage derived as described above. *para*-Depside linkage has been detected but not confirmed (Gross, *loc. cit.*). The observation that the lichen metabolites, *meta*- and *para*- depsides, interconvert easily and that the *meta* -isomer is more stable(J.A.Elix *et al.*, Aus.J.Chem., 1986, **39**, 613) may have a bearing on gallotannin biosynthesis. Ellagitannins form the next clearly identifiable group in which (S)-hexahydroxy -diphenoyl [(S)-HHDP] esters are elaborated by oxidative coupling of vicinal galloyl ester groups (4,6 and 2,3) attached to the D-glucopyranose in its most stable $^4C_1$ conformation. The third grouping is based on a further metabolic variant in which oxidative coupling of adjacent galloyl groups occurs 1,6; 2,4 or 3,6 in the D-glucopyranose precursor which has adopted the less favourable $^1C_4$ conformation. It is noteworthy that the 3,6-coupled congeners display (R)-HHDP stereochemistry.These three classes are summarised below.

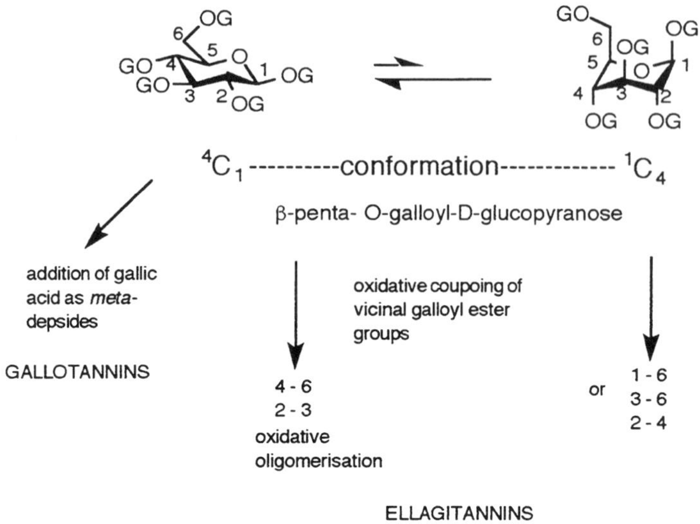

Overall patterns of metabolism of β-penta-O-galloyl- D-glucose in higher plants

(e) Synthesis

There has been a number of approaches towards the synthesis of tannins. All have remained incomplete but for the two now reported. The partial synthesis of elaeocarpusin (14),was reported (T.Okuda *et al.*, Heterocycles, 1986, **24**, 1841). Geraniin (11), when treated with L-ascorbic acid under simulated physiological conditions (pH 4), was converted to elaeocarpusin (14). It is considered that this is how geraniin and perhaps other tannins, are solubilised in water - a long standing anomoly.

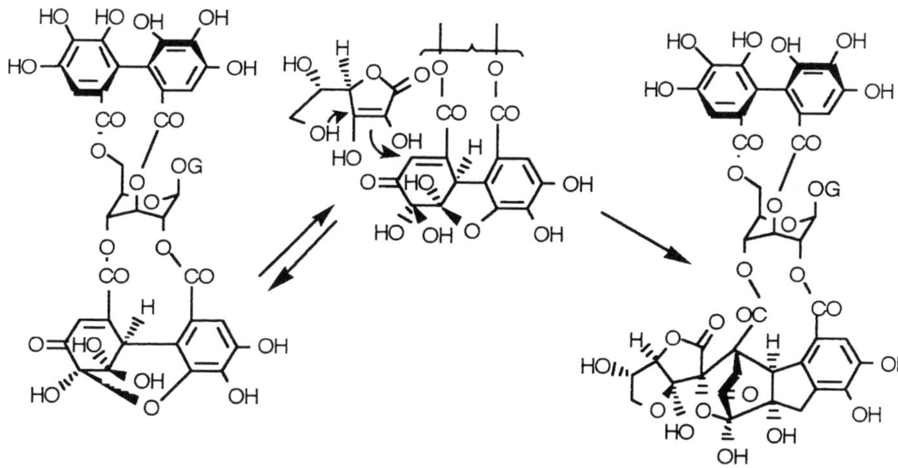

(11) (14)

The first total synthesis of an ellagitannin , tellimagrandin I (16), was reported recently (K.S. Feldman *et al.*, J.Am.Chem.Soc., 1994, **116**, 1742; 3357). The precursor (15) was prepared in three steps from D-glucose. When subjected to lead tetraacetate oxidation, only 4,6-coupling ensued. Debenzylation provided homochiral tellimagrandin. The enantiospeciific formation of the (S)-HHDP moiety had been postulated by Haslam(1989), who had earlier considered that the inherent conformational preferences of the precursor galloylated glucose, would translate directly into the naturally occurring HHDP sterochemistry upon oxidative bond formation.

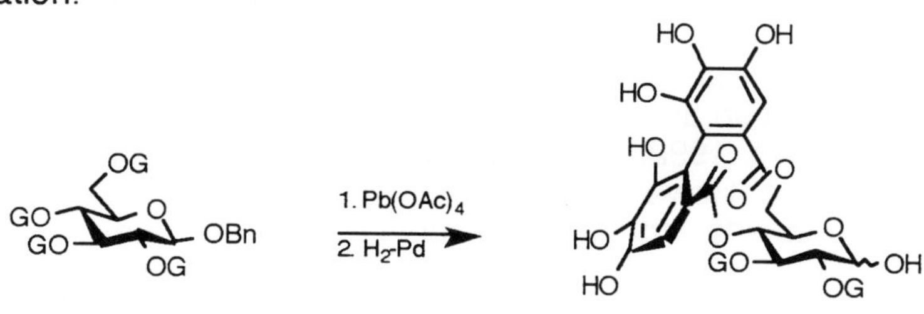

(15) (16)

(f) Individual tannins

Terchebulin (**17**) is a new ellagitannin which was isolated from the fruits of *Terminalia chebula* RETZ., *Combretaceae* (T.C.Lin *et al.*, Chem. Pharm. Bull.Tokyo, 1990,**38**,3004). It possesses the novel tetraphenylcarboxylic acid, terchebulic acid. This was determined by isolation following saponification with sodium methoxide. **Terflavin A** (18) was also isolated from the leaves of *T.chebula*. This finding provides biogenetic evidence that the terchebulic acid moiety is derived by an oxidative carbon-oxygen coupling of adjacent flavogallonic and gallic acid esters.

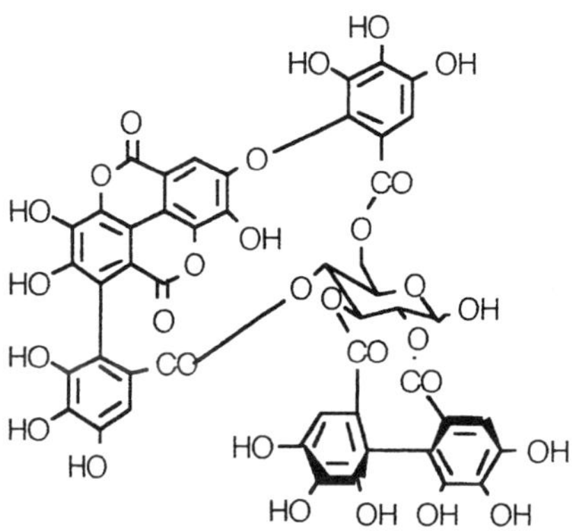

(17)

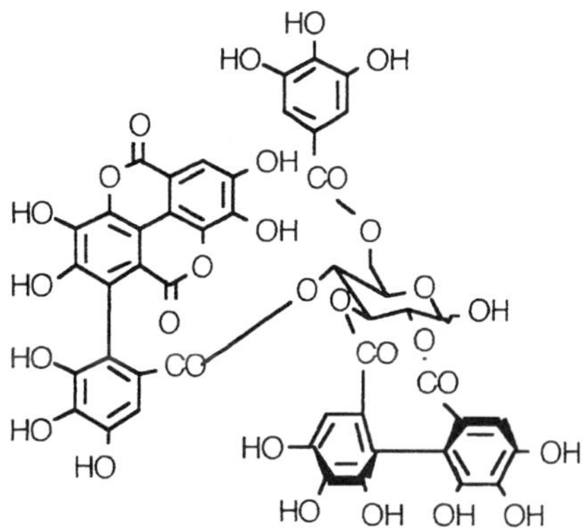

(18)

Nobotanin I (19). Three new hydrolysable tannin dimers were isolated from the leaves of *Heterocentron roseum* (Melastomataceae), and their structures elucidated on the basis of chemical degradation and nmr spectra (T.Yoshida *et al.*, Chem. Pharm. Bull. Tokyo, 1992, **40**, 66). Nobotanins G, H and I are structurally related. However, nobotanin (19) is a novel dimer possessing a depside - forming valoneoyl group. This same functionality had been detected earlier in a number of ellagitannins, *e.g.* prostatin C (T. Hatano *et al* ., Heterocycles, 1990,**31**,1221) and praecoxins C and D (*idem*, Chem.Pharm.Bull. Tokyo, 1991,**39**, 1689).

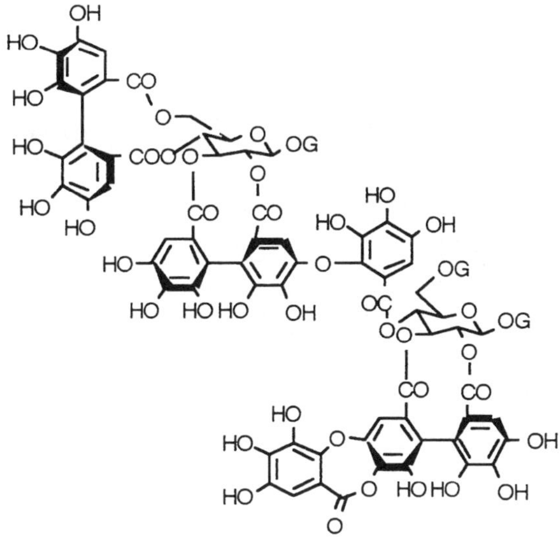

(19)

Hirsunin (**2 0**) This novel ellagitannin was isolated from the leaves of *Alnus hirsuta* var. *microphylla*. The structure consists of a diarylheptanoid glycoside (oregonin) and the ellagitannin (praecoxin A) moieties connected through an ester linkage (M.W.Lee *et al.*, Phytochemistry, 1992,**3 1**, 967).

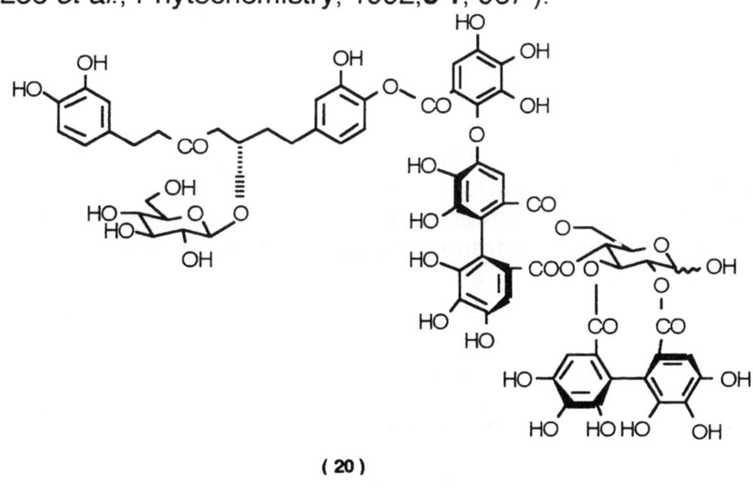

(20)

Phyllanthusiin-D(21) This unusual ellagitannin was isolated from the biologically active polar fraction of *Phyllanthus amarus* (L.Y. Foo *et al.*, Phytochemistry 1992, **3 1**, 711). Its structure was established as 1-galloyl-2,4-(acetonyl-dehydrohexahydroxydiphenoyl) -3,6-hexahydroxydiphenoyl-glucopyranoside and its conformation was determined as $^1C_4$ by chemical and spectroscopic methods.

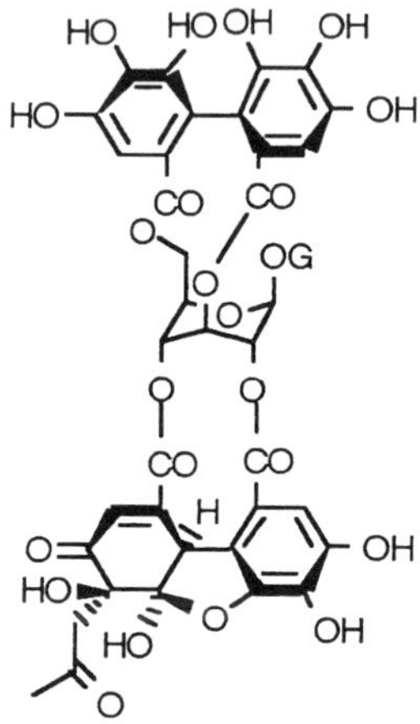

(21)

Woodfordin-D (22) and Oenothein-A (23) Two new trimeric hydrolyzable tannins, woodfordin-D (22) and oenothein-A (23) were isolated from the dried flowers of *Woodfordia fruticosa* (T.Yoshida *et al.*, Chem.Pharm.Bull. Tokyo, 1991, **3 9**, 1157; *idem.,ibid.*, 1992, **4 0**, 2023). Their macrocyclic structures , which have a novel constituent unit (the woodfordinoyl group) connecting the monomers , have been elucidated on the basis of spectral and chemical evidence. Oenothein-A (23) was also isolated from the leaves of *Oenothera biennis*. Both tannins exhibit antitumour activity.

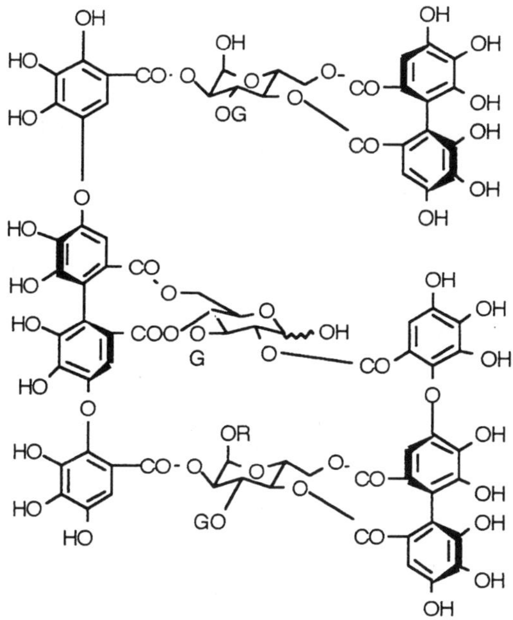

(22) R = G
(23) R = H

Alienanins A(24) **and B**(25). Two novel hydrolyzable tannins, alienanins A and B, were isolated from the leaves of *Quercus aliena* Blume(Fugaceae) (G. Nonaka *et al.*, Chem.Pharm.Bull. Tokyo, 1991,**39**, 884). Their structures were elucidated on the basis of chemical and spectroscopic evidence and shown to be dimeric C-glycosidic ellagitannins in which the hexahydroxydiphenoyl ester group is linked to the C-glycosidic C-1 atom through a carbon to carbon linkage.

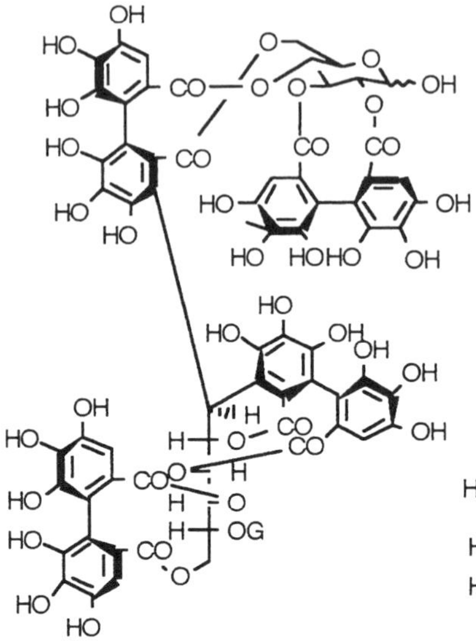

(24)

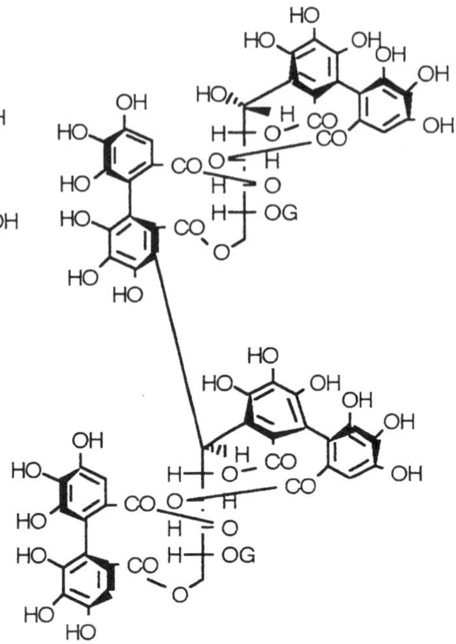

(25)

3. Lignans

(a) Introduction, biosynthesis

There has been increased activity in this area due to the varied and often powerful biological activities of lignans and neolignans (W.D.MacRae and G.H.N.Towers, Phytochemistry,1984, **23**,1207). For example, podophyllotoxin and others of the podophyllum family have been used in cancer chemotherapy (Chem.in Britain, **1981**, 457), whereas enterolactone may be an example of a new type of hormone-controlling cell growth (R.S.Ward, Chem.Rev.,1982, **11**,75) and natural 2,5-bisaryltetrahydrofuran lignans exhibit platelet-activating factor antagonists (T.Biftu and R.Stevenson, Phytother.Res., 1987,**1**,97).

The chemistry of lignans has been reviewed (D.C.Ayres and J.D.Loike, "Lignans, Chemical, Biological and Clinical Properties", Cambridge University Press, 1990; O.R. Gottlieb in J.W. Rowe, ed.,

"Natural Products of Woody Plants", Verlag - Chemie, 1989; G.M. Massanet *et al.*, Fitoterapia,1989,**60**,3; A.Pelter, Recent Adv. Phytochem., 1986, **20**, 201; M.N. Zaprometov, Cell Cult.Somatic Cell Genet.Plants, 1988,**5**,89). As in previous editions, only the chemistry of lignans will be reviewed. Neolignans are not included. Readers are directed to Gottlieb (*loc.cit.*).

Biosynthesis and steroselectivity in biosynthesis of lignans have been reviewed (B.Davin *et al.*, Recent Adv.Phytochem., 1992, **26**, 325; Ayres and Lioke *loco. cit.*.; N.G.Lewis and L.B.Davin in R.W. Hemingway and P.E.Laks,eds. 'Plant Polyphenols' Plenum Press, 1992, resp.).

The first example of a new class of lignan has been reported. It was isolated from Douglas-fir bark (L.Y.Foo and J.Karchesy, J.Chem.Soc., Chem.Commun., **1989**, 217). It is a biphenyl-linked pinoresinol-dihydroquercetin and is named pseudosuganol.

(b) Individual lignans

(i) 1,4-Diarylbutanes

Anolignans A(1) and **B**(2) are new dibenzylbutadiene lignans isolated from *Anogeissus acuminata*. They exhibited HIV-1 reverse-transcriptase inhibitory activity. Compound (2) showed weak activity alone but in the presence of (1), the activity was high. This synergistic effect was also observed by (2) on (1). Another new lignan , **anolignan C**(3) was also isolated from this plant. These three substances were either weakly cytotoxic or noncytotoxic towards various cancer cell lines.(A.M.Rimando *et al.*, J. Nat. Prod.,1994, **57**,896). The structures were established by spectroscopic methods, especially ID and 2D nmr.

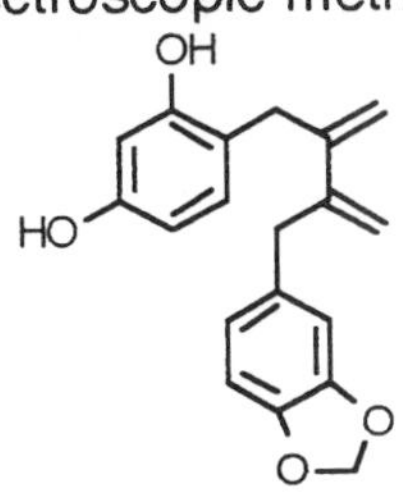

(1)

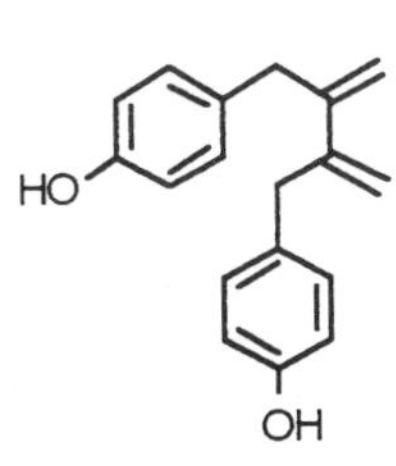

(2)

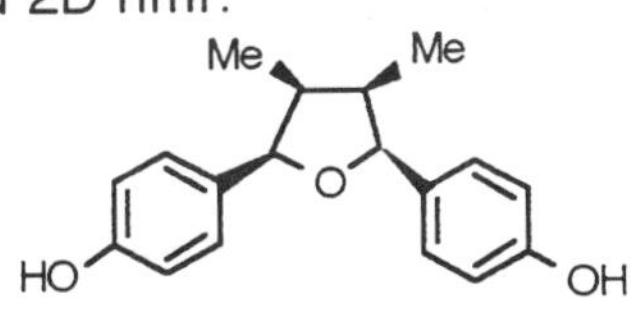

(3)

(±)-9,9'-O-Diferuloylsecoisolariciresinol (4) was one of two new compounds among the sixteen compounds isolated from *Machilus thunbergii*. The structure of (4) was determined by means of spectroscopic studies and chemical transformations (H. Karikome *et al.*, Phytochemistry, 1991,**3 0**,315).

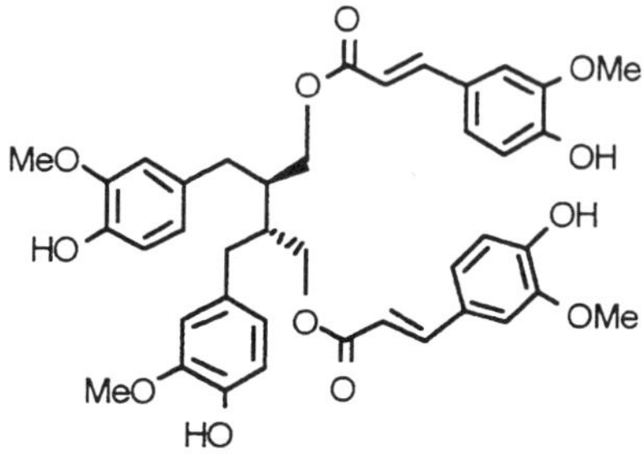

(4)

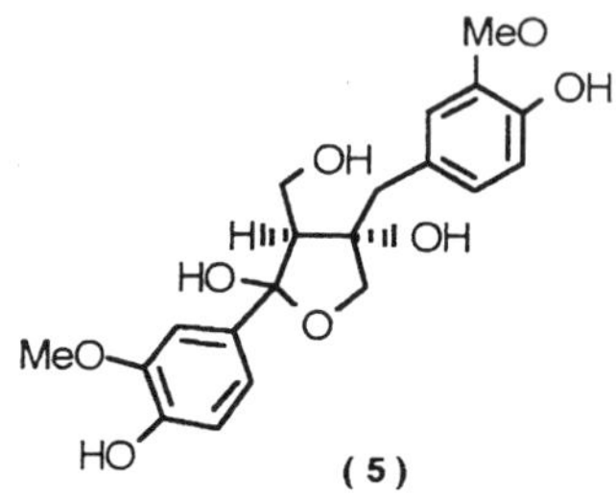

(5)

The 2-phenyl-3-methyl-4-benzyltetrahydrofuran subgroup of lignans is well established (cf. Gottlieb, *loc cit.*). The oxidation level may vary, for example, the new lignan, **2-hydroxyolivil** (5), isolated from the aerial parts of *Carduus assoi,* is highly oxygenated (I. Fernandez *et al.*, Phytochemistry, 1991, **3 0**,1030).

A short, stereoselective synthesis of (±)-dihydrosesamin (6) has been reported. Radical cyclisation is the key step. The overall synthesis occurs in good yield (G.Maiti *et al.*, Tetrahedron Letters, 1994,**3 5**,3985).

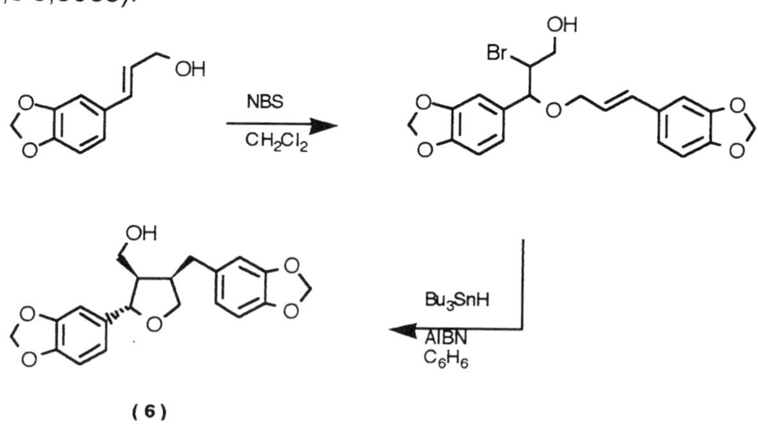

(6)

Actifolin (7) has been isolated together with two known lignans piperitol and sesamin, from the leaves of *A.longifolia*. This is the first report of lignans isolated from the genus *Actinodaphne*. (H.Tanaka *et al.*, Phytochemistry,1989,**28**,952). The structure was established on the basis of chemical and spectroscopic evidence.

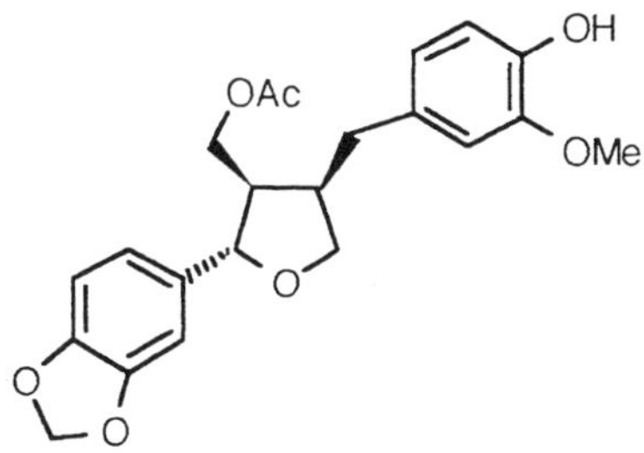

(7)

(ii) Dibenzylbutyrolactones

Phenylalanine and ferulic acid are biosynthetic precursors of the dibenzylactone lignan, arctigenin and the furofurans, phillygenin and epipinoresinol. Detailed investigation supports the hypothesis that the lignans arise by oxidative coupling of two coniferyl alcohol units rather than from two ferulic acid or coniferaldhyde units (M.M.A. Rahman *et al.*, Phytochemistry,1990,**29**,1841). *Forsythia intermedia* cell-free extracts were examined for their abillity to catalyse the enantioselective and regiospecific O-methylation of matairesinol. Regiospecific glucosylation first occurs to give matairesinoside. Subsequent methylation affords arctiin, which is then converted into arctigenin *via* the action of a β-glucosidase (S.Ozawa *et al.*, Phytochemistry, 1993,**32**,643).

The oxidative intramolecular aryl-aryl coupling of dibenzylbutanolides (butyrolactones) to aryltetralin lactones has been achieved using cell-free extracts from *Catharanthus roseus* (J.P. Kutney *et al.*, Can.J.Chem.,1992,**70**,2115). This latter transformation has been achieved synthetically by a number groups (R.C. Cambie *et al.*, Aust.J. Chem., 1988,**41**,897; M.Tanoguchi *et al.*, Chem. Pharm. Bull. Tokyo 1991, **39**,1873; H.Ishibashi *et al.*, Heterocycles 1991, **32**, 1279; *idem*, Tetrahedron, 1993, **49**,4173; A.S. Krauss *et al.*, Aust. J. Chem. 1991, **44**, 1307).

The total synthesis of optically active lignans has been reviewed (E.Brown, Trends Org.Chem., 1991,**2**,59).The asymmetric synthesis of homochiral dibenzylbutyrolactones has been achieved by conjugate addition to a chiral butenolide(R.S.Ward *et al.*, Tetrahedron-Asymmetry, 1990,**1**,857; B.L.Feringa *et al.*, J.Org.Chem., 1994,**59**,5999). A novel asymmetric synthesis (T.Honda *et al.*, J. Chem. Soc.Perkin Trans.1, **1994**, 1043) which involved enantioselective deprotonation as key step, is outlined below.

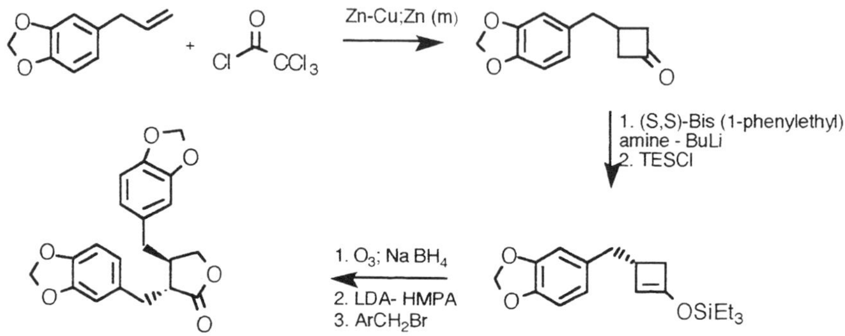

Sesquipinsapol C (8) a novel sesquilignan, was isolated as its acetoxyl derivative from the wood of *Abies pinsapo*. Its structure was determined by spectroscopic methods and chemical transformation (A.F.Barrero *et al.*, J.Nat.Prod., 1994,**57**,713).

Diarctigenin (9) , is a novel lignan dimer. It was isolated from the seeds of *Arctium-lappa* (B.H.Han, Phytochemistry, 1994, **37**,1161). Its structure was elucidated as *bis* -5,5'-arctigenin by spectral analysis and chemical modification.

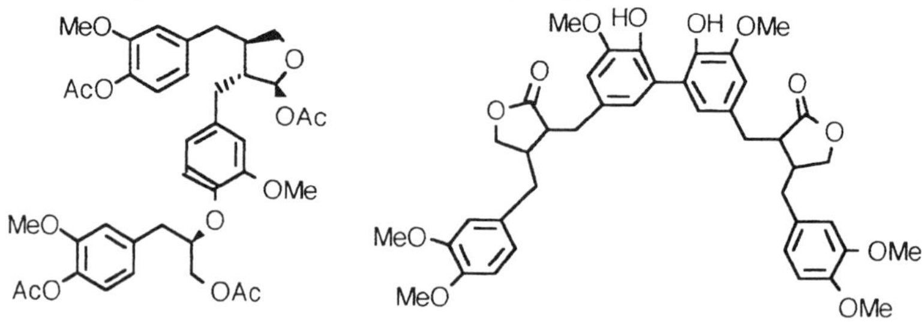

(8) (9) X

Trachelosiaside(10) was isolated from *Trachelospermum asiaticum*. It was found to be the first naturally occurring C-glucosidic lignan (I.Yamauchi *et al.*, Chem.Pharm.Bull.Tokyo, 1986,**34**,4340). The structure is matairesinol 5-C-1β-D-glucoside.

(+)-**Calocedrin** (11) was first isolated from the wood of *Calocedrus formosana.* (J.M.Fang *et al.*, Phytochemistry, 1985, **24**, 1863). The lignan had an unusual dihydroanhydride structure. Interestingly, when reduced, the hibalactone formed is optically inactive.

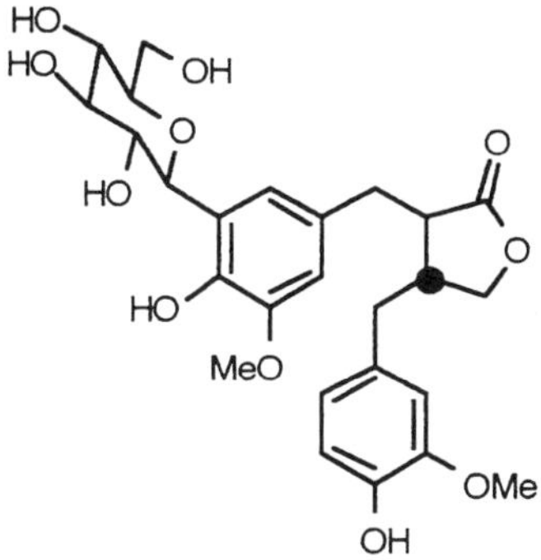

(10)

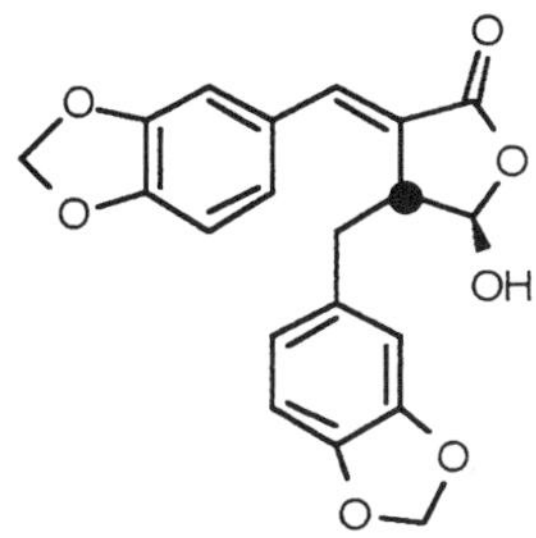

(11)

(iii) 2,5-Diaryl-3,4-dimethyltetrahydrofuranoids

all-cis-**2,5-Di-(*p*-hydroxyphenyl)-3,4-dimethyltetrahydrofuran**(12) was isolated from the roots of *Krameria interior* (Z.A.Dominguez *et al.*, Phytochemistry, 1990, **29**, 2655).

Aristolignin (13) and (-)**zuonin-A**(14) were isolated from *Aristolochia chilensis*.. These and related tetrahydrofuranoids were classified into six stereochemical groups(A.Urzua *et al.*, Phytochemistry, 1987,**26**,1509).

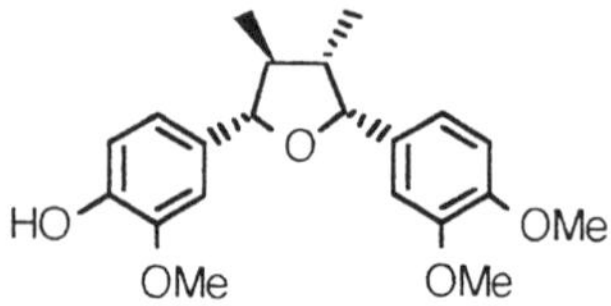

(13)

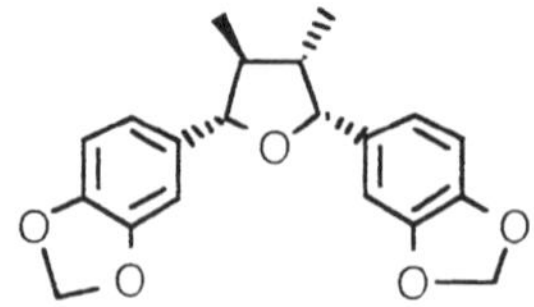

(14)

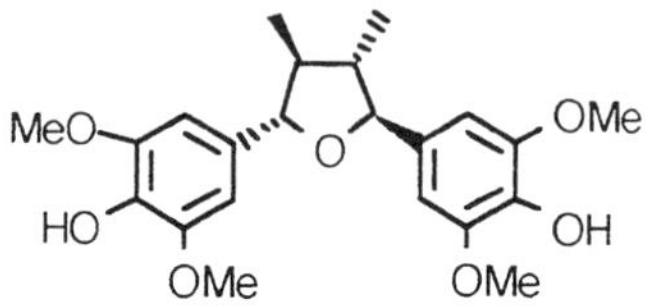

(15)

(-)-di-de-O-Methylgrandisin (15), was isolated from the leaves of *Virola pavonis* (P.H.Ferri and L.E.S.Barata, Phytochemistry, 1991,**30**, 4204).

(iv) Pinoresinol and lariciresinol types : the furofurans

The details of stereo- and enantio-selectivity in the course of lignan biosynthesis are finally unravelling. The coupling of achiral coniferyl moieties to (+)-pinoresinol either in *Forsythia - suspensa* or with (+)-pinoresinol synthase and coupling to (-) -pinoresinol in cell free extracts of *Forsythia-intermedia* , have been reported (D.G.I. Kingston *et al.*, J.Chem.Commun., **1990**, 1405; N.G.Lewis *et al.*, Phytochemistry, 1992,**31**,3875; *idem*, Tetrahedron Lett., 1994, **35**, 4731). Specific de-O-methylation of the *para*-methoxyl groups of (+)-magnolin and (+)-yangabin occurs in rat (A.M. Miyazawa *et al.*, Phytochemistry,1993,**32**,1421). This effect is also observed with *Aspergillus-niger* (*idem,ibid.*,1994,**35**,1191).

The solution conformations of furofurans were determined by means of the lanthanide shift technique which utilises a 1,2,3-trimethoxyaryl arrangement as a bidentate coordination site for the NMR shift reagent (O.Hofer, *et al.*, Monatsh. Chem., 1992,**123**,105).

Numerous syntheses of furofurans have already been achieved. There has been a wide spectrum of approaches: *via* dibenzylbutyrolactones (B.L.Feringa *et al.*, Tetrahedron Lett., 1991, **32**, 3239), vinylbutyrolactones (F. Ishibashi *et al.*, Chem.Lett., 1986, **10**,1771), the Claisen rearrangement (D.W.Knight *et al.*,

J.Chem. Soc. Chem. Commun., **1991**,1641), radical annulation (G.Maiti *et al.*, Tetrahedron Lett., 1994,**35**,6731), photchemistry, (K.Orito *et al.*, Tetrahedron Lett., 1991,**32**,5999), intramolecular Mukayama reaction (D.A.Whiting *et al.*, J. Chem. Soc. Perkin Trans.1, **1992**, 185), and oxidative coupling (F.Chioccara *et al.*, Acta Chem. Scand., 1993, **47,** 610). The first enantiocontrolled synthesis from tartaric acid has been described and involves a hetero Diels -Alder reaction (S.Takano *et al.*, J.Chem.Soc. Chem. Commun., **1988**, 189).

Individual furofuran lignans

Sambocolignoside(16). Sambacolignoside was isolated from *Jasminum sambac* (L.) Ait. This is the first example of a lignan - linked secoiridoid glucoside (T.Tanahashi *et al.*, Chem. Pharm. Bull. Tokyo, 1987, **35**, 5032).

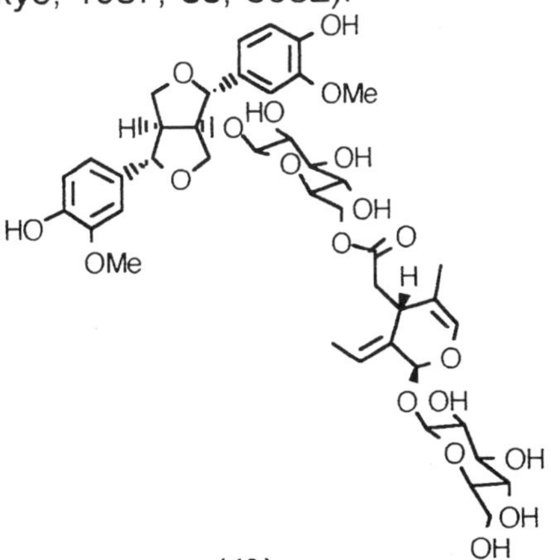

(16)

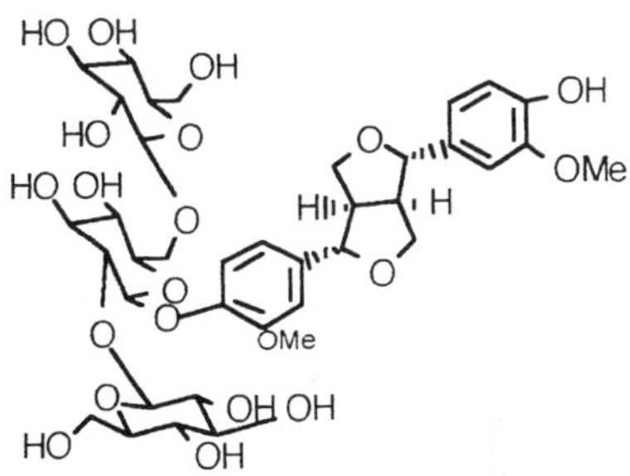

(17)

KP$_3$ (17) is a novel triglucoside that was isolated from sesame seed. The compound had branched (1->2)-and (1->6)-glucosidic linkages. It showed antioxidative activity (H.Katsuzaki *et al.*, Heterocycles, 1993,**36**,933). Later, other related lignans were isolated from sesame seed (*idem*, Phytochemistry, 1994,**35**,773).

Syringaresinol tetra-glycoside(18) was one of a number of novel glycosides isolated from *Albizziae Cortex*. The tetraglycoside and the accompanying triglycoside were the first examples of this type of lignan derivative(J.Kinjo *et al.*, Chem.Pharm.Bull.Tokyo, 1991, **3 9**, 1623).

An unnamed **lignan** (19) with novel ether functionality was isolated from the stems and leaves of *Zanthozylium petiolare.* This lignan was identified by spectroscopic methods (M.S.P.Arruda *et al.*, Phytochemistry,1994,**3 6**,1303).

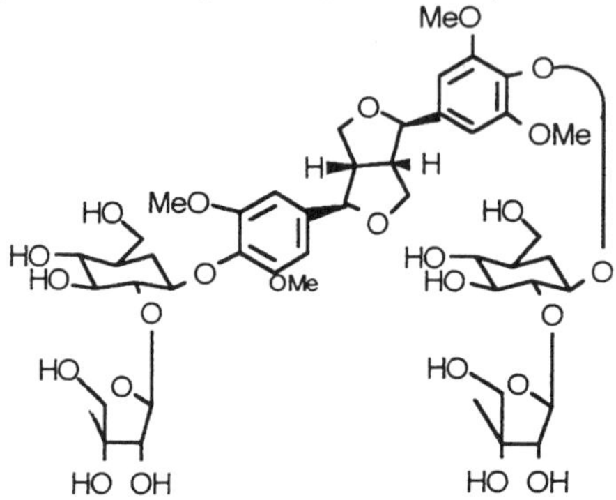

(18)

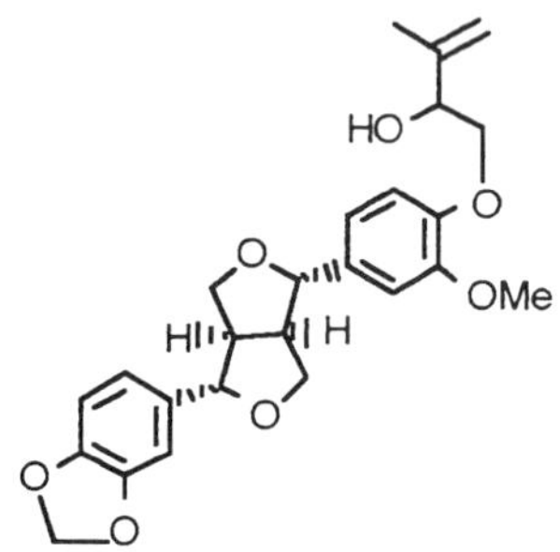

(19)

(v) Aryl tetralins

Two lignan glycoconjugates, isolariciresinol-beta-4'-O-glucopyranoside and tentively, seco-isolariciresinol-beta-D-glucoside were detected in reisling wine by fabms. These two lignans represent a new category of wine phenols (V.A. Marinos *et al.*, Phytochemistry, 1992,**3 1**,4307). Evaluation of synthetic analogues of novel lignans isolated from *Arnebis euchroma* , led to the discovery of three synthetic lignans which act as selective inhibitors of topoisomerase II. They do not induce DNA breakage in either cellular or *in vitro* assay systems.(K.Yoshiki *et al.*, BioMed. Chem. Lett.,1995, **5**,905). One example of the latter is the following:

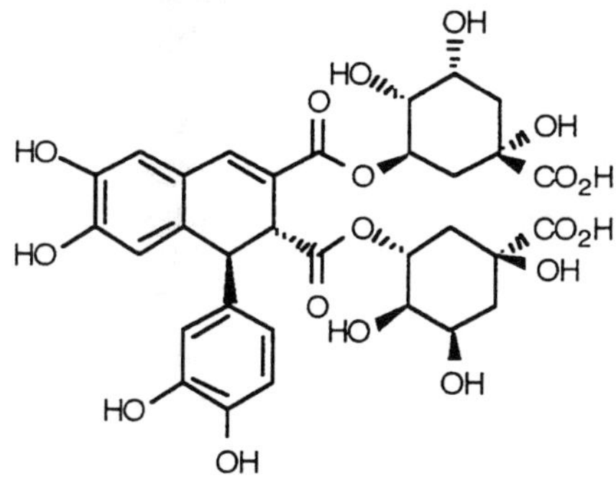

Synthesis of aryltetralin lignans has been achieved by BF_3 -catalysed dimerisation of methyl cinnamates (B.Botta *et al.*,Chem. Pharm. Bull. Tokyo, 1990, **3 8**,3238) and by intramolecular non-phenolic oxidative aryl-aryl coupling of 1,4-diaryl butanes (A.S. Krauss and W.C.Taylor, Aust.J.Chem., 1992,**4 5**,935 ; A.Pelter, R.S. Ward and R.R.Rao, Tetrahedron ,1985, **4 1**, 2933). Other routes involve the cyclisation of dibenzylbutyrolactones (J.M.Trujillo *et al.*, Tetrahedron, 1986,**4 2**,3899; E. Brown *et al.*,Tetrahedron 1992, **4 8**, 687), cycloaddition reactions of benzoisofurans (R.Rodrigo *et al.*, J.Org.Chem,.1985,**5 0**,5902), orthoquinodimethanes (J.L.Charlton *et al.*,Tetrahedron -Asymmetry,1993,**4**,6451; J.LCharlton and M.M. Alauddin, J. Org. Chem., 1986,**5 1**,3490), 1-phenyl-2-benzopyrones (D.W. Jones, J.Chem.Soc.Perkin Trans.I, **1994**,399), and intramolecular cycloaddition (R.Stephenson and J.V.Weber, J. Nat. Prod., 1991,**5 4**,310).

Cerberlignan L(20) Following the isolation of sesqui-, di-, sester-, and tri-lignans from the stems of *C.manghas,* five complex lignans , cerberalignans J (olivil-4.0.8''-*threo* - guaiacylglycerol), K (olivil-5',5''-cycloolivil), L (20) (olivil-5.5''-cycloolivil), M (olivil-5.5''-olivil-4'''.0.8''''- *threo* -guaiacylglycerol), and N (olivil-5'.5''-olivil-4'''.0.8''''- *threo* -guaiacylglycerol) were isolated and their structures elucidated by spectral methods (F. Abe *et al.*, Phytochemistry, 1989, **2 8**, 3473).

Caffeic tetramers(21),(22).Two caffeic tetramers were isolated from the dried root of *Macrotomia euchroma.* The structures were shown to have enantiomeric 1,2-dihydro-1-(3',4'-dihydroxyphenyl)naphthalene-2,3-carboxylic acid (caffeic acid dimer) moieties linked to two molecules of 3-(3',4'-dihydroxyphenyl)-(R)-lactic acid (M. Nishizawa *et al.*, Phytochemistry,1990,**29**,2645).

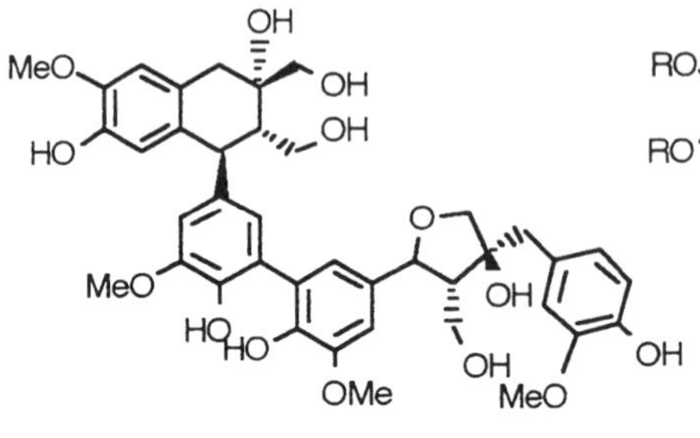

(20)

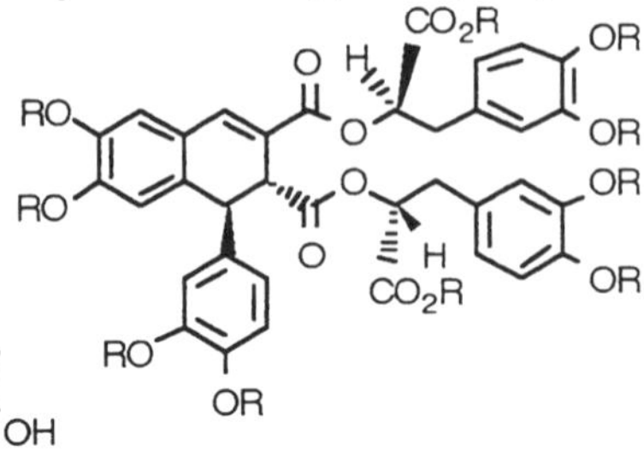

(21) : R=H
(22) : R=Me

Peperomin A(23), **B**(24), **and C**(25) are three novel lignans having the unusual seco structure and were isolated from the fresh whole plant of *P.japonica* (*Piperaceae*). The structures were established by spectroscopic and crystallographic evidence. They are constituents of Formosan antitumour folk medicine(C.M.Chen *et al.*, Heterocycles ,1989,**29**,411).

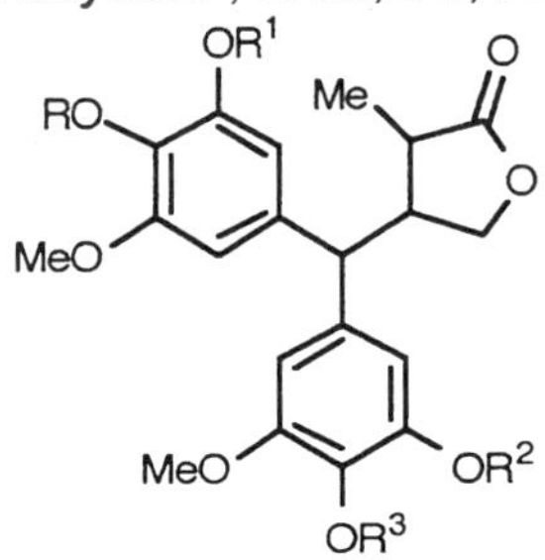

(23) : $R R^1 = R^2 R^3 = CH_2$
(24) : $R = R^1 = Me, R^2R^3 = CH_2$
(25) : $R - R^3 = Me$

(vi) 1-Arylnaphthalenes

A wide range of synthetic methodologies have been developed for the synthesis of this class of lignan. Oxidative coupling of phenolic cinnamates with silver (I) has been successful (S.Maeda *et al.*, Chem.Pharm. Bull., Tokyo,1994,**42**,2506). Other routes involve aryl cyanohydrins,(T.Ogiko *et al.,* Synlett., **1992**,651); bis -benzal maleic anhydride (A.S.R. Anjaneyulu *et al.*, Indian J.Chem.Sect.B, 1994, **33**,839), butenolide(D.C.Harrowven, Tetrahedron Lett., 1991, **32**, 3735), cyclopropylcarbinols (S.Seko *et al.*., Tetrahedron Lett., 1990, **31**, 6883); cycloadditions involving *ortho*-quinodimethanes

(K.Kobayashi *et al.*, J.Chem.Soc.Perkin Trans.1, **1992,** 3111 ; S. Takano *et al.*, Tetrahedron Lett.,1985, **26**, 1655); phenylallyl-propiolates (L.Revesz *et al.*, Helv.Chim.Acta, 1988, **71**, 1697) and arylpropargyl arylpropiolates (R.Stevenson *et al.*, J. Nat. Prod., 1991, **54**, 1687).

Several ring C aromatized analogues of podophyllotoxin were synthesized for testing against human DNA topoisomerase II. The results indicate that aromatization of ring C gave rise to no inhibition of this enzyme at 200μg / M. A comparison of the cytotoxicity among these compounds also demonstrates that a free hydroxyl group at C-4 contributes to significant cytotoxicity (S.A. Beers *et al.*, J.Nat.Prod., 1988, **51**,901).

Phyllanthostatin A(26).The unusual cytostatic (PS ED_{50} 4 μg/mL) lignan ester (26) was isolated from the Central American tree *Pbyllanbus acuminatus.* The lignan was isolated and purified by means of high speed countercurrent distribution and hplc. Its structure was determined by 2D nmr. It slowly converted to justicidin B (G.R. Pettit *et al.*, J.Nat.Prod., 1988,**51**, 1104).

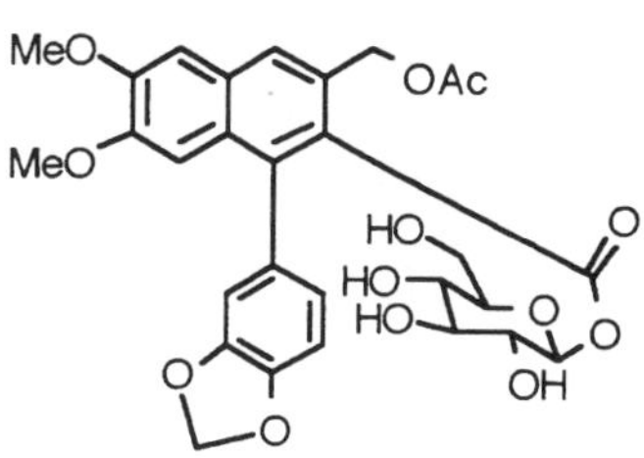

(26)

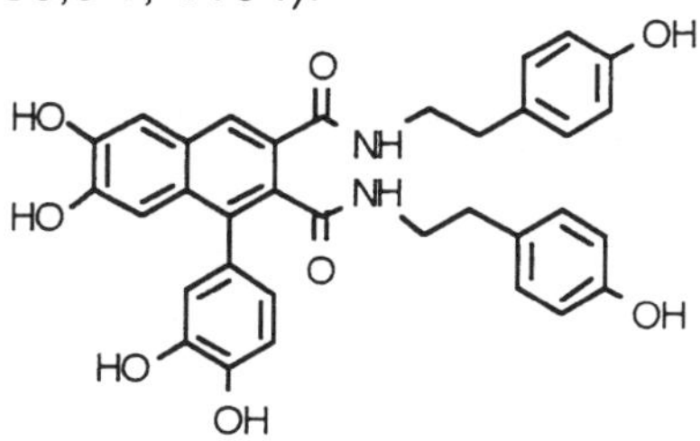

(27)

Cannabisin-A(27) is a new lignan amide which was isolated from the fruits of *Cannabis sativa.* The structure was established on the basis of chemical and spectral evidence (I.Sakakibara *et al.*, Phytochemistry, 1991, **30**, 3013).

(vii) Podophyllum lignans

Due to the anti-cancer activity of the podophyllotoxin derivatives etoposide and teniposide, there has been a major drive to investigate all aspects of this group of lignans, including synthetic analogues. A sensitive immunoassay for naturally occurring podopyllotoxin has been developed.The assay allows direct quantitation of podophyllotoxin in less than 0.2mg of plant tissue (K.J. Yoo and J.R.Porter, J.Nat.Prod.,1993,**56,**715).

It has been reported that the action of etoposide as an antitumor agent is mediated through its interaction with DNA topoișomerase II, which results in DNA breakage inside the cell. A wide range of synthetic and some natural products were examined and compared (Y.C.Cheng and K.H.Lee, J.Med.Chem., 1986, **29**, 1547). 4-Amino derivatives of etoposide are more powerful than the parent (Y.L.Zhang *et al.*, J.Med Chem., 1994, **37**,446; Z.Q.Wang *et al.*, *ibid.*,1992,**35**,871). Other 4-, 2"- O-, 3"-O-, and 2",3"-di-O-substituted etoposides (H.F. Hansen *et al.*, Acta Chem.Scand., 1993,**47**, 1190; T.Ohnuma *et al.*, Chem. Pharm. Bull. Tokyo, 1992,**40**, 1783) and desoxypodophyllotoxin derivatives (T. Terada *et al.*, Chem. Pharm. Bull. Tokyo, 1992,**40**,2720) have been investigated for biological activity. Several podophyllotoxin - related lignans possess significant antifungal activity against a number of filamentous fungi (D.P.Figgitt *et al.*, Biochem. Biophys.Res.Commun., 1989,**160**,257).

The biosynthesis of the podophyllum lignans has been investigated by the research group of P.M.Dewick (Phytochemistry, 1984, **23**, 1029, 1037; 1986,**25**, 2089, 2093; 1991,**30**,1489). The biotransformation of appropriate dibenzylbutyrolactones with cell cultures of *Podophyllum peltatum* is being developed as a 'biological factory' for the continuous synthesis of *podophyllum* related lignans
(J.P. Kutney *et al.*, Heterocycles, 1993,**36**,13).

There is a very extensive literature relating to the synthesis of the podophyllum lignans much of which has been reviewed (D.A.Whiting, Nat.Prod.Rep.,1985,**2**,191; *idem, ibid.*, 1987, **4**, 499; 1990,**7**, 349; R.S.Ward , Tetrahedron, 1990,**46**,5029; *idem*, Synthesis, **1992**,719). An interesting approach with fine stereocontrol involves a transitory quinonemethide produced by photoenolisation which is induced to undergo intramolecular tandem cycloaddition by a tethered dienophile (G.A. Kraus and Y.S.Wu, J.Org.Chem., 1992, **57,** 2922). The corresponding intermolecular varient has been accomplished (J.L. Charlton and K.Koh, J.Org.Chem.,1992,**57**,1514).

Thuriferic acid (28) This novel lignan was extracted from the leaves of *J.thurifera* (Cupressaceae). Its structure was confirmed by synthesis and 1D- and 2D-nmr (A.San Feliciano *et al.*,Tetrahedron, 1988, **44**, 7255). The new lignan **deoxypicropodophyllotoxin** (29) was also isolated from the same source (*idem*,Phytochemistry,1989,**28**,2863).

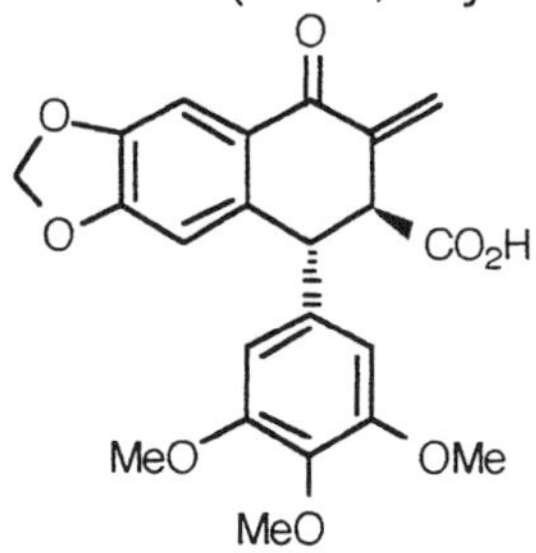

(28)

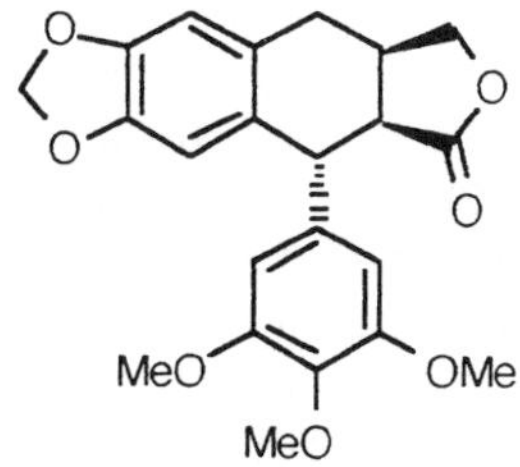

(29)

beta-Hydroxy-4'- demethyldesoxypodophyllotoxin(30). The stems of *Casearia clerkei* have yielded this new lignan (K. Shaari and P.G. Waterman, J.Nat.Prod., 1994,**57**,720).

Hyptinin(31).This new lignan was isolated from the mixe-Indian medicinal plant *Hyptis-verticillata*. It was fully characterised by spectroscopic methods (M.Kuhnt *et al.*, Phytochemistry, 1994, **36**, 485).

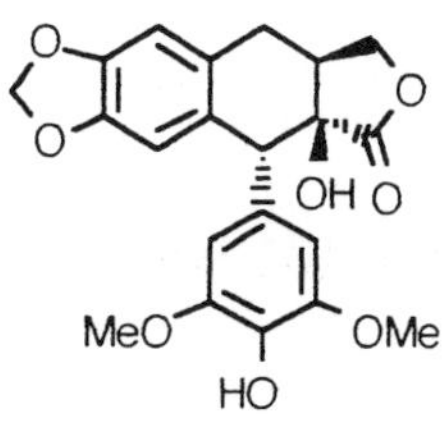

(30)

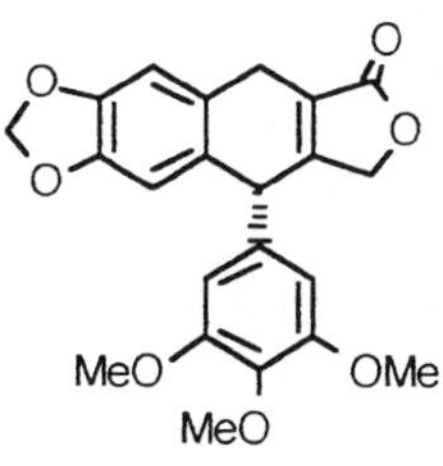

(31)

(viii) Dibenzocyclooctadiene lignans

This group of lignans is important because they exhibit a broad range of biological activities including antileukemic activity. Additonally, they have intrinsically interesting chemical structures.

There is now an extensive literature dealing with the synthesis of these natural products, a number of which are asymmetric syntheses. Many new examples of this group of lignans have been discovered , the majority of which are closely related to known examples. A number of novel examples have been reported recently. The cytotoxic activities of optically pure and racemic steganacin and analogues against KB cells in culture and the inhibitor activity of cilia regeneration in *Tetrahymena* were studied (K.Tomioka *et al*., J.Med Chem., 1991, **34**, 54). An extensive study of mass spectrometric fragmentation of dibenzocyclooctadiene lignans from *Schisandraceae* plants has been reported (H.B. Zhai *et al*., Yaoxue Xuebao, 1990, **25**, 110).

Synthesis has focussed almost exclusively on ring closure by phenolic and non-phenolic oxidation of 1,4-diarylbutanes and dibenzylbutyrolactones (A. Pelter *et al*., Tetrahedron-Asymmetry, 1992, **3**, 239; M.Tanaka *et al*., *i bid*., 1993, **4**, 605; *idem*, Heterocycles, 1994,**37**,739; A.R.Carroll *et al*., Aust.J.Chem., 1994, **47**, 937; S.Tobinaga *et al*., Chem. Pharm. Bull. Tokyo, 1994, **42**, 438; R.Dhal *et al*., Tetrahedron , 1994, **50**, 1153). A notable exception is the asymmetric total synthesis of (-) schizandrin and (-) isoschizandrin by an initial biaryl coupling followed by sequential build - up of the cyclooctadiene ring (A.M.Warshawsky and A.I.Meyers, J.Amer.Chem.Soc.,1990,**112**,8090).

Kadsulignans E (32)-**G** (34). These three new lignans with a spirodienone structure were isolated from the root bark of *Kadsura longipedunculata*. Their structures were elucidated by 2D ^{1}H -^{13}C long range COSY experiments and chemical conversions (J.S. Liu and M.F.Huang, Phytochemistry,1992,**31**,957).

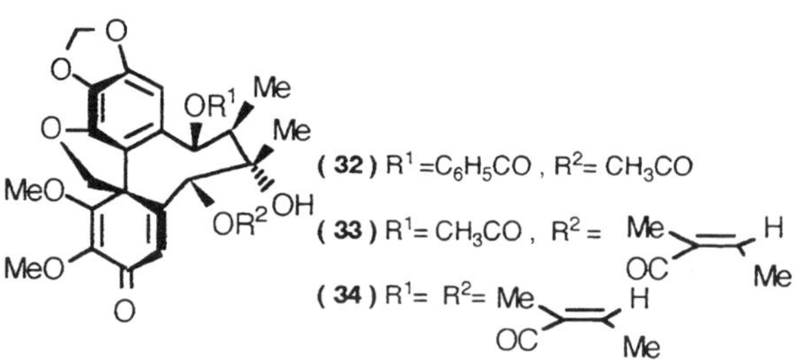

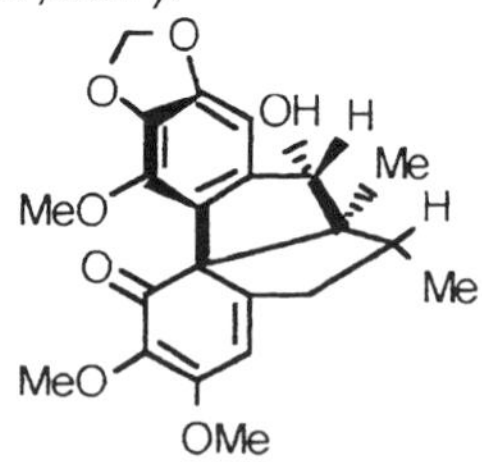

(35)

Kadsulignan-K (**heteroclitin-G**) (35). It was named heteroclitin-G by M.Hattori *et al.*, (Chem. Pharm. Bull.Tokyo, 1992, **40**, 1510) who isolated it from the stem of *Kadsua heteroclita* and established this novel benzobicyclooöctane skeleton by x-ray crystallography. J.S.Liu *et al.*, (Phytochemistry,1992,**31**,1379) isolated it from the seeds of *Kadsura* sp. The structure, including absolute configuration, was elucidated by the latter group using 2D ^{1}H -^{13}C long range COSY spectra and chemical conversions. They named it kadsulignan-K.

4.Lignin

(i) Structure

The chemistry and biosynthesis of lignin has been briefly reviewed (D.C. Ayres and J.D. Loike, 'Lignans Chemical ,Biological and Clinical Properties', Cambridge ,1990). Pyrolysis - GC-MS pyrograms of an *-iter alia* - angiosperm wood was particularly valuable for screening and differentiating *p*-coumaric acid and ferulic acid derivatives in tissues (J.Ralph and R.D.Hatfield, J.Agr.Food Chem., 1991,**39**,1426). The relative distribution of carbon-carbon and diaryl ether condensed bonds in spruce lignins was investigated by thioacidolysis followed by Raney nickel desuphurisation. GC and GC-MS analysis of the main lignin-derived dimeric degradation products, revealed that condensed bonds were not homogeneously distributed within lignin fractions. The β-1 diarylpropane lignin structures whose relative importance is still an open question, were comparatively more abundant in the water soluble lignin fragment. Chemical methods have been devised for the quantitative determination of cinnamic acids participating in ester- ether bridges between cell wall polymers. It has been calculated that there are10 ferulic acid ester - ether bridges for every 100 C_6-C_3 lignin monomers(T.B.T.Lam *et al.*, Phytochemistry, 1992,**31**,1179).

Fractionation, size exclusion chromatography, thioacidolysis and C-13 NMR of lignin released by cultures of *Picea- abies* showed that it is intermediate in structure between milled wood lignin and synthetic lignins produced by dehydration of coniferyl alcohol. It is a highly branched polymer with a large proportion of inter unit carbon-carbon bonds. This indicates that when lignins are formed outside the carbohydrate matrix, the structure is changed to a large extent (G. Brunow *et al.*, Phytochemistry,1993,**32**,845).

NMR methods can be used to delineate detailed structural and regiochemical information on the plant cell wall and elucidate biochemical incorporation pathways. It was unambiguously determined that *p*-coumaric acid is incorporated into maize lignin exclusively at the γ - position.The implication is that lignin acylation occurs by enzymatically controlled pre-acylation of lignin monomers which are subsequently incorporated into the lignin polymer *via* oxidative coupling (J. Ralph *et al.*, J.Amer.Chem.Soc.,1994, **116**, 9448).

(ii) Delignification

Loss of anthraquinone (AQ) in alkaline pulping is partly attributed to irreversible reactions of (AQ) or (AQ) derived compounds with lignin fragments. The formation of 3-(4-hydroxy-3-methoxyphenyl) benz[*d,e* anthracen-7-one, 3-(4-hydroxy-3,5-dimethoxyphenyl)benz[*d,e*]anthracen-7-one and3-(4-hydroxyphenyl) benz[*d,e*]anthracen-7-one, in the reactions of isolated lignins have been studied (P.Kafhunen and G.Brunow, Acta Chem.Scand., 1991, **45**, 945). Mechanochemical reactions of a lignin model compound veratrylglycerol-β-(α'-methyl)syringylalcohol ether were investigated using a ceramic ball mill or a vibration ball mill. Homolytic $C_{\alpha}-C_{\beta}$ cleavage occurred to form a pair of carbon radicals during the milling. The radicals subsequently gave rise to a range of products which were monitored (A.H.Wu *et al.*, Ach-Models Chemistry 1994, **131**, 571).

The anodic oxidation in methanol of some lignin model compounds was studied. Cyclic voltammetry and preparative scale electrolyses were employed. Rapid cleavage of the C_{α}-C_{β} bonds in the first - formed radical cations was proposed (P.C.B. Page *et al.*, Tetrahedron, 1992,**48**,7221; P. Hapiot *et al.*, New J. Chem., 1993, **17**, 211 ; V.L.Pardini *et al.*, J.Org.Chem., 1991,**56**,7305).

Using 2D-nmr techniques the incorporation of feruloyl esters in coniferylalcohol dehydration polymers indicated that the degree to which hydroxy acids are involved in the lignification of forages may be significantly underestimated (J.Ralph *et al.*, J.Chem.Soc. Perkin Trans.1, **1992**, 2961; 2971).

5.Humic acids

Determination of humic acids and related substances in the environment has been reviewed (J.Lawrence, Anal.Trace Org. Aquat. Environ., **1989**, 313). Mn(IV) and Fe(III) oxides play an important catalytical role in the synthesis of humic acids from phenolic compounds by short-range ordered oxides [H.Shindo, Humic Subs. Global Environ. Implic. Hum. Health, Proc. Int. Meet., Int. Humic Subst. Soc. 6th,1992 (pub. **1994**)361].

Many techniques, both physical and chemical have been applied to the determination of the structure of humic acids. Humic acid varies in structure depending on the source. This is now fully recognised. For example, N.Senesi *et al.* (Soil Sci.,1991,**152**,259) examined fifty samples of humic acids and fulvic acids. The following includes most of the techniques recently employed for investigating humic acids : FTIR, pyrolysis-soft ionizaton MS, synchronous excitation fluorescence spectroscopy, solid state C-13 NMR spectrocopy, analytical pyrolysis, x-ray diffraction, transmission and scanning electron microscopy, proton NMR, one and 2D-NMR, dipolar dephasing NMR, cross polarisation / magic angle C-13 NMR, EPR, and ozonation.

Using direct pyrolysis-field ionization mass spectrometry (Py-FIMS), seven humic acids were examined by H.R.Schulten and M.Schnitzer (Sci Total Environ.,1992,**117-8**, 27). The major building blocks of all humic acids were found to be phenolics, lignin monomers and dimers, in addition to mono- and polysaccharides and N-components, C_{12}-C_{26} alkanes, C_{11}-C_{33} fatty acids, and C_{34}-C_{48} alkyl monoesters. Extensive Curie-point pyrolysis-mass spectrometric analysis indicated that much of the structural network of humic substances consist of aromatic rings which are joined by long-chain alkyl structures to form a flexible network (H.R. Schulten *et al.*, Naturwissenschaften, 1991,**78**, 311). C.Saiz-Jimenez *et al.*, (J.Anal. Appl.Pyrolysis, 1987,**11**,3567) also noted the presence of polysaccharides and proteins. G.M.Lantukh *et al.* (Khim. Tekhnol. Vody, 1988,**10**,536) detected pyrimidine and hydroxypurine derivatives which they considered were indicative of the biological origin of the humic acid. Sh.Zh. Zhorobekova *et al.*, (Khim. Tverd. Topl. (Moscow) **1987**,30) isolated and charcterised 16 amino acids from the humic acids isolated from indigenous peats and brown coals. ^{13}C NMR studies (P.G.Hatcher *et al.*, Geochim.Cosmochim. Acta, 1989, **53**,125) of volcanic ash soils revealed that the humic acids were highly aromatic. They concluded that these humic acids contained benzene di- and tri-carboxylic acids interconnected by biphenyl linkages.and that this matrix was probably characteristic of the core that survived degradation.

The compexation of metal cations with humic acids remains a highly active area of soil research. The retention and depletion of metals from arable soil and the effect of acid rain has been studied. From IR and ESR complexation studies of Cu^{2+} and Fe^{3+} it was found that two principal classes of binding site are available for compexation. One type was a weak surface site which is water stable but easily disrupted by proton exchange, and another resulting in highly stable inner sphere complexes resistant to extensive proton exchange (N.Senesi *et al.*, Sci.Total Environ., 1986, **55**, 351).

Compexation studies have been extended to zinc and lead (A. Steinbrich *et al.*, Rocz.Glebozn.1986,**37**,333), vanadyl and uranyl ions (S.Goncalves *et al.*, Talanta, 1987,**34**,839) and thorium (W.P.Cacheris *et al.*, Radiochim.Acta 1987,**42**,185). A number of aquahumate complexes of iron (III), copper (II) and manganese (II) were identified using ESR (A.T.Pilipenko *et al.*, Dokl. Akad. Nauk Ukr.SSR, Ser.B: Geol.,Khim. Biol. Nauki.,**1987**,57).

The effect of pH on the precipitation of heavy metal-humic acid complexes has been reported (M.Swiderska-Broz, Environ. Prot. Eng. 1987,**13**,29) as has the fixation of Zn-65 and Ru-106 by humic acid in polluted waters (D.Huljev *et al.*, Radio. Iugosl., 1989, **23**,169). Using diffuse reflectance FTIR (DRIFT), relative concentrations of functional groups were found to be fairly constant regardless of sample concentration(J.Niemeyer *et al.*, Soil Sci.Soc.Am.J.,1992,**56**,135). Therefore, changes in the relative concentration of functional groups could be measured during the humification process.

Humic acids retains organic compounds (J.C.Del Rio *et al.*, Sci. Total Environ,1988,**81-82,**373). Repeated extraction of lignite humic acids with organic solvents led to the isolation of fatty acids, dicarboxylic acids, long chain ω-hydroxyacids , n-alkanols and alkanes.

Second Supplements to the 2nd Edition of Rodd's Chemistry of Carbon Compounds, Vol.III D/E/F (Partial), edited by M. Sainsbury

Chapter 17

DIHYDRIC AND POLYHYDRIC ARALKANOLS AND THEIR OXIDATION PRODUCTS WITH FUNCTIONAL GROUPS IN SEPARATE SIDE CHAINS

J. CHARALAMBOUS AND E. G. EVAGOROU

1. Side-chain halogenated alkylbenzenes

(a) Bis- and tris-(halogenomethyl)benzenes

As previously reviewed there are several methods for the preparation of di- and tri-substituted halomethyl benzenes [J. Charalambous, Ist Supplement, 2nd Ed. C. C. C.]. The reaction of thionyl bromide with 1,2-, 1,3-, 1,4-benzene dicarboxaldehydes yields the corresponding dibromo compounds $(BrCH_2)_2C_6H_4$ [S. D. Saraf, *Synth Commun.*, 1983, **13**, 7]. Treatment of 1,3-$(HOCH_2)_2C_6H_4$ with an aqueous solution of HCl at 130°C under pressure affords 1,3-$(ClCH_2)_2C_6H_4$ in excellent yield [T. Okawa, CA, 1991, 114:184960q]. Partial bromodehydroxylation of dimethanols (**1**) (R = OH; R^1 = F, Cl) with aqueous HBr in toluene gives (**1**) (R = Br) in high yields [A. T. Costello and D. J. Milner, *Synth. Commun.*, 1987, **17**, 219].

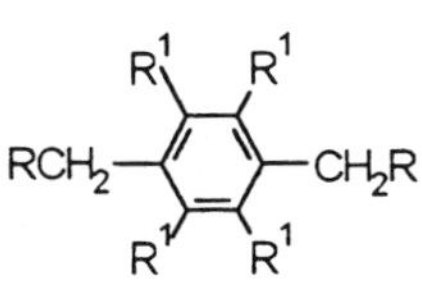

(**1**)

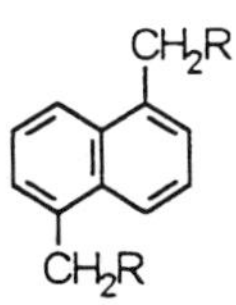

(**2**)

1,2-, 1,3- and 1,4-bis(trichlorosilylmethyl)benzenes and related compounds with methyl ring substituents are obtainable from the reaction of bis(chloromethyl)benzenes with trichlorosilane and isopropylamine [M. Motevalli, D. L. Ou *et al.*, *J. Organomet. Chem.*, 1993, **445**, 35]. 1,2-, 1,3- and 1,4-$(BrCH_2)_2C_6H_4$ can be directly converted in high yields into methyl and iso-propyl phthalates yields using silver(II) oxide in methanol or isopropanol respectively [B. Ortiz *et al.*, *Synth. Commun.*, 1993, **23**, 749]. Treatment of 1,2-, 1,3- and 1,4-$(XCH_2)_2C_6H_4$ (X = Cl, Br, I) with cinnamyl alcohol in the

presence of Bu_4NBr and NaOH at 80°C gives *ca.* 90% of the corresponding bis(cinnamyloxymethyl) benzene [K. Ida, CA, 1993, 118: 233633d].
Carbonylation of 1,4-$(ClCH_2)_2C_6H_4$ in methanol containing catalytic amounts of $Co_2(CO)_8$ in the presence of a base under 1 at. of CO yields 1,4-$(MeO_2C)_2C_6H_4$ [M. I. Terekhova *et al.*, *Zh. Org. Khim.*, 1991, **27**, 2368]. Carboalkoxylation of 1,4-$(ClCH_2)_2C_6H_4$ with ROH (R = Me, Et, Me_2CH, Ph) and carbon dioxide in the presence of $PdCl_2(PPh_3)_2$ as catalyst and N,N-dicyclohexylmethylamine gives the corresponding esters 1,4-$(RO_2CCH_2)_2(C_6H_4)$; similar behaviour is shown by other bis(chloromethyl)arenes *e.g.* (**2**) [T. Kobayashi, F. Abe, M. Tanaka, *J. Chem. Catal.*, 1988, **45**, 91]. Reaction of 1,4-$(XCH_2)_2C_6H_4$ (X = Cl or Br) with CO, $Fe(CO)_5$, a phase transfer catalyst and $CaCO_3$ yields 1,4-$(HO_2CCH_2)_2C_6H_4$ and 4-$MeC_6H_4CH_2CO_2H$ the former being the major product [S. C. Shim. *et al.*, *Bull. Korean Chem., Soc.*, 1988, **9**, 61].
The phosphine oxide 1,4-$(Bz_2P(O)CH_2)_2C_6H_4$ can be obtained by the reaction of 1,4-bis(bromomethyl)benzene and $(Bz)_2P(H)O$ in a two phase system containing a hydrocarbon solvent and aqueous NaOH [E. N. Tsvench and A. V. Kharitonov, *Zh. Obshch. Khim.*, 1985, **55**, 2481]. Di(bromomethyl)benzenes, 1,2-, 1,3-, and 1,4-$(BrCH_2)_2C_6H_4$, react with benzothiazoles to give the corresponding 3-(bromomethylbenzyl)benzothiazolium bromides [P. Chabrecek, V. Sutoris and A. Gaplovsky, *Chem. Pap.*, 1990, **44**, 209].
Reductive debromination of 1,2-$(BrCH_2)_2C_6H_4$ with $SnCl_2$-$Pd(PPh_3)_4$-pyridine-THF gives 1,2-quinodimethane *in situ*, which undergoes stereoselective [4+2] cycloaddition with *Z*- and *E*-$MeO_2CCH{:}CHCO_2Me$ to give 63 and 72% *cis* and *trans* methoxycarbonyltetralins (**3**) respectively. A similar reaction of 1,2-$(BrCH_2)_2C_6H_4$ with $RR^1C{:}CH_2$ (R = H, Me; R^1 = MeO_2C, $PhSO_2$, CN) gives up to 48% of the tetralins (**4**). Double cycloadditon of 1,2-$(BrCH_2)_2C_6H_4$ with *Z*-$MeO_2CCH{:}CHCO_2Me$ gives 31% of adduct (**5**) [Y. Masuyama and Y. Haratani, *Chem. Express*, 1988, **3**, 127].

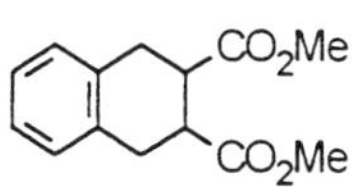

(3)

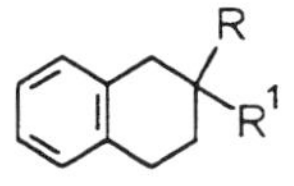

(4)

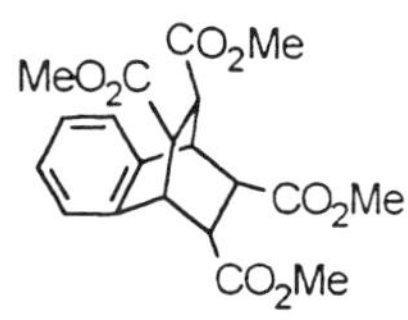

(5)

Sodium iodide mediated debromination of tetrakis- and hexakis-(dibromomethyl)arenes gives bromodicyclobutanobenzenes (**6**) and/or tetra(bromomethylene)benzenes (**7**) [M. K. Shepherd, *J. Chem. Soc., Perkin Trans. 1*, 1988, 961]. Cycloaddition of 1,2-quinonedimethane, generated *in situ* from 1,2-$(BrCH_2)_2C_6H_4$ and NaI with acepleiadylene-5,8-dione affords (**8**) [T. Josuke, T. Shrinri, E. Seiji and M. Kenichi, *J. Chem. Soc., Perkin

Trans. 1, 1988, 1541]. 1,2-$(BrCH_2)_2C_6H_4$ on treatment with Zn in dioxane under sonication affords 1,2-quinonedimethane which can undergo cycloadditon with 1,4-cyclopentenedione to give benzo[f]ninhydrin [R. Heffner, J. E. Safaryn and M. M. Joullie, *Tetrahedron Lett.*, 1987, **28**, 6539]. Di-Grignard reagents derived from various di- and tri-halomethylbenzenes (halo = Cl or Br) or 1,8-$(ClCH_2)_2C_{10}H_6$ have been prepared in yields of up to 96% by reactions with magnesium anthracene in THF at 20°C [S. Harvey *et al.*, *J. Org. Chem.*, 1988, **53**, 3134]. Symmetrical dibenzylbenzenes (**9**) (R = Me, OMe, CH_2OH, CHO, OH, CH_2 Br) can be isolated by the cross-couplings of 2-R-C_6H_4MgBr with 1,2-$(BrCH_2)_2C_6H_4$ in the presence of CuI [W. Y. Lee *et al.*, *Bull. Korean Chem. Soc.*, 1989, **10**, 387].

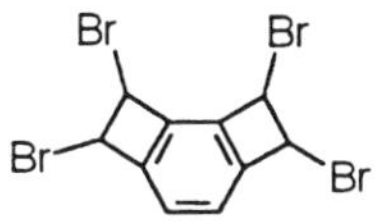

(6)

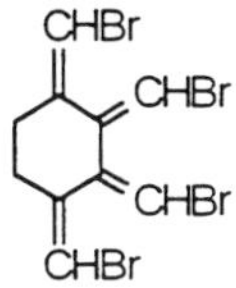

(7)

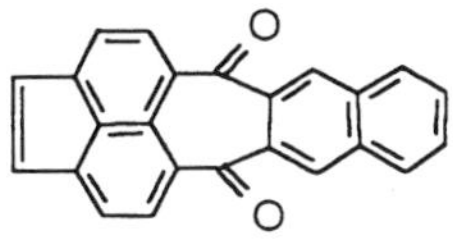

(8)

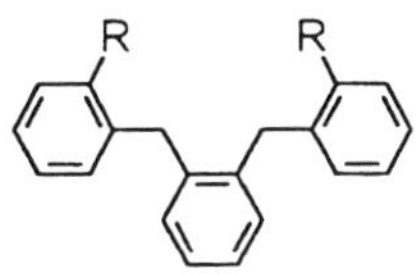

(9)

The reaction of 1,2-$(BrCH_2)_2C_6H_4$ with RNH_2 (R = Me, Et, Pr, Me_2CH, Me_3C, Me_3CCH_2), followed by MnO_2 oxidation, affords the isoindoline N-oxides (**10**) the reaction of which with Ac_2O in the presence of Et_3N provides an efficient and general route to 2-alkyl-2*H*-isoindoles (**11**) [R. P. Kreher *et al.*, *Chem. Ztg.*, 1988, **112**, 85]. The 2-aryl-2*H*-isoindole (**11**) (R = Ph) is obtained similarly [R. P. Kreher, U. Feldhoff, J. Seubert and D. Schmitt, *Chem. Ztg.*, 1987, **111**, 155]. The base-catalysed cyclisation of 1,2-bis(dihalomethyl)benzenes with diethyl(acetylamino)malonate, followed by decarboxylation and amide cleavage, provides a convenient route to 1,2,3,4-tetrahydroisoquinoline-3-carboxylic acids and esters [B. O. T. Kammermeier, U. Lerch and C. Sommer, *Synthesis*, 1992, 1157]. A direct synthesis of 2,2-dichloro-2-silaindan (**12**) consists of reacting metallic silicon with 1,2-$(ClCH_2)C_6H_4$ in the presence of Cu catalyst at 260-320°C [I. N. Jung, S. H. Yeon and J. S. Han, *Bull. Korean Chem. Soc.*, 1993, **14**, 315].

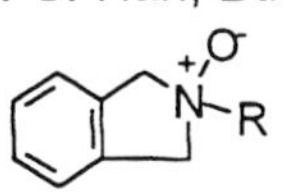

(**10**)

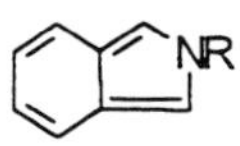

(**11**)

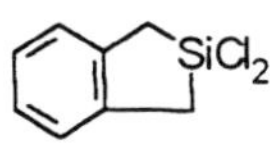

(**12**)

All isomeric dihalides $(XCH_2)_2C_6H_4$ (X= Cl or Br) and their derivatives have been utilised extensively for the synthesis of a variety of cyclic and macrocyclic compounds as illustrated in Table 1.

Table 1: Synthesis of Cyclic and Macrocyclic Compounds from Halogenomethylbenzenes

Type of compound (Ref.)	Reactants	Product
Cyclophane (1)	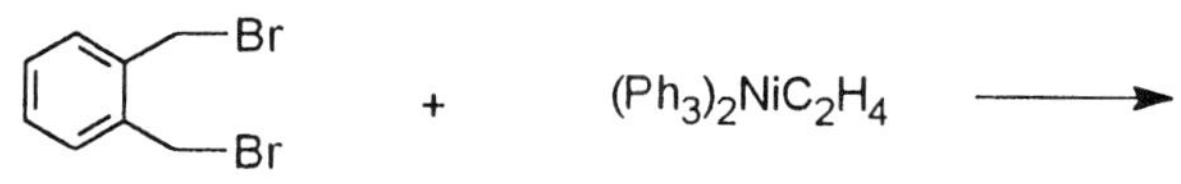	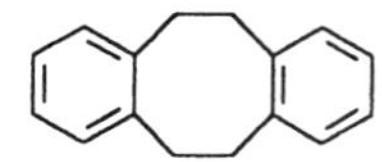
Dithiacyclophane (2)	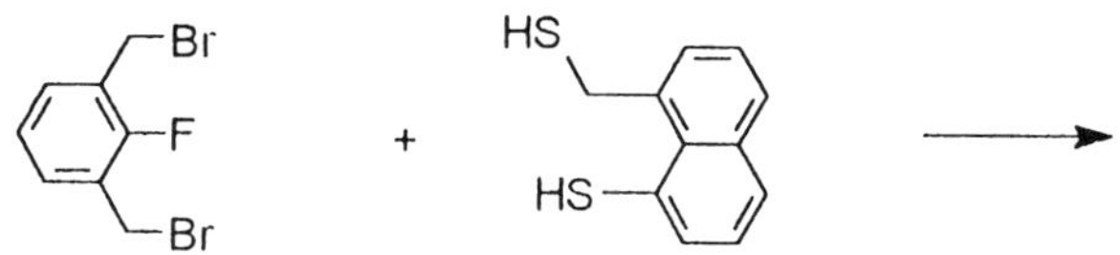	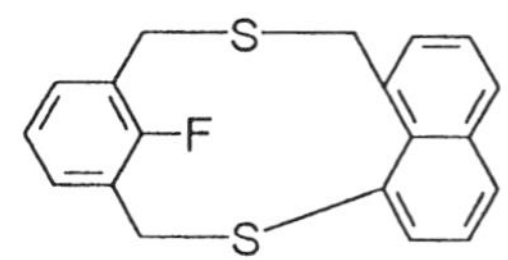
Crown ether (3)	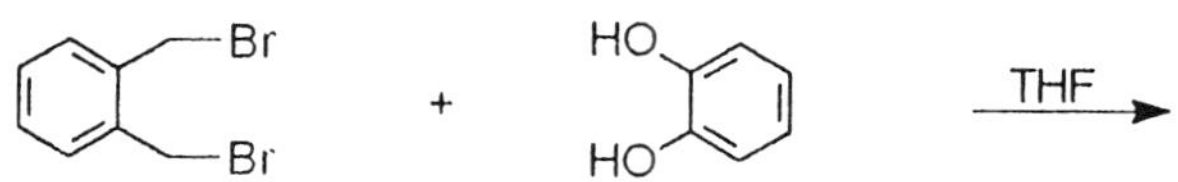	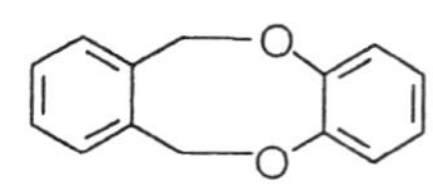
Aza macrocycle (4)	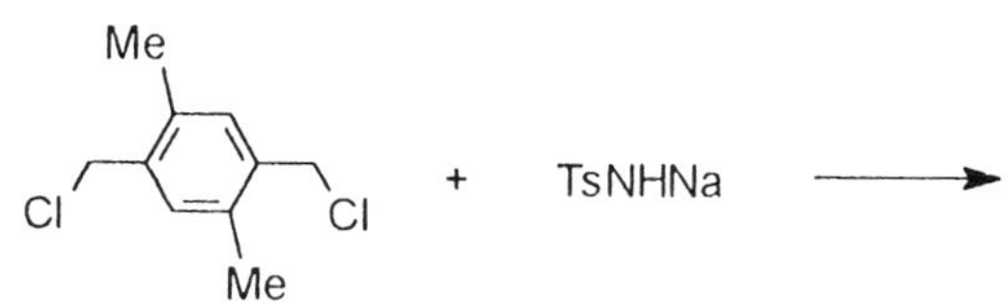	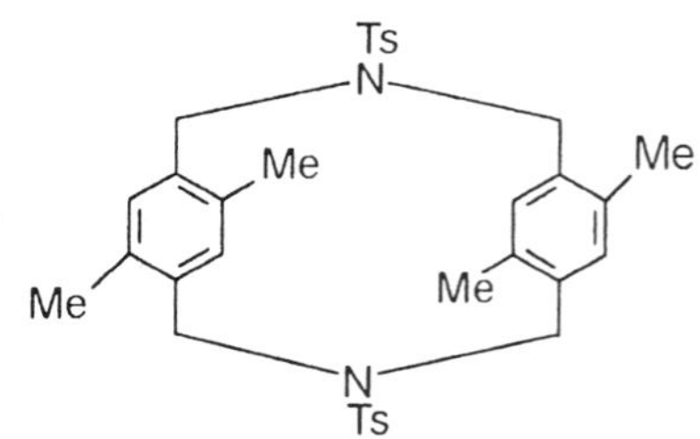

Table 1 cont'd.

Cyclophane (5)

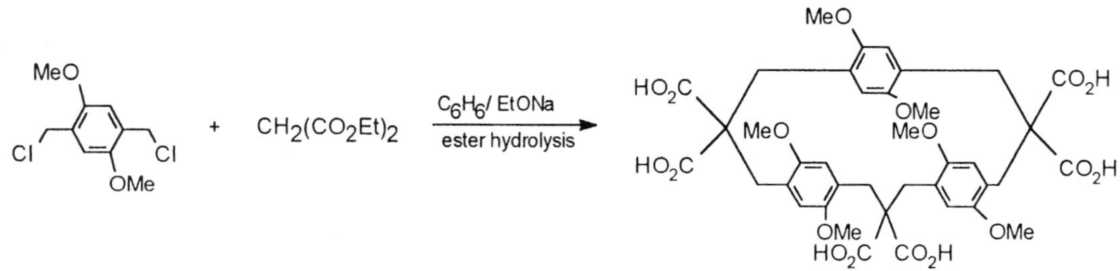

Crown ether (6)

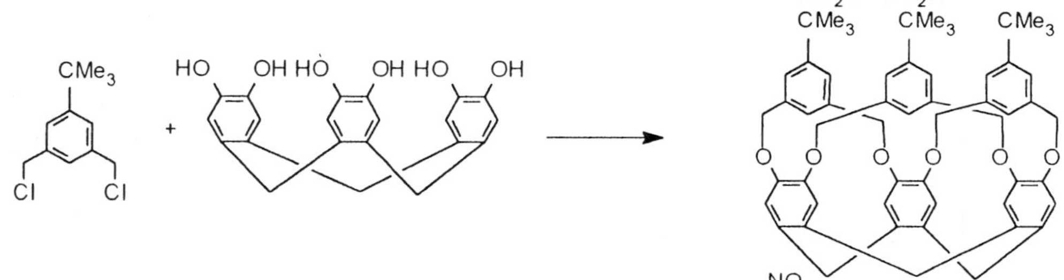

Cyclophane (7)

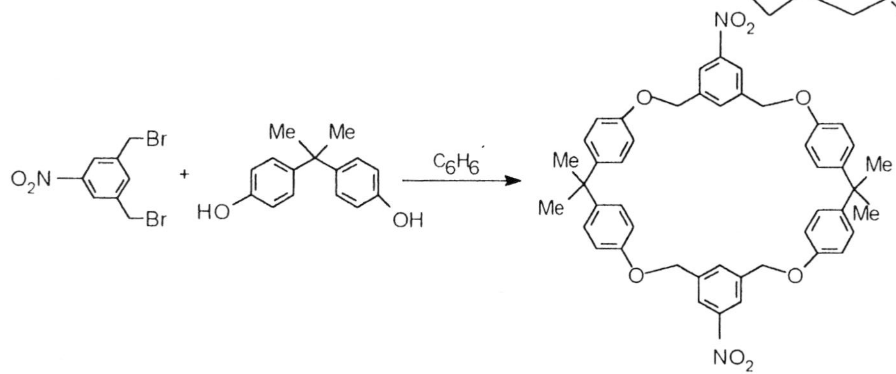

References for Table 1

1. E. Uhlig, B. Hipler, *Tetrahedron Lett.*, 1984, **25**, 5871.
2. Y. H. Lai, S. Y. Wong and D. H. Chang, *Tetrahedron*, 1993, **49**, 669.
3. A. J. Poss and M. S. Smyth, *Synth. Commun.*, 1989, **19**, 3363.
4. F. Bottino *et al.*, *J. Org. Chem.*, 1988, **53**, 3521.
5. X. Xu, Z. Xie and Y. Wang, **CA**, 105:172022m.
6. D. J. Cram *et al.*, *J. Chem. Soc., Chem. Commun.*, 1988, 407.
7. K. Sago *et al.*, *J. Am. Chem. Soc.*, 1986, **108**, 1996.

2. Di- and poly-hydric aralkanols, thiols and amines

(a) Aralkanols

1,2-Bis(hydroxymethyl)benzene can be prepared conveniently and in good yield by the reduction of phthalic acid with $NaBH_4$ and I_2 [J. V. Kanth and M. Perriasamy, *J. Org. Chem.*, 1991, 56, 5964], with $Zn(BH_4)_2$ in 1,2-dimethoxyethane in the presence of trifluoroacetic acid [B. C. Ranu and A. R. Das, *J. Chem. Soc., Perkin Trans. 1*, 1992, 1561] or with SmI_2 by the addition of a base in the presence of a protic solvent at room temperature [Y. Kaniochi and T Kudo, *Bull Chem. Soc. Jpn.*, 1992, **65**, 3049]. This compound and its isomers have also been obtained by reduction of the corresponding benzene dicarboxaldehyde with Al-$NiCl_2.6H_2O$ in THF [B. K. Sarmah and N. C. Barna, *Tetrahedron*, 1991, **47**, 8587] or with sodium diethyldihydroaluminate in THF/toluene [N. M. Moon, Y. S. Shon and J. H. Ahn, *Bull. Korean Chem. Soc.*, 1992, **13**, 199]. Sodium derivatives of phenolic di- and tri-hydroxymethylbenzenes *e.g.* 2,4,6-$(HOCH_2)_3C_6H_2ONa$ have been obtained by the reactions of RC_6H_4OH (R = H, 2-Me, 3-Me) with HCHO in butanol [K. Tsutomu, T. Saburo and T. Akio, *Kagaku to Kogyo (Osaka)*, 1987, **61**, 363].

The selective oxidation of 1,2-, 1,3- or 1,4-bis(hydroxymethyl)benzenes with γ-MnO_2 in THF leads to the corresponding hydroxyaldehyde and dialdehyde plus, in the case of the 1,2- isomer, to 1-hydroxy-1,3-dihydroisobenzofuran and phthalide [I. Constantinides and R. S. Macomber, *J. Org. Chem.*, 1992, **57**, 6063]. Other oxidising agents which have been used to convert bis(hydroxymethyl)benzenes to their corresponding aldehydes or ketones include tetrakis(pyridine)silver dichromate, $Py_4Ag_2Cr_2O_7$ [H. Firouzabadi, A. Sardarian and H. Gharibi, *Synth. Commun.*, 1984, **14**, 89] and polyvinyl supported silver dichromate [B. Tamami and D. Mohadjer, *Polymer*, 1991, **32**, 2666]. For the oxidation of 1,2-bis(hydroxymethyl)benzenes to phthalides see section 5a.

2,6-Bis(hydroxymethyl)phenols are converted to 2-hydroxyisophthaldehydes [P. Taniguchi, *J. Chem. Soc. Jpn.*, 1984, **57**, 2683]. EtMgBr can be used to convert 4-substituted-2,6-bis(hydroxymethyl)phenols to their magnesium salts which on treatment with paraformaldehyde gives the corresponding 5-substituted-2-hydroxy-3-hydroxymethyl benzaldehydes (75-88%) [Y. Hu and H. Hu, *Huaxue Xuebao*, 1987, **45**, 613]. Further oxidation of the second hydroxymethyl group in these products is prevented by the presence of formyl groups. Formation of 4-substituted-2,6-diformylphenols is achieved by condensing substituted-2-hydroxy-3-hydroxymethyl benzaldehyde with 1,2-diaminoethane in the presence of $Cu(OAc)_2$. Oxidation of the resulting complex with $K_2Cr_2O_7$/DMSO gives the required dicarbaldehyde (Scheme 1) [Y. Hu and H. Hu, *Synthesis*, 1991, 325].

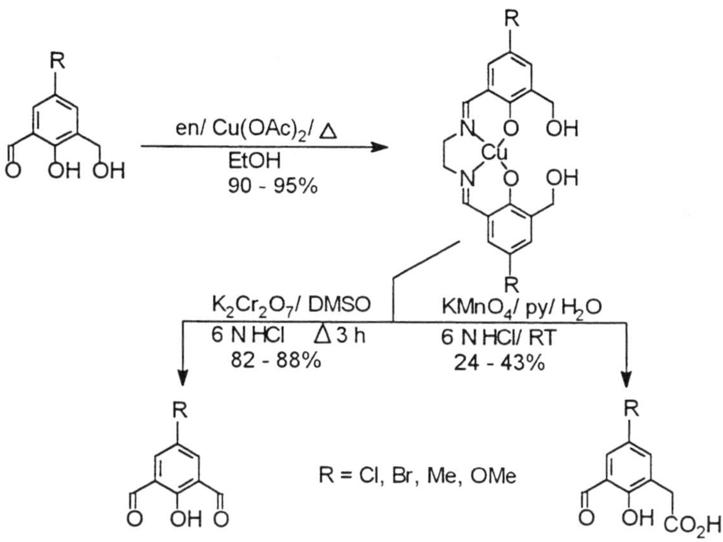

Scheme 1

Selective hydrogenolysis of (**13**) (R = CH_2OH) over Raney nickel in alkaline solution at 20°C affords the corresponding monohydrogenolysis products (**13**) (R = CH_3) [C. Behrens, M. Egholm and O. Bucharolt, *Synthesis*, 1992, 1235]. For a kinetic study of the acetylation of 1,2-, 1,3- and 1,4-$(HOCH_2)C_6H_4$ with acetic anhydride/ pyridine in THF see I. Constantinides, M. Lourdes-Guerra and R. S. Macomber [*J. Phys. Org. Chem.*, 1990, **3**, 789]. The 1,2-diol reacts with carbonyl compounds under heterogeneous conditions in the presence of a sulphonated charcoal catalyst to give 1,3-dioxacycloheptanes (**14**) [H. K Patney, *Tetrahedron Lett.*, 1991, **32**, 413] and on xanthation affords 2-$HOCH_2C_6H_4CH_2OCS_2Me$ and 1,2-$(MeS_2COCH_2)_2C_6H_4$

[B. Kryczka and G. Descotes, *Bull. Pol. Acad. Sci., Chem.*, 1985, **33**, 475].

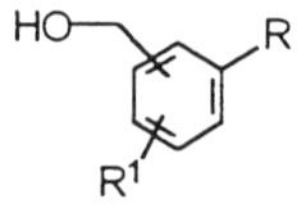

(**13**)

R = H, 4-CH_2OH

R^1 = 5-CN, 5-NH_2, 3-CH_2OH

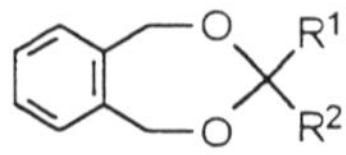

(**14**)

The phosphonate ester (**15**) is formed on treatment of 1,2-$(HOCH_2)_2C_6H_4$ with phenyldichlorophosphate in ethyl acetate containing anhydrous Et_3N [W. N. Setzer *et al.*, *J. Org. Chem.*, 1992, **57**, 2812]. The seven-membered heterocycle adopts an asymmetric twist-conformation [I. A. Litvinov *et al.*, *Izv. Akad. Nauk SSSR, Ser. Khim,* 1991, 214]. The reaction of 1,2-$(HOCH_2)_2C_6H_4$ with $Sb(OR)_3$ causes the formation of (**16**) (R = Me, Et, Me_3C) in 74-98% yield, these exist as polymeric species in the solid state but in solution are monomeric or dimeric [B. A. Arbuzov *et al.*, *ibid*, 1987, 423]. The reaction of the diol with Me_3Al in the molar ratio 2:3 in diethyl ether leads to the formation of (**17**), where the central Al atom is five-co-ordinated [S. Pasynkiewicz and W. Ziemkoska, *J. Organomet. Chem.*, 1992, **423**, 1].

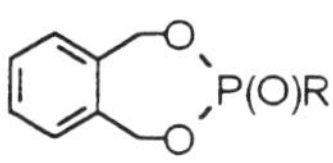

(**15**)

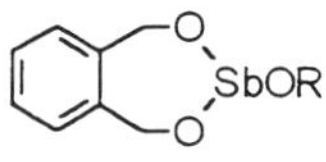

(**16**)

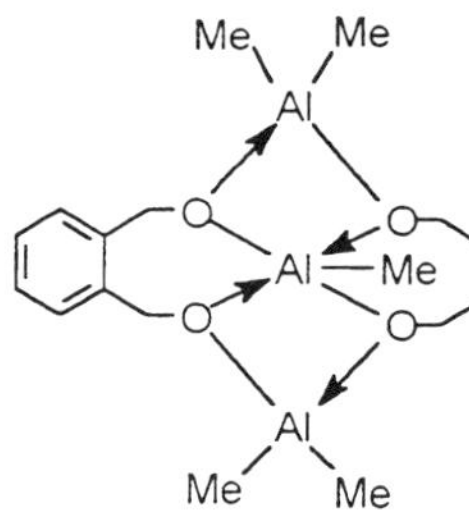

(**17**)

The condensation reaction of bis(hydroxymethyl)benzenes with a dihalide or ditosyl compound provides convenient routes towards the synthesis of cyclic products, examples being the reaction of 1,2-bis(hydroxymethyl)benzene and $PrCH[O(CH_2)_2Cl]_2$ yielding (**18**). In the case of the tosyl derivative the reaction affords a much higher yield of the macrocyclic product [J. S. Bradshaw *et al.*, *Tetrahedron*, 1987, **43**, 4271]. The reaction of 2,6-bis(hydroxymethyl)-4-methylphenol and benzylamine in toluene at 135°C affords (**19**) [H. Takemura , *Tetrahedron Lett.*, 1992, **33**, 5775]. The crown ethers (**20**) are formed from the reactions of 2,6-bis(hydroxymethyl)-4-methylphenol with either 1,2-dibromoethane or 1,2-bis(bromomethyl)benzene in THF/NaH under inert atmosphere [T. I. Richards *et al.*, *J. Chem. Soc., Perkin Trans 1*, 1987, 2765].

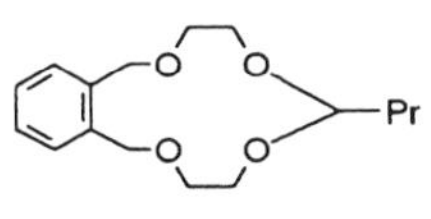

(18)

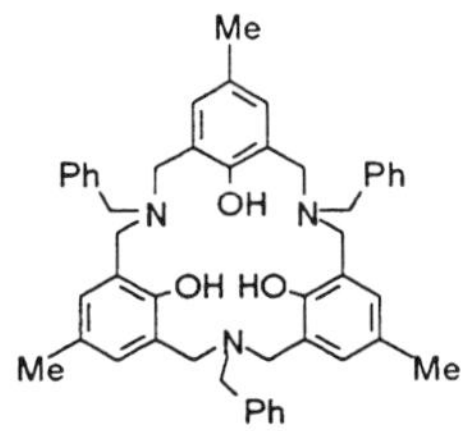

(19)

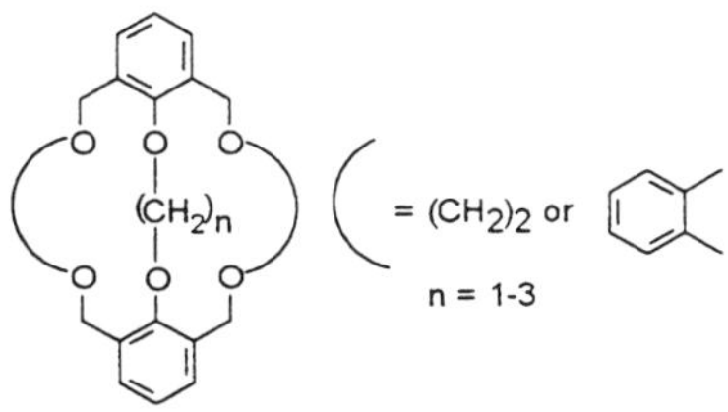

(20)

(b) Thiols

A simple one pot reaction between 1,4-bis(hydroxyalkyl)benzenes (**21**) (X = O) and 2,4-bis(4-methoxyphenyl)-1,3,2,4-dithiadiphosphetane 2,4-disulphide (Lawesson's reagent, LR) affords the corresponding dithiols (**21**) (X = S). Similarly, treatment of 1,2-bis(hydroxymethyl)benzenes (**22**) or 1,2-(mercaptomethyl)(hydroxymethyl)benzenes (**23**) with LR gives 1,3-dihydrobenzo[*c*]thiophenes (**24**) [T. Nishio, *J. Chem. Soc., Perkin Trans 1*, 1993, 1113].

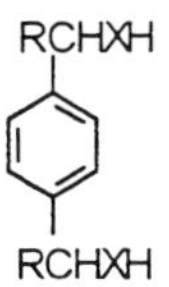

(21)

R = Me, Ph

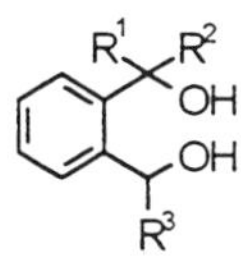

(22)

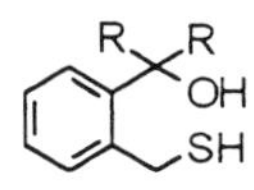

(23)

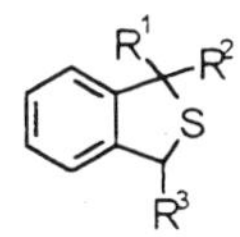

(24)

R^1 = Me, Ph, Ph, Ph, *p*-MeC_6H_4
R^2 = Me, H, Ph, H, H
R^3 = H, H, H, Ph, *p*-MeC_6H_4

1,2-, 1,3- and 1,4-bis(mercaptomethyl)benzenes have been utilised extensively for the preparation of cyclophanes, cappedophanes and other cyclic or acyclic sulphur containing compounds. The general synthesis involves the condensation of the thiol with a dihalo compound, although dialcohols, ditosylates and other compounds also produce similar results. At high dilution 1,2-$(HSCH_2)_2C_6H_4$ reacts with 1,3-dichloroacetone in Cs_2CO_3/ DMF to yield a mixture of (**25**) and (**26**). The same reaction in

NaOMe/ MeOH affords (**27**) in low yield as well as (**25**) and (**26**). The dithioacetal (**27**) is formed as the overall product on replacing 1,3-dichloroacetone with 1,1-dichloroacetone [J-J. Chui, R. S. Grewal, H. Hart and D. L. Ward, *J. Org. Chem.*, 1993, **58**, 1553].

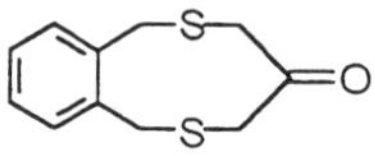

(**25**)

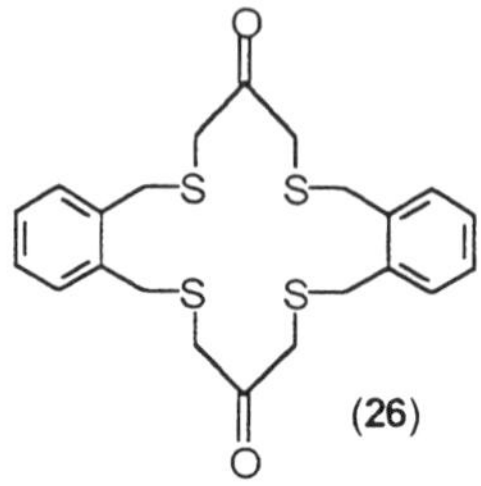

(**26**)

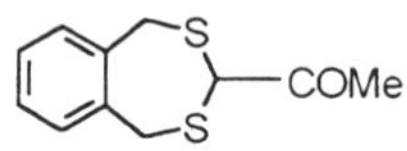

(**27**)

The synthesis of (**28**) is achieved by the cyclocondensation of $ClCH_2CHClOEt$ with substituted 1,2-bis(mercaptomethyl)benzenes [I. M. Nasyrov, N. Mukhamedzhanar and S. Sh. Shukurov, *Org. Khim. Geterotsikl, Soedin.,* 1985, **7**, 997].

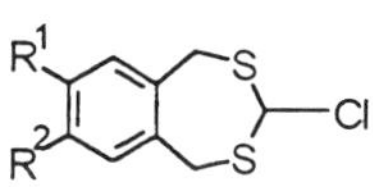

(**28**)

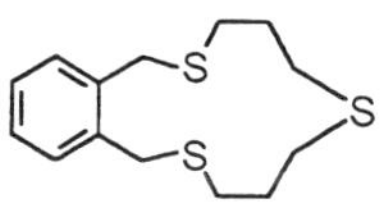

(**29**)

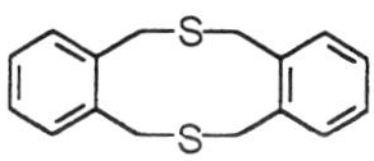

(**30**)

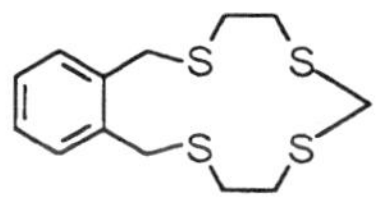

(**31**)

R^1 = H, Me, O
CH_2
CH_2
R^2 = H, Me, O

Other medium sized rings containing multiple sulfur atoms such as (**29**) and (**30**) are isolated when 1,2-$(HSCH_2)_2C_6H_4$ is reacted with $Br(CH_2)_2S(CH_2)_2Br$ and 1,2-$(BrCH_2)_2C_6H_4$ respectively in methanol in the presence of boric acid and a base [J. J. H. Edema *et al.*, *Tetrahedron,* 1993, **49**, 4355]. The tetrathia compound (**31**) is obtained from the di cesium salt of 1,3-dithiaheptane-1,7-dithiol and 1,2-$(BrCH_2)_2C_6H_4$. Araliphanes of the type (**32**) are obtained using similar cyclocondensation reactions [J. Dohm, M. Neiger, K. Rissanen, F. Vögtle, *Chem. Ber.*, 1991, **124**, 915] and bicyclo araliphanes (**33**) are formed on reacting 1,3,5-$(HSCH_2)_3C_6H_3$ and $(BrCH_2CH_2)_2CHCH_2Br$ [R. A. Pascal, C. G. Winans and D. Van Engen, *J. Am. Chem. Soc.*, 1989, **111**, 3007].

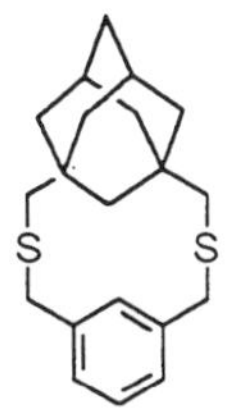

(32)

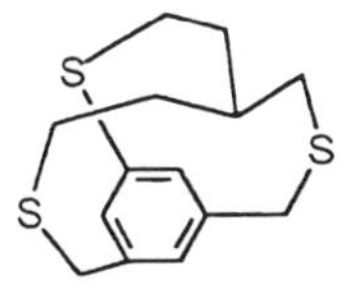

(33)

1,4-$(HSCH_2)_2C_6H_4$ undergoes cyclisation with 1,2-$(BrCH_2)_2C_6H_4$ to give selectively the [2+2] macrocycle (**34**) whereas the reaction of 1,2-$(HSCH_2)_2C_6H_4$ with 1,4-$(BrCH_2)_2C_6H_4$ gives a mixture of (**34**) and the [1+1] product (**35**) [G. Bodwell, L. Ernst and H. Hopf, *Chem. Ber.*, 1989, **122**, 1013]. In contrast [1+1] cyclocondensation products are the result of the reaction of 1,4-$(HSCH_2)_2C_6H_4$ and 1,4-$(BrCH_2)_2C_6H_4$ [Z. Shi, *Xibei Daxue Xuebao, Ziran Kexueban*, 1990, **20**, 49]. The formation of the [2+2] product (**36**) is illustrated in Scheme 2 [T. Yamato *et al.*, *Chem. Express.*, 1990, **5**, 777].

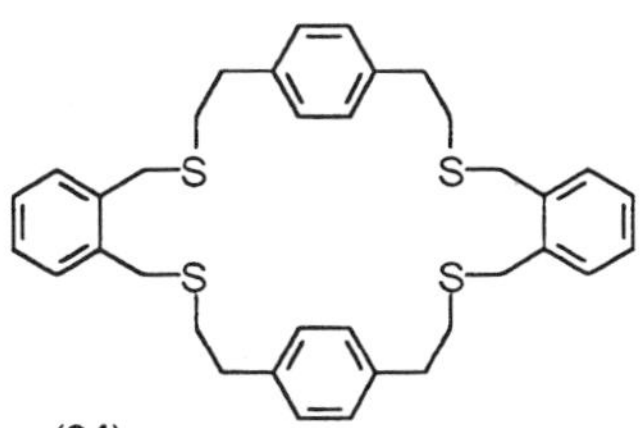

(34)

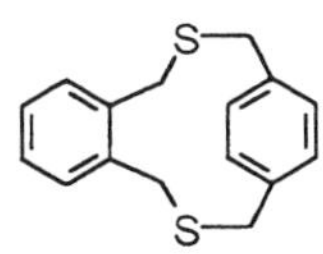

(35)

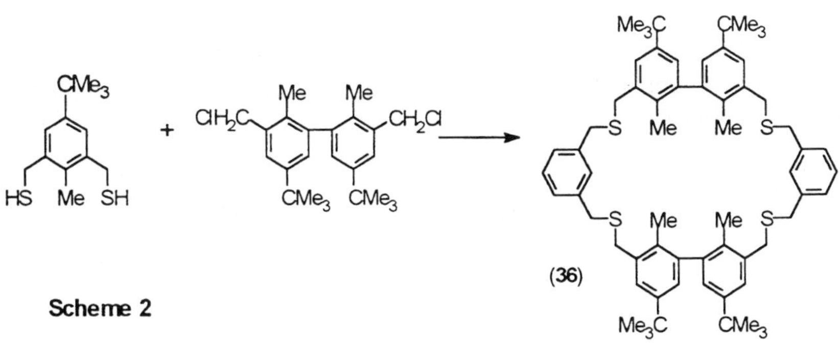

(36)

Scheme 2

A modification to the general procedure for the preparation of *ortho-ortho, ortho-meta* and *meta-meta* dithia[3,3]phanes from bis(halomethyl)- and bis(mercaptomethyl)-benzenes requires the addition of benzene and NaOH in 96% ethanol over ½ h to the reaction mixture [P. Kus, *Pol. J. Chem.*, 1989, **63**, 285]. Using this method (**30**) is obtained in 76% yield. By employing the same procedure with naphthalene based cyclophanes, the isolation of (**37**), (**38**) and (**39**) has been achieved [P. Kus, *Pol. J. Chem.*, 1991, **65**, 1611].

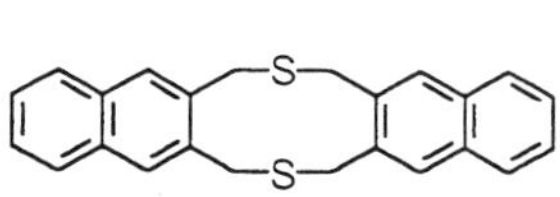

(**37**)

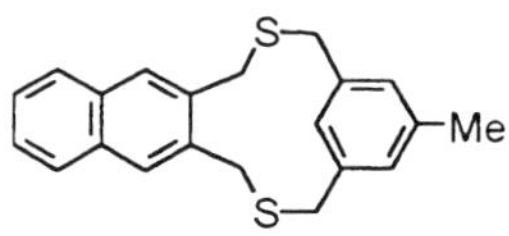

(**38**)

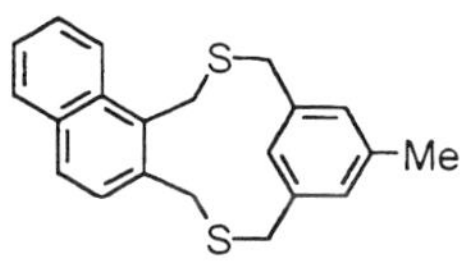

(**39**)

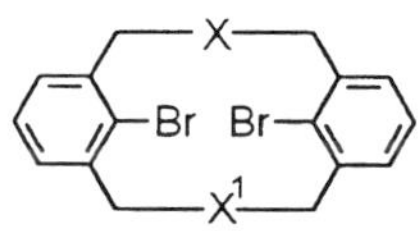

(**40**)

In the absence of the dihalo compound the dithio reactant may undergo oxidative coupling in the presence of iodine to form *syn* and *anti* (**40**) ($X = X^1 = S_2$). A reaction of the anti-derivative with $(Et_2N)_3P$, affords (**40**) ($X = X^1 = S$) while (**40**) ($X = S$, $X^1 = S_2$) is isolated on reacting the *syn* isomer with this reagent [H. Fujihara, J. Chiu, N. Furukawa, *Chem. Lett.,* 1991, **1**, 141].

The use of 1,3- or 1,4-$(HSCH_2)_2C_6H_4$ or 1,3- or 1,4-$(BrCH_2)_2C_6H_4$ allows the introduction of substituent groups within the macrocycle. Such examples are shown in Table 2:

Table 2

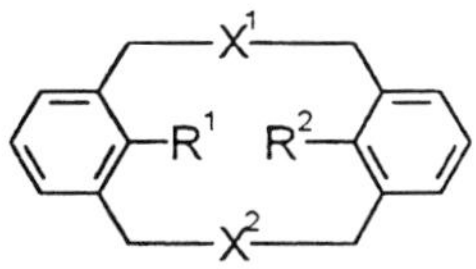

(**41**)

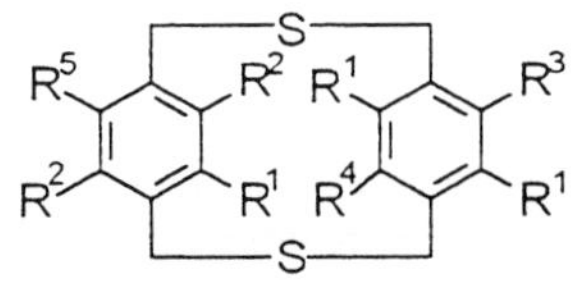

(**42**)

X^1	X^2	R^1	R^2	Ref.
S	SS	Br	Br	1
SS	SS	Br	Br	1
S	S	CH_2OMe	Me	2
S	S	F,Cl,MeO	F,Cl,MeO	8
S	S	Me	Me	9

R^1	R^2	R^3	R^4	R^5	Ref.
H	F	H	H	H	3
H	CN	H	H	H	4
H	H	Ph	Ph,H	H	10
H	H	Ph	H	PH	10

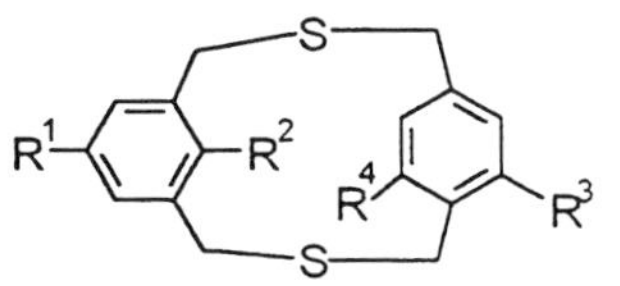

(**43**)

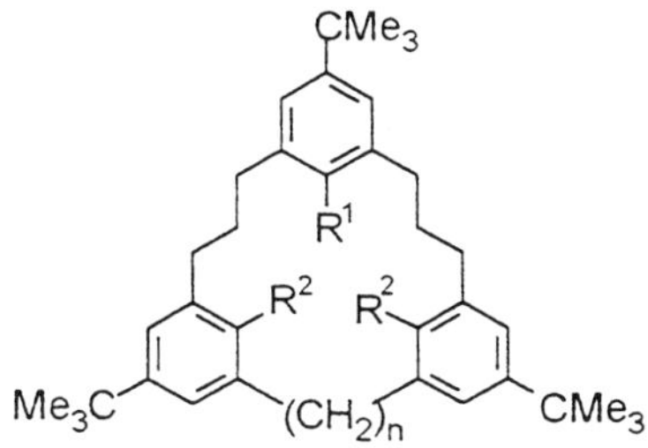

(**44**)

R^1	R^2	R^3	R^4	Ref.
H	H	H	Me	6
H	H	F	F	3
CMe_3	H,Me, OMe,Br	H	H	7

R^1 = Me, Et, Pr, OMe
R_2= Me, OMe
n = 1,2
Ref: 5

References:

1. H. Fujihara, J. -J. Chui and N. Furukawa, *Chem. Lett.*, 1991, **1**, 141.
2. Y. H. Lai and Q. M. Tan, *Aust. J. Chem.*, 1990, **43**, 1783.
3. R. Filler *et al.*, *J. Fluorine Chem.*, 1986, **30**, 399.
4. H. A. Staab, P. Wahl and K. Y. Kay, *Chem. Ber.*, 1987, **120**, 541.
5. A. Tsuge *et al.*, *Chem. Lett.*, 1992, **3**, 345.
6. Y. H. Lai and A. H. T. Yap, *J. Chem. Soc., Perkin Trans 2*, 1993, 703.
7. T. Yamato *et al.*, *J. Chem. Soc., Perkin Trans 1*, 1992, 2675.
8. T. Yamato, A. Miyazawa and M. Tashiro, *J. Org. Chem.*,1992, **57**, 266.
9. T. Yamato *et al., Chem. Express*, 1991, **6**, 285.
10. L. Czuchajowski and A. Zemanek, *Pol. J. Chem.*, 1990, **64**, 499.

For the synthesis of *anti*-(**41**) (X^1=X^2=S, R^1=CH_2OMe, R^2=Me), the benzylic (methoxymethyl) functional group is introduced after the cyclocondensation reaction. This maximises the overall yield. Oxidation of (**42**) (R^2=CN, R^1=R^3=R^4=R^5=H), with MCPBA or AcOOH gives the corresponding disulphones. Wittig rearrangement of (**41**) (X^1 = X^2 = S, R^1 = R^2 = F) by BuLi, followed by methyl-ation with MeI, gives bis(methylthio) derivatives (**45**). Quaternisation by $(MeO)_2CH^+BF_4^-$ and treatment with Me_3COK/ THF forms (**46**).

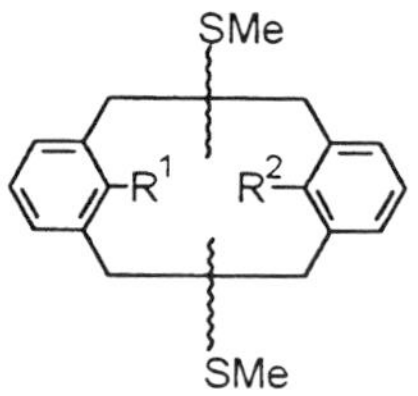

(**45**) R^1 = R^2 = F, Cl, OMe

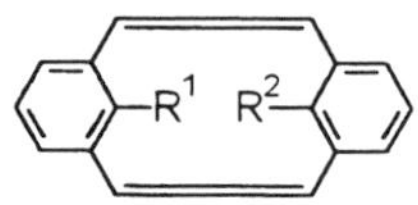

(**46**)

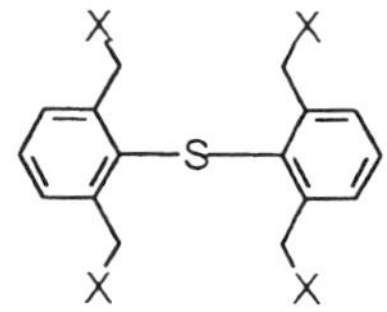

(**47**)

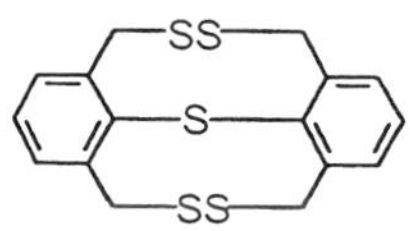

(**48**)

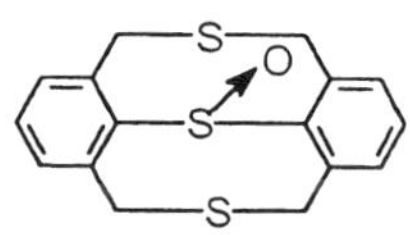

(**49**)

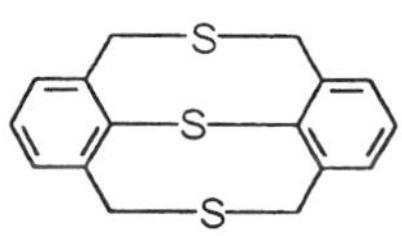

(**50**)

The reaction of bis[2,6-bis(bromomethyl)phenyl] sulphide (**47**) (X = Br) with thiourea, NaOH and HCl leads to the sulphide (**47**, X = SH). A high dilution reaction of the latter in chloroform in the presence of iodine and triethylamine affords (**48**) [H. Fujihara, J. J. Chui, N.Furukawa, *Tetrahedron Lett.*, 1990, **30**, 7441]. On dissolving (**48**) in conc. H_2SO_4 the monosulphoxide (**49**) is obtained in 48% yield after hydrolysis of the H_2SO_4 solution. Similar results are achieved when $NOPF_6$ is used instead of initial product [H. Fujihara, J. -J. Chui, N.Furukawa, *J. Am. Chem. Soc.*, 1988, **110**, 1280]. When (**49**) is irradiated in trimethylphosphite at 15°C under high pressure, the tris-sulphide (**50**) results in 82% yield. The X-ray structure of (**50**) is described in: F. Iwasaki, N. Toyoda, N. Yamazaki, H. Fujimara and N. Funikawa, *Acta Crystallogr.,Section C: Cryst. Struct. Commun.*, 1990, C**46**, 2154.

When two or more cyclophanes are linked together by a bridging group the resulting structure is known as a cuppedophane or as a cappedophane. Examples are (**51**), from the reaction of 1,3-$(HSCH_2)_2C_6H_4$ with tetrakis(bromomethyl)terphenyl (**52**) (R=H), in KOH/EtOH) in 43% yield [T. K. Vinod and H. Hart, *J. Am. Chem. Soc.*, 1988, **110**, 6574], (**53**) from the treatment of tetrakis(bromomethyl)terphenyl (**52**) (R=H) with 1 equivalent of

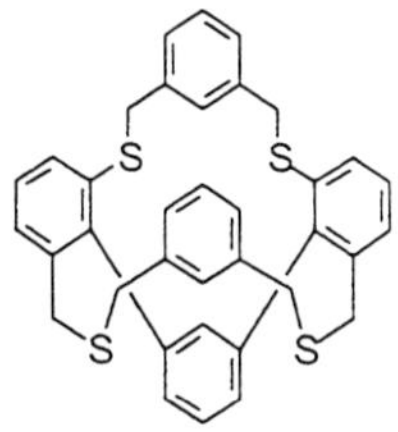

(**51**)

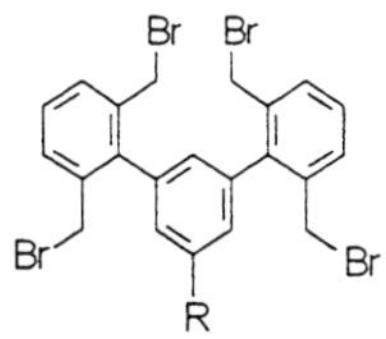

(**52**) R = H, CO_2H

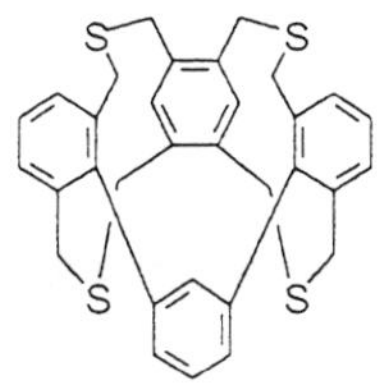

(**53**)

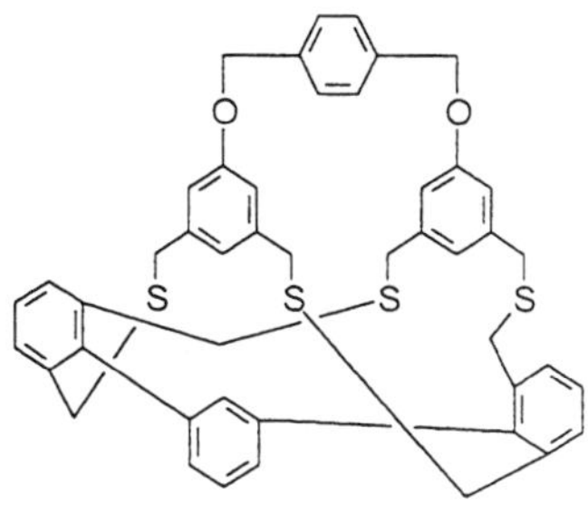

(**54**)

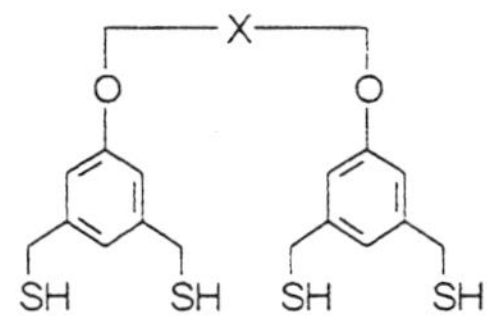

(**55**)

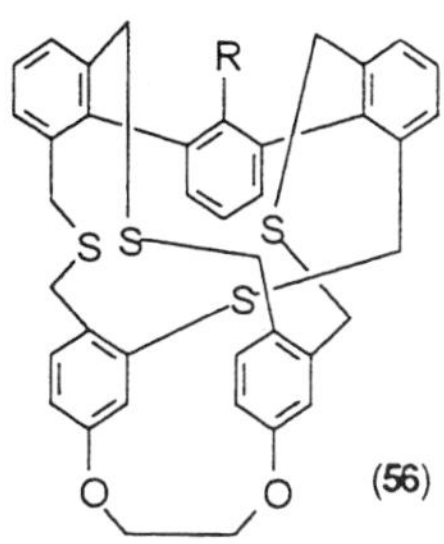

(**56**)

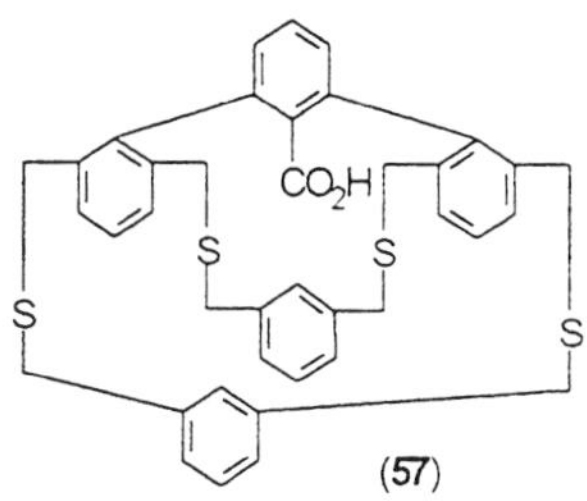

(**57**)

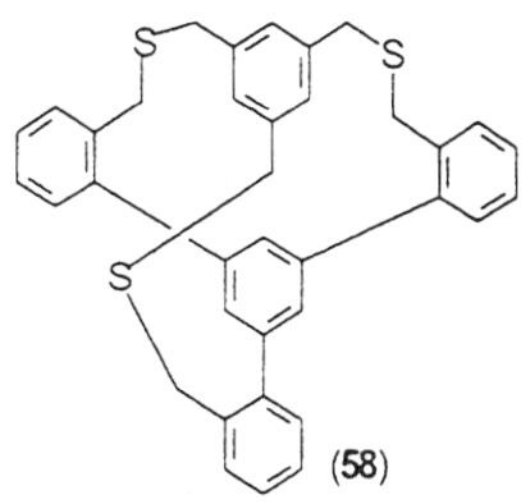

(**58**)

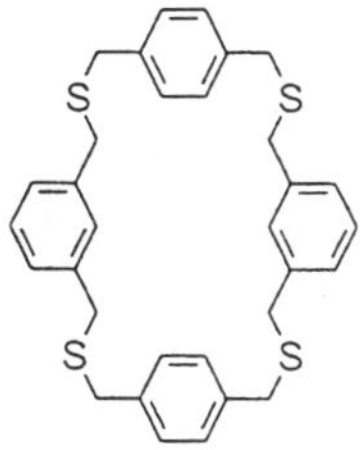

(**59**)

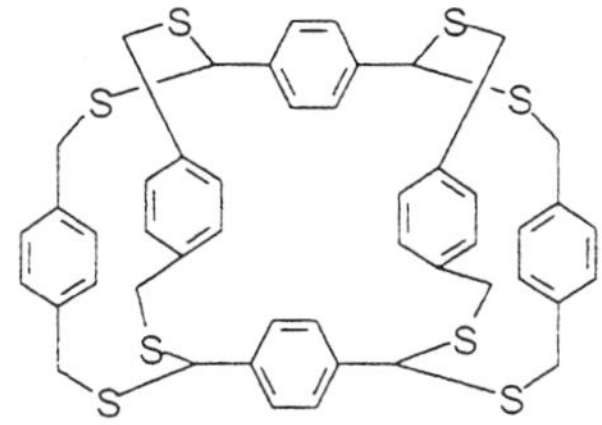

(**60**)

1,2,4,5-tetrakis(mercaptomethyl)benzene [T. K. Vinod and H. Hart, *J. Org. Chem.*, 1990, **58**, 881] and (**54**) (R= Ph) from the reaction of (**52**) (R=Ph) with (**55**) (X=C_6H_4-1,4). When R=H or Br the self filled conformer (**56**) (R=H, Br; X= C_6H_4) is the major product [T. K. Vinod and H. Hart, *J. Org. Chem.*, 1991, **56**, 5630]. Other recent examples include the synthesis of (**57**) from the reaction of 1,3-$(HSCH_2)_2C_6H_4$ and (**52**) (R=CO_2H) [U. Lueng *et al.*, *Chem. Ber.*, 1991, **124**, 397], (**58**) and (**59**) [J. J. Chui, H, Hart and D. Ward, *J. Org. Chem.*, 1993, **58**, 964] and (**61**) [T. Takenmura, K. Kozawa, T. Uchida and N. Mori, *Chem. Lett.*, 1984, **10**, 1839].
The inclusion of ferrocene, both external as the cyclophanes (**61**) (X = 1,2-, 1,3-, 1,4-$SCH_2C_6H_4$) [M. Hisatome *et al.*, *Tetrahedron Lett.*, 1983, **24**, 5757; *Bull. Chem. Soc., Jpn.*, 1987, **60**, 2953] and as part of the ring (**62**) (X = 1,2-, 1,3-, 1,4-$SCH_2C_6H_4$) [M. Hisatome *et al.*, *Organometallics*, 1987, **6**, 1498] is noted during the reactions of ferrocene dimethanol with 1,2-, 1,3-, 1,4-$(HSCH_2)_2C_6H_4$.

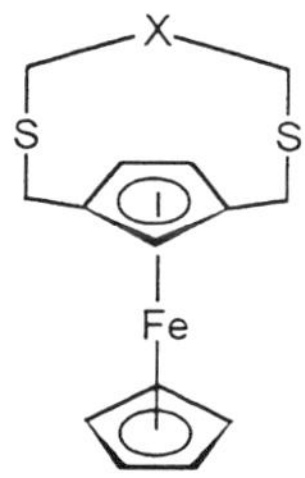

(**61**)

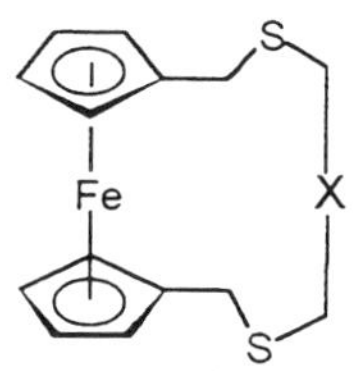

(**62**)

Compounds (**63**) (X = P, R = Ph) and (**63**) (X = As, R = Cl, Br, Ph, 1,4-BrC_6H_4, 1,4-$O_2NC_6H_4$ are prepared by the reaction of bis(methylenethiol)-benzene with $PhPCl_2$ and AsR_3 or $RAs(OMe)_2$ respectively [R. P. Arshinova *et al.*, *Zh. Obshch. Khim.*, 1988, **58**, 2417; R. R. Shagidullin, I. Kh. Shakirov, P. N. Sobolev and N. A. Chdaeva, *Izv. Akad. Nauk SSSR, Ser. Khim.*, 1988, **1**, 182].

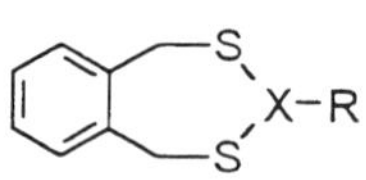

(**63**)

(c) *Amines*

Several 1,3-bis(trialkylaminomethyl)-, 1,3-bis(trialkylaminoethyl)- and 1,3,5-tris(trialkylaminomethyl)-benzene bromides and iodides (*e.g.* **64**) have been obtained by heating the corresponding haloalkylbenzene with R_2NH (R = Me or Et) in benzene [H. G. Löhr *et al.*, *Chem. Ber.*, 1984, **117**, 1487]. Reaction of 1,3-$(H_2NCH_2)_2C_6H_4$ with hexamethylenetetramine in water in the presence of conc. HCl at 100°C affords 1,3-$(HCO)_2C_6H_4$ [E. A. Meier, CA, 1989, **111**, 232302p].

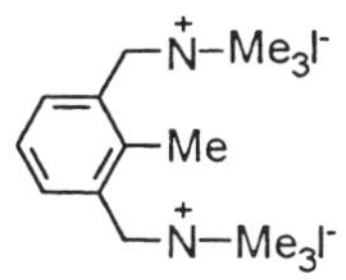

(**64**)

3. Hydroxyalkylarene carbaldehydes and hydroxyalkyl aryl ketones

The $Pd(Ph_3)_4$ catalysed reaction of 3-iodobenzyl alcohol with tributyltin hydride at 50°C under 1 at. CO gives 3-hydroxymethylbenzaldehyde in high yield. Similarly, 2-iodobenzyl alcohol affords a mixture of the open and cyclic forms of 2-hydroxymethylbenzaldehyde (**65**) ($R^1 = R^2$ = H) [V. Baillargeon and J. K. Stille, *J. Am. Chem. Soc.*, 1983, **105**, 7175].
4-Hydroxymethylbenzaldehyde can be prepared in excellent yield from 4-$MeO_2CC_6H_4CHO$ *via* protection with lithium 2-[N-methyl-N-(2-pyridyl)]amide, followed by trimethylsilation and reduction with $LiAlH_4$ [D. L. Comins and J. D. Brown, *Tetrahedron Lett.*, 1981, **22**, 4213] and by partial catalytic hydrogenation of 1,4-$(HOC)_2C_6H_4$ [J. Wild, B. Janssen and N. Goetz, CA, 1989, **110**, 17287u]. Alcohols of type (**66**) (R = saturated or unsaturated alkyl, aryl) can be obtained from the reaction of 1,4-$(HOC)_2C_6H_4$ with RMgX (X = halo) [S. Miyata, N. Yonezawa and A. Murao, CA, 1992, **116**, 6230k]. Partial oxidation of 4-substituted 2,6-bis(hydroxymethyl)phenols with activated MnO_2 in $CHCl_3$ gives (**67**) (R = Cl, Br, Me, OMe) in 40-70% yield [Y. Hu and H. Hu, *Synthesis*, 1991, 325]. See also section 2a, Scheme 1.

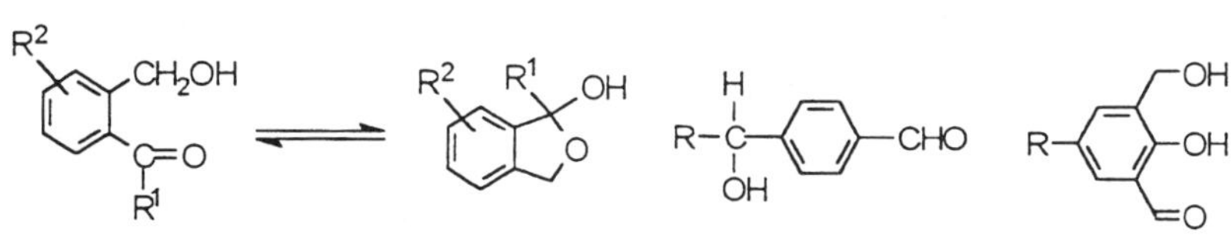

(65a) (65b) (66) (67)

Reaction of methyl 2-(dimethoxymethyl)benzoate with $LiAlH_4$ gives 2-(dimethoxymethyl)benzyl alcohol, which is a protected form of 2-(hydroxymethyl)benzaldehyde. 2-(Dimethoxymethyl)benzyl alcohol is readily converted to alkoxyphthalan and as an isobenzofuran precursor is useful in synthesis [J. G. Smith and P. W. Dibble, *J. Org. Chem.*, 1983, **48**, 5361]. The compounds (**65**) (R^1 = H, R^2 = 6-NO_2, 6-Br) can be obtained by reduction of the corresponding phthalide with diisobutylaluminium hydride and (**65**) (R^1 = H, R^2 = 7-OMe) by lithiation of 3-methoxybenzaldehyde. All compounds except (**65**) (R^1 = H, R^2 = 6-Br) have open structures in the solid state. In solution the open structure predominates in all cases [G. Wulff and G. Wolf, *Chem. Ber.*, 1986, **119**, 1876]. The hydroxy group of 4-(2-hydroxyethyl)benzaldehyde can be selectively reduced using trialkylborane and triflic acid [G. A. Olah and A. H. Wu, *Synthesis*, 1991, **5**, 407].
2-Acetylbenzyl alcohol together with its cyclic form can be obtained by reduction of the ketalised methyl ester of 2-acetylbenzoic acid with $LiAlH_4$ [E. Berner, *Acta Chem. Scand. Ser. B,* 1982, **B36**, 729]. 4-$MeCOC_6H_4CH_2CH_2OH$ results in excellent yield from the oxidation of 4-$MeCH(OH)C_6H_4CH_2CH_2OH$ with cetyltrimethylammonium permanganate [R. Rathore, V. Bushan and S. Chandrasekaran, *Chem. Lett.*, 1984, 2131]. The hydroxy group of 4-(2-hydroxyethyl) benzaldehyde can be selectively reduced using trialkylborane and triflic acid [G. A. Olah, A. H. Wu, *Synthesis*, 1991, 407].

4. Arene- di and poly-carboxaldehydes

(a) Arenedicarboxaldehydes

A facile synthesis of aromatic 1,4-dicarboxaldehydes involves the direct anodic oxidation of 1,4-substituted toluenes. The yields are influenced by the solvent used [I. Nishiguchi and T. Hirashima, *J. Org. Chem.*, 1985, **50**, 219]. 1,4-$(HOC)_2C_6H_4$ can be obtained by the gas phase oxidation of toluene [J. S. Yoo *et al.*, *Appl. Catal.,* A 1993, **105**, 83]. The $FeCl_3$ catalysed reaction of 1,3-$(Cl_2CH)_2C_6H_4$) and $(Me_3Si)_2O$ affords 1,3-$(HOC)_2C_6H_4$ [T. Nakano, M. Kayama and Y. Nagai, *Nippon Nagaku Kaishi*, 1985, 1504]. By using polymer supported H_2CrO_4 1,3- or 1,4-$(XCH_2)_2C_6H_4$, (X = Br, Cl, H) are readily converted to the corresponding dicarboxaldehydes. In contrast 1,2-$(BrCH_2)_2C_6H_4$ gives the phthalide and not the 1,2-dialdehyde [H. K. Lee, B. H. Chang and H. R. Kim, *Pollimo*, 1985, **9**, 377]. 1,2-$(HOC)_2C_6H_4$ has been prepared by sonicating PhBr and $Me_2NCH_2CH_2NMeCHO$ in lithium-containing THF to give 1,2-$C_6H_5CH(OLi)NMeCH_2CH_2NMe$, which on treatment with BuLi gives 1,2-$LiC_6H_4CH(OLi)NMeCH_2CH_2NMe$. Condensation of the latter with DMF or MeI, followed by hydrolysis affords the dicarboxaldehyde in 62% yield [J. Einhorn and J. L. Luche, *Tetrahedron Lett.*, 1986, **27**, 1793].

Under Vilsmeier reaction conditions 1,4-dihydroxybenzoic acids, such as 2-, 3- and 4-methyl-1,4-dihydroxybenzoic acids, undergo a one-pot conversion to benzene-mono-, di- and tri-carboxaldehydes. Similarly, naphthalene-1,3-dicarboxaldehyde is obtained in 91% yield from 1,4-dihydro-1-naphthoic acid [B. Raju and G. S. K. Rao, *Synthesis*, 1987, 197]. The Vilsmeier reaction of 4-methyl-2- and 3-cyclohexenone gives 63% 4,3,5-Cl$(HOC)_2C_6H_2$Me whereas the reaction of 2- and 6-methyl-2-cyclohexenone results in 60% 2,3,5-Cl$(HOC)_2C_6H_2$Me [B. Raju and G. S. K. Rao, *Indian J. Chem., Sect. B*, 1987, **26B**, 177]. Similarly 2-hexen-4-one gives 2,3,5-Cl$(HCO)_2C_6H_2$Me [M. Sreenivasulu and G. S. K. Rao, *ibid*, 1989, **28B**, 494]. Sequential reaction of 2,4,6-$Me_3C_6H_2$CHO with PCl_5 and Cl_2CHOMe/ $TiCl_4$, followed by HCl in EtOH gives 2,4,6-$Me_3C_6H(CHO)_2$-1,3 [A. P. Yakubov *et al.*, *Tetrahedron*, 1993, **49**, 3397].

Benzenedicarbaldehydes have been obtained by sequential treatment of benzenedicarboxylic acids with 9-borabicyclo[3.3.1]nonane, followed by treatment with lithium tris(diethylamino)aluminium hydride in the presence of pyridine [J. S. Cha *et al.*, *Org. Prep. Proced. Int.*, 1992, **24**, 327; J. S. S. Cha, *Bull. Korean Chem. Soc.*, 1992, **13**, 670]. Treatment of 1,3-$(PhO_2C)_2C_6H_4$ with 4-$MeSCH_2SO_2C_6H_4$Me in the presence of NaH gives 4-PhCOCH(SMe)$SO_2C_6H_4$Me, which on reduction with $NaBH_4$ and treatment with base gives the dialdehyde [K. Ogura, N. Yahata, K. Takahashi and H. Iida, *Tetrahedron Lett.*, 1983, **24**, 5761].

Condensation of 1,2-$(HOC)_2C_6H_4$ with 1,2-$(HOCH_2)_2C_6H_4$ or $HOCH_2CMe_2CH_2OH$ in dichloromethane containing anhydrous $CuSO_4$ results in the monoacetals (**68**) [R = 1,2-$(CH_2)_2C_6H_4$, $CH_2CMe_2CH_2$] which can be reduced by $NaBH_4$ to give the hydroxymethylbenzaldehyde acetals (**68**) (R = CH_2OH; X = 1,2-$(CH_2)_2C_6H_4$, $CH_2CMe_2CH_2$) [J. G. Smith and G. Kruger, *J. Org. Chem.*, 1985, **50**, 5759].

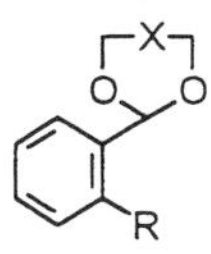

(68)

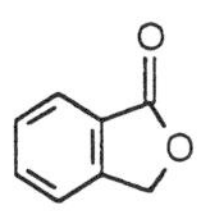

(69)

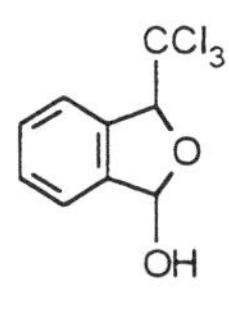

(70)

1,2-Benzenedicarboxaldehyde undergoes intramolecular Cannizzaro-type reaction in acetone catalysed by compounds of the type [Rh(diphosphine)$(Me_2CO)_2$]ClO_4 to give the γ-lactone (**69**) in excellent yield; 2,3-naphthalenedicarboxaldehyde behaves similarly [S. H. Bergens, D. P. Fairlie and B. Bosnich, *Organometallics*, 1990, **9**, 566]. Oxidation of the 1,2-benzene dicarboxaldehyde by excess sodium perborate in acetic acid affords benzene-1,2-dicarboxylic acid [A. McKillop and D. Kemp, *Tetrahedron*, 1989, **45**, 3299]. Treatment of the same aldehyde with PhCONHOH in benzene containing 4-$MeC_6H_4SO_3H$ yields 2-formylbenzonitrile in 72% [D. Geffken, *Arch. Pharm.*, 1985, **318**, 1142]. 1,2- and 1,4-$(HOC)_2C_6H_4$ react at 20°C with trichloroacetic acid in DMSO to yield insoluble products corresponding to the addition of trichloromethyl anion to carbonyl groups. The addition of one CCl_3^- ion to 1,2-isomers leads to the formation of the cyclised phthalan (**70**) [W. N. Wassef and N. M. Ghobrial, *J. Chem. Soc., Perkin Trans 2*, 1989, 251].

Reaction of 1,3-$(HOC)_2C_6H_4$ with ethylcyanoacetate in the presence of piperidine at room temperature yields 1,3-bis(β-cyano-β-ethoxycarbonylvinyl)benzene [T. Teshirogi, CA, 1988, **108**, 76054q]. Reaction of the enamine (**71**) (formed *in situ* from isoquinoline and sodium triethylborohydride) with 1,4-$(HOC)_2C_6H_4$ produces the 4,4'-coupled bis(isoquinoline) (**72**) in good yield which is readily oxidised to the diketone with MnO_2 [B. Guilani *et al.*, *J. Heterocycl. Chem.*, 1990, **27**, 1007]. The bisstilbene (**73**) is obtainable from the reaction of 2-methylbenzothiazole with 1,4-$(HOC)_2C_6H_4$ in the presence of $PhCH_2N^+Et_3Cl^-$ [S. B. Lokhande and D. W. Rangnekar, *Indian J. Chem. Sect. B*, 1986, **25B**, 485]. 1,2-$(HOC)_2C_6H_4$ on heating with MeONa in methanol affords the indandione (**74**) which exists in equilibrium with the pyran (**75**) [J. Dusemund and E. Kroeger, *Arch. Pharm.,* 1987, **320**, 617]. The indene derivatives (**76**) (R = 4-ClC_6H_4; 2-, 3-, 4-pyridyl; 2,4-$HO(MeO)C_6H_3$; 2,4-$(MeO)_2C_6H_3$; 2-thienyl) can be prepared in 54 - 65% yields by the reaction of the phthalaldehyde with RCOMe in 15% aq. NaOH [K. Samula and B. Cichy, *Acta Pol. Pharm.*, 1985, **42**, 256]. Darzens condensation of 1,4-$(HOC)_2C_6H_4$ with the α-chloroketones $RC(O)CH_2Cl$ (R = Me, Ph) in 50% aqueous dioxane with NaOH affords the mono-oxiranes (**77**) (R = Me, Ph, OH); the dioxiranes are also obtainable when R = Ph or OEt [W. N. Wassef and M. El-Berkey, *J. Chem. Research(S)*, 1990, 402] Benzylidine acetophenones *e.g.* (**78**) can be readily prepared by condensation of 1,2- or 1,4-$(HOC)C_6H_4$ with 4-RC_6H_4COMe (R = NO_2, Cl, NH_2) in the presence of $(4\text{-}MeC_6H_4)_2TeO$ in DMSO [J. Shao *et al.*, *Org. Prep. Proced. Int.*, 1993, **25**, 581] and by condensation of the 1,4-isomer with 4-RC_6H_4COMe (R = H, F, MeO) or 2-acetylthiophene [N. Elrayyes and A. J. A. Al-Johary, *J. Chem. Eng. Data,* 1985, **30**, 500]. Treatment of 1,2-$(HOC)_2C_6H_4$ and its 1,3- and 1,4-isomers with 1,3-bis(dimethylamino) trimethinium perchlorate in AcOH in the presence of acid affords the respective arylidene malonaldehydes $(AcO)_2CHC_6H_4CH{=}C(CHO)_2$ [L. A. Yanovskaya *et al.*, *Collect. Czech. Chem. Commun.*, 1985, **50**, 2270]. In the presence of ammonium acetate, the reaction of 4-$RC_6H_4CH_2SO_2CH_2CO_2Me$ (R = H, Me, Cl) with 1,3- and 1,4-$(HOC)_2C_6H_4$ affords the respective methoxycarbonyl styryl benzyl sulphones $C_6H_4[CH{:}C(CO_2Me)SO_2CH_2C_6H_4R\text{-}4]_2$ [M. V. R. Reddy *et al.*, *Sulfur Lett.*, 1990, **10**, 211].

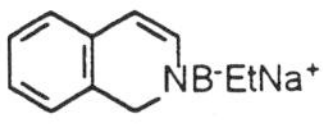

(71)

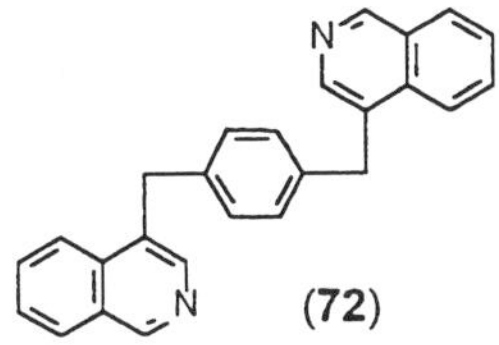

(72)

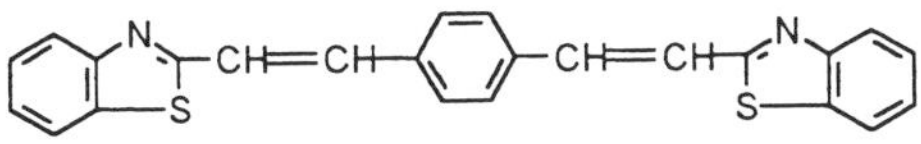

(73)

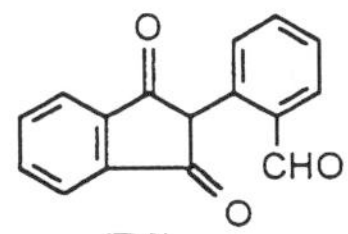

(74)

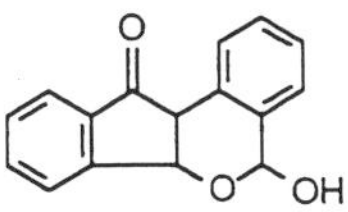

(75)

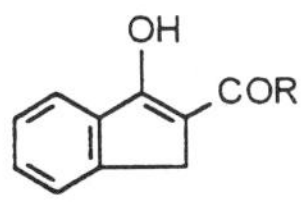

(76)

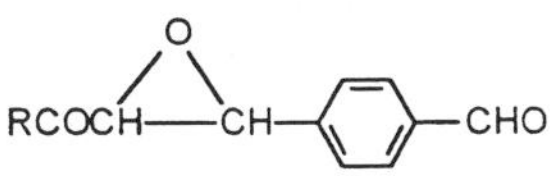

(77)

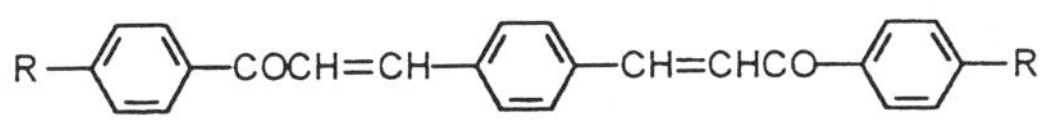

(78)

Isoindolin-1-ones have been isolated in good yields by reacting 1,2-$(HOC)_2C_6H_4$ with α-amino acids or their methyl esters, arylamines, heterocyclic amines and less efficiently aryl substituted aliphatic amines in the presence of acetic acid. [R. Grigg, H. Q. N. Gunaratne and V. Sridharan, *J. Chem. Soc., Chem. Commun.*, 1985, 1183]. Treatment of the aldehyde successively with aqueous $NaHSO_3$ followed by RNH_2 (R = Me, $PhCH_2CH_2$, Bu, 4-$MeOC_6H_4$) and KCN gives the cyanoisoindoles (**79**) in 33-86% yield. The yields are much higher when the reactions are carried out under a nitrogen atmosphere [K. Takahashi, K. Suenobu and K. Ogura, *Chem. Lett.*, 1985, 1487]. The ylidene (**81**) and its isomers are among the products of the reaction of 2-$H_2NC_6H_4C_6H_4NH_2$-2 with 1,2$(HOC)_2C_6H_4$ [D. M. Hall *et al.*, *J. Chem. Soc., Perkin Trans 1*, 1987, 1763]. Cyclocondensation of 1,2-$(HOC)C_6H_4$ with $RC_6H_4NHNH_2$ (R = 4-MeO, 4-Me, H, 4-Br, 2-Me, 2-Br) gives hydroxydihydrophthalazines (**82**) in 70 - 99% yields, which dimerise in acetic acid [R. N. Butler, A. M. Gillan, P. McArdle and D. Cunningham, *J. Chem. Soc., Chem. Commun.*, 1987, 1016].

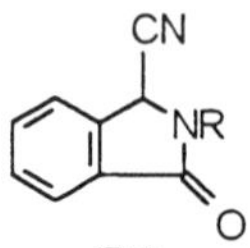

(**79**)

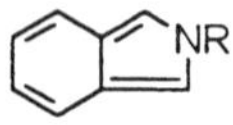

(**80**)

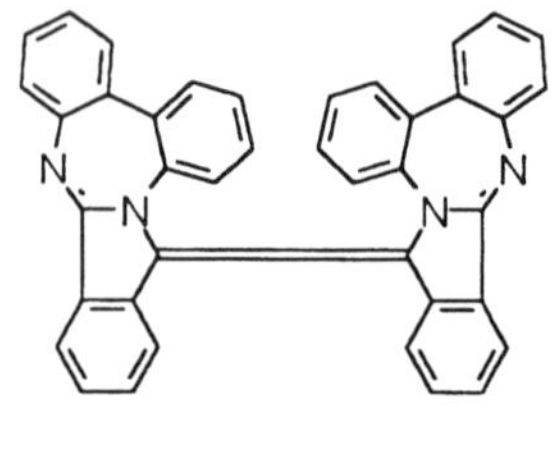

(**81**)

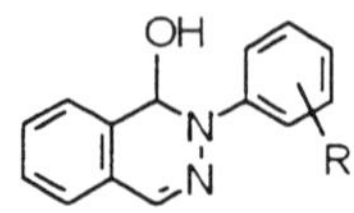

(**82**)

The reaction of PhN_3 with PPh_3 in dichloromethane followed by treatment with 1,2-$(HOC)_2C_6H_4$ gives (**80**) (R = Ph) in 22% yield, together with 17% isobenzofuranylisoindoline (**83**). Whereas a similar reaction starting with $PhCOCH_2N_3$ affords the isoquinoline (**84**) (R = Ph) [T. Aubert, M. Farnier and R. Guilard, *Can. J. Chem.,* 1990, **68**, 642]. The condensation of active methyl groups in toluene with 1,4-$(HOC)_2C_6H_4$ in aqueous medium at room temperature, using $PhCH_2N^+Et_3Cl^-$ as a phase transfer catalyst, affords bis(stilbenes) and bis(styryls), *e.g.* the reaction of 2-methylbenzothiazole and 1,4-$(HOC)_2C_6H_4$ in the absence of $PhCH_2N^+Et_3Cl^-$ gives the benzothiazole (**85**) [S. B. Lokhande and D. W. Rangnekar, *Indian J. Chem., Sect. B*, 1986, **25B**, 485]. The Wittig reaction of 1,4-$(HOC)_2C_6H_4$ with equimolar Ph_3P:CHOMe or $Ph_2P(O)CH_2OMe$ gives 4-$HOCC_6H_4CH_2OH$ and, with excess of these reagents, 1,4-$(CH{=}CHOMe)_2C_6H_4$ [C. Capa, F. Sanchez-Ferrando and M. Tristan-Polo, *Nouv. J. Chim.*, 1985, **9**, 493]. The 1,2- and 1,3 substituted compounds $R_2C_6H_4$ (R = $(CF_3)_2C{=}CH$) can be obtained from the reaction of $(F_3C)_2CCl_2$ with P_3Ph and the respective benzene dicarboxaldehyde [M. Hanack and C. Korhummel, *Synthesis*, 1987, 944]. Reaction of 1,4-$(HOC)_2C_6H_4$ C_6F_5CH:PPh_3 gives 1,4-$(C_6F_5CH_2CH_2)_2C_6H_4$ [R. Filler and J. E. Gadomski, *J. Fluorine Chem.*, 1990, **47**, 175].

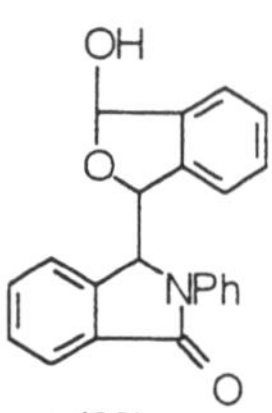

(83)

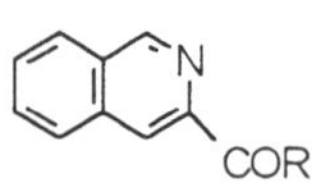

(84)

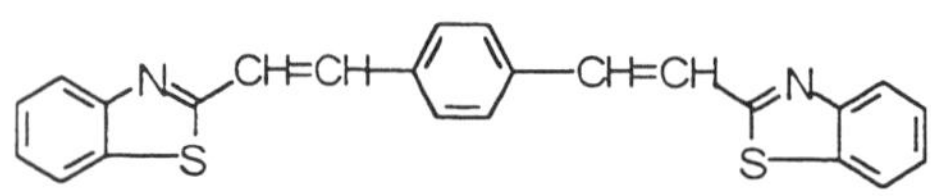

(85)

Substituted 1-naphthols have been prepared regioselectively in good to excellent yields by the reaction of 1,2-$(HOC)_2C_6H_4$ with Ta (or Nb) alkyne complexes, produced by the treatment of acetylene with low valent Ta (or Nb) [derived by Zn reduction of $TaCl_5$ (or $NbCl_5$)] [Y. Kataoka *et al.*, *Tetrahedron Lett.*, 1990, **31**, 369].

1,2-, 1,3- and 1,4-Benzene dicarboxaldehydes have been used extensively for the synthesis of various macrocycles and other types of cyclic compounds *e.g.* **86** - **88** [Z. Li, Y. Lin and C. Chi, *Heterocycles*, 1992, **34**, 1729], (**89**) [M. G. B. Drew *et al.*, *J. Chem. Soc., Dalton Trans.*, 1992, 11]. See also Section 2b for the synthesis of cyclophanes. The direct cyclocondensation of isophthaladehyde and 3-azapentane-1,5-diamine in acetonitrile leads to the isolation of an 18-membered imidazoline containing macrocycle (**90**) [H. Adams *et al.*, *J. Inclusion Phenom., Mol. Recognit. Chem.*, 1991, **11**, 63] instead of the expected 24-membered hexaaza Schiff base (**91**). The ring contraction occurs as a consequence of the internal nucleophilic addition of the secondary amine functions across adjacent azomethine bonds. Isolation of the non-contracted macrocycle (**91**) is possible as its dicopper(II) complex on complexation of (**90**) [R. Menif *et al.*, *Inorg. Chem.*, 1990, **29**, 4723].

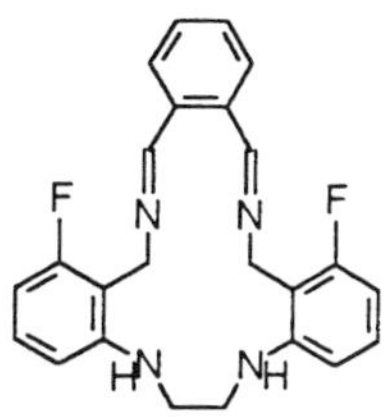

(**86**)

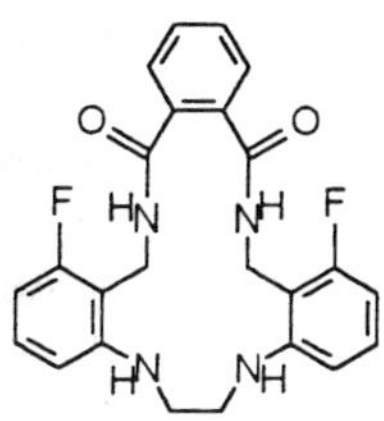

(**87**)

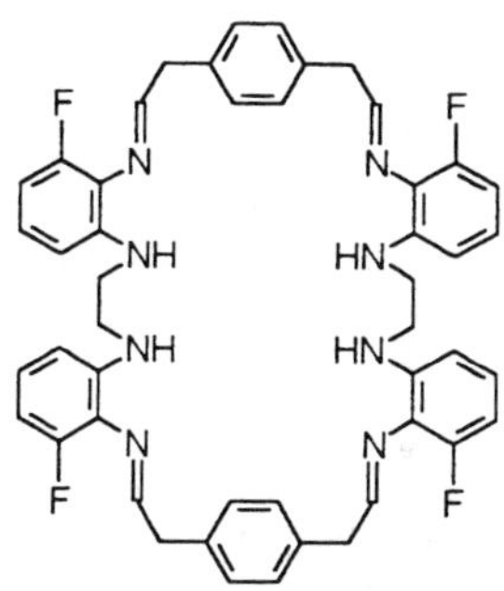

(**88**)

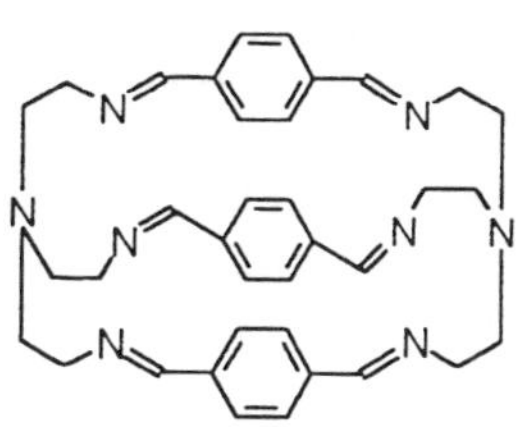

(**89**)

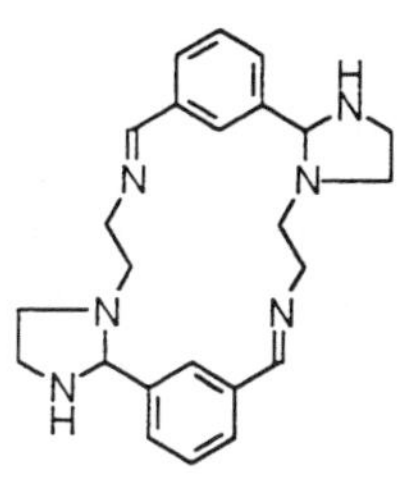

(**90**)

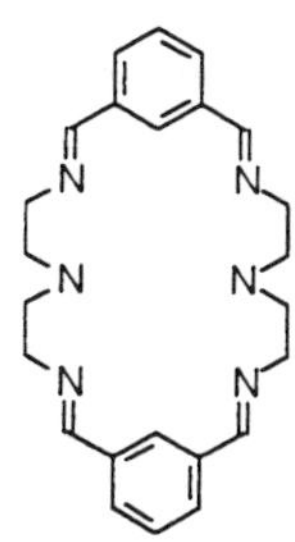

(**91**)

(b) Phenolic benzenedicarboxaldehydes

A one pot synthesis of 4-substituted 2,6-diformylphenols (**92**) (R = Cl, Br, Me, OMe) is illustrated in Scheme 1 (Section 2a). Another route to (**92**) (R = Me), which is also applicable for the synthesis of (**92**) (R = CMe_3), involves the oxidation of the respective 4-R-2,6-bis(hydroxymethyl)phenol with activated MnO_2 [S. Taniguchi, *Bull. Chem. Soc. Jpn.*, 1984, **57**, 2683]. The analogue (**92**) (R = CF_3) has been prepared by a multistage process from 4-BrC_6H_4OH [J. Leroy *et al.*, *J. Fluorine Chem.*, 1988, **40**, 23]. 2,4-Diformylphenol and its 2,6-isomer are obtained together with monoformyl compounds by the Reimer-Tiemann formylation of phenol in a weakly hydrated solid-liquid medium [A. Thoer, G. Denis, M. Delmas, A. Gaset, *Synth. Commun.*, 1988, **18**, 2095]. The dicarboxaldehydes 2,3,5-$HO(HOC)_2C_6H_2(OR)$ (R = Me, Et) can be prepared by heating the respective mandelic acid 2,3,5-$HO[HO_2C(HO)CH]_2C_6H_2OR$ in aqueous alkali in the presence of Cu, Ni, Co or Zn [CA, 1981, 96:68597n].

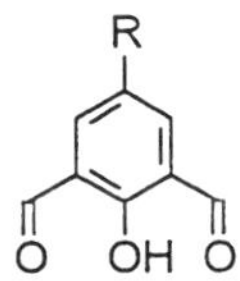

(**92**)

The [2+2] cyclocondensation of 2,6-diformylphenol and a diamine has been utilised extensively for the synthesis of various macrocycles. As shown in Scheme 3 the product, *e.g.* (**93**), can be isolated by the direct approach (Table 3). In some cases where this is not possible the presence of a metal ion ensures that a [2+2] macrocyclic entity predominates over a mixture of polymeric products such as [2+2], [3+3], [4+4], *etc.*, which are extremely difficult to separate. The isolation of a monometallic (**94**) or dimetallic (**95**) complex depends on the nature of the metal ion *i.e.* its size (ionic radius), properties and the counterion as well as the nature of the diamine. For example, in the case of (**94**) (M^{n+} = Pb^{2+}, X = $CH_2CH(OH)CH_2$) the metal co-ordinates to only one compartment of the macrocycle as opposed to co-ordinating symmetrically in the centre *i.e.* co-ordination occurs between the metal and the two phenolic oxygen atoms and only two of the four nitrogen atoms, specifically the two nitrogen atoms from the same diamine [S. S. Tandon *et al.*, *Inorg. Chem.*, 1992, **31**, 4635]. This is also observed for the much larger metals such as the lanthanide metals when the diamine chain is relatively short *e.g.* (**94**) (X = $CH_2CH_2NHCH_2CH_2$, M^{n+} = La^{3+} - Lu^{3+}) [P. Guerriero *et al.*, *Inorg. Chem. Acta*, 1987, **129**, 127]. When the diamine chain is much longer *e.g.* X = $CH_2CH_2NHCH_2CH_2NHCH_2CH_2$ or

Table 3

Macro-cycle	X	R	M^{n+}	counter-ion	Ref
93	$CH_2CH_2CH_2$	Cl, Me	-	-	1,2
	$CH_2CH(Me)CH_2$	Cl, Me	-	-	1
	$CH_2CH_2NHCH_2CH_2$	Cl, Me	-	-	3
	C_6H_4	Me	-	-	3
	$CH_2CH_2OCH_2CH_2OCH_2CH_2$	Cl	-	-	4
94	$CH_2CH(OH)CH_2$	Me, CMe	Pb^{2+}	ClO_4	5
95	$CH_2CH(OH)CH_2$	Me, CMe	Pb^{2+}	ClO_4	6
	$CH_2CH(OH)CH_2$	Me, CMe	Mn^{2+}, Ni^{2+}	OAc	7
	$CH_2CH(OH)CH_2$	Me, CMe	Cu^{2+}	NO_3	8
	$(CH_2)_2N(Me)(CH_2)_2N(Me)(CH_2)_2$	Me	Pb^{2+}	ClO_4	9,10
	$CH_2CH_2CH_2$	Me	Mn^{2+}	NO_3	11
	$CH_2CH_2CH_2$	Me	Pb^{2+}	ClO_4	12
	$CH_2CH_2OCH_2CH_2OCH_2CH_2$	Me	La^{3+},Gd^{3+}	NO_3	13
	$[(CH_2)_2N(Me)(CH_2)_2N(Me)(CH_2)_2$	Me	La^{3+},Dy^{3+}, Eu^{3+},Gd^{3+}	NO_3	14
	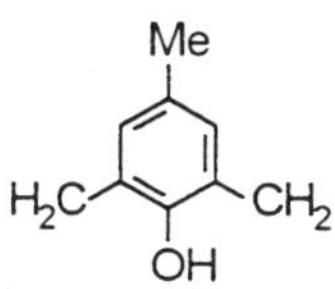	Me	Mn^{2+}	ClO_4	15

References for Table 3:

1. P. Guerriero *et al.*, *Rapid Commun. Mass Spectrom.*, 1992, **6**, 741.
2. A. J. Atkins, A. J. Blake and M. Schroder, *J. Chem. Soc. Chem. Commun.*, 1993, 353.
3. A. Aguiari *et al.*, *Inorg. Chim. Acta*, 1992, **202**, 157.
4. P. Guerriero *et al.*, *J. Chem. Soc., Dalton Trans.*, 1990, 647.
5. S. S. Tandon *et al.*, *Inorg. Chem.*, 1992, **31**, 4635.
6. A. J. Downard, V. McKee and S. S. Tandon, *Inorg. Chim. Acta*, 1990, **173**, 181.
7. S. S. Tandon *et al.*, *Inorg. Chem.*, 1993, **32**, 4621.
8. S. S. Tandon and V. Mckee, *J. Chem. Soc., Dalton Trans.*, 1989, 19.
9. H. Sakiyama *et al.*, *Chem. Lett.*, 1992, 1021.
10. Y. Ikawa, T. Nagata and K. Maruyama, *Chem. Lett.*, 1993, 1049.
11. M. Mikuriya, K. Nakadera and T. Tokii, *Inorg. Chim. Acta*, 1992, **194**, 129.
12. M. Tadokoro *et al.*, *Bull. Chem. Soc. Jpn.*, 1990, **63**, 3337.
13. I. Kahwa *et al.*, *J. Chem. Soc., Chem. Commun.*, 1989, 1531.
14. I. A. Kahwa, F. R. Fronczek and J. Selbin, *Inorg. Chim. Acta*, 1988, **148**, 273.
15. M. Mikuriya *et al.*, *Inorg. Chim. Acta*, 1989, **156**, 21.

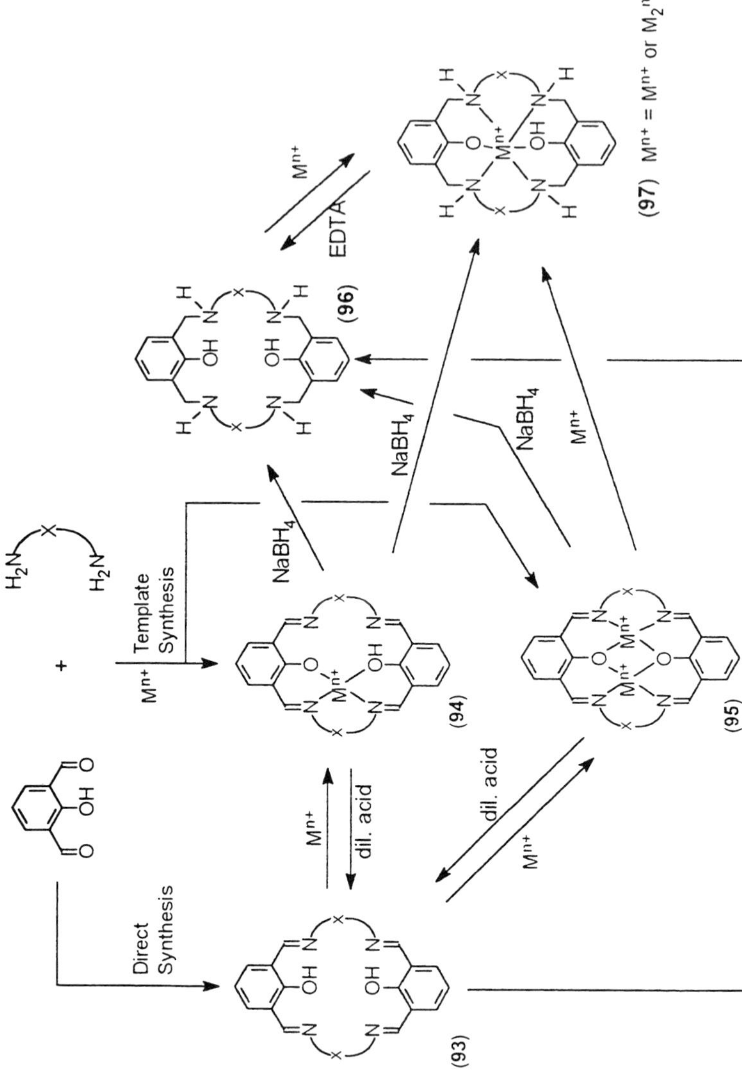

Direct Synthesis
Template Synthesis
M^{n+}
dil. acid
$NaBH_4$
EDTA
(93)
(94)
(95)
(96)
(97) $M^{n+} = M^{n+}$ or M_2^{n+}

Scheme 3

$CH_2CH_2OCH_2CH_2OCH_2CH_2$ dinuclear lanthanide metal complexes of (**95**) are formed [I. A. Kahwa *et al.*, *Inorg. Chim. Acta*, 1986, **118**, 179]. Where (**93**) cannot be obtained by direct demetallation of (**94**) or (**95**) demetallation is achieved by reduction with $NaBH_4$ thereby forming the free reduced macrocycle (**96**) [P. Guerriero *et al.*, *Org. Mass Spectrom.*, 1992, **27**, 231] or its corresponding metal complex (**97**) *e.g.* X = $(CH_2)_3$, M^{n+} = Pb^{2+}. The metal is removed from (**97**) with the use of dilute H_2SO_4 or EDTA [S. K. Mandal and K. Nag, *J. Org. Chem.*, 1986, **51**, 3900] forming free (**96**). Mass spectrometry of the free macrocycle (**93**) is detailed by P. Guerriero *et al.*, [*Org. Mass Spectrom.*, 1992, **27**, 231]. Macrocyclic compounds arising from (**92**) are listed in Table 3.

(c) Benzene polycarboxaldehydes

The major use of 1,3,5-$(HOC)_3C_6H_3$ is in the synthesis of bicyclophanes. For example the Wittig reaction of aromatic bisphosphonium salts with 1,3,5-$(HOC)_3C_6H_3$, followed by hydrogenation affords the bicyclophane (**98**) [H. E. Hoegberg and O. Wennerstroem, *Acta Chem., Scand., Ser. B*, 1982, **B36**, 6614; F. Vögtle, J. Gross, C. Steel and M. Nieger, *Angew. Chem.*, 1992, **104**, 1112]. Condensation of 1,3,5-$(HOC)_3C_6H_3$ with $Me_2C(NHOH)CMe_2NHOH$, followed by oxidation with PbO_2 gave the stable triradical (**99**) [L. Dulog and J. S. Kim, *Angew. Chem.*, 1990, **102**, 403].

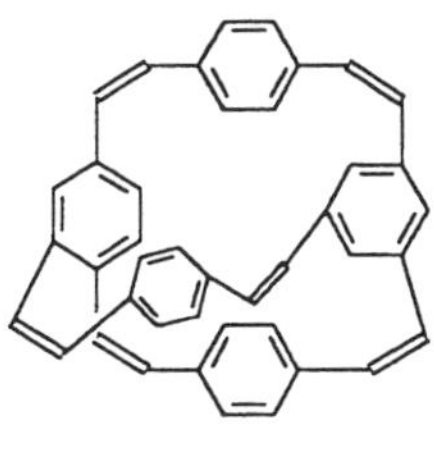

(**98**)

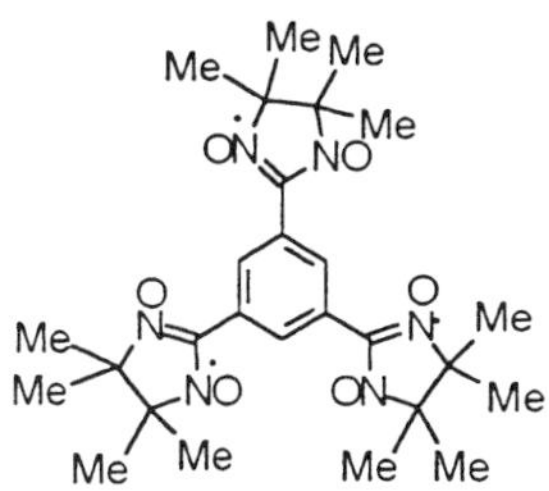

(**99**)

The tris(sydnonyl)benzenes (**100**) (n = 1-8, R = H, Ph) are obtained from 1,3,5-$(HOC)_3C_6H_3$, $R(CH_2)_nNH_2$ and KCN [K. Rehse and A. Martens, *Arch. Pharm. (Weinheim, Ger.)*, 1993, **326**, 217]. The diacetal of 1,3,5-$(HOC)_3C_6H_3$ (**101**) is formed from the reaction of 1,3-$(HOC)_2C_6H_3$-Br-5 with neopentyl alcohol in the presence of TsOH, CH_2Cl_2, followed by reflux in THF in the presence of magnesium. The aldehyde (**101**) has been utilised for the synthesis of 1,3,5-triporphyrinylbenzene [A. Osuka, B.-L. Liu and K.

Maruayama, *J. Org. Chem.*, 1993, **58**, 3582]. Wittig or Wittig-Horner reactions between 1,2,4,5,-$(HOC)_4C_6H_2$ and phosphorus ylides, derived from 1,3-dithiols, affords compounds of the type (**102**) [M. Salle *et al.*, *Tetrahedron Lett.*, 1991, **32**, 2897].

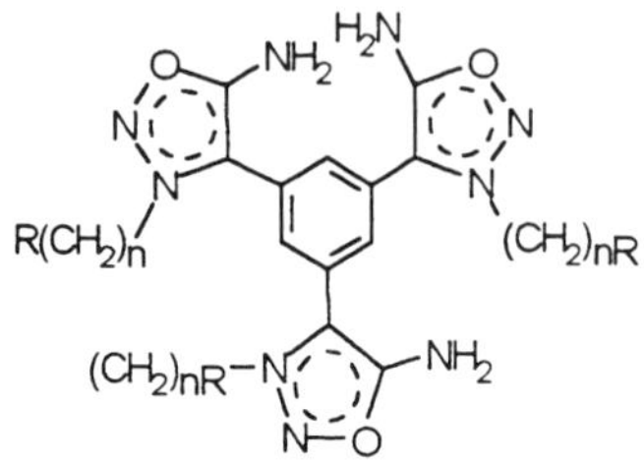

(**100**)

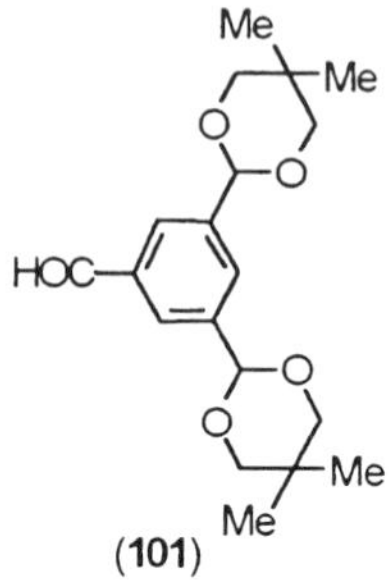

(**101**)

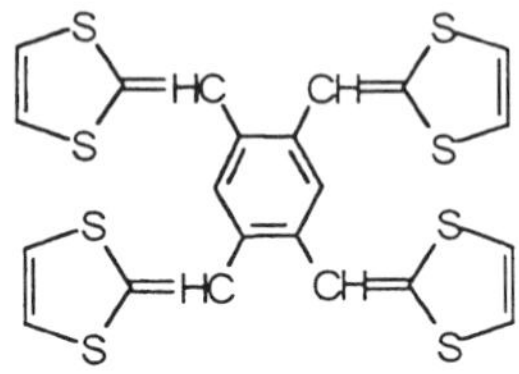

(**102**)

5. Phthalides and Related Compounds

(a) Phthalides

Phthalides can be obtained by the oxidative lactonisation of 1,2-bis(hydroxymethyl)benzenes (Table 4):

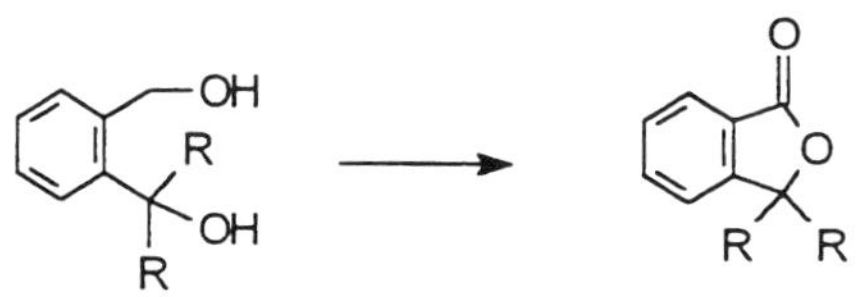

Table 4

R	Reagent/ Conditions	Yield(%)	Ref.
R=H,Me	NIS/AgOAc/Δ/dark	*ca.* 85	T. R. Beebe *et al., J. Org. Chem.*, 1987, **52**, 5472.
R=H	$IrH_5[(Me_2CH)_3P]_2$	96	Y. Lin *et al., J. Organomet. Chem.*, 1992, **429**, 269.
R=H	$3ClC_6H_4CO_2H/HCl/DMF$	-	H. R. Kim., *et al., Bull. Korean Chem.Soc.*, 1992, **13**, 579.
R=H	$NIS/MeOH/K_2CO_3$	-	C. E. MacDonald *et al., Microchem. J.*,1993, **47**, 115.
R=H	$RuH_2(PPh_3)_4/(Me)_2CO$	90	S. Murashi *et al., J. Org. Chem.*, 1987, **52**, 4319.
R=H	AcOOH/EtOAc/NaBr	74	T. Morimoto *et al., Bull.Chem. Soc. Jpn.*, 1992, **65**, 703

6,7-Dimethoxyphthalide can be obtained in excellent yield by treatment of the lithiated derivative of 3,4-dimethoxybenzyl α-ethoxyethyl ether with carbon dioxide. 6,7-(Methylenedioxy)phthalide can also be prepared in a similar fashion [E. Napolitano *et al.*, *J. Org. Chem.*, 1983, **48**, 3653]. Phthalides of the type (**103**) (R^1 = H, R^2 = aryl; R = R^1 = Et) are obtained from the reaction of 3-bromophthalide with amines [M. I. Shevchuk, S. T. Shpak and A. F. Tolochko, *Zh. Obshch. Khim.*, 1987, **57**, 799]. Reaction of 2-acetyl- and 2-benzoyl-benzoic acids with diethylamino sulphur trifluoride gives (**104**) (X = F; R = Me, Ph) in excellent yields [T. B. Patrick and Y.-F. Poon, *Tetrahedron Lett.*, 1984, **25**, 1019]. A mixture of phthalides (**105**) and (**106**) results from the hydrolysis of 4-R-1,2-$(BrCH_2)_2C_6H_3$ (R = NO_2, Bz) with 95% H_2SO_4 in the presence of air [W. Guo and Y. F. Duann, *Org. Prep. Proceed. Int.*, 1990, **22**, 85].

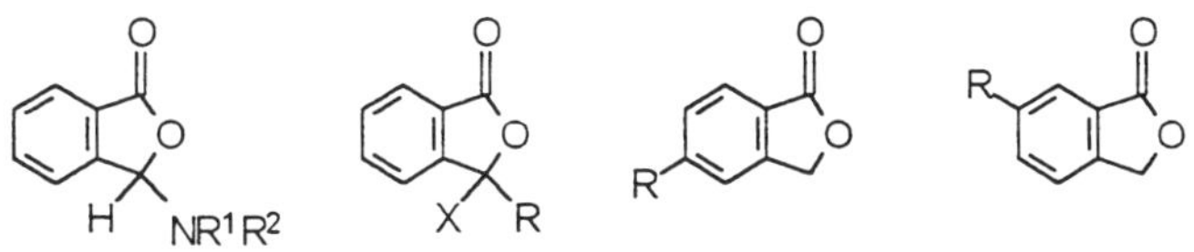

(103) (104) (105) (106)

Phthalic anhydride reacts with PhLi to give 3,3-di(phenyl)phthalide [M. T. Rahman and S. K. Nahar, *J. Organomet. Chem.*, 1992, **425**, 210] and trichloroacetic acid in DMSO to give the lactone (**107**) (R=H). 2,3-Naphthalic and 3-nitrophthalic anhydrides behave similarly, the latter affording a 3:1 mixture of (**107**) (R=3-NO_2) and (**107**) (R=6-NO_2) [W. N. Wassef and N. Ghobrial, *J. Chem. Res., (S)*, 1990, 326].

The Wittig reaction of phthalaldehydic acid with the phosphoranes $Ph_3P{:}CRCO_2Et$, (R = H or Br) [N. S. Narasimhan and S. S. Kusurkar, *Indian J. Chem.*, Sect. B 1983, **22B**, 345] and $Me_2C{:}CHCH_2C({:}PPh_3)CO_2R$, (R = Me or Et) [M. D. Chordia, N. S. Narasimhan and S. Nurani, *J. Chem. Soc. Perkin Trans 1*, 1991, 371] affords phthalides of the type (**108**) and (**109**) respectively. 3-(Arylsulphonyl)phthalides can be prepared by the reactions of phthalaldehydic acid with *p*-$RC_6H_4SO_2H$, (R = H, MeO) in the presence of boron trifluoride [V. S. K. Murty *et al.*, *Synth. Commun.*, 1990, **20**, 1711]. Condensation of phthalaldehydic acid with benzene in concentrated sulphuric acid gives (**110**) (R^1 = R^2 = H) in high yield. Similar reactions with PhR (R = HO, MeO) give the *para* regioisomers (**110**) (R^1 = H; R^2 = HO, MeO) whereas PhR (R = Cl, Br, Me) lead to mixtures of *ortho* and *para* regiosomers [R. Al-Hamdany, J. M. Al-Rawi and S. Ibrahim, *J. Prakt. Chem.*, 1987, **329**, 126].

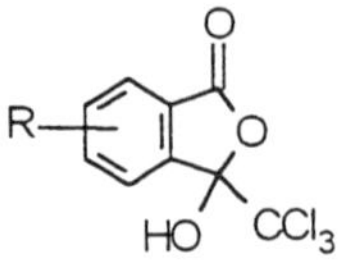

(107)

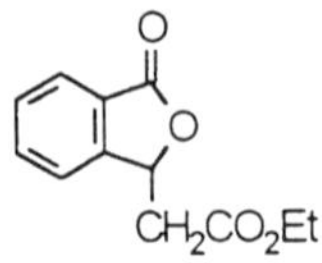

(108)

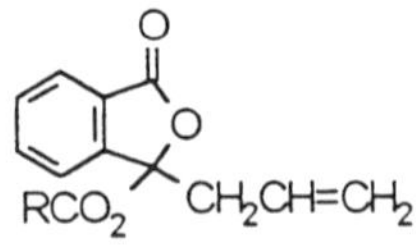

(109)

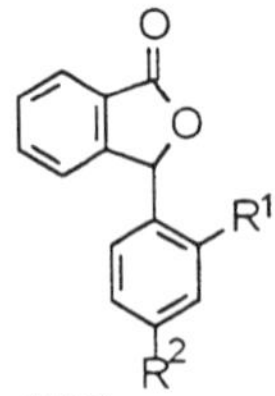

(110)

Reaction of phthaldehydic acid with RCH_2PPh_3Br (R = alkyl) affords 2-$HO_2CC_6H_4CN$:CHR which on iodolactonization gives 3-iodophthalides. Reaction of the latter with sodium acetate gives 3-alkylidenephthalides [R. S. Mali *et al.*, *Indian J. Chem.*, Sect. B 1990, **29B**, 319]. 3-Ylidenephthalides can be obtained by the reaction of phthalic anhydrides with phosphoranes [A. Allahbad and D. W. Knight, *J. Chem. Soc., Perkin Trans 1*, 1982, 1985; M. D. Chordia, N. S. Narasimhan and N. S. Nurani, *J. Chem. Soc Perkin Trans 1*, 1991, 371]. When an unsymmetrically substituted phthalic anhydride is used this route leads to mixtures of regioisomers. However, regiospecific synthesis of 3-ylidenephthalides (**111**) can be accomplished using the phosphonate reagents (**112**) which are obtainable from phthalaldehydic acids and dimethyl phosphite. Wittig-Horner type condensation of (**112**), with aromatic aldehydes affords isomeric mixtures, which can be converted quantitatively to the more stable (2)-isomers by opening of the lactone ring, followed by recyclisation of the intermediate deoxybenzoin-2-carboxylic acid (**113**). Reaction of the phosphonate (**112**) ($R^1 = R^2 = H$) with n-butanol affords the natural product (**111**) ($R^1 = R^2 = H$; R = n-Pr) [E. Napolitano *et al.*, *Synthesis*, 1985, 38].

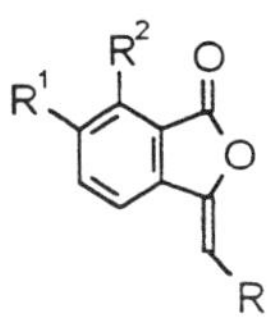

(111)

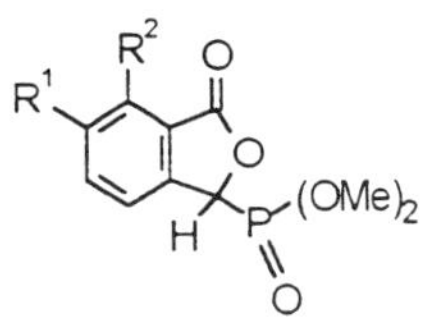

(112)

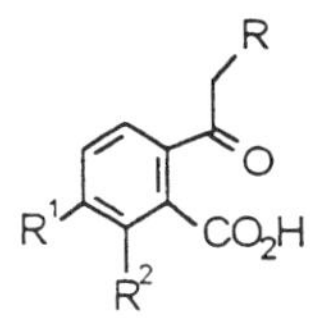

(113)

R =

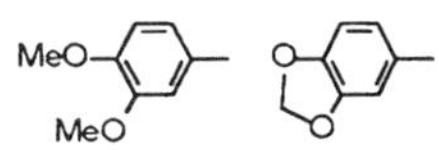

R^1	R^2
OMe	OMe
—O—CH_2—O—	
H	H

The kinetics of the equilibration of 3-hydroxyphthalide (**114**) and 2-formylbenzoic acid suggest that the phthalide and the anion interconvert. This process is catalysed in the ring-opening direction by the addition of water and general bases [R. A. McClelland and P. E. Soerengen, *Can. J. Chem.,* 1986, **64**, 1196].

$$(\mathbf{114}) \rightleftharpoons 2\text{-HOOC}_6\text{H}_4\text{CO}_2^- \rightleftharpoons 2\text{-HOOC}_6\text{H}_4\text{CO}_2\text{H}$$

(**114**) = 3-hydroxyphthalide

The mechanism of the alkaline hydrolysis of a wide range of phthalides of the type (**115**) and of analogous naphthalides and phenanthralides, involves attack by the hydroxy anion at the carbonyl group, followed by rapid ring fission to form the carboxylate anion of the corresponding acid as the product [F. Anvia *et al.*, *J. Chem. Soc., Perkin Trans 2*, 1990, 1809].

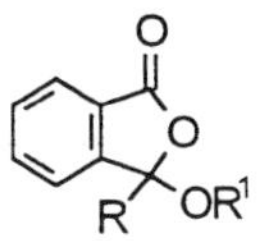

R = H, Me, Ph
R^1 = Me, Ph, 3-substituted phenyl

(**115**)

O-Ethylation of the 3-methyl- or 3-phenyl-phthalides by diethoxycarbenium tetrafluoroborate, followed by borohydride reduction gives mixtures of the corresponding *cis*- and *trans*-1,3-dihydroisofurans [D. Tobia and B. Rickborn, *J. Org. Chem.*, 1986, **51**, 3849]. Reaction of phthalide (**116**) (R = R^1 = H, R^2 = H, NO_2; R = Me, Ph, R^1 = R^2 = H; R^1 = R^2 = H, R^1 = MeO, NO_2; R = H, R^1 = R^2 = MeO) with 3,4-$R^3{}_2C_6H_3CH_2CH_2NH_2$, (R = H, MeO) gives (**117**); the aminolysis is influenced by the substituents of (**116**) [B. Suedkamp and W. Meise, *Arch. Pharm.* (Weinheim, Ger.), 1985, **318**, 640]. The course of the reactions of the phthalides (**104**), (R = H, Ph; X = Cl, I, OH) with pyrrolidine, piperidine, morpholine, or imidazole to give the corresponding (**118**) (R = H, Ph; $R^1{}_2N$ = pyrrolidino, piperidino, morpholino, or imidazolyl) and/or (**119**) (R, $R^1{}_2N$ = same) depends on the amine nucleophilicity and the leaving group ability of X in (**104**). The better leaving group containing (**104**) also tends to favour the formation of (**119**) [K. B. Sloan and S. A. M. Koch, *J. Org. Chem.*, 1983, **48**, 635].

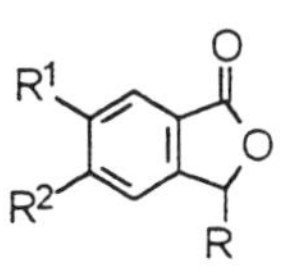

(**116**)

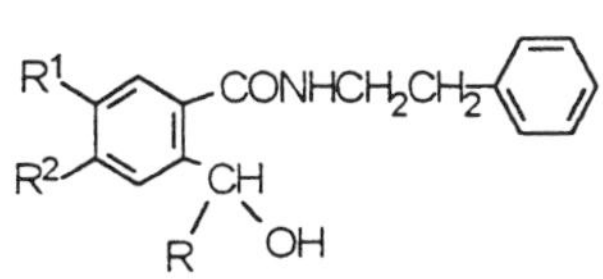

(**117**)

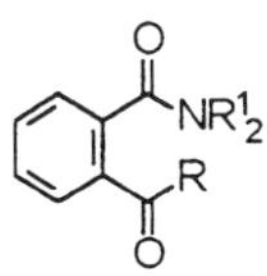

(**118**)

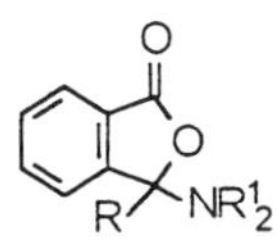

(**119**)

1,3-Dipolar cycloadditions of aromatic nitrile oxides to 3-methylenephthalide or to (*E*)-3-benzylidenephthalides (**120**) affords spiro heterocycles (**121**), which can be converted in high yield to 2-(3-aryl-5-isoxazolyl)benzoates (**122**) (X = H or alkyl) (Scheme 4). (*Z*)-3-Benzylidenephthalides and 3-ethylidenephthalides do not form analogous adducts probably due to steric effects [R. K. Howe *et al.*, *J. Org. Chem.*, 1983, **48**, 4590; 1985, **50**, 903].

Scheme 4

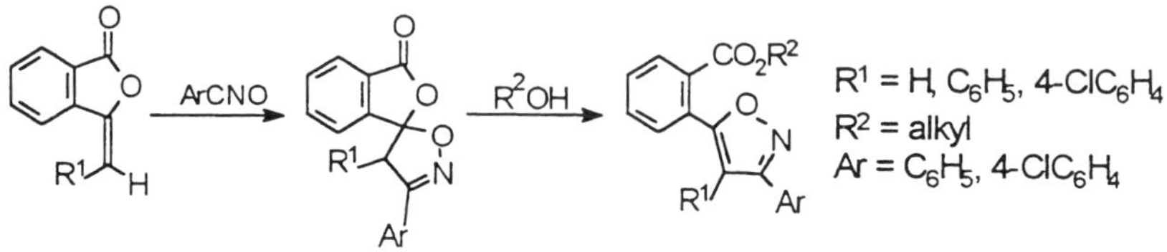

(120) (121) (122)

(b) Nitrogen and sulphur analogues of phthalides

Thionation of phthalaldehyde or naphthalene 2,3-dicarboxaldehyde with Lawesson's reagent provides an efficient route to the dithiaphthalide (**123**) and 2,3-dithianaphthalide (**124**) respectively [P. N. Nugara *et al.*, *Heterocycles*, 1991, **32**, 1559]. Reaction of 2-acetylbenzoic acid with $PhNH_2$, $3\text{-}F_3CC_6H_4NH_2$ and Bz_2NH_2 gives the corresponding phthalimidines (**125**) (R = Ph, $3\text{-}CF_3C_6H_4$, Bz). These react readily with aromatic nitrile oxides to give the spiroheterocycles (**126**) (Ar = *e.g.* Ph, $3\text{-}CF_3C_6H_4$) [R. K. Howe and B. R. Shelton, *J. Org. Chem.*, 1990, **55**, 4603]. Reductive decarboxylation of N-acyloxyphthalimides through a radical chain process is possible with visible light irradiation using $Ru(bpy)_3Cl_2$ as a sensitizer in the presence of BANH and t-BuSH [K. Okada *et al.*, *Tetrahedron Lett.*, 1992, **33**, 7377].

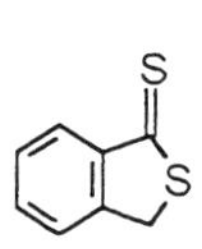

(123)

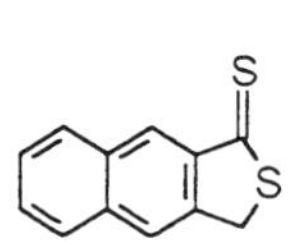

(124)

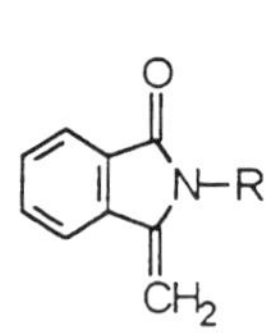

(125)

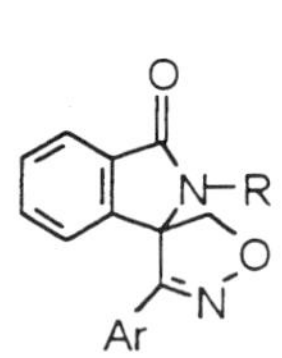

(126)

6. *Oxoaralkane carboxylic acids*

(a) Aldehydic Acids

Phthalaldehydic acids (**127**) with nitro, ethoxycarbonyl, chloro, bromo and methoxy substituents have been prepared by decarboxylation of the 2-carboxyphenyl pyvuric acids or by hydroxylation or ring-opening of phthalides *via* their bromo derivatives. Reaction of the phenylhydrazine derivatives of these phthalaldehydic acids affords 2-phenyl-1(2H)-phthalazinones [A. Sugimato *et al.*, *Chem. Pharm. Bull.*, 1985, **33**, 2809]. Reaction of phthalaldehydic acid with 2-aminopyridine and potassium cyanide affords pyrido[2',1':2,3]imidazo[4,5-c]isoquinolin-5(6*H*)-one [J. P. Paolini *et al.*, *J. Heterocycl. Chem.*, 1987, **24**, 549], whereas reaction with piperazine-2,5-dione gives a mixture of its analogue (**128**) and the acid (**129**) [M. L. Baron, I. D. Rae and P. M. Simmonds, *Aust. J. Chem.*, 1982, **35**, 2567]. Reaction of phthalaldehydic acid with 2,4,5-(HO)RR^{1}C$_6$H$_2$COMe (R = R^1 = H; R = OMe, R^1 = H; R = H, R^1 = OMe) affords 2,4,5-(HO)RR^{1}C$_6$H$_2$COCH:CHC$_6$H$_4$CO$_2$H-2 which rapidly isomerise to (**130**) by acid or heating and can be readily converted to 2'-carboxyflavones [N. Abe, CA, 1990, **112**, 7324y]. The corresponding reaction of 4-HOCC$_6$H$_4$CO$_2$H provides a route to 4'-carboxyflavones [K. Sato, G. Saitoh and N. Abe, CA, 1990, **112**, 7323x].

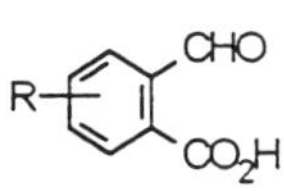

R = NO$_2$, Cl, Br, MeO, CO$_2$Et

(**127**)

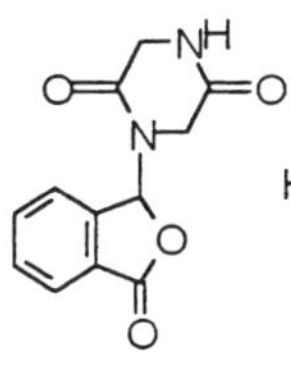

(**128**)

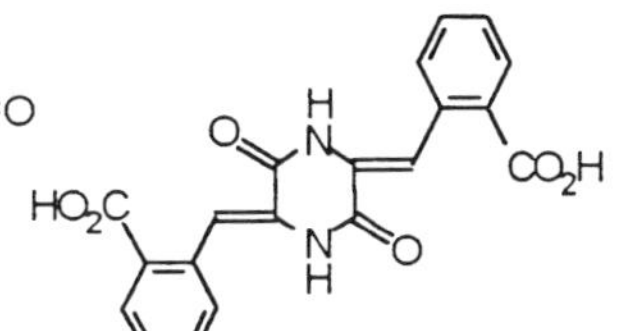

(**129**)

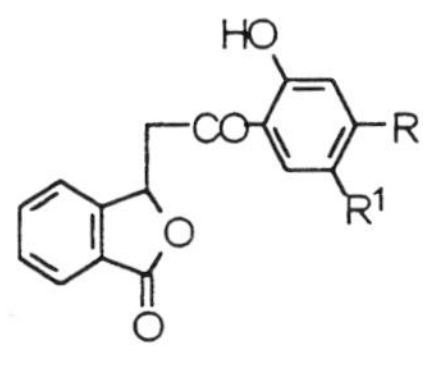

(**130**)

Reaction of phthaldehydic acid with anisole in the presence of polyphosphoric acid provides a convenient route to 1,3-di-*p*-anisylisobenzofurans [R. Al-Hamdany, M. T. Ayoub, R. Al-Bayatti, *J. Chem. Res. Synops.*, 1989, 820] whilst condensation of 1,4-(HOC)$_2$C$_6$H$_4$ with 5,6-diamino-1,3-dipropyluracil provides a facile route to the uracil (**131**) [J. Perumattam, *Synth. Commun.*, 1989, **19**, 3367]. The isocoumarins (**132**) (R = H, Br, Cl, MeO; R^1 = Ph, Pr, β-naphthyl) can be obtained by the condensation of phthalaldehydic acids with R^{1}COCH$_2$R^2 in the presence of 1,5-diazabicyclo[5.4.0]undec-5-ene [C. Bhakta, *J. Indian Chem., Sect. B*, 1985, **24B**, 428]. Cyclocondensation of the methyl esters of phthalaldehydic acid with RNH$_2$ and R^{1}SH in the presence of 4-MeC$_6$H$_4$SO$_3$H affords the isoindolinones (**133**) (*e.g.* R = aryl, benzyl, CH$_2$CO$_2$H; R^1 = CH$_2$CO$_2$H) [R.

Liebl and R. Handte, *Liebigs Ann. Chem.*, 1985, 1093].

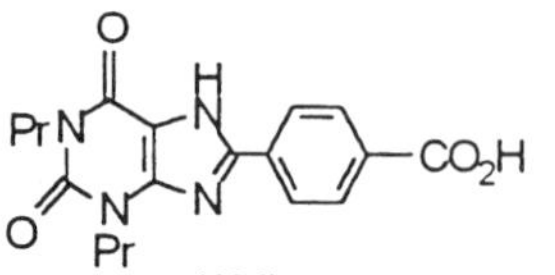

(131)

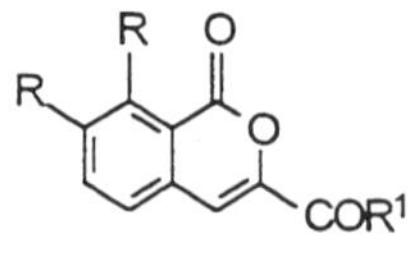

(132)

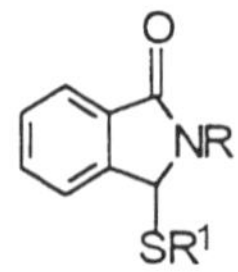

(133)

Selective acetalisation of the formyl group of 4-$HO_2CC_6H_4CHO$ can be achieved effectively using 1,2-ethanediol and alumina as catalyst [Y. Kamitori *et al.*, *Tetrahedron Lett.*, 1985, **26**, 4767]. Selective thioacetylisation is also possible with 1,3-propanedithiol and silica gel treated with thionyl chloride as catalyst. [Y. Kamitori *et al., J. Org. Chem.,* 1986, **51**, 1427].
A Reformatsky type reaction of 4-$RO_2CC_6H_4CHO$, (R = H, Me or Me_3C), with propargyl bromide and Zn-Cu couple provides a route to the acetylenic carbinols (**134**) (R^1 = OH), which can be converted to their iodo and bromo derivatives. 4-NCC_6H_4CHO behaves similarly [M. V. Papadopoulou, *Chem. Ber.,* 1989, **122**, 2017].

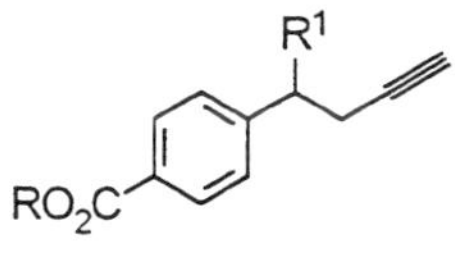

(134)

4-Formylbenzoic acid, 4-$HO_2CC_6H_4CHO$, is readily oxidised by sodium perborate to terephthalic acid [A. McKillop and D. Kemp, *Tetrahedron*, 1989, **45**, 3299] but undergoes ring nitration without oxidation of the formyl group on treatment with dinitrogen pentaoxide or nitronium trifluoromethane sulphonate in nitric acid [R. B. Moody and R. Willmer, *J. Chem. Soc., Perkin Trans* 2, 1992, 229]. Reduction of the acid with samarium iodide/base in the presence of protic solvents is rapid and leads to 4-toluic acid as the major product [Y. Kamochi and T. Kudo, *Bull. Chem. Soc. Jpn.*, 1992, **65**, 3049]. Reduction of 4-$RO_2CC_6H_4CHO$, (R = H or Me) using 1,1,3,3-tetramethyldisiloxane in combination with $XSiMe_3$, (X = Cl, Br) leads to 4-$RO_2CC_6H_4CH_2OSiMe_3$ in excellent yields [J. M. Aizpurua, B. Lecea and C. Palomo, *Can. J. Chem.*, 1986, **64**, 2342].
For the use of phthaldehydic acids in the synthesis of phthalides see section 5a. For the infrared and Raman spectra of 4-$HO_2CC_6H_4CHO$ see P. Venkoji [*Spectrochim. Acta*, Part A, 1986, **42A**, 1307] and for a NMR study of proton (deuteron) transfer of hydrogen bonds in 4-$HO_2CC_6H_4CHO$ see T. Agaki *et al. [Chem. Phys. Lett.*, 1987, **139**, 331].

(b) *Acylarene-carboxylic acids*

4-Acetylbenzoic acid, 4-$MeCOC_6H_4CO_2H$, is readily obtained in good yield by the palladium catalysed carbonylation of 4-$MeCOC_6H_4N_2BF_4$ with carbon monoxide and silyl hydrides or polymethylhydrosiloxane at room temperature [K. Kiyoshi *et al.*, *J. Organomet. Chem.*, 1984, **270**, 283]. 4-Acetylphenol, as its triflate, can be converted by potassium cyanide in the presence of catalytic $Ni(PPh_3)_4$ to 4-$MeCOC_6H_4CN$ [M. R. J. Chambers and D. A. Widdowson, *J. Chem. Soc., Perkin Trans 1*, 1989, 13657]. Reactions of phthalic anhydride with aryl coppermagnesium reagents afford 2-aroylbenzoic acids (aryl = 4-tolyl, 4-$MeOC_6H_4$, 4-ClC_6H_4) in high yield [M. T. Rahman and S. K. Nahar, *J. Organomet. Chem.*, 1992, **425**, 201]. Phenyl 2-acylbenzoates (**135**) can be prepared by coupling the respective acid and phenol with N,N-dicyclohexylcarbodiimide. Alkaline hydrolysis of these esters is first order both in ester and hydroxide anion concentration [F. Anvia and K. Bowden, *J. Chem. Soc., Perkin Trans 2*, 1990, 1805].

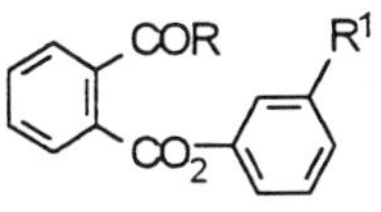

R = Me; R^1 = H, Me, Br, NO_2

R = Ph; R^1 = H, Me, Cl, NO_2

(135)

Reactions of 2-acylbenzoic acids sodium azide, diphenylphosphoryl azide and trimethylsilyl azide leads mainly to the azido derivatives (**136**) of the ring tautomer of the acids. Similar behaviour is shown by the 2-acylbenzamides. Thermal decompositions of (**136**) occur at low temperature to give one or two of the rearrangement products (**137**) - (**139**) depending on the substituents and stereoelectronic factors [H. Takeuchi and S. Eguchi, *J. Chem. Soc., Perkin Trans 2,* 1988, 2149]. Photochemical excitation of 4-$MeCOC_6H_4CON_3$ with UV light leads to nitrene formation. Thus, in tertiary butyl alcohol it gives the hydroxamate 4-$MeCOC_6H_4CONHOCMe_3$ whilst in benzene solution containing *cis*- or *trans*-4-methyl-2-pentene aziridines are formed with total retention of configuration [M. E. Sigman, T. Autrey and G. B. Schuster, *J. Am. Chem. Soc.*, 1988, **110**, 4297].

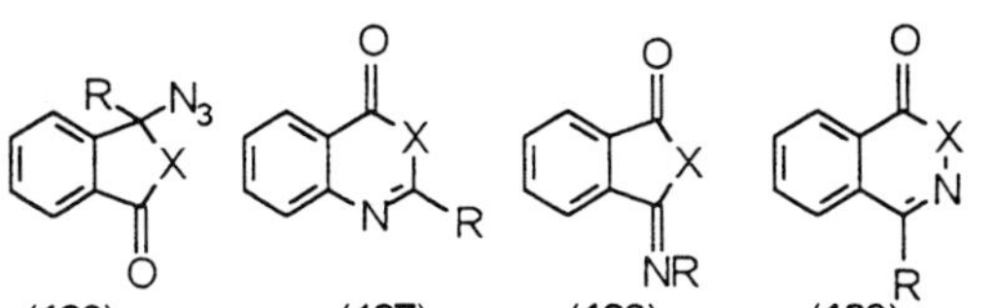

(136) (137) (138) (139)

X = O, NMe; R = H, Me, Ph

X = NPh; R = Me, Ph

Reaction of 2-$MeCOC_6H_4CO_2H$ with α-amino acids affords oxazolidinones (**140**) (R = Me, CH_2Ph, $CHMe_2$, CHMeEt) [M. K. Mao and R. K. Webber, *J. Chem. Soc., Chem. Commun.*, 1990, 679] and cyclocondensation of 2-$BrCH_2COC_6H_4CO_2Et$ with 2-NCC_6H_4XH (X = O, NR; R = 2-$COC_6H_4NO_2$, Bz) in the presence of sodium hydride gives azepinones (**141**) [G. Viti *et al.*, *J. Heterocycl. Chem.*, 1991, **28**, 379]

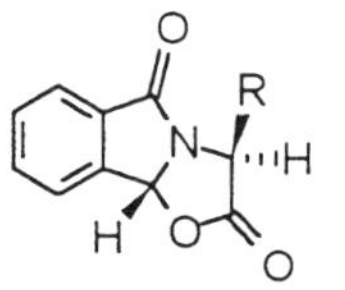

(**140**)

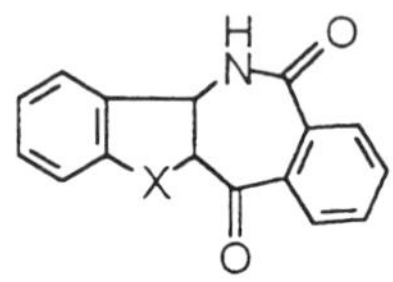

(**141**)

7. *Arene-di, -tri and -poly -carboxylic acids*

(a) Phthalic acid and its derivatives

Phthalic acid (benzene-1,2-dicarboxylic acid) (**142**) (R = H) is obtained in high yield by the oxidation of 1-naphthol, 1,2- or 1,4- naphthoquinone, β-tetralone and also of related compounds with potassium superoxide in DMF in the presence of 18-crown-6 ether. Naphthalene-, anthracene- and phenanthrenedicarboxylic acids can be prepared similarly [C. Soteriou, W. Lee and R. W. Giese, *J. Org. Chem.*, 1990, **55**, 2159]. Phthalic acid has also been obtained by oxidation of 1,2-$(HOC)_2C_6H_4$ with sodium perborate [A. McKillop and D. Kemp, *Tetrahedron*, 1989, **45**, 3299] and by palladium catalysed carbonylation of 2-iodobenzoic acid [N. A. Bumagin, K. V. Nikitin and I. P. Beletskaya, J. *Organomet. Chem.,* 1988, **358**, 563]. Phthalic acid can be converted to its dimethyl ester in high yield by treatment with methanol in the presence of $CuCl_2$ [T. L. Ho, *Synth. Commun.*, 1989, **19**, 2897]. The mechanism of oxidation of phthalic acid to oxalic acid and CO_2 has been investigated [S. Ichinose and A. Okuwaki, *Bull. Chem. Soc. Jpn.*, 1990, **63**, 159]. Carboxylic acids, including phthalic acid, are readily transformed to the corresponding aldehydes by lithium tris(dialkylamino)aluminium hydrides [J. S. Cha, *Bull. Korean Chem. Soc.*, 1992, **13**, 670]. Electrochemical reduction of phthalic acids (**142**) (*e.g.* R = H, alkyl) in the presence of aqueous acids affords 1,2-dihydrophthalic acids [I. Nishiguchi *et al.*, CA, 1993, **118**, 147307u]. The isomeric acid chlorides obtained by treating the half methyl esters of 3-methoxyphthalic acids (**143**) with thionyl chloride undergo facile interconversion [D. N. Gupta, P. Hodge and P. N. Hurley, *J. Chem. Soc., Perkin Trans 1*, 1989, 391].

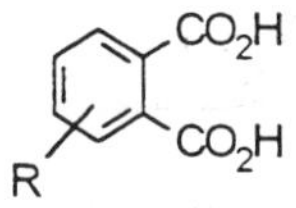

(142)

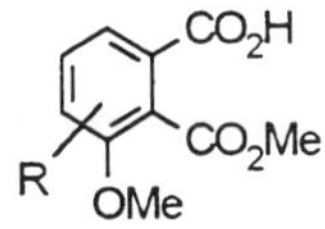

(143)

For the use of phthalic anhydride in the synthesis of phthalides and 2-aroylbenzoic acids see Sections 5a and 6 respectively. The vapour phase chlorination of phthalic anhydride proceeds with a high degree of regio selectivity yielding the 3- and 4- chlorophthalic anhydrides in 5:11 molar ratio and only small amounts of dichlorinated products. Similarly chlorination of 3-chlorophthalic anhydride yields 3,5-dichlorophthalic anhydride [J. W. Verbicky, Jr and L. Williams, *J. Org. Chem.*, 1893, **48,** 2465]. Treatment of tetrachlorophthalic anhydride with zinc and aqueous sodium hydroxide leads to nearly exclusive formation of 3,4,6-trichlorophthalic acid, which on further dehydrochlorination with 2M NaOH affords 3,6 - dichloro - and 3-chlorophthalic acids [L. B. Fertel, N. J. O'Reilly and K. M. Callaghan, *J. Org. Chem.*, 1993, **58**, 261].

(b) Isophthalic acid and its derivatives

Isophthalic acid (benzene-1,3-dicarboxylic acid) can be prepared in excellent yield from 1,3-$(HOC)_2C_6H_4$ by oxidation with sodium perborate in acetic acid [A. McKillop and D. Kemp, *Tetrahedron,* 1989, **45**, 3299]. Isophthalonitrile is obtained by the nickel catalysed oxidative dehydrogenation of 1,3-$(H_2NCH_2)_2C_6H_4$ with $K_2S_2O_8$ [S. Yamazaki and Y. Yamazaki, *Bull. Chem. Soc. Jpn.*, 1990, **63**, 301] and can be quantitatively hydrolysed to the diamide with basic hydrogen peroxide in DMSO (A. R. Katrinzky, B. Pilarsky and L. Urogdi, *Synthesis,* 1989, 949].

(c) Terephthalic acid and its derivatives

Terephthalic acid (benzene-1,4-dicarboxylic acid) has been obtained in high yield by the oxidation of 4-RC_6H_4CHO (R=CHO, CO_2H or CN) with sodium perborate in acetic acid [A. McKillop and D. Kemp, *Tetrahedron*, 1989, 3299]. Terephthalamide is obtained by hydration of terephthalonitrile in NH_3/H_2O under optimal conditions [Y. G. Abdullaev *et al.*, *Zh. Prikl. Khim. (Leningrad)*, 1989, **62**, 1401] and by the action of sodium percarbonate [G.

W. Kabalka, S. M. Deshpande, P. P. Wadgaonkar and N. Chatla, *Synth. Commun.*, 1990, **20**, 1445] or of sodium perborate in aqueous methanol [A. McKillop and D. Kemp, *Tetrahedron,* 1989, **45**, 3299] on the nitrile.
Terephthalonitrile has been prepared by the oxidation of the dimethylhydrazone of terepthalaldehyde with 3-chlorobenzoic acid [S. B. Said, J. Skarzewski and J. Mlochowski, *Synthesis*, 1989, 223].

(d) Di and poly-carboxylic acids with carboxyl groups in separate side chains

1,2-Benzenediacetic acid has been obtained together with 4-$MeC_6H_4CH_2CO_2H$ by treating 1,2-$(XCH_2)_2C_6H_4$ (X=Cl or Br) with CO, $Fe(CO)_5$ and a phase transfer catalyst [C. Shim *et al.*, *Bull. Korean Chem. Soc.*, 1988, **9**, 61]. 1,3-Benzodioxole-5,6-diacetic acid has been obtained from the corresponding 5,6-dichloromethyl compound *via* the dinitrile [D. H. Kim, *J. Heterocyclic Chem.*, 1992, **29**, 11]. Tetracarboxylic acids 1,3-$R_2C_6H_4$ (R = $CH(CO_2H)CH_2CO_2H$, $CH(CH_2CO_2H)_2$) are obtained from 1,3-$(HOC)_2C_6H_4$ by the reaction with *e.g.* ethyl cyanoacetate [T. Teshirogi, CA, 1988, **108**, 76054q].
Perkin condensation of 1,4-benzenediacetic acid with $PhCOCO_2K$ gives phenylene bis(maleic anhydride) [E. K. Fields *et al.*, *J. Org. Chem.,* 1990, **55**, 5165]. Reduction of diethyl terephthalates affords 1,4-$(HOCH_2)_2C_6H_4$ [C. Campa, F. Sanchez-Ferrandon and M. Tristan-Polo, *Nouv. J. Chim.*, 1985, **9**, 493]

(e) Phthalimides and related compounds

A wide range of N-substituted phthalimides has been prepared by the palladium catalysed carbonylation and coupling of 1,2-dihaloaromatics (bromides and iodides) with primary amines (*e.g.* $PhNH_2$, 4-$NCC_6H_4NH_2$, $C_8H_{17}NH_2$) under pressure. [R. J. Perry and S. R. Turner, *J. Org. Chem.*, 1991, **56**, 6573]. The synthesis of N-glycinylphthalimide by the reaction of phthalic anhydride and glycine can be achieved efficiently by microwave irradiation [S. S. Bari *et al.*, *J. Chem. Educ.*, 1992, **69**, 939]. N-Allylphthalimide is obtained in good yield *via* a Mitsunobu reaction from phthalimide allyl alcohol, thiaphenyl phosphine and diethyl azidocarboxylate. [G. J. Hitchings, M. Helliwell and J. M. Vernon, *J. Chem. Soc. Perkin Trans 1*, 1990, 83] and N-acyloxyphthalimides can be readily prepared from N-hydroxyphthalimide and carboxylic acids [T. W. Green in *Protective Groups in Organic Synthesis*, John Wiley, New York, 1981, p.183].
Reaction of 3-fluoro-N-methylphthalimide with pyrrolidine, piperidine and morpholine in refluxing methanol affords the corresponding 3-amino-N-methylphthalimide in good yield. The kinetics of the reaction between 3-

fluoro-N-methylphthalimide and the secondary amines in methanol and in aprotic solvents have been investigated [L. R. Caswell and M. E. Goldsmith, *J. Org. Chem.*, 1989, **54**, 5101].
Phthalimide hydroperoxides such as (**144a**) have been prepared by singlet oxygenation of (**144b**) [I. Sato, M. Takayama and T. Matsuura, *J. Am. Chem. Soc.*, 1990, **112**, 883], whereas hydroxyperoxides such as (**144c**) by the CF_3SO_3H catalysed action of H_2O_2 on the corresponding dimethyl ketal [A. G. Myers, M. M. Fundy and P. A. Lindstrom, Jr., *Tetrahedron Lett.*, 1988, **29**, 5609]. Such and other phthalimide hydroperoxides are efficient photochemical hydroxyl radical generators and effective DNA cleaving agents [I. Sato, M. Takayama and T. Matsuura, *J. Am. Chem. Soc.*, 1990, **112**, 883].

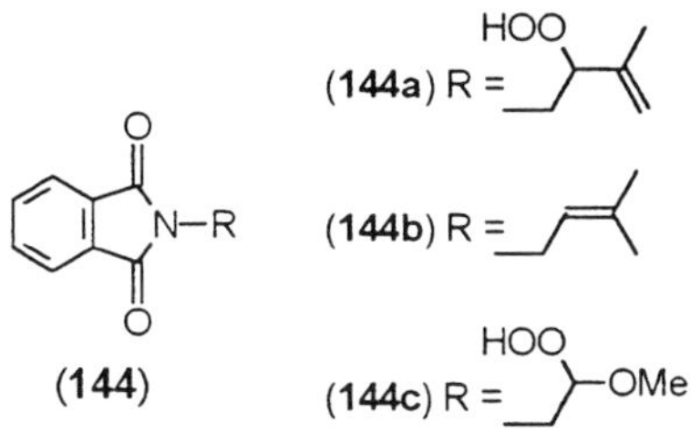

(**144**)

Irradiation of N-acyloxyphthalimides in the presence of 1,4-diazabicyclo[2.2.2]octane provides a route for the radical mediated decarboxylation of carboxylic acids [K. Okada, K. Okamoto and M. Oda, *J. Chem. Soc., Chem. Commun.*, 1989, 1636; *J. Am. Chem. Soc.*, 1988, **110**, 8736]. Such phthalimides can also undergo photosensitized decarboxylative Michael addition to electron-deficient alkenes [K. Okada *et al.*, *J. Am. Chem. Soc.*, 1991, **113**, 9401].

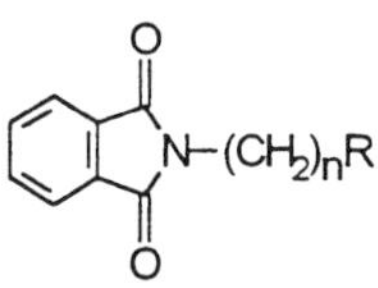

(**145**)

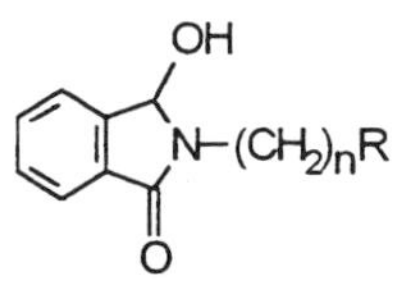

(**146**)

Reduction or Grignard addition to the N-substituted phthalimides (**145**) affords the α-hydroxylactams (**146**) [G. J. Hitching, *J. Chem. Soc. Perkin Trans 1*, 1990, 83]. For a mass spectrometric study of the deamination reactions of phthalamides see E. V. Ganin, B. V. Rozynov and V. I. Nitikin [*Ukr. Khim. Zh.*, 1989, **55**, 302].

Second Supplements to the 2nd Edition of Rodd's Chemistry of Carbon Compounds, Vol.III D/E/F (Partial), edited by M. Sainsbury

Chapter 18

DIHYDRIC AND POLYHYDRIC ARALKANOLS AND THEIR OXIDATION PRODUCTS

Hak-Fun CHOW

1. Aralkanediols and aminoaralkanols

(a) Aralkanediols and their derivatives

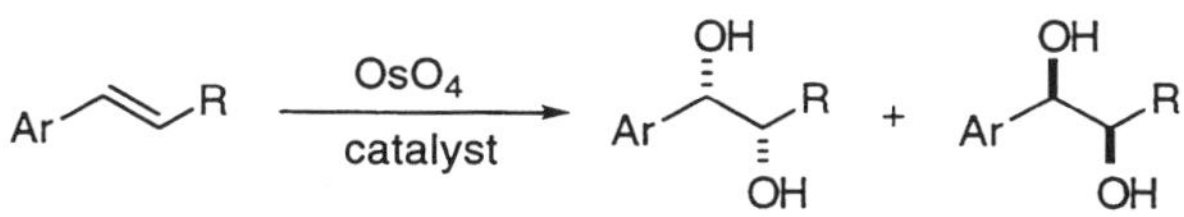

The past decade witnessed a widespread interest in the enantioselective functionalization of substituted olefins. In particular, aralkane-*threo*-1,2-diols can be synthesized in good to excellent optical purity *via* a chiral amine mediated catalytic osmium tetraoxide dihydroxylation of phenyl substituted olefins. For example, Corey (E. J. Corey *et al.*, J. Am. Chem. Soc., 1989, **111**, 9243) used the chiral diamine **1** as a ligand to generate optically active aralkane-1,2-diols in good yields (81 - 95%) and excellent enantioselectivity (92 - 93% e.e.). Tomioka employed the C_2 symmetrical diamine **2** to achieve reaction enantioselectivity of up to 90 - 99% e.e. (K. Tomioka, M. Nakajima and K. Koga, J. Am. Chem. Soc., 1987, **109**, 6123). Sharpless utilized the naturally occurring alkaloid derivatives **3** and **4** to give 1,2-diols with moderate enantioselectivity (33 - 88% e.e.) (S. G. Hentges and K. B. Sharpless, J. Am. Chem. Soc., 1980, **102**, 4263; E. N. Jacobsen *et al.*, J. Am. Chem. Soc., 1988, **110**, 1968; K. B. Sharpless *et al.*, J. Org. Chem., 1991, **56**, 4585). Later, the phthalazine ligand **5** was synthesized and this discovery leads to the development of an excellent enantioselective dihydroxylation procedure which is applicable to a wide range of olefinic substrates (K. B. Sharpless *et al.*, J. Org. Chem., 1992, **57**, 2768). Jacobsen, on the other hand, used the (salen)manganese complexes **6** and iodosylmesitylene to achieve moderate reaction enantioselectivity (W. Zhang *et al.*, J. Am. Chem. Soc., 1990, **112**, 2801).

Interestingly, the globular protein bovine serum albumin, a protein known to have strong affinity for a wide range of metals, had also been used as an enantioselective catalyst in the osmium tetraoxide dihydroxylation reactions of phenyl substituted alkenes with moderate success (6 - 68% e.e.) (T. Kokubo *et al.*, J. Chem. Soc., Chem. Commun., 1983, 769).

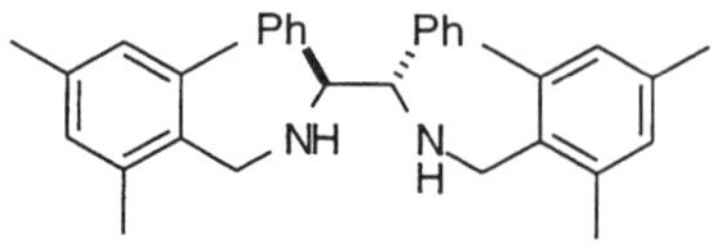
1

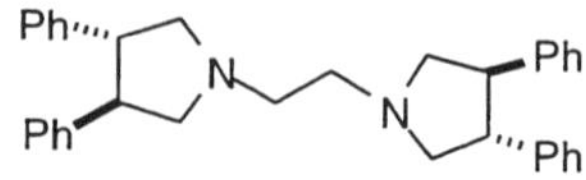
2

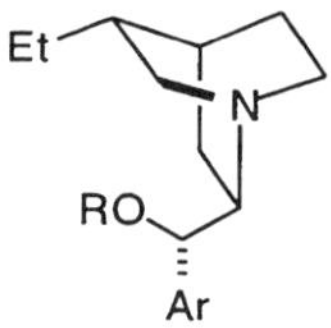
3

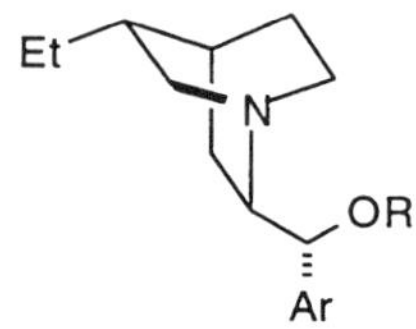
4

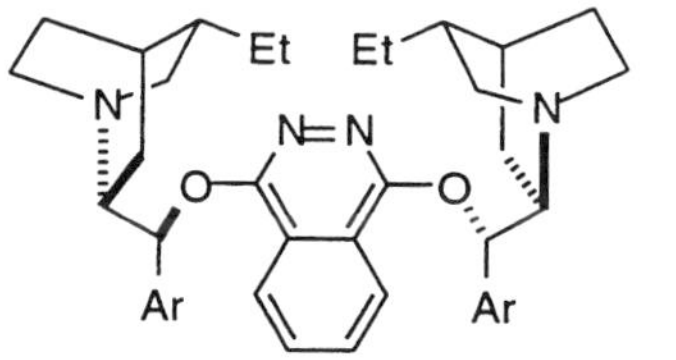
5

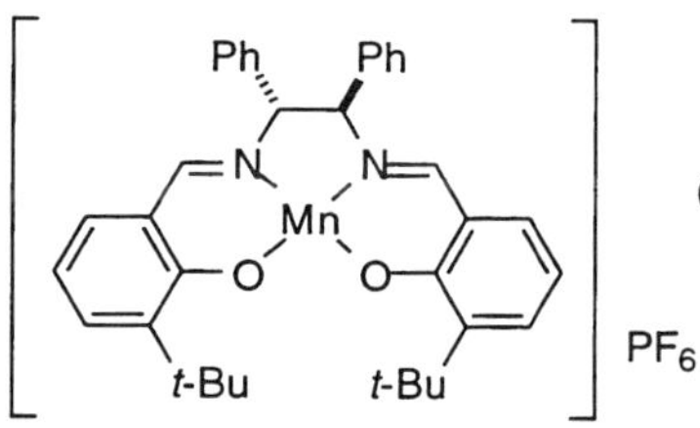
6

Reduction of aromatic α-hydroxy ketones **7** with zinc borohydride gives preferentially the *erythro*-1,2-diols **8** (*erythro* : *threo* = 3 : 1). However, a reversal of stereoselectivity can be achieved in the reduction of the protected aromatic α-(*tert*-butyldiphenylsilyloxy) ketones **9** with sodium

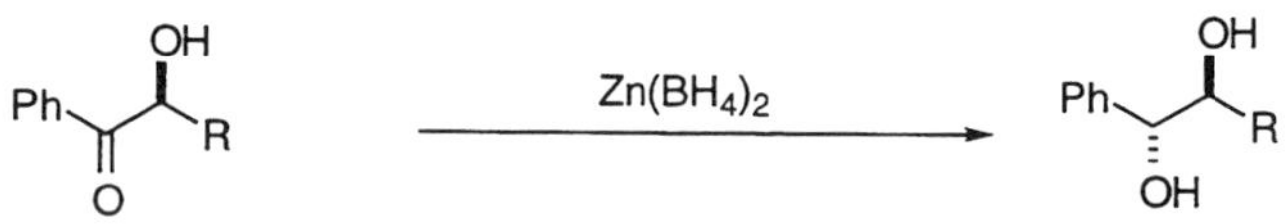

7 8

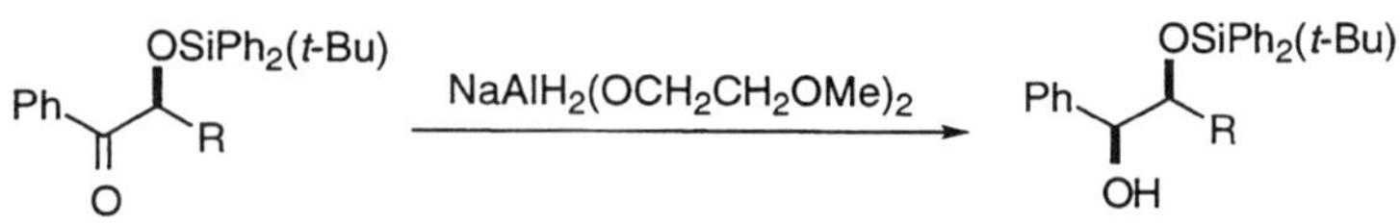

9 10

bis(2-methoxyethoxy)aluminum hydride. In this case the *threo*-isomers **10** are isolated as the major product (*erythro* : *threo* = 2 : 3) (T. Nakata, T. Tanaka and T. Oishi, Tetrahedron Lett., 1983, **24**, 2653).

The phenyl substituted *erythro*-1,2-diol **11** can be prepared by using organosilicon chemistry (K. Tamao, E. Nakajo and Y. Ito, J. Org. Chem., 1987, **52**, 957). Deprotonation of allylsilane **12** with butyllithium followed by metal exchange with zinc chloride gives the allylzinc derivative **13**. Reaction of the organozinc intermediate **13** with benzaldehyde followed by oxidative cleavage of the carbon-silicon bond then affords the diol **11** in 80% yield.

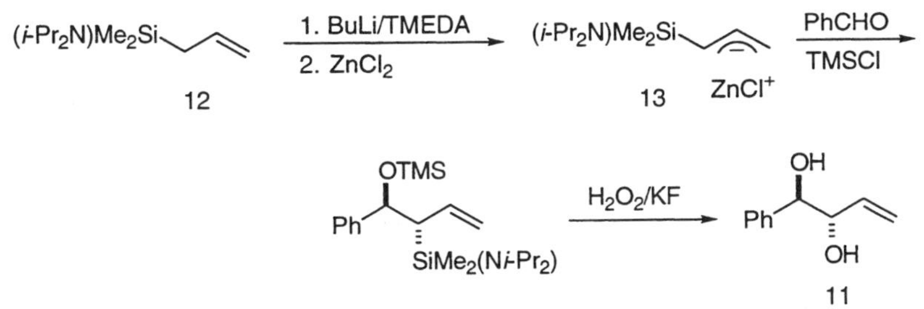

Treatment of benzylic *N,N*-dialkylcarbamates **14** with butyllithium in the presence of TMEDA followed by quenching of the anion with aldehydes or ketones gives aralkane-1,2-diol derivatives **15** in 82 - 93% yields. Hydrolysis of the carbamates **15** then affords the aralkane-1,2-diols (D. Hoppe and A. Brönneke, Synthesis, 1982, 1045).

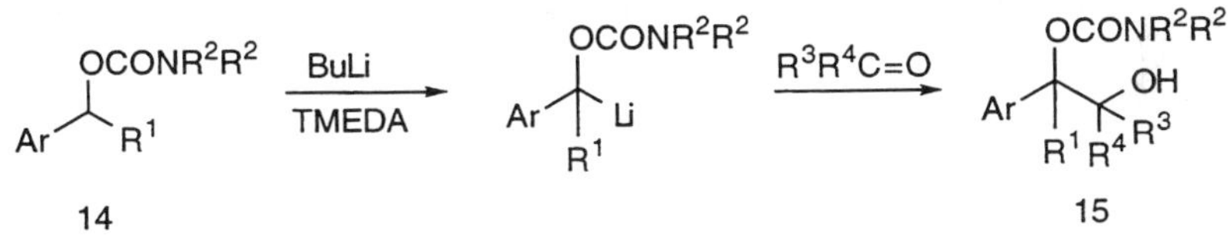

The formation of hydroxyacetates in the reaction of styrene with lead (IV), cobalt (III) and manganese (III) salts in acetic acid has been demonstrated to involve the intermediacy of alkene radical-cations (J. A. Norman, C. B. Thomas and M. J. Burrow, J. Chem. Soc., Perkin Trans. 1, 1985, 1087).

Optically active aromatic *syn*-1,3-diols **16** can be synthesized in one step from the asymmetric aldol reaction of aldehydes with the silyl enol ether of acetophenone **17** followed by the asymmetric borane reduction. Both reactions are catalyzed by the chiral borane **18** (Y. Kaneko, T. Matsuo and S.-i. Kiyooka, Tetrahedron Lett., 1994, **35**, 4107).

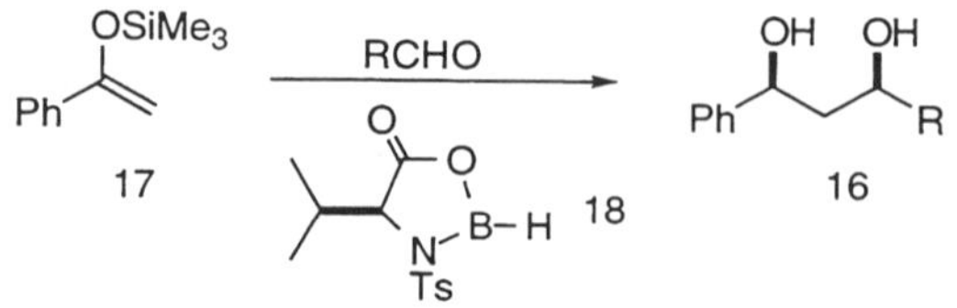

(b) Aminoaralkanols and their derivatives

Homochiral aminoalcohols and diamines have played an important role in asymmetric organic reactions and bioorganic synthesis. They also possess interesting biological activity and are important compounds in the pharmaceutical industry. (B. G. Main and H. Tucker, in Chapter 10 - Beta Blockers, Medicinal Chemistry, 2nd ed., ed. C. R. Ganelin and S. M. Roberts, Academic Press, London, 1993, p. 187). An efficient synthetic route to the optically active aryl β-aminoalcohols **19** and aryl α,β-diamines **20** has recently been reported. The aryl substituted diols **21**, available in both enantiomerically pure forms through the asymmetric dihydroxylation of aryl substituted olefins, can be converted into the corresponding cyclic sulfates **22** in good yields. These reactive cyclic sulfates **22** are readily opened by a variety of nucleophiles to give optically active aminoethanols **19** in good yields (B. B. Lohray, Y. Gao and K. B. Sharpless, Tetrahedron Lett., 1989, **30**, 2623). In a similar approach, the cyclic sulfite of the diol can also be synthesized and converted into optically active β-aminoethanols **19** and α,β-diamines **20** (B. B. Lohray and J. R. Ahuja, J. Chem. Soc., Chem. Commun., 1991, 95).

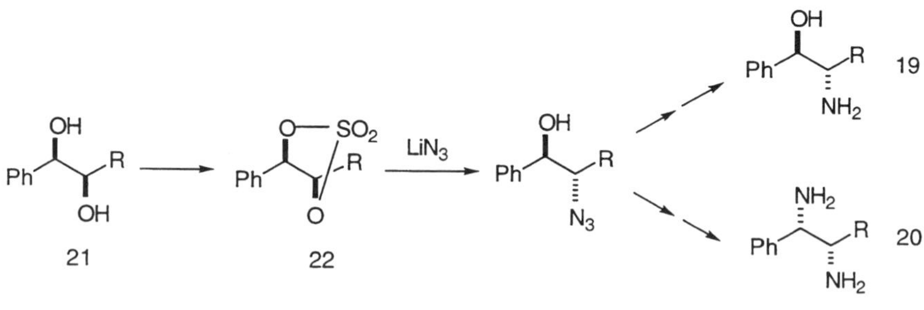

An alternative synthesis of optically active aryl β-aminoalcohols involves a dipeptide - (*S,S*)-cyclo(phenylalanylhistidyl) - catalyzed enantioselective cyanohydrin formation of 4-anisaldehyde (R. F. C. Brown, W. R. Jackson and T. D. McCarthy, Tetrahedron: Asymmetry, 1993, **4**, 205). The optical induction of this reaction is >99%.

Solladié-Cavallo recently reported a method for the synthesis of racemate aryl β-aminoalcohols **23**. Reaction of *N*-methyl-*N*-substituted formamidine **24** with *tert*-butyl lithium gives the carbanion **25**, which reacts with aromatic aldehydes to give aryl substituted aminoalcohols **23** in 60 - 95% yield (A. Solladié-Cavallo and M. Bencheqroun, Tetrahedron Lett., 1990, **31**, 2157).

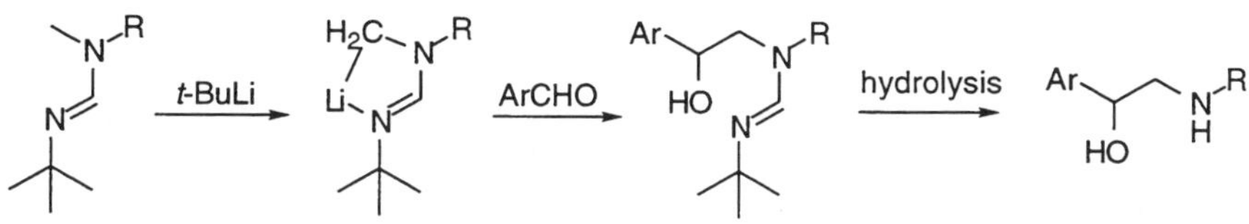

24 25 23

Aryl substituted β-aminoalcohols are important neurotransmitters or β-blockers. As a result, there are numerous literature reporting on the total synthesis of these drug molecules. The adrenergic drug, (*R*)-phenylephrine hydrochloride **26** can be obtained in 85% e.e. through an asymmetric hydrogenation reaction catalyzed by the hydroxyproline derived diphosphine **27** - rhodium complex (H. Takeda *et al.,* Tetrahedron Lett., 1989, **30**, 367). Optically active ephedrine analogs **28** can be prepared in 80% yield *via* a diastereoselective trimethylsilyl cyanide addition to the aldehyde moiety of the chiral arenechromium tricarbonyl complexes **29**. Subsequent Grignard

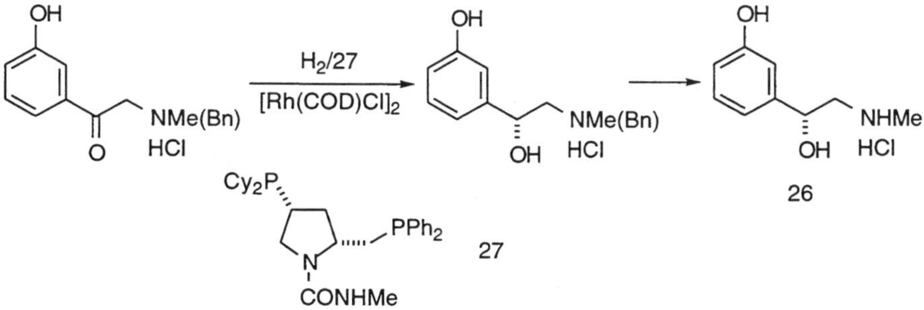

26

27

addition and hydride reduction to the nitrile functionality then furnishes the β-aminoalcohols **28** (A. Solladie-Cavallo *et al.*, Chem. Lett., 1987, 1583).

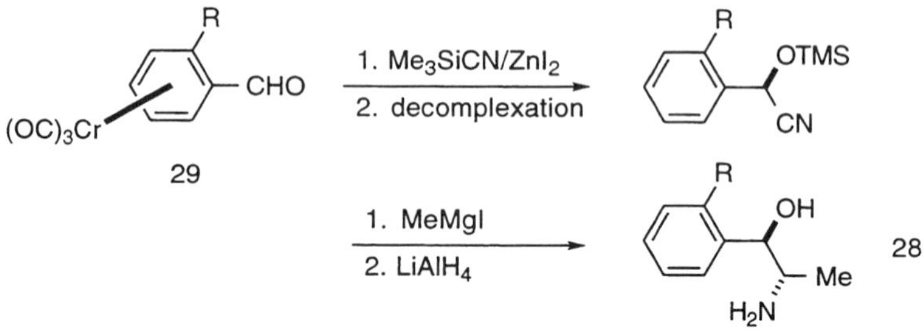

An extensive investigation on derivatizing the β-adrenergic drug isoproterenol **30** had been carried out in order to search for analogs with more potent and useful pharmacological activity (A. B. Reitz *et al.*, J. Med. Chem., 1985, **28**, 634 and references cited therein). A benzyl carbamate derivative **31** (n = 5, R^1 = Me, R^2 = $NHCO_2Bn$) containing a branched, seven-carbon spacer group was found to be 40 times more potent than isoproterenol in the *in vitro* assay.

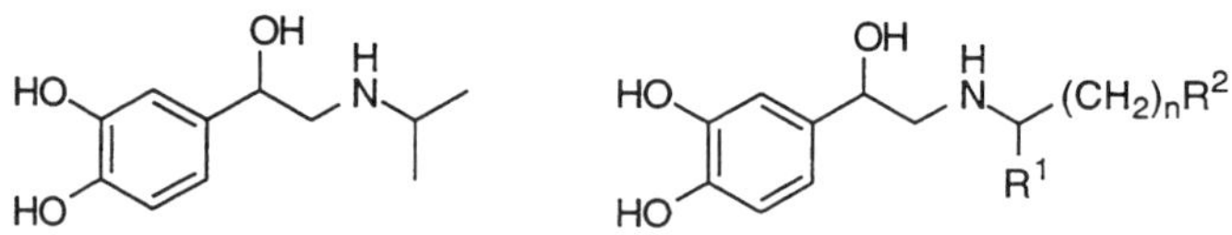

30 31

A facile one-pot preparation of optically active *N*-substituted phenylglycinols and 2-phenyloxazolidinones from phenylglycine has been reported (A. Correa, J. N. Denis and A. E. Greene, Synth. Commun., 1991, **21**, 1). Chiral oxazolidinones are important and useful chiral auxiliaries for asymmetric aldol condensations (D. A. Evans, J. Bartroli and T. L. Shih, J. Am. Chem. Soc., 1981, **103**, 2127).

Optically active aryl β-aminoalcohols are very useful in asymmetric synthesis. For example, (*R*)-phenylglycinol **32** can be used as a chiral auxiliary for the synthesis of optically active aliphatic and aromatic (*S*)-α-amino acids **33** (T. Inaba *et al.*, Bull. Chem. Soc. Jpn., 1992, **65**, 2359). The previous asymmetric Strecker reaction employing (*S*)-α-methylbenzylamine gives only aliphatic amino acids with lower optical purities. This modified procedure relies on the equilibration of the diastereomeric α-amino carbonitriles **34a** and **34b** (4 : 1) in refluxing methanol at 80°C. (*R*)-phenylglycinol has also been utilized to synthesize homochiral proline

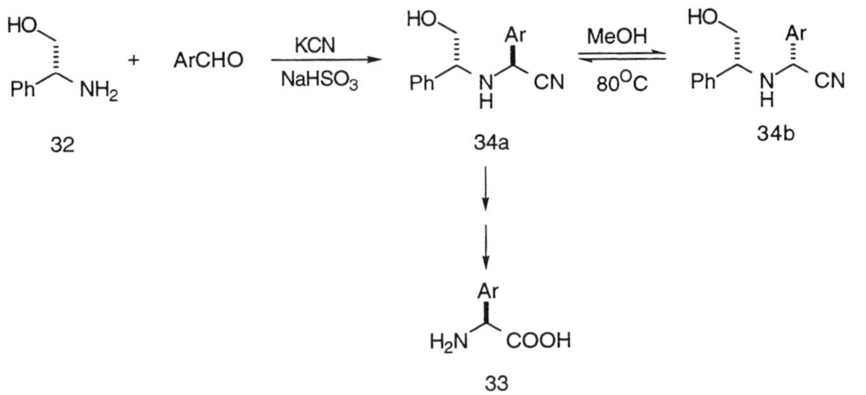

derivatives **35** and **36** (C. Agami *et al.*, Tetrahedron, 1993, **49**, 7239). The key reaction step is an aza-Cope rearrangement of an ene-iminium intermediate **37** generated from the reaction between the *N*-substituted phenylglycinol and glyoxal.

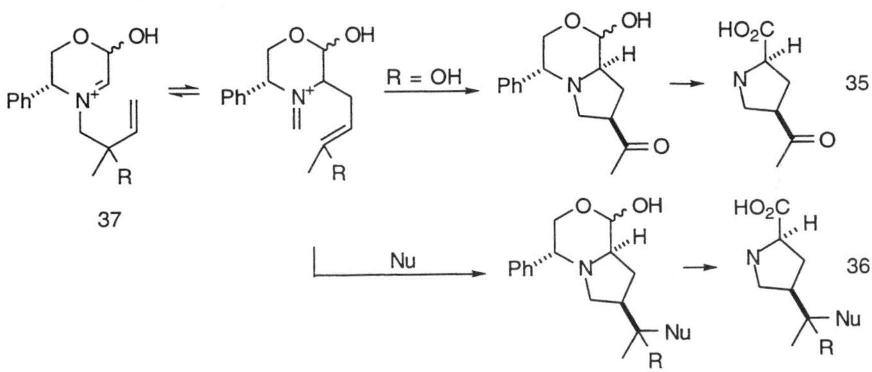

The antipode (*S*)-phenylglycinol has been used as a non-recoverable chiral auxiliary for the synthesis of optically active primary amines **38** with good optical purity (80 - 94% e.e.) (Z.-Y. Chang and R. M. Coates, J. Org. Chem., 1990, **55**, 3475). In this case the key process is the addition of a Grignard reagent to either *C*-aryl- or *C*-alkyl-*N*-(α-phenyl-β-(benzyloxy)-

ethyl)nitrones **39**. The only drawback of this synthetic route is that the phenylglycinol has to be oxidatively cleaved to release the desired product **38**.

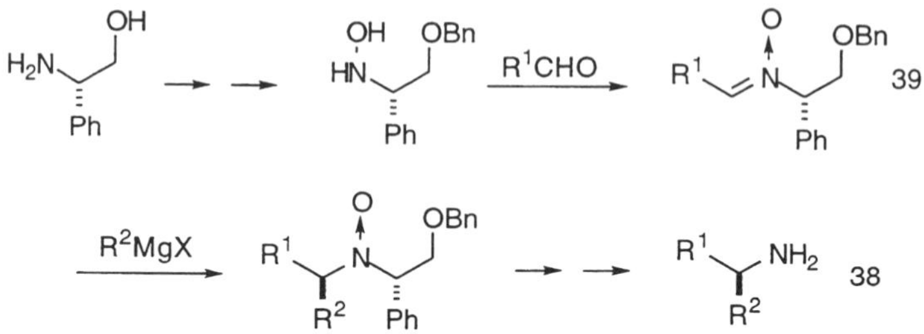

N-Methyl-(*R*)-phenylglycinol has been employed as a chiral auxiliary for the synthesis of both (*R*)- and (*S*)-*N*-methyl-α-aminoacid derivatives (C. Agami *et al.*, Tetrahedron, 1991, **47**, 4343). Treatment of the optically active morpholine derivative **40** with organocopper reagents gives (*S*)-α-aminoacids **41**. On the other hand, reaction of the heterocycle **40** with organozinc reagents gives (*R*)-α-aminoacids **42**. The optical induction of this reaction is high, with e.e. up to 96%.

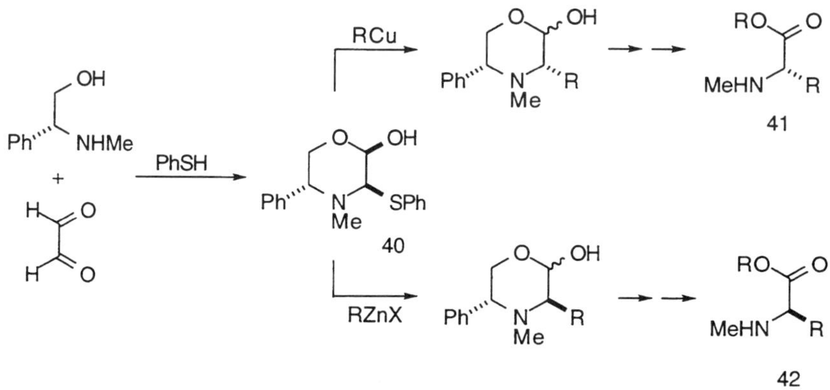

Ephedrine, another member of the aryl β-aminoalcohol family, can be coupled to various α,β-unsaturated carboxylic acid functionalities to give the amides **43**, which undergo Michael addition reaction with an excess of Grignard reagents. Hydrolysis of the primary addition products **44** then affords the optically active β,β-disubstituted carboxylic acids **45** with good enantioselectivity (79 - 99% e.e.) (T. Mukaiyama and M. Iwasawa, Chem. Lett., 1981, 913).

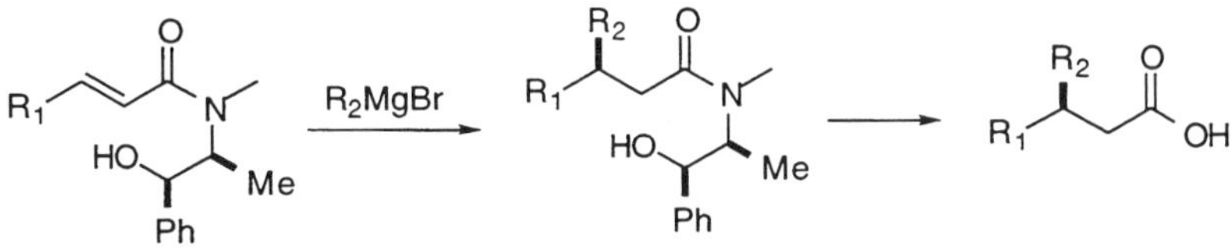

43 44 45

γ-Aminoalcohols **46** can be synthesized in two steps *via* a [3 + 2] cycloaddition reaction between phenyl nitrile oxide and monosubstituted alkenes (V. Jäger and V. Buß, Liebigs Ann. Chem., 1980, 101). Reduction of the cycloadducts with lithium aluminum hydride furnishes the desired compound **46** with good stereo- and regio-selectivity. These phenyl

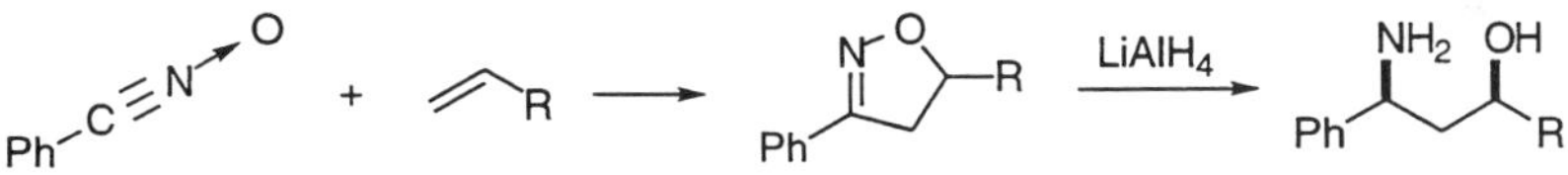

46

substituted γ-*N*-methyl-aminoalcohols **47** can undergo intramolecular cyclization to give *N*-methylazetidines **48** with triphenylphosphine and triethylamine in carbon tetrachloride (V. Stoilova, L. S. Trifonov and A. S. Orahovats, Synthesis, 1979, 105).

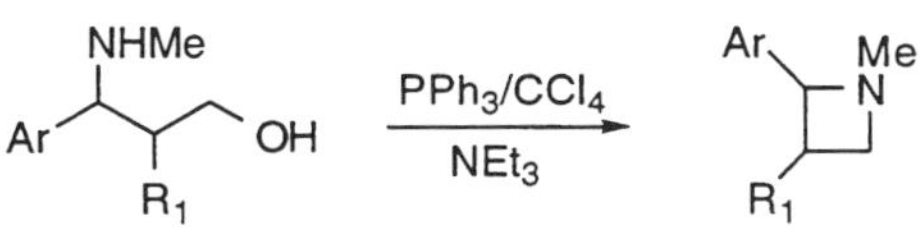

47 48

2. Hydroxycarbonyl and aminocarbonyl derivatives of aralkanes

(a) 2-Hydroxyaralkanones and their derivatives

2-Hydroxyaralkanones **49** are usually prepared *via* the direct oxidation of the corresponding aralkanones. Thus, the lithium enolates (**50**, R^3 = Li) or silyl enol ethers (**50**, R^3 = $SiMe_3$) of aralkanones react with

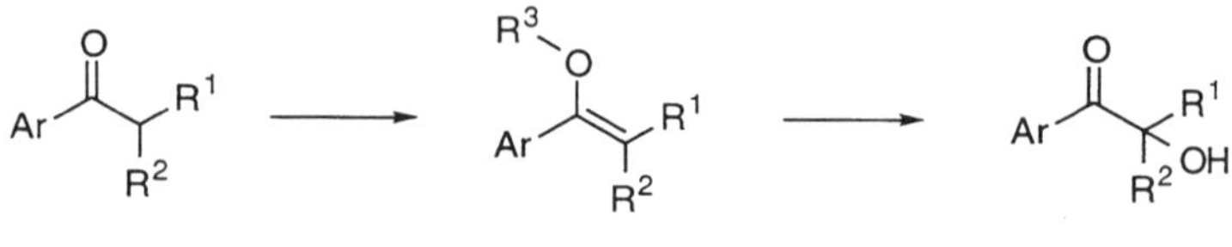

50 49

dimethyldioxirane to give 2-hydroxyaralkanones in high yields (77 - 82%) (K. R. Guertin and T. H. Chan, Tetrahedron Lett., 1991, **32**, 715; H. K. Chenault and S. J. Danishefsky, J. Org. Chem., 1989, **54**, 4249; W. Adam, L. Hadjiarapoglou and X. Wang, Tetrahedron Lett., 1989, **30**, 6497). Other oxidants include CrO_2Cl_2 (T. V. Lee and J. Toczek, Tetrahedron Lett., 1982, **23**, 2917), oxaziridine (F. A. Davis and A. C. Sheppard, J. Org. Chem., 1987, **52**, 954), osmium tetraoxide/*N*-methylmorpholine-*N*-oxide (J. P. McCormick, W. Tomasik and M. W. Johnson, Tetrahedron Lett., 1981, **22**, 607), iodosobenzene in the presence of boron trifluoride etherate (R. M. Moriarty, M. P. Duncan and O. Prakash, J. Chem. Soc., Perkin Trans. 1, 1987, 1781) or *tert*-butyl hypochlorite in the presence of palladium tetrakis-(triphenylphosphine) (T. Nakatsuka and T. Mukaiyama, Chem. Lett., 1982, 369).

Of special interest is that silyl enol ethers can also be smoothly oxygenated to α-hydroxycarbonyl compounds **51** by molecular oxygen in the presence of an aldehyde and a catalytic amount of nickel (II) complex (T. Takai *et al.*, Chem. Lett., 1991, 281). Molecular oxygen is used as the oxidant and the aldehyde being the reductant in this reaction.

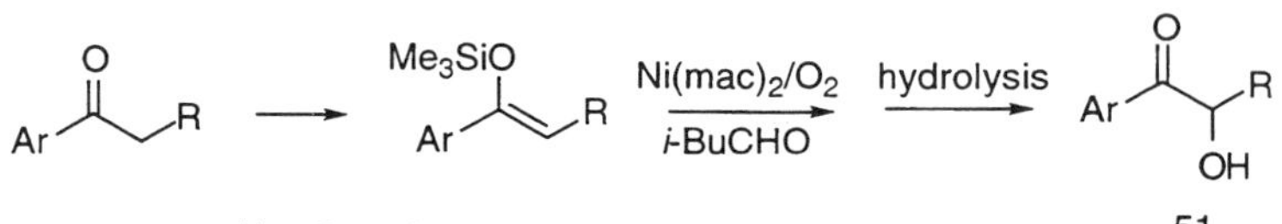

mac-H = 3-methyl-2,4-pentanedione

51

If chiral (camphorylsulfonyl)oxaziridine **52** is used as the oxidant, α-hydroxyketones with optical purity of 69 - 95% e.e. can be obtained (F. A. Davis and M. S. Haque, J. Org. Chem., 1986, **51**, 4083). The enantio-selectivity of the hydroxylation depends on the nature of the metal cation and the presence of cosolvent such as HMPA. Other chiral sulfonyl oxaziridine **53** has also been employed, but the optical purity of the product **49** is low (7 - 11% e.e.) (F. A. Davis and A. C. Sheppard, J. Org. Chem., 1987, **52**, 954).

52

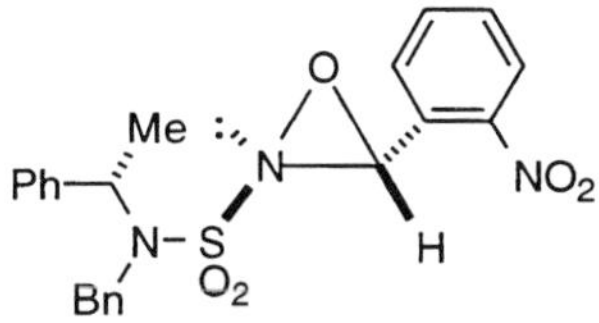

53

The silyl enol ether of aromatic ketones, can also be converted to the corresponding α-trifluoromethanesulfonyloxy ketones **54** with trimethylsilyltrifluoromethanesulfonate/iodosobenzene in good yields (53 - 77%) (R. M. Moriarty *et al.*, Tetrahedron Lett., 1989, **30**, 667). The triflate group can then be displaced by a wide variety of oxygen and nitrogen nucleophiles to afford α-hydroxy and α-aminoketones respectively.

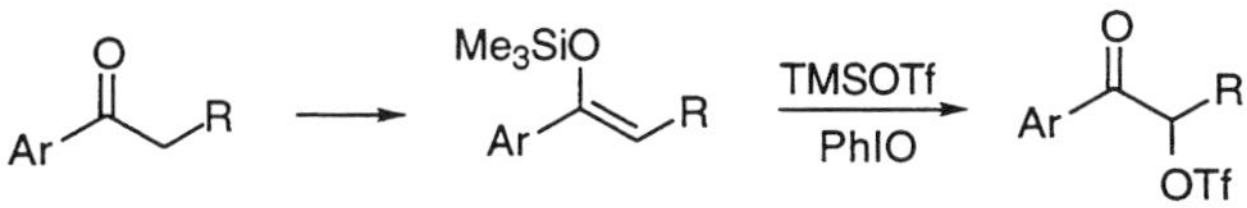

54

Optically active α-hydroxy-α-phenylacetals (46 - 100% e.e.) can be prepared from the reduction of α-keto-α-phenylacetals using microorganisms (P. Ferraboschi et al., Tetrahedron: Asymmetry, 1993, **4**, 1931).

Treatment of aromatic aldehydes with the sodium salt of tetraethyl methanediphosphonate **55** in benzene affords the olefins **56** in good (75 - 85) yields. Dihydroxylation of alkenes **56** with *N*-methylmorpholine *N*-oxide in the presence of osmium tetraoxide gives the diols **57**, which undergo base-catalyzed C-P bond cleavage and tautomerization to give α-hydroxy-

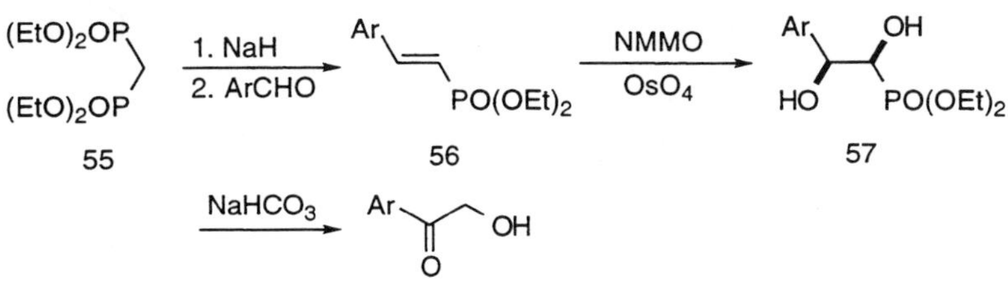

55 56 57

58

ketones **58** in good yields (W. Waszkuc, T. Janecki and R. Bodalski, Synthesis, 1984, 1025).

β-Aryl vinyl sulfoxides **59** undergo an additive Pummerer rearrangement with trifluoroacetic anhydride to produce 1,2-bis(trifluoroacetoxy)-thioethers **60** in excellent yields. Basic methanolysis of **60** then gives the aryl substituted α-hydroxyketones **61** in good yields (D. Craig, K. Daniels, and A. R. MacKenzie, Tetrahedron, 1993, **49**, 11263).

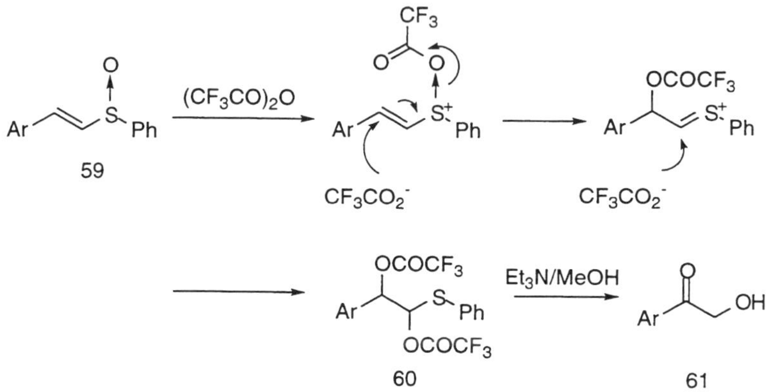

Styrene oxide and stilbene epoxide can be oxidatively opened in trifluoromethanesulfonic acid and DMSO/triethylamine to give α-hydroxyacetophenone and benzoin respectively in moderate yields (B. M. Trost and M. J. Fray, Tetrahedron Lett., 1988, **29**, 2163). The key step involves the DMSO opening of a protonated epoxide.

1-Trimethylsilylmethanol is used *via* its dilithiated carbonate in a one-pot procedure to hydroxymethylate benzoyl chlorides, benzoates or benz-

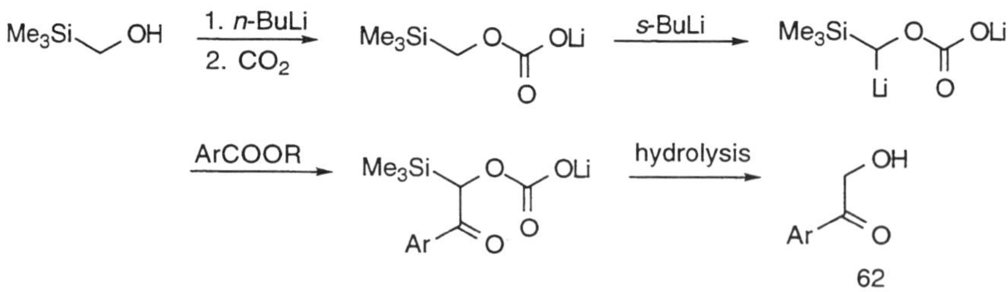

amides to give 2-hydroxyaralkanones **62** in 60 - 68% yields (A. R. Katritzky and S. Sengupta, Tetrahedron Lett., 1987, **28**, 1847).

(b) 3-Hydroxyaralkanones and their derivatives

Titanium enolates **63** derived from aromatic ketones are easily accessible by quenching the lithium enolate with titanium halides. Aldol condensation of **63** with aldehydes gives *erythro*-3-hydroxyaralkanones **64** irrespective of the starting enolate geometry (M. T. Reetz and R. Peter, Tetrahedron Lett., 1981, **22**, 4691).

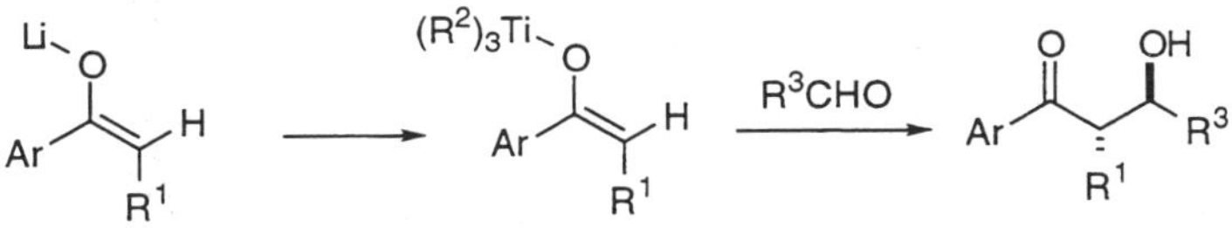

63 64

(*S*)-Gingerol **65**, the major pungent principle of *Zingiber officinale* Roscoe, was synthesized. The key step is a diastereoselective cycloaddition of a nitrile oxide **66** to a chiral iron complexed triene **67** (T. Le Gall, J.-P. Lellouche and J.-P. Beaucourt, Tetrahedron Lett., 1989, **30**, 6521).

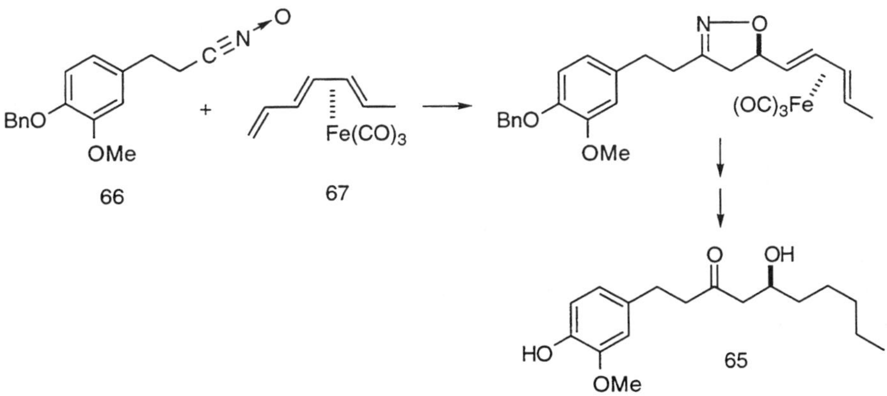

(c) 2-Aminoaralkanones and their derivatives

2-Aminoaralkanones are important intermediates for the construction of nitrogen containing heterocycles and also have considerable value as intermediates for the synthesis of adrenergic β-ethanolamine derivatives.

α-Aminoketones **68** can be synthesized from substituted phenacyl bromides **69** by reaction with sodium diformylamide in acetonitrile. The formyl groups can then be cleaved readily with hydrochloric acid (H. Yinglin and H. Hongwen, Synthesis, 1990, 615).

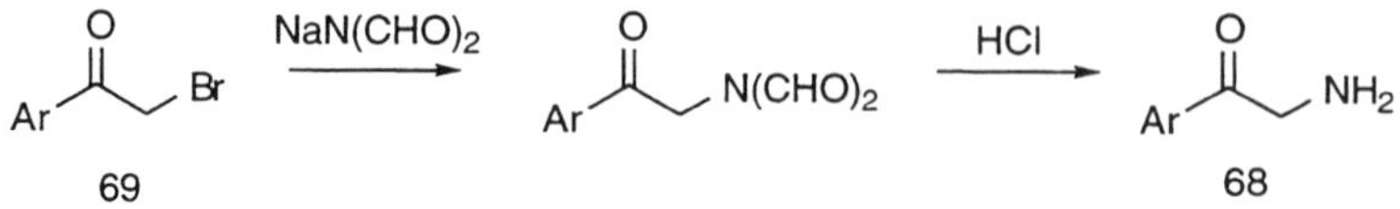

2-Aminoaralkanones can also be synthesized by Friedel-Craft acylation. Reaction of arenes with the α-*N*-(trifluoroacetamido)- or α-*N*-(acetamido) acid chloride **70** in the presence of aluminum chloride gives α-aminoaralkanones **71** in 22 - 77% yield (J. E. Nordlander *et al.,* J. Org. Chem., 1984, **49**, 4107; D. E. McClure *et al.,* J. Org. Chem., 1981, **46**, 2431). Selective reduction of the ketone function leads to a wide variety of

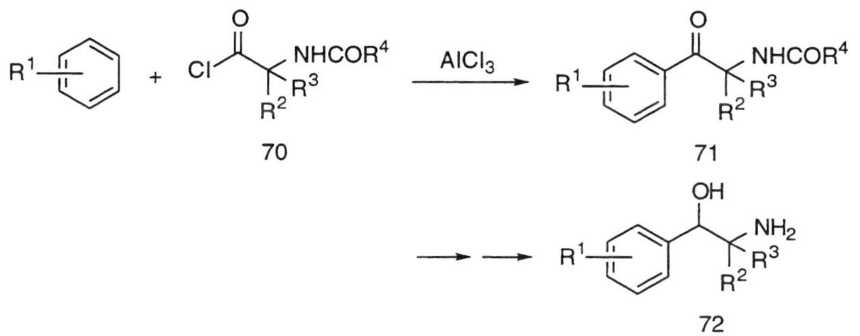

β-aminoalcohols **72**. Similarly, treatment of arenes with bis(*N*-methoxy-carbonylamino)methyl methyl ketone **73** in the presence of acid gives β-aminoketone derivatives **74** in good yields (D. Ben-Ishai and D. Denenmark, Heterocycles, 1985, **23**, 1353).

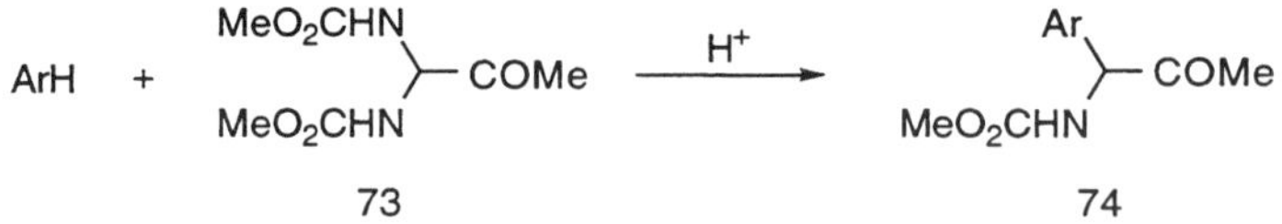

Hydrogenation of α-nitroacetophenones **75** with sulfur poisoned platinum catalyst results in the selective reduction of the nitro group into an amino functionality to give α-aminoketones **76** in 73 - 98% yield (R. Tamura, D. Oda and H. Kurokawa, Tetrahedron Lett., 1986, **27**, 5759).

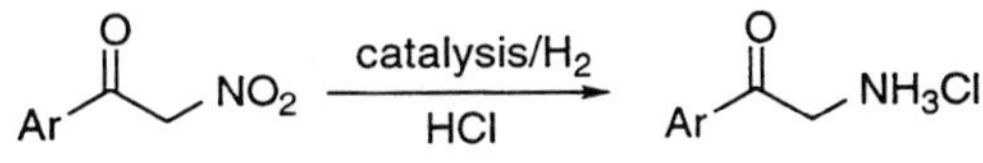

75 76

Silyl enol ethers **77** react with ethyl azidoformate to give the aziridine intermediates **78**, which undergo ring opening to give ethyl *N*-phenacylcarbamates **79** on contact with silica gel (S. Lociuro, L. Pellacani and P. A. Tardella, Tetrahedron Lett., 1983, **24**, 593).

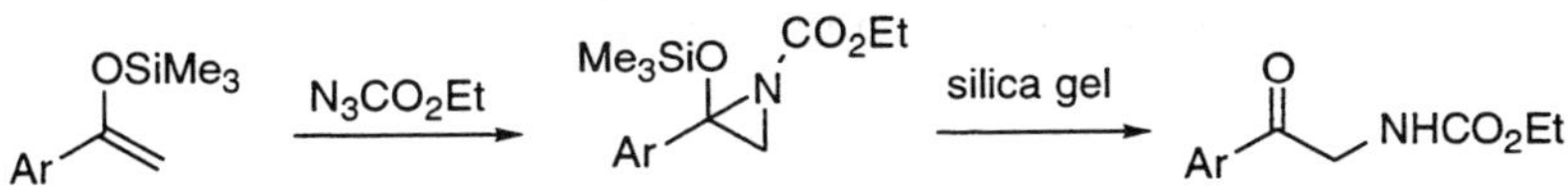

77 78 79

Phenacylamides **80** can now be prepared (M. R. Shipton, Synlett, 1992, 491) by heating the hydrochloride salt of α-amino ketones **81** in neat anhydride in the presence of a catalytic amount of DMF (82 - 99% yield). No diacylated products is formed, which is often a side reaction in the Schotten-Baumann procedure. Previous attempts in treating the hydrochloride salt **81** with base in the presence of air led to the formation of pyrazine derivatives.

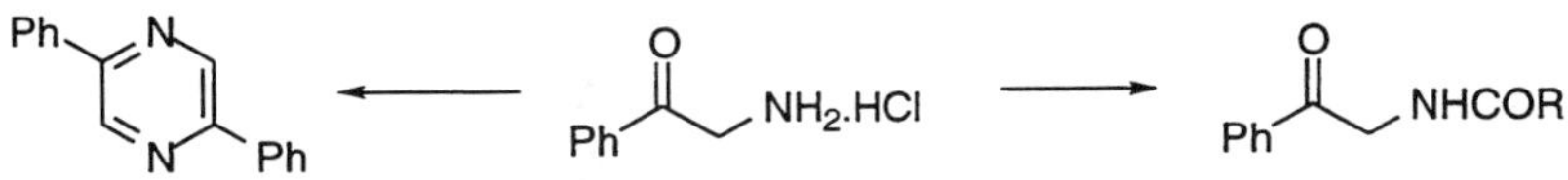

81 80

α-*N*-Formyl ketones **82** can also be α-*C*-alkylated to give, after deformylation, α-alkyl-α-aminoketones **83** (J. Ackrell *et al.*, J. Org. Chem., 1986, **51**, 3374).

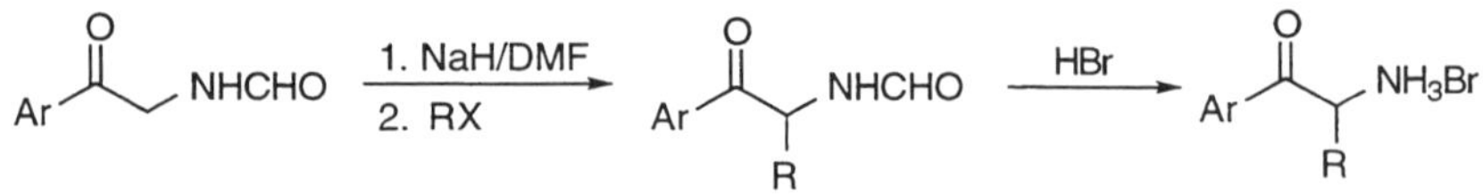

82 83

(d) Polyhydric aralkanones and their derivatives

α-Dicarbonyl compounds react with activated tin to form intermediate tin (II) enediolates **84**. The latter compounds undergo aldol reaction with aldehydes affording α,β-dihydroxy ketones **85** as a mixture of diastereoisomers (T. Mukaiyama, J. Kato and M. Yamaguchi, Chem. Lett., 1982,

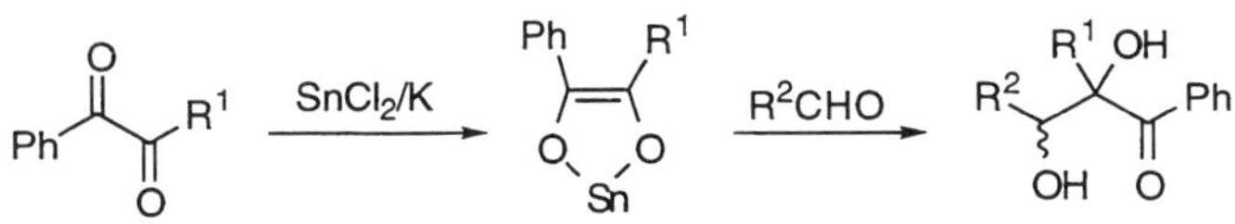

84 85

1291). Alternatively, α,β-dihydroxy ketones **86** can also be prepared in stereoselective *threo*-configuration *via* the corresponding boron enediolate starting from α-hydroxyacetophenone **87** (T. Mukaiyama and M. Yamaguchi, Chem. Lett., 1982, 509).

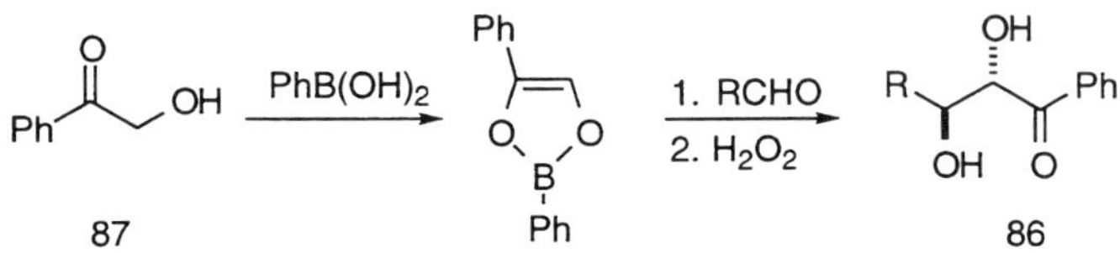

87 86

3. Hydroxy- and amino-aralkanecarboxylic acids and their derivatives

(a) 2-Aryl-2-hydroxycarboxylic acids and their derivatives

Optically active (*R*)-mandelic acid **88** can be prepared in 94% e.e. *via* a diastereoselective potassium tributylborohydride reduction of 8-phenylmenthyl phenylglyoxylate **89** by using 8-phenylmenthol as the chiral auxiliary (A. Solladié and M. Bencheqroun, Tetrahedron: Asymmetry, 1991, **2**, 1165).

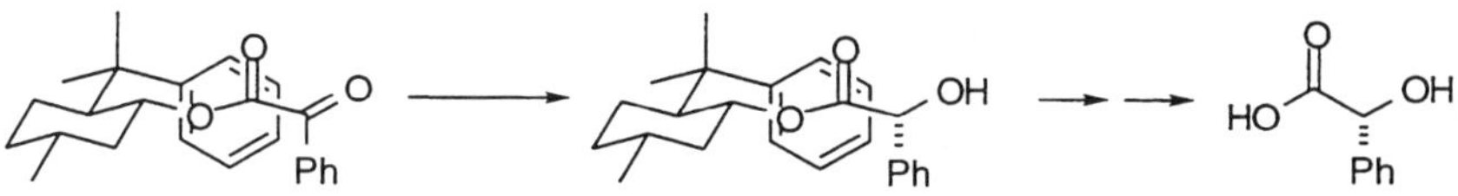

89 88

Acetalization of (*S*)-mandelic acid with aldehydes gives *cis*-1,3-dioxolan-4-one **90** as the major product. Subsequent alkylation, followed by hydrolytic cleavage of the product **91** affords (*S*)-α-hydroxy acids **92** with good enantioselectivity (70 - 100% e.e.) (G. Fráter, U. Müller and W. Günther, Tetrahedron Lett., 1981, **22**, 4221).

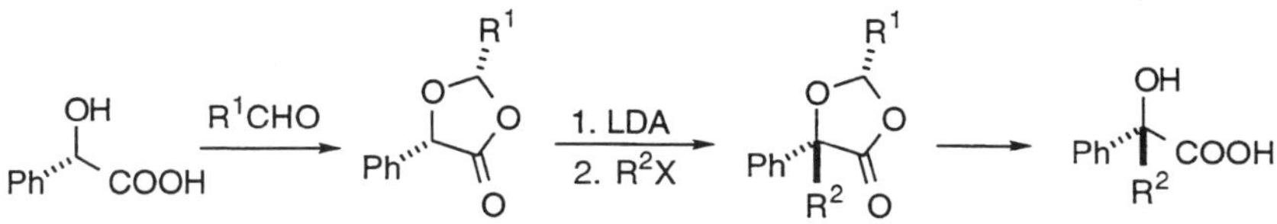

90 91 92

Phenyl ketones **93** can be transformed to α-hydroxyphenylacetamides **94** by cyanosilylation of the carbonyl group and subsequent hydrolysis of the nitrile moiety with acid (G. L. Grunewald, W. J. Brouillette and J. A. Finney, Tetrahedron Lett., 1980, **21**, 1219).

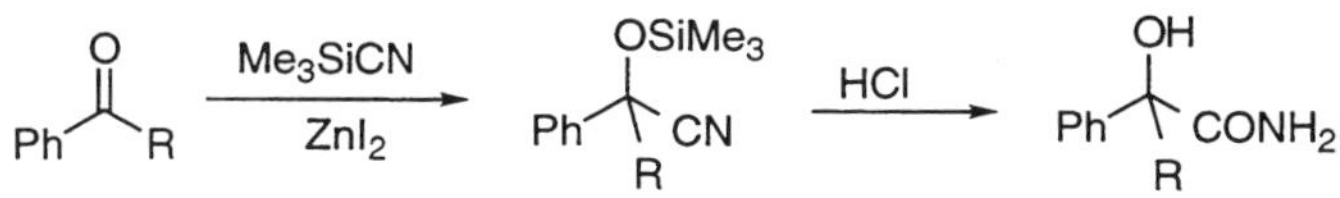

93 94

(b) 3-Aryl-3-hydroxycarboxylic acids and their derivatives

The enantioselective aldol reactions of ketone lithium enolates with aromatic aldehydes mediated by chiral lithium amides have been studied (A.

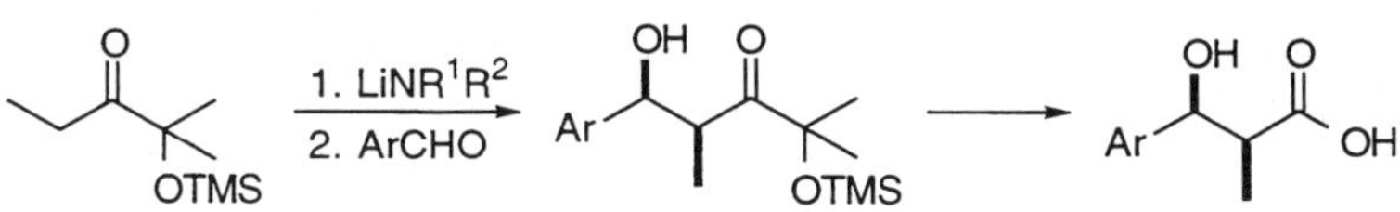

95 96

Ando and T. Shioiri, Tetrahedron, 1989, **45**, 4969). The resulting aldol adducts **95** can be obtained in good *erythro* selectivity but the optical yields are moderate (10 - 77% e.e.). 3-Aryl-3-hydroxycarboxylic acids **96** can be obtained *via* oxidative cleavage of the α-hydroxyketone moiety in **95**.

(c) 2-Amino-2-arylcarboxylic acids (α-amino acids) and their derivatives

Both α- and β-amino acids are important natural products and numerous methods for their synthesis have been developed in the past. Please also refer to section 1b for more details.

(*S*)-Phenylglycine can be obtained in 80% yield and 82% e.e. *via* the addition of phenyl Grignard reagent to the chiral electrophilic glycine equivalent α-bromo-*N*-Boc-glycine (-)-phenylmenthyl ester **97** (P. Ermert *et al.*, Tetrahedron Lett., 1988, **29**, 1265).

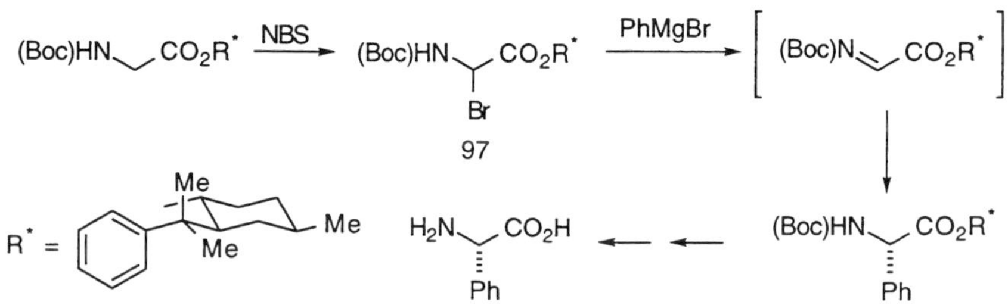

(*R*)-Phenylglycine can be converted to a wide variety of chiral *N*- and *N,N*-substituted diamines (G. Buono, C. Triantaphylides and G. Peiffer, Synthesis, 1982, 1030). This preparative method avoids the potential problems of racemization by using the acyl azide **98** as an intermediate. Displacement of the azide with amines proceeds without racemization to give various diamides in good yields. Reduction of the diamides with lithium aluminum hydride affords the diamines **99** in good yields. These optically active diamines are very useful ligands for the preparation of chiral bis-

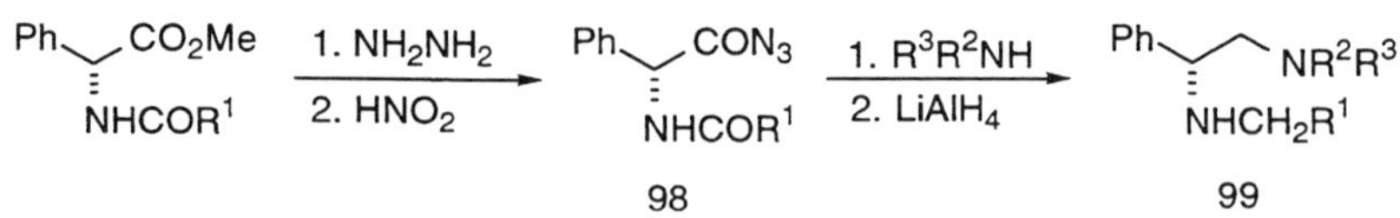

diaminophosphines which have been successfully employed in asymmetric homogeneous catalysis.

(d) 2-Amino-3-arylcarboxylic acids and their derivatives

The asymmetric hydrogenation of 2-acylaminocinnamic acid derivatives **100** using transition metal complexes with chiral diphosphine ligands has been studied extensively. Of particular significant is the use of optically active BINAP (BINAP = 2,2'-bis(diarylphosphino)-1,1'-binaphthyl) ligands as catalysts, sufficiently high asymmetric induction (>90% e.e.) could be achieved in these reactions to give nearly optically pure 2-amino-3-arylcarboxylic acids **101** (R. Noyori and H. Takaya, Acc. Chem. Res., 1990, **23**, 345; H. Kawano *et al.*, J. Chem. Soc., Perkin Trans. 1, 1989, 1571).

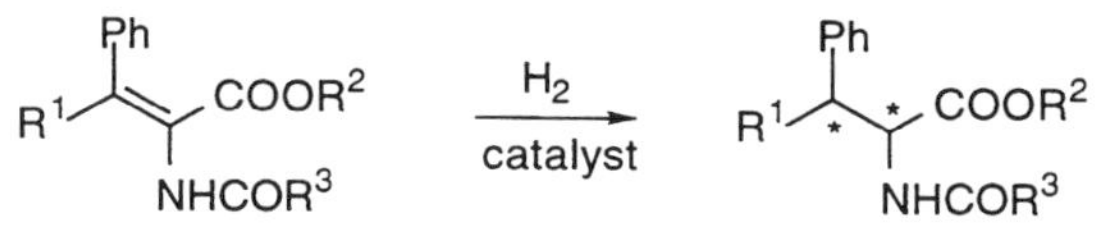

100 101

(e) 3-Amino-3-arylcarboxylic acids (β-amino acids) and their derivatives

β-Amino acids are important natural products found in biologically active peptides as well as useful chiral building blocks in the synthesis of β-lactam antibiotics. They have also been used to prepare peptide analogs with increased potency and enzymatic stability. For these reasons the enantio-selective synthesis of β-amino acids has attracted considerable interest.

Optically active β-amino-β-phenyl acid (*R*)-**102** can be obtained in 90% e.e. via a (*R*)-BINAP-ruthenium (II) catalyzed hydrogenation of (*E*)-β-acylamino-β-arylacrylic acid (*E*)-**103a** (W. D. Lubell, M. Kitamura and R. Noyori, Tetrahedron: Asymmetry, 1991, **2**, 543). Hydrogenation of the (*Z*)-isomer with the same catalyst, however, gives (*R*)-**102** with poor optical purity (9% e.e.), although the (*Z*)-isomer is much more reactive towards hydrogenation. When (*R*)-BINAP-rhodium (I) complex is used as the catalyst, both the (*E*)- and (*Z*)-isomers produce the (*R*)-enantiomer with modest enantioselectivity (45 - 60% e.e.).

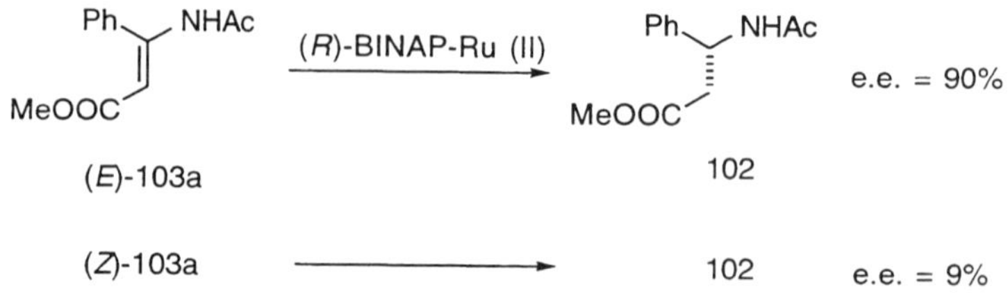

Reaction of ethyl tributylstannylacetate **104** with (*R*)-2-aryl-1,3-oxazolidines **105** under Lewis acid catalyzed conditions yields the (1*R*,1*R*')-amino alcohols **106** in 91 - 99 % diastereoselectivity (M. K. Mokhallalati, M.-J. Wu and L. N. Pridgen, Tetrahedron Lett., 1993, **34**, 47). Oxidative cleavage of the amino alcohols **106** then gives aromatic (*R*)-β-amino esters **107**.

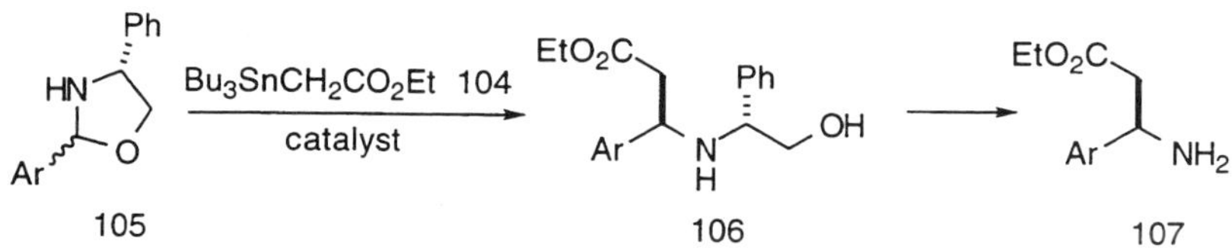

3-Amino-4-phenylbutanoic acid **108** has been synthesized in enantiomerically pure form from (*S*)-asparagine **109** (P. Gmeiner, Tetrahedron Lett., 1990, **31**, 5717). The key step relies on a nucleophilic substitution of the mesylate leaving group with phenylcuprate.

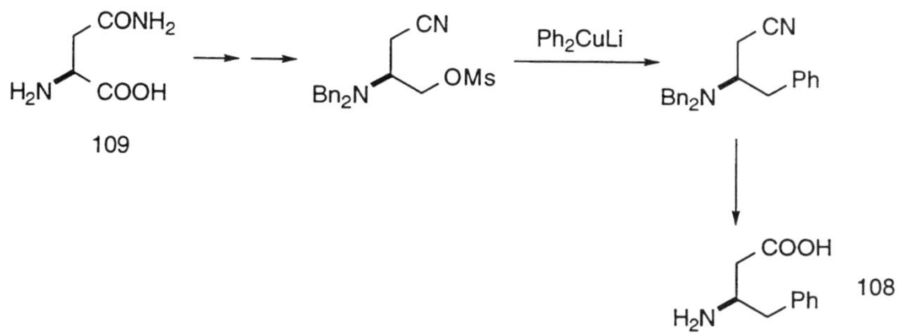

Ethyl acrylate adds to *N*-(tosyl)imines **110** of aromatic aldehydes in the presence of DABCO as a catalyst to give ethyl 3-aryl-2-methylene-3-tosylaminopropanoates **111** in good yields (P. Perlmutter and C. C. Teo, Tetrahedron Lett., 1984, **25**, 5951).

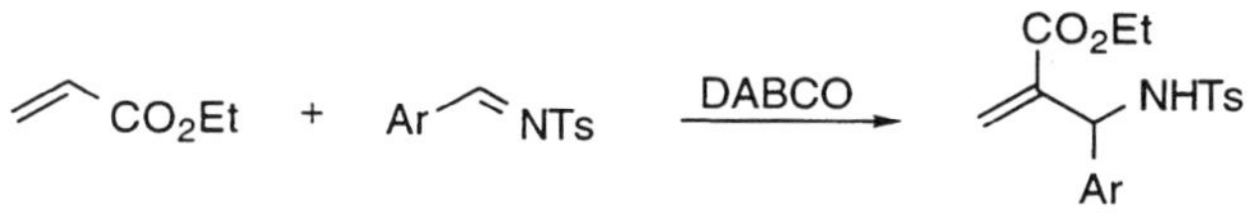

110 111

(f) Polyhydric aralkanecarboxylic acids and their derivatives

Reaction of isocyanoacetate **112** with aromatic aldehydes in the presence of the cationic gold complex **113** and chiral ferrocenylphosphine ligands **114** bearing a tertiary amino group at the terminal position of a pendant chain gives optically active 5-aryl-2-oxazoline-4-carboxylates, which can be transformed into aromatic (2*S*,3*R*)-β-hydroxy-α-amino acids **115** with good to excellent optical purity (86 - 96% e.e.) (T. Hayashi, M. Sawamura and Y. Ito, Tetrahedron, 1992, **48**, 1999).

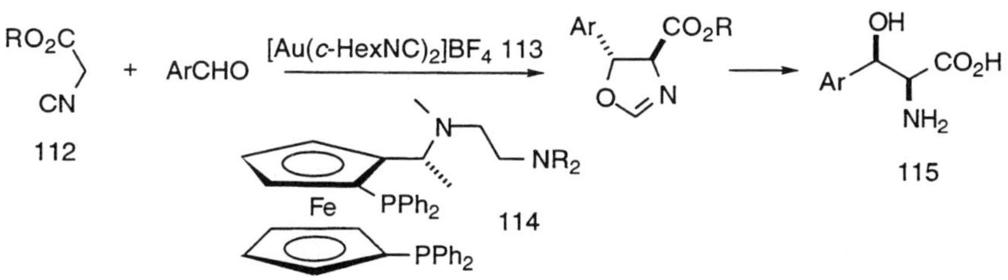

112 114 115

(2*S*,3*R*)-β-Hydroxy-α-amino acids can also be obtained enantioselectively from the aldol reaction between the chiral imine **116** and aromatic aldehydes (T. Nakatsuka, T. Miwa and T. Mukaiyama, Chem. Lett., 1981, 282). Hydrolysis of the initial adducts then gives the amino acids **117** with moderate optical purity and diastereoselectivity.

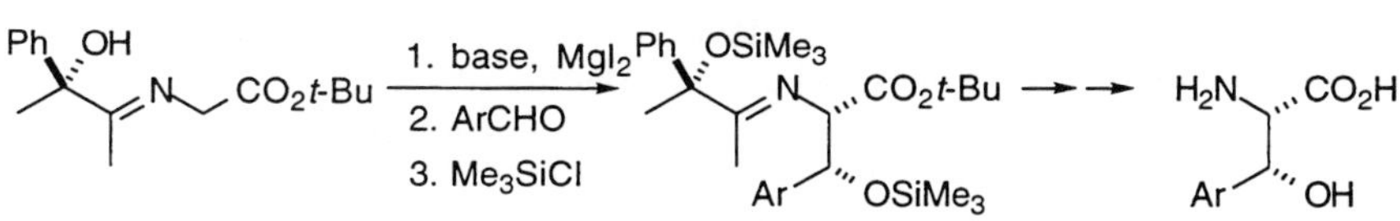

116 117

The C-13 side chain of the antitumor agent taxol, *N*-benzoyl-(2*R*,3*S*)-3-phenylisoserine **118**, has been synthesized *via* the α-hydroxylation of the homochiral aromatic β-amino acid **119** (F. A. Davis, R. T. Reddy and R. E. Reddy, J. Org. Chem., 1992, **57**, 6387). Treatment of **119** with 2 equivalents of lithium diisopropylamide followed by (+)-(camphorylsulfonyl)-oxaziridine gives **118** as the major diastereoisomer in 58% yield. Interestingly, using the (-)-(camphorylsulfonyl)oxaziridine enantiomer does not alter the diastereoselectivity of the reaction.

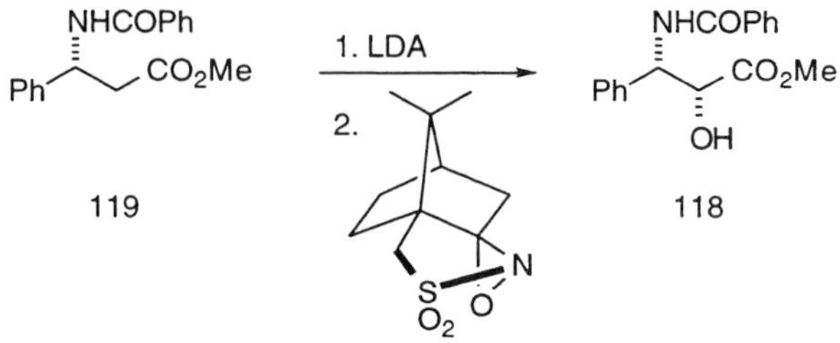

119 118

N-(Methoxycarbonyl)amino phenylglycine **120** in methanol solution containing sodium chloride undergoes electrochemical methoxylation at the α-position to furnish α-amino-α-hydroxyacid derivative **121** (T. Shono, Y. Matsumura and K. Inoue, J. Org. Chem., 1983, **48**, 1388).

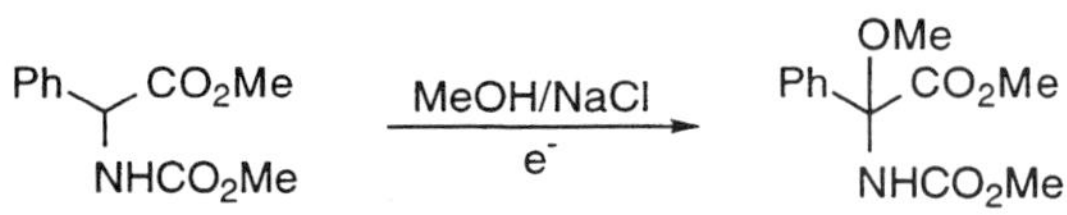

120 121

4. Dicarbonyl derivatives of aralkanes

(a) Aralkane-1,2-diones

A versatile method has been developed to synthesize unsymmetrical aralkane-1,2-diones (T. Yamamori and I. Adachi, Tetrahedron Lett., 1980, **21**, 1747). 2-Substituted 2-lithio-1,3-dithianes **122** add to nitrile oxides **123** to give the oximes **124**, which may be demasked to the unsymmetrical 1,2-diketones **125** with trifluoroacetic acid and aqueous formaldehye. The yields of the reaction vary from 40 - 91%.

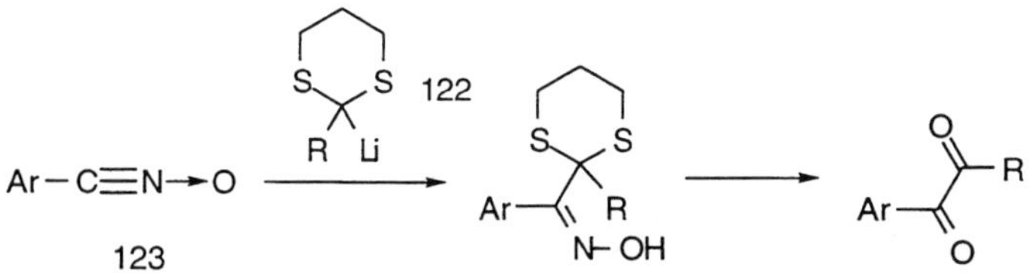

123 124 125

2-*Tert*-Butyloxycarbonylethyl-5-oxo-2-trifluoromethyl-2*H*,4*H*-1,3-oxazoles **126**, prepared from α-amino acids, trifluoroacetic anhydride and *tert*-butyl acrylate in the presence of triethylamine, react with phenyllithium to give the hemiacetals **127**. Subsequent treatment with formic acid gives the α-diketones **128** in 70 - 77% yield (J. Leyendecker, U. Niewöhner and W. Steglich, Tetrahedron Lett., 1983, **24**, 2375).

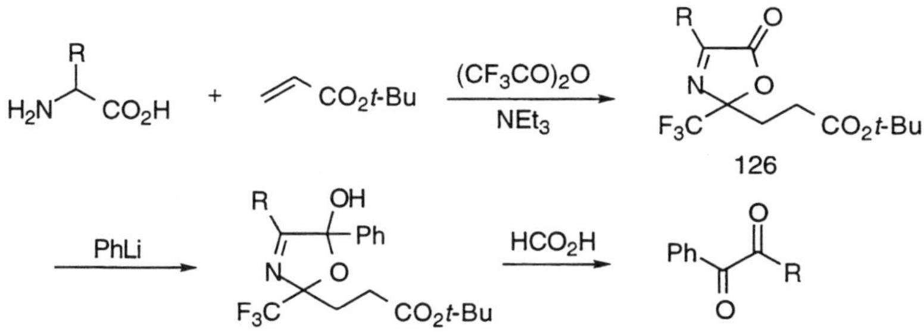

127 128

(b) Aralkane-1,3-diones

1-Phenyl-1,3-butanedione can be prepared *via* Na_2PdCl_4 catalyzed oxidation of 4-phenyl-3-buten-2-one in the presence of *tert*-butyl hydroperoxide in aqueous acetic acid (J. Tsuji, H. Nagashima and K. Hori, Chem. Lett., 1980, 257).

Silyl enol ether **129** reacts with 1-bromo-1-phenylthio-1-trimethylsilyl methane **130** in the presence of zinc bromide to give β-phenylthio-β-trimethylsilyl ketone **131** (D. J. Ager, Tetrahedron Lett., 1983, **24**, 419). This adduct can be converted into 3-phenylpropan-3-onal **132** by oxidation with MCPBA, thermal sila-Pummerer rearrangement, and hydrolysis.

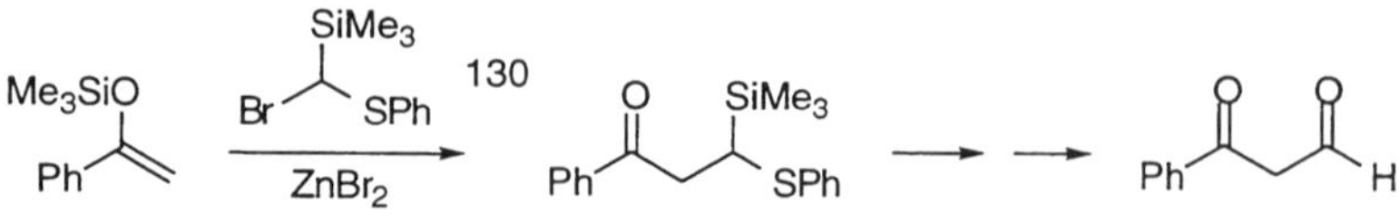

129 131 132

(c) Aralkane-1,4-diones

1,4-Diphenyl-1,4-butanedione **133**, has been prepared by thiazolium salt-catalyzed addition of benzaldehyde to vinylsulphone **134** (H. Stetter and H.-J. Bender, Chem. Ber., 1981, **114**, 1226).

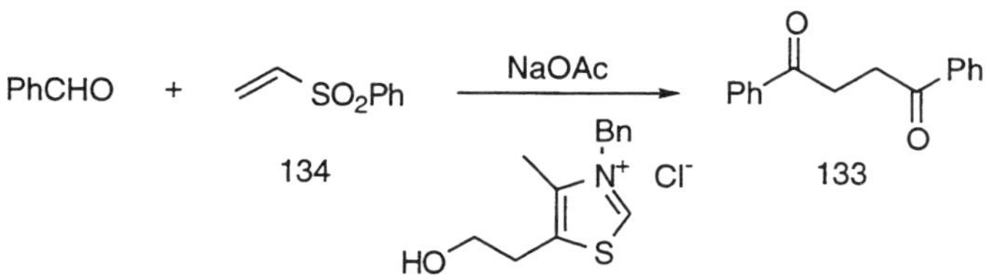

Substituted 1,4-diketones can also be prepared *via* direct nucleophilic acylation of α,β-unsaturated ketones (S. Hünig and G. Wehner, Chem. Ber., 1980, **113**, 302; Chem Ber., 1980, **113**, 324). Thus, lithated *O*-(trimethylsilyl)cyanohydrin **135** reacts with unsaturated ketones **136** in ether to give exclusively the 1,4-addition products **137**, which can be transformed into 1,4-diketones **138** upon hydrolysis. The dithiane anion **139** also reacts with unsaturated ketones **136** in 1,4-fashion in THF/HMPA mixture (M. El-Bouz and L. Wartski, Tetrahedron Lett., 1980, **21**, 2897), subsequent cleavage of the dithiane moiety then again generates the 1,4-diketones.

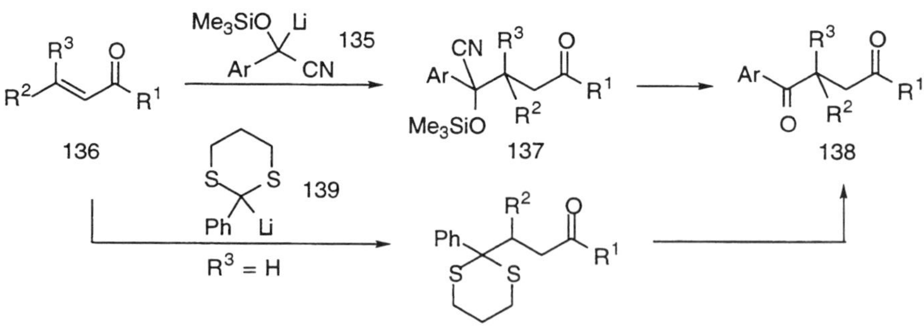

5. *Oxoaralkanecarboxylic acids and their derivatives*

(a) *Aryl 2-oxocarboxylic acids (α-ketoacids) and their derivatives*

Oxidation of mandelic acid with *tert*-butyl hydroperoxide in the presence of ruthenium catalysis gives ethyl phenylglyoxylate (M. Tanaka, T. Kobayashi and T. Sakakura, Angew. Chem., Int. Ed. Engl., 1984, **23**, 518). It can also be prepared in 80% yield from the reaction of phenyllithium with triethoxyacetonitrile **140** in ether (G. P. Axiotis, Tetrahedron Lett., 1981, **22**, 1509). On the other hand, treatment of alkyl α-oxo-1*H*-

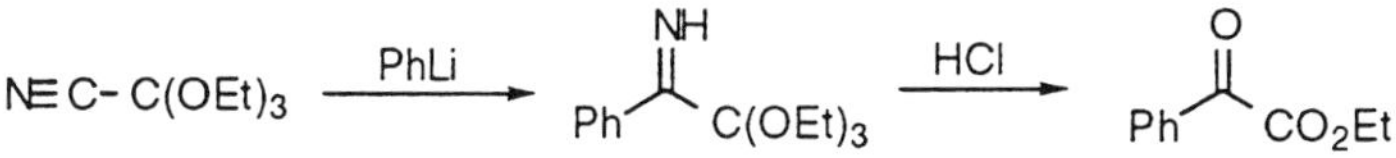

140

imidazol-1-acetates **141** with aryl Grignard reagents gives a wide variety of aryl α-keto acids **142** in good yields (J. S. Nimitz and H. S. Mosher, J. Org. Chem., 1981, **46**, 211).

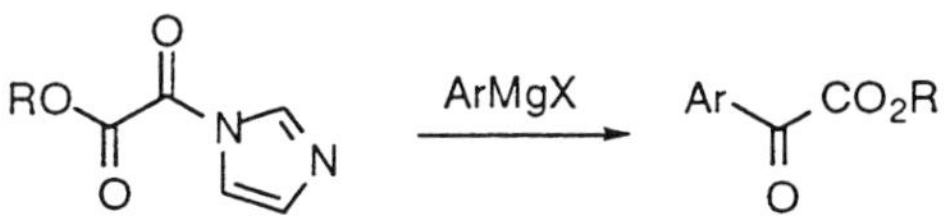

141 142

Reaction of α,β-epoxyesters **143** with hydrofluoric acid and pyridine gives the 3-fluoro-2-hydroxyalkanoic esters **144** (A. Ourari, R. Condom and R. Guedj, Can. J. Chem., 1982, **60**, 2707). Oxidation of **144** with Jones reagent affords fluoro-substituted phenylpyruvates **145** in 58 - 87% yield.

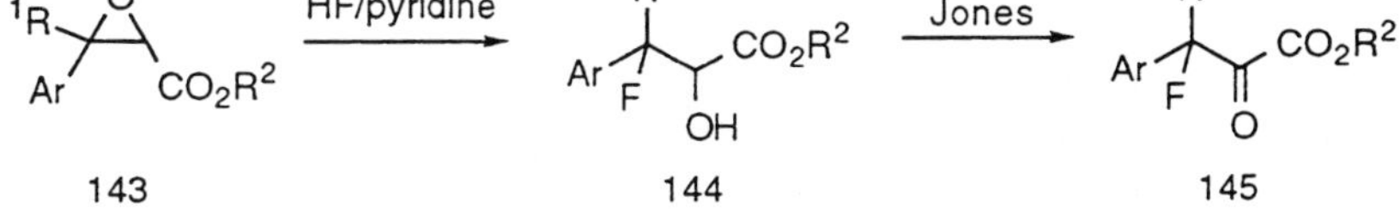

143 144 145

Arylglyoxylates **146** can be converted into the corresponding α-(*N*-methylsilyl)imino esters **147** on treatment with lithium hexamethyldisilazane followed by chlorotrimethylsilane (Y. Matsuda *et al*., J. Chem. Soc., Perkin

Trans. 1, 1989, 279). The silylated imines **147** can be reduced to the corresponding α-amino acid esters **148** by sodium cyanoborohydride. On the other hand, dimerization of **2** in methanol affords 2-alkoxycarbonyl-imidazol-4(2*H*)-ones **149**.

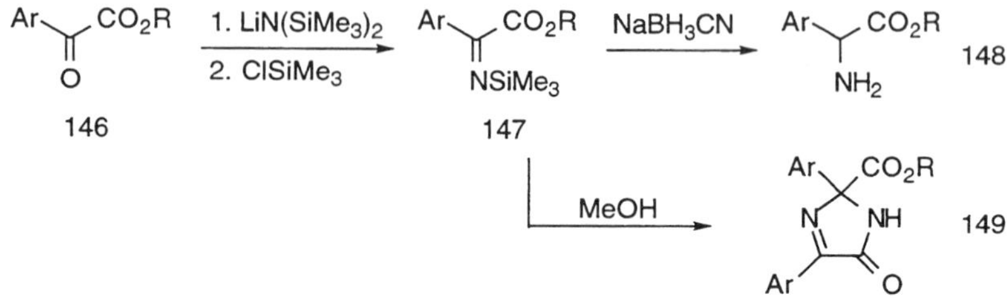

Phenylglyoxylic acid undergoes oxidative decarboxylation to give benzoic acid in the presence of diethyl persulphoxide (W. Ando, H. Miyazaki and T. Akasaka, Tetrahedron Lett., 1982, **23**, 2655).

(b) Aryl 3-oxocarboxylic acids (β-ketoacids) and their derivatives

Ethyl benzoylacetate **150** can be converted into the corresponding β-amino acid **151** by reductive amination (M. Furukawa *et al.*, Chem. Pharm. Bull., 1979, **27**, 2223). In the presence of a chiral amine, optically active β-amino acid can be obtained, although the optical purity is low (7 - 28% e.e.).

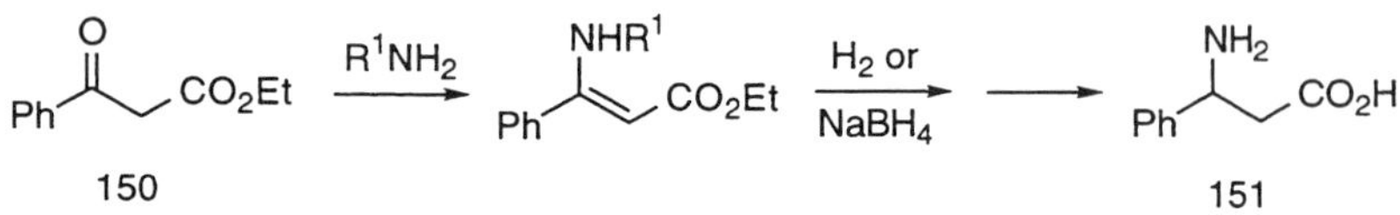

(c) Aryl 4-oxocarboxylic acids (γ-ketoacids) and their derivatives

Zinc bromide catalyzed coupling of the silyl enol ether **152** of aromatic ketones with methyl 1-chloro-1-phenylthioacetate gives γ-ketoesters **153** (T. V. Lee and J. O. Okonkwo, Tetrahedron Lett., 1983, **24**, 323). The phenylthio moiety can be oxidatively eliminated to give the aryl substituted γ-keto-α,β-unsaturated esters **154**. Application of the same methodology to 1-arylbut-2-en-1-ones **155** gives the corresponding homologs **156**.

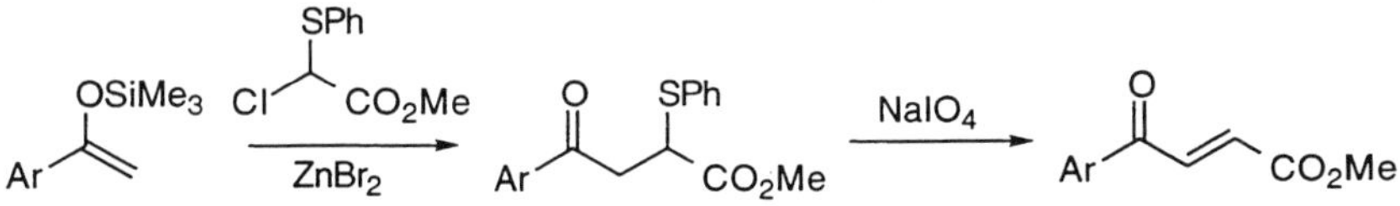

152 153 154

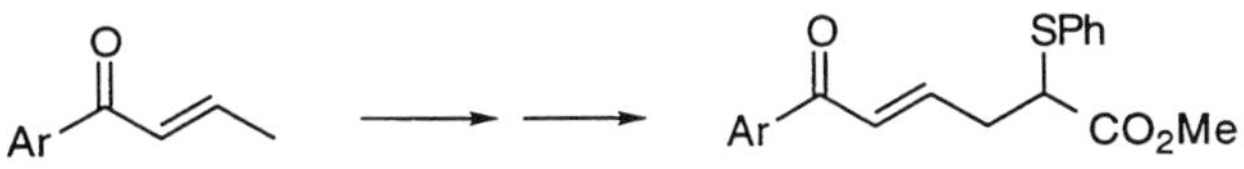

155 156

Reaction of the silyl enol ether of acetophenone with ethyl diazoacetate gives the cyclopropane **157**, which can be cleaved to give the γ-ketoacid **158** (H. Kunz and M. Lindig, Chem. Ber., 1983, **116**, 220).

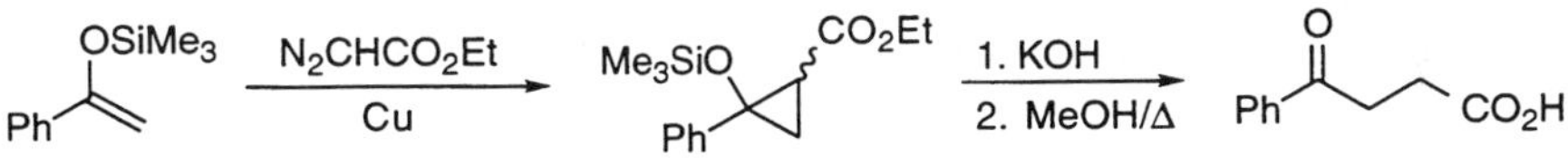

157 158

Alternatively, 3-trimethylsilyloxy-2-alkenoate **159**, prepared from 3-oxo-alkanoate with sodium hydride and chlorotrimethylsilane, also reacts with ethyl diazoacetate in the presence of copper (II) salts to afford cyclopropane **160**. Subsequent *in situ* hydrolysis leads to γ-ketoester **161** (K. Saigo *et al.,* Synth. Commun., 1984, **14**, 787).

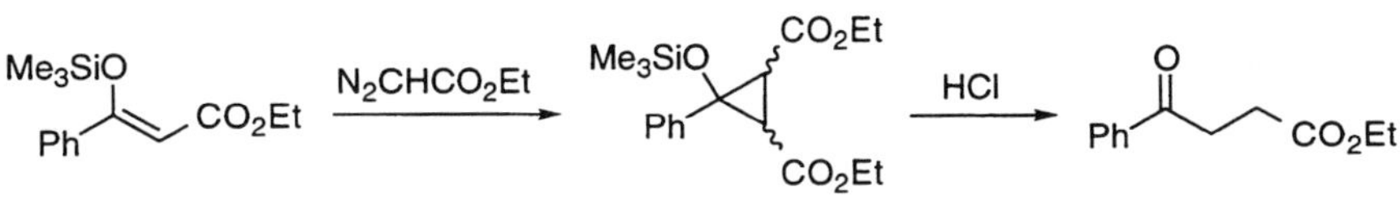

159 160 161

6. Aralkanedicarboxylic acids and their derivatives

2-Arylglutaramides **162** and 3-arylglutaramides **163** can be converted into aryl substituted 1,3-diaminopropanes **164** and **165** respectively by a double Hofmann rearrangement. The resulting diamines can be transformed into aryl substituted 2-amino-3,4,5,6-tetrahydropyrimidines **166**. These tetrahydropyrimidines have been shown to possess potent antidepressant activity (K. Weinhardt, M. B. Wallach and M. Marx, J. Med. Chem., 1985,

28, 694). The most active compound in this series (**166**, Ar = 5-Ph, R^1 = H, R^2 = CO_2Me) is more potent than the tricyclic antidepressant standards.

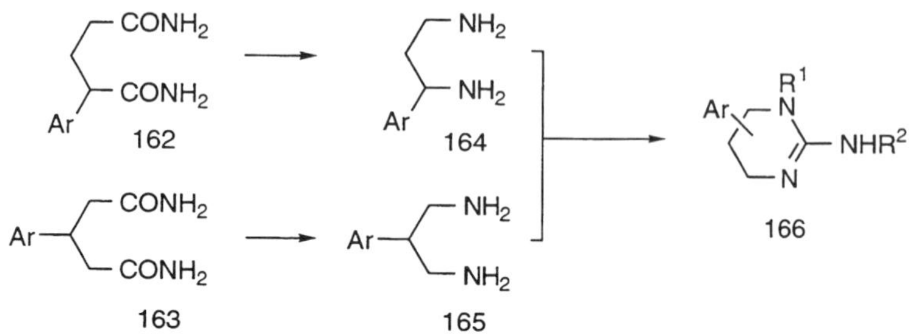

Dimethyl (*S*)-phenylsuccinate **167** can be prepared in 93% e.e. and 40% yield from a palladium catalyzed carbonylation of styrene in the presence of a chiral biphenyl ligand **168** (S. C. A. Nefkens, M. Sperrle and G. Consiglio, Angew. Chem., Int. Ed. Engl., 1993, **32**, 1719).

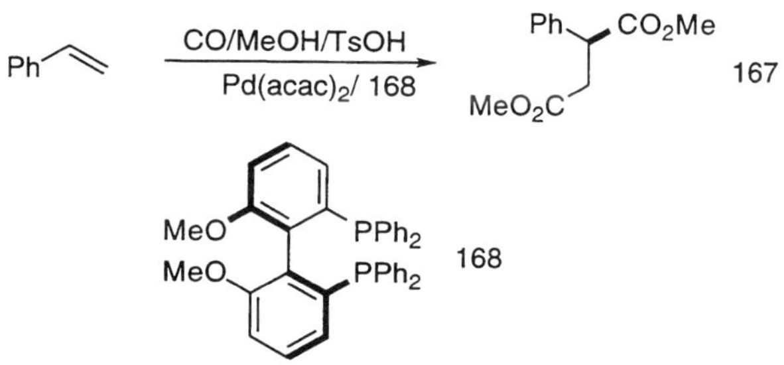

Phenyl substituted 1,1-bis(ethoxycarbonyl)cyclopropanes **169** can be obtained by reacting phenyl substituted olefins with diethyl bromomalonate in the presence of DBU and a catalytic amount of Cu (II) (N. Kawabata, S. Yanao and J. Toshida, Bull. Chem. Soc. Jpn., 1982, **55**, 2687). Depending on the substitutent on the olefin, the yield of the reaction varies from 27 - 87%.

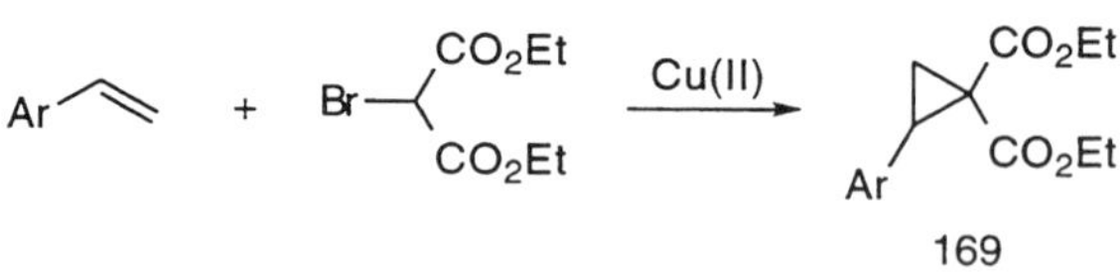

Second Supplements to the 2nd Edition of Rodd's Chemistry of Carbon Compounds, Vol.III D/E/F (Partial), edited by M. Sainsbury

Chapter 19

BENZENE DERIVATIVES WITH ONE OR MORE UNSATURATED SIDE CHAINS (JANUARY 1980-JANUARY 1995)

KARL J. HALE

1. Arylalkenes

(a) Synthesis of styrenes and derivatives

(i) Nickel (0)-catalysed olefination of benzylic cyclic dithioacetals with Grignard reagents (Scheme 1) is a convenient method for obtaining functionalised styrenes from masked benzaldehyde derivatives in a single step (T.-Y. Luh *et al., J. Org. Chem.*, **1991**, *56*, 4035). It is believed that a catalytic

Scheme 1

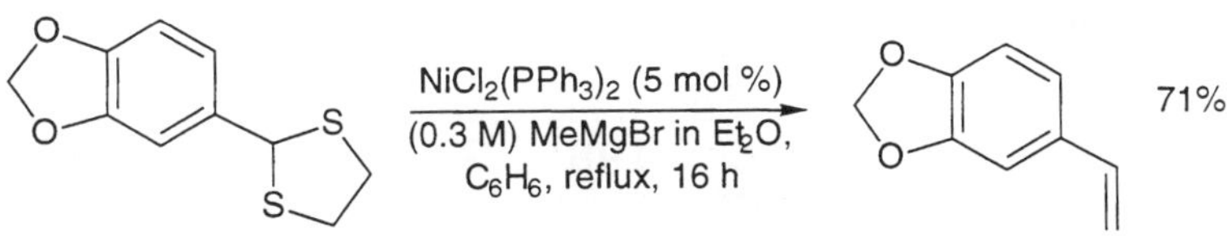

cycle operates in these reactions (Scheme 2) in which a Ni(0) complex oxidatively adds to one of the carbon-sulfur bonds in the dithiolane to produce a nickel(II)-sulfur coordinated intermediate **I**. The latter then associates with the alkyl group of the Grignard reagent and participates in a reductive elimination to give **II**. The sulfide anion in **II** has the capacity to further complex with the Ni (0) catalyst and entice it to engage in oxidative addition to the remaining carbon-sulfur bond to give **III**. *Syn* β-hydride elimination is then believed to occur in **III** to generate the styrene. The active Ni(0) catalyst is regenerated after a reductive elimination mediated by the Grignard reagent.

Scheme 2

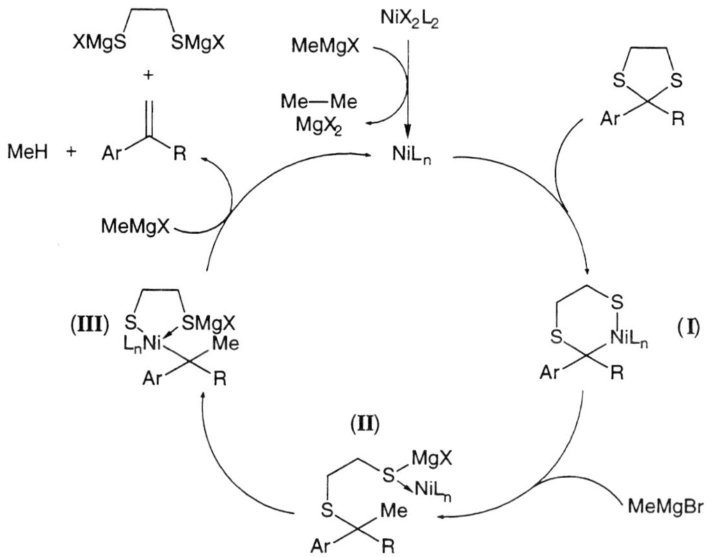

(ii) The Heck arylation reaction (R.F. Heck, *Org. React.*, **1982**, *27*, 345; R.F. Heck, in *Palladium Reagents In Organic Synthesis*, Academic Press, London, **1985**; G.D. Daves, Jr., and A. Hallberg, *Chem. Rev.*, **1989**, *89*, 1433) remains a popular choice amongst organic chemists for the preparation of substituted styrenes under mild conditions. It requires only catalytic quantities of Pd(II) salts to proceed and couples alkenes to aryl halides or triflates, or alternatively simple styrenes to vinyl halides or triflates. The Heck reaction is thought to proceed through the catalytic cycle shown in Scheme 3. Initially, a Pd(II) salt is reduced *in situ* to a Pd(0) species by a triarylphosphine ligand or the alkene itself. The Pd(0) intermediate then oxidatively adds to the aryl or vinyl halide/triflate to produce **IV**, which subsequently complexes with the donor alkene to give **V** prior to engaging in C-C bond formation. The latter step produces **VI** which spontaneously undergoes *syn* β-hydride elimination to form the coupled styrene and the hydrido palladium species **VII**. Reaction of **VII** with mild base finally regenerates the Pd(0) complex and propagates the catalytic cycle.

Scheme 3

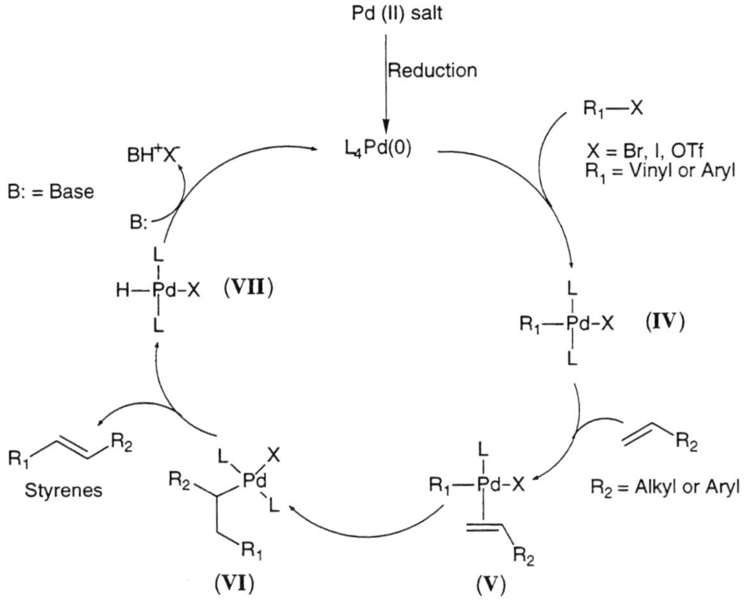

The awesome power of the Heck reaction for the assembly of structurally complex styrene derivatives is well illustrated by the example shown in Scheme 4, where a vinylic tetrabromide is coupled to styrene, using catalytic $Pd(OAc)_2$, tetra-*n*-butylammonium bromide, and potassium carbonate in DMF at 70°C (O. Reiser, S. Reichow, and A. de Meijere, *Angew. Chem. Int. Edn. Engl.*, **1987**, *26*, 1277).

Scheme 4

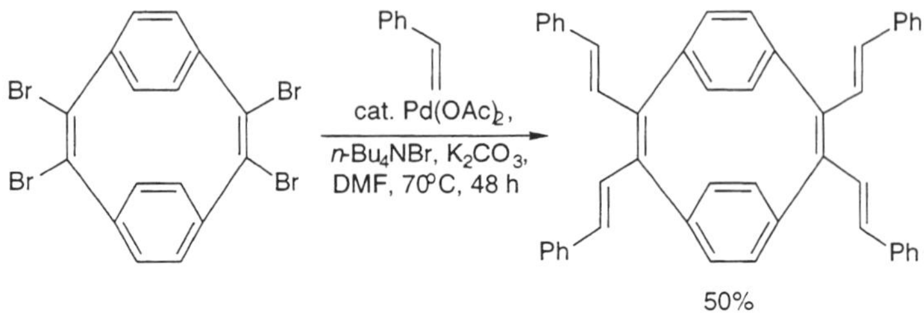

Catalytic Heck reactions have also been conducted between polymer-bound styrenes (Wang resin) and aryl bromides (Scheme 5) (K.-L. Yu, M.S. Deshpande, and D.M. Vyas, *Tetrahedron Lett.*, **1994**, *35*, 8919). This approach to styrenes may prove of particular value for combinatorial organic synthesis in which chemical libraries of compounds are prepared for random biological screening.

Scheme 5

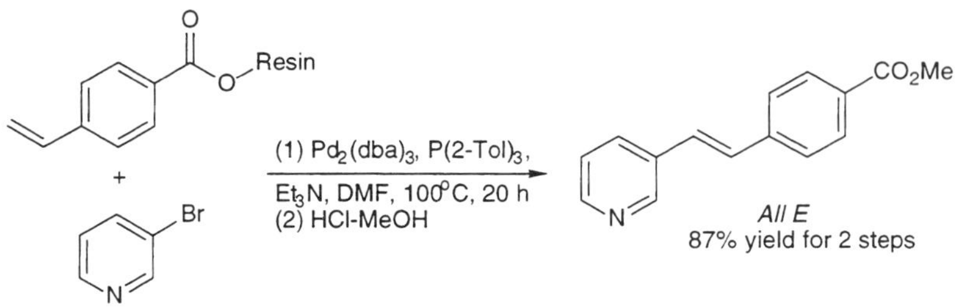

Intramolecular variants of the Heck reaction are proving especially powerful tools for styryl ring annulation, as evidenced by the palladium (0)-catalysed internal arylation of the iodoarene shown in Scheme 6. This is particularly noteworthy since the substrate contains both an activated aziridine and a mixed allylic hydroxylamino acetal (K.F. McClure and S.J. Danishefsky, *J. Am. Chem. Soc.*, **1993**, *115*, 6094; K.F. McClure, S.J. Danishefsky, and G.K. Shulte, *J. Org. Chem.*, **1994**, *59*, 355).

Scheme 6

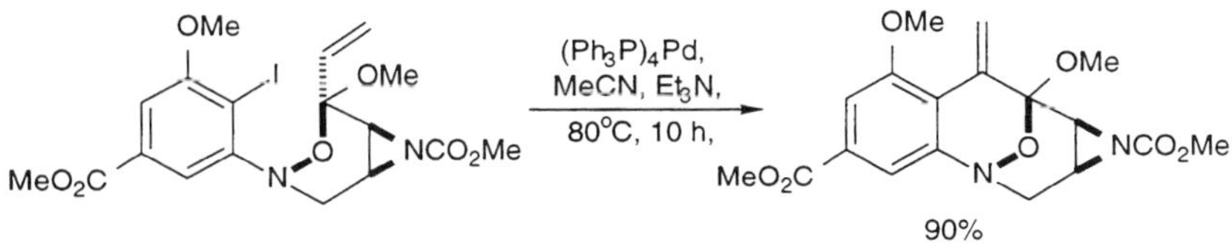

L. Tietze and T. Grote (*J. Org. Chem.*, **1994**, *59*, 192) have also successfully performed the intramolecular Heck reaction shown in Scheme 7 using catalytic $Pd(OAc)_2$, *n*-Bu_4NCl, and NaOAc in *N,N*-dimethyformamide at 70 ^oC (as recommended by R.C. Larock and S. Babu, *Tetrahedron Lett.*, **1987**, *28*, 5291). Interestingly, when catalytic quantities of $(Ph_3P)_4Pd$ were used to initiate this Pd(0) cyclisation, the indole predominated in a 7:1 mixture of these two products, formed in 80% yield.

Scheme 7

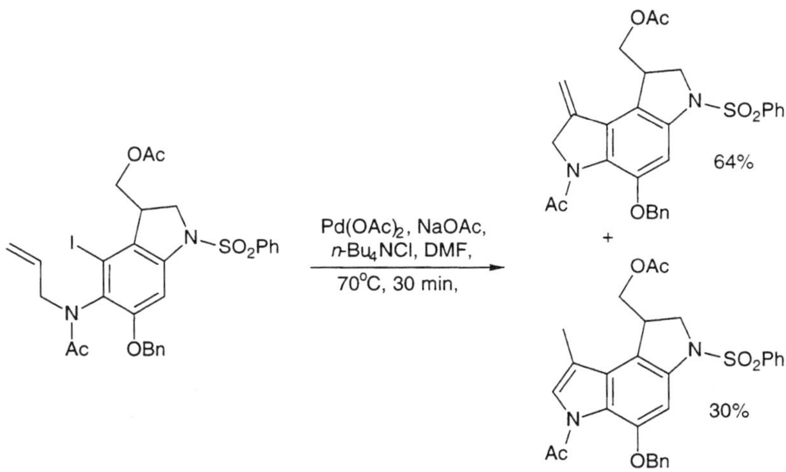

(iii) Vinyl boronates readily cross-couple with aryl iodides in the presence of catalytic quantities of $(Ph_3P)_4Pd$ with retention of parent alkene geometry (Scheme 8) (N. Miyaura, M. Satoh, and A. Suzuki, *Tetrahedron Lett.*, **1986**, *27*, 3745). These reactions are believed to operate via a catalytic cycle in which there is oxidative addition of the Pd(0) complex to the aryl iodide, transfer of the vinyl group from boron to palladium, and finally reductive elimination.

Scheme 8

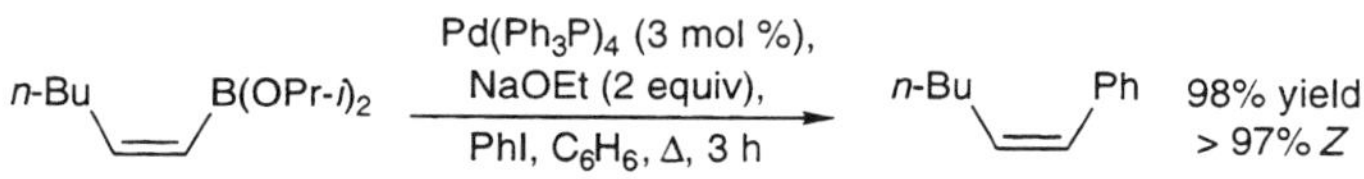

The latter step instigates release of the styrene from the Pd (II) complex to regenerate the Pd(0) catalyst for propagation of the cross-coupling cycle.

(iv) Monosubstituted alkynes undergo smooth thioboration in THF at 50 °C when mixed with 9-phenylthio-9-BBN and a catalytic quantity of $(Ph_3P)_4Pd$. The resulting vinyl boranes cross-couple *in situ* with aryl iodides to give trisubstituted β-thio

Scheme 9

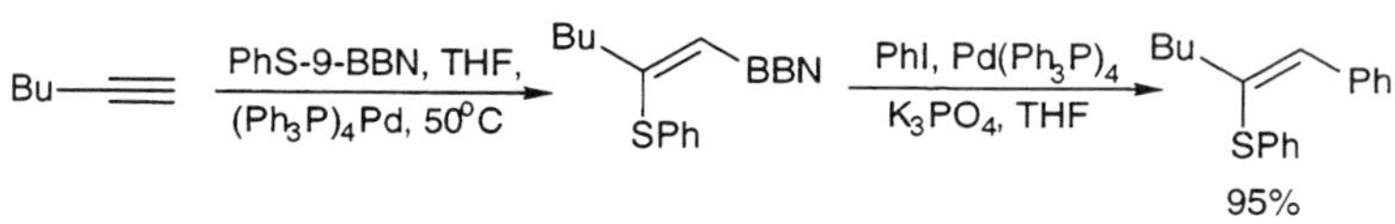

phenylstyrenes with complete preservation of stereochemistry in the original double bond (Scheme 9) (T. Ishiyama, K. Nishijima, N. Miyaura, and A. Suzuki, *J. Am. Chem. Soc.*, **1993**, *115*, 7219).

(v) A stereoselective route to trisubstituted styryl homoallylic alcohols is provided by the nickel-catalysed coupling of aryl Grignard reagents with 5-alkyl-2,3-dihydrofurans (P.J. Kocienski *et al.*, *J. Chem. Soc. Perkin Trans. 1*, **1992**, 3419). One very good example of this procedure being applied is the reaction between 5-methyl-2,3-dihydrofuran and phenylmagnesium bromide in the presence of a catalytic quantity of $(Ph_3P)_2NiCl_2$. This led to the (*Z*)-homoallylic alcohol being formed in 85% yield and 96% geometrical purity (Scheme 10).

Scheme 10

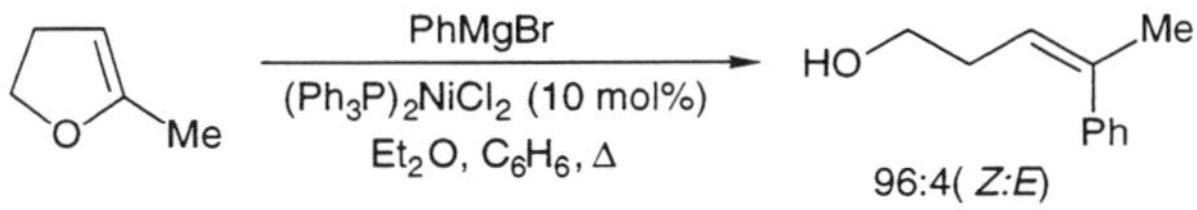

(vi) (*Z*)-β-Phenyltellurostyrenes can be stereospecifically constructed by the *trans*-addition of phenyltellurium hydride to phenylacetylenes (Scheme 11). Treatment of these adducts with Grignard reagents and catalytic quantities of $(Ph_3P)_2NiCl_2$

Scheme 11

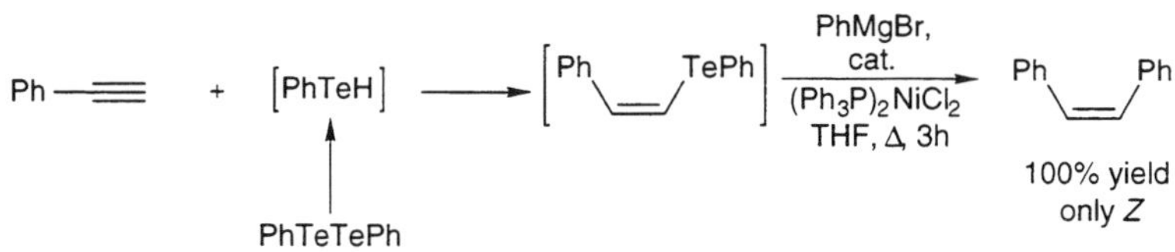

leads exclusively to β-alkylated styrenes with *Z*-olefin geometry (S. Uemura and S. Fukuzawa, *Tetrahedron Lett.*, **1982**, *23*, 1181). (*Z*)-β-Phenyltellurostyrenes will also undergo replacement when treated with lithium dialkylcuprates at 0°C or room temperature to produce stereodefined β-alkylated styrenes (Scheme 12) (A. Ogawa *et al., J. Org. Chem.*, **1994**, *59*, 1600). In the case of trisubstituted α,β-(phenyltelluro)styrenes, it is possible to regioselectively substitute the β-tellurophenyl group without scrambling the starting alkene geometry (Scheme 12).

Scheme 12

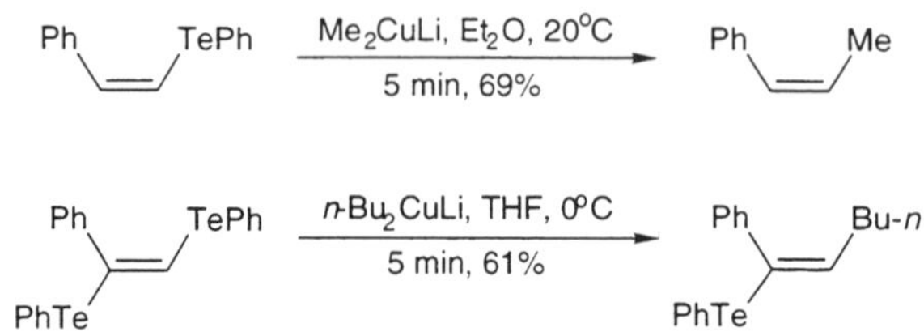

(vii) 1,1-Dihaloalkenes can often be selectively monocoupled with equimolar amounts of arylzinc chlorides at ambient temperature in the presence of 1 mol % of [$PdCl_2$-$Ph_2P(CH_2)_4PPh_2$] to provide the corresponding α-halo styrenes in good yield (Scheme 13) (A. Minato, *J. Org. Chem.*, **1991**, *56*, 4052). Treatment of 1,1-dibromoalkenes with an excess of the arylzinc reagent under the same reaction conditions usually leads to dicoupled products being isolated in good yield. 1,1-Dichloroalkenes fail to give dicoupling products even when a large excess of the arylzinc reagent is employed.

Scheme 13

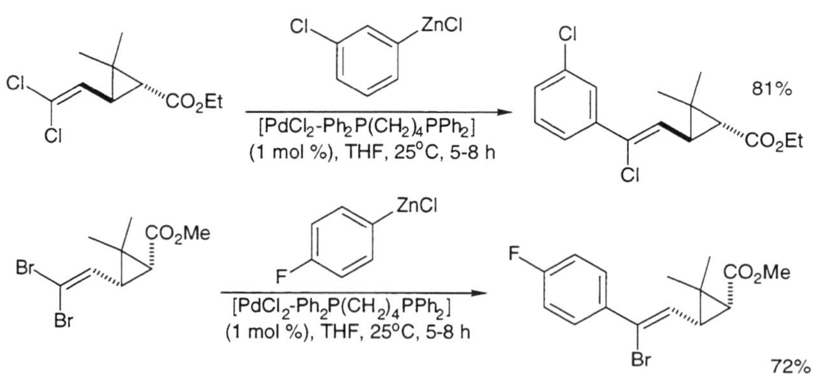

(viii) Palladium (0)-catalysed allylic substitution of β-substituted styrenes that contain a leaving group at the γ-position offers unique possibilities for the preparation of functionalised styrenes. For instance, treatment of the styryltosylate found in Scheme 14 with sodium azide, 15-crown-5, and 5 mol % of bis(dibenzylideneacetone)palladium (0) [$Pd(dba)_2$] in the presence of 10 mol % triphenylphosphine led to the γ-azido-γ-cyclopropylstyrene being isolated in 79% yield (P. Aufranc *et al.*, *Tetrahedron Lett.*, **1993**, *34*, 4193).

Scheme 14

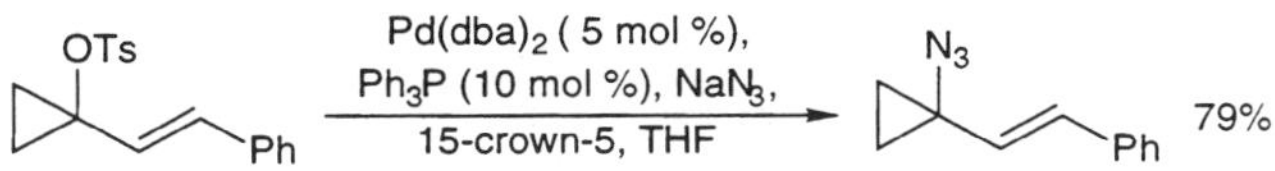

A. Pfaltz and coworkers (*Tetrahedron*, **1992**, *48*, 2143) have observed that (allyl)palladium (II) complexes of chiral aza-semicorrins are superb catalysts for the enantioselective allylic alkylation of racemic (*E*)-1,3-diphenyl-3-acetoxy-1-propene with dimethyl malonate and *N,O*-bis-(trimethylsilyl)acetamide (Scheme 15).

Scheme 15

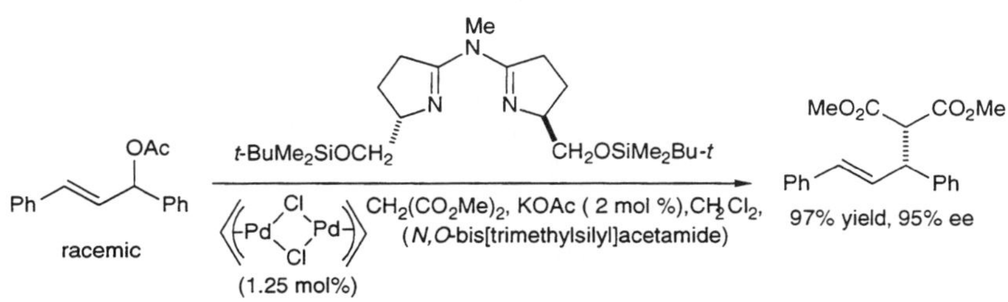

C.G. Frost and J.M.J. Williams (*Tetrahedron:Asymmetry*, **1993**, *4*, 1785) have also examined this reaction with the structurally-related chiral oxazoline ligand shown in scheme 16 and obtained the enantiomeric product in 86% yield and 80% ee.

Scheme 16

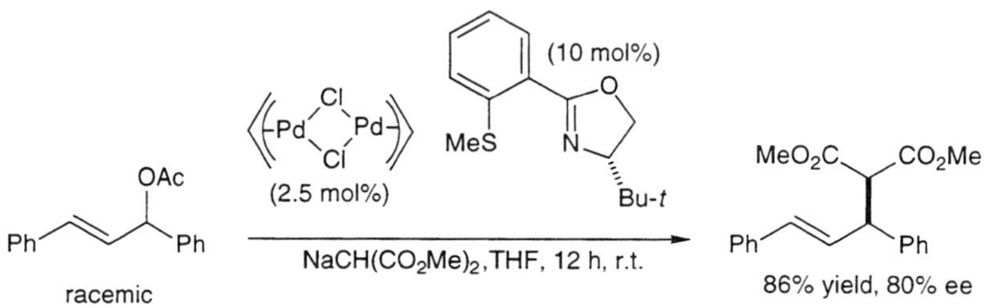

A chiral phosphine ligand of similar lineage has been developed by J. Sprinz and G. Helmchen (*Tetrahedron Lett.*, **1993**, *34*, 1769) (Scheme 17). It leads to even higher ee in this alkylation reaction; the chiral γ-substituted styrene being produced in 99% yield and 97.5% ee.

Scheme 17

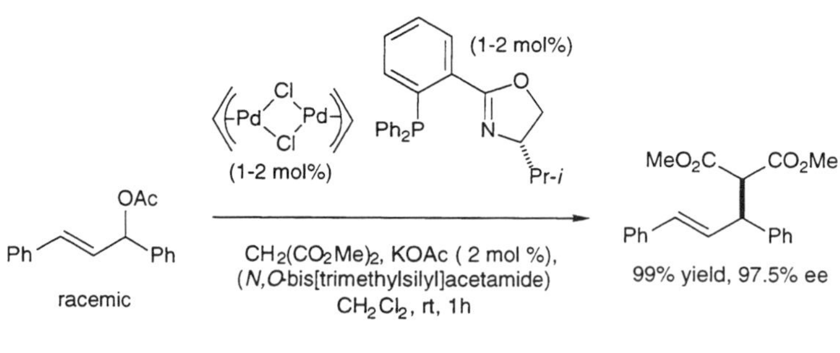

T. Hayashi *et al.*, (*Pure & Appl. Chem.*, **1988**, *60*, 7; *J. Chem. Soc. Chem. Comm.*, **1986**, 1090) have discussed the issues that determine the regio- and stereochemical outcome of the palladium-catalysed asymmetric allylic substitution of racemic 1-[(*E*)-1-styryl]-2-methylpropyl acetate mediated by the chiral ferrocenylphosphine ligand shown in Scheme 18. When carried out to 80% product conversion, with 0.5 equivalents of sodium acetylacetonate in THF at 40°C for 3 days, the reaction delivered optically pure (*R*)-acetate in 20% yield, along with 30% of the γ-alkylation product in 94% ee, and 48% of the α-alkylation product, formed in 16% ee (Scheme 18).

Scheme 18

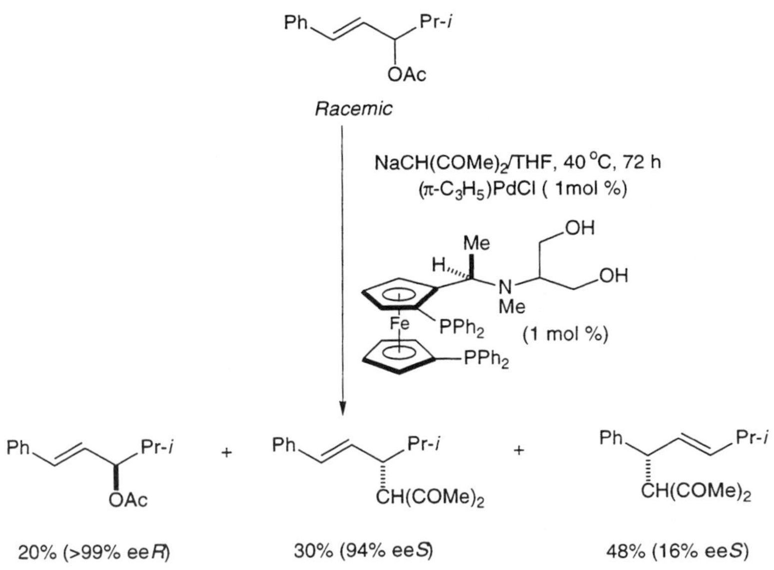

Results after 80% chemical conversion

(ix) W.E. Crowe and Z.J. Zhang (*J. Am. Chem. Soc.*, **1993**, *115*, 10998) have disclosed a highly selective method for assembling (*E*)-α,β-disubstituted styrenes via cross-metathesis

Scheme 19

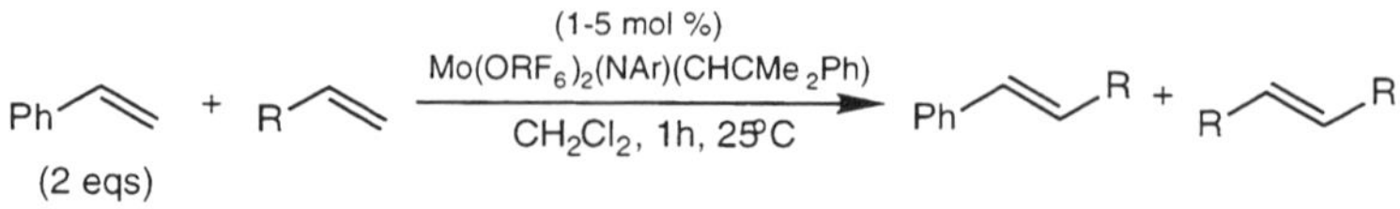

R	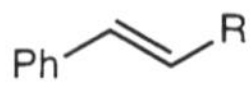 *(%)*	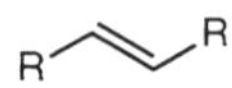 *(%)*
n-Hexyl	89	2
$CH_2CH_2C(=CH_2)CH_3$	86	2
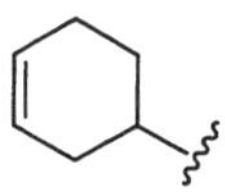	60	15
$CH_2CH_2CH_2Br$	90	4
$CH_2CH_2CH(Br)CH_3$	84	8
CH_2CH_2Br	50	42
$CH_2CH_2COCH_3$	11	Not determined
$CH_2CH_2CO_2Bn$	67	20
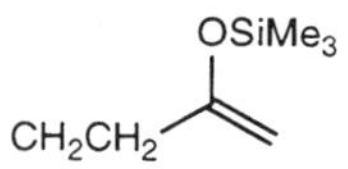	66	Not determined
$CH_2CH_2CH_2OBn$	85	2

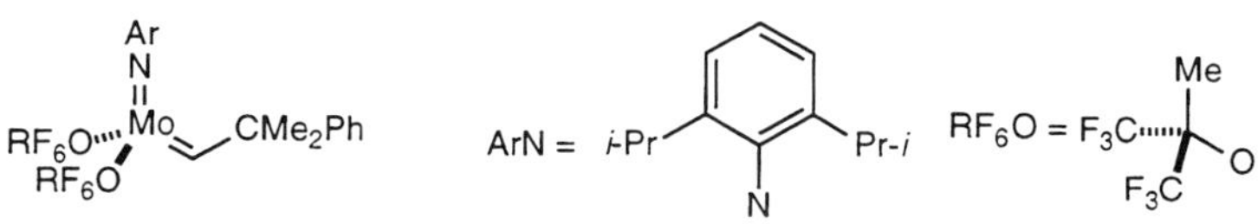

between terminal styrenes and terminal alkyl-substituted olefins (Scheme 19). The reactions are mediated by catalytic quantities of the complex [$Mo(ORF_6)_2(NAr)(CHCMe_2Ph)$], first prepared by Schrock *et al.* (*Organometallics*, **1987**, *6*, 1373;

J. Am. Chem. Soc., **1990**, *112*, 3875). As a rule, an excess of the terminal styrene component is employed. This does not usually cause problems since styrenes show little tendency towards undergoing self-metathesis under these conditions. The high (*E*)-selectivity observed in this cross-metathesis process, along with the formation of ethene as the sole reaction by-product, make it a transformation of quite considerable synthetic utility. Mechanistically, the (*E*)-selectivity is believed to arise from a cycloreversion reaction of a *trans*-α,β-disubstituted metalla-cyclobutane intermediate.

(x) A spectacular route to highly functionalised β-substituted styrenes has been developed by J. E. Baldwin and D.R. Kelly (*J. Chem. Soc. Chem. Comm.*, **1985**, 682). It is based on the intermolecular addition of an alkyl free radical to the β-position of a β-stannylstyrene followed by the β-elimination of a tri-*n*-butylstannyl radical (Scheme 20). In a typical experiment, an alkyl bromide is heated at 86 ^{o}C in toluene with β-tri-*n*-butylstannylstyrene (2 equiv) for 24-36 hours, with azobisisobutyronitrile (AIBN) (16 mol %) being added every 12 hours. The resulting β-alkylated (*E*)-styrenes are isolated in yields that vary between 22 and 82%. Some particularly meritorious features of this process are its operation at neutral pH, and its compatibility with hydroxyl groups in the starting alkyl halide.

Scheme 20

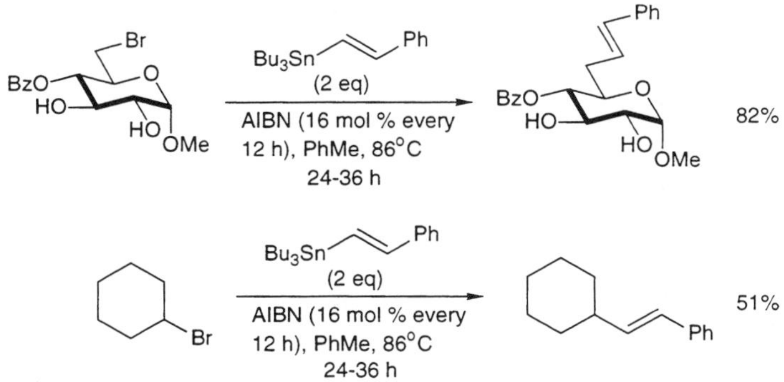

Scheme 20 (continued)

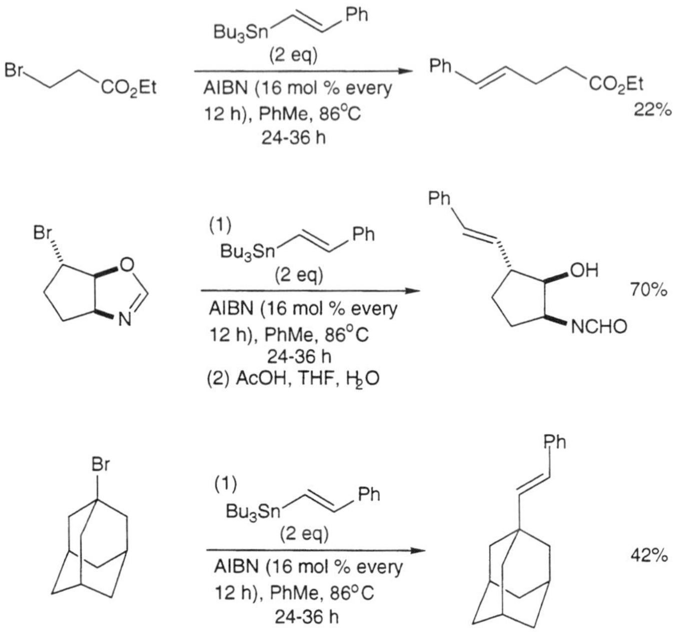

G.A. Russell, H. Tashtoush, and P. Ngoviwatchai (*J. Am. Chem. Soc.*, **1984**, *106*, 4622) have also found that alkyl radicals generated thermally or photochemically from alkylchloromercurials will add regioselectively to the β-position of β-tributylstannylstyrene and instigate loss of $Bu_3Sn\cdot$. A chain reaction

Scheme 21

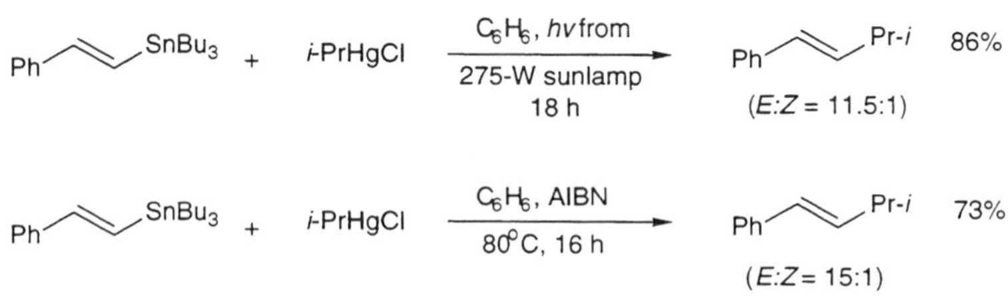

then ensues, with $Bu_3Sn\cdot$ attacking the RHgCl to regenerate the alkyl radical by either an electron transfer or a halogen atom transfer process. The method clearly provides a very versatile route to a wide range of (*E*)-β-substituted styrenes (Scheme 21).

(xi) Cross-coupling of (*E*)-β-styryl bromides with bromozinc ester enolates can be achieved with retention of alkene stereochemistry if carried out in the presence of a $(Ph_3P)_4Pd$ catalyst (Scheme 22). The reaction complements earlier work

Scheme 22

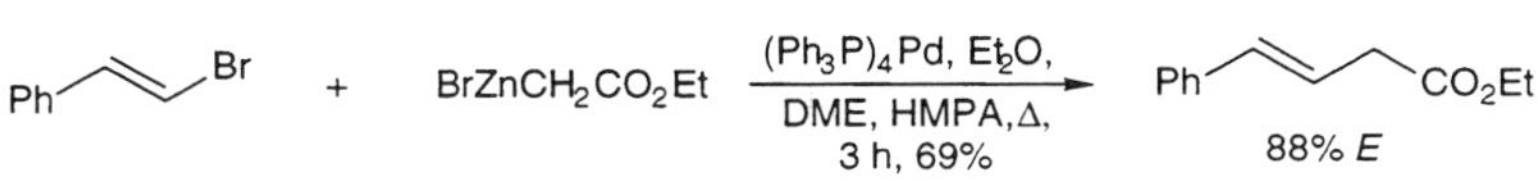

by A.A. Millard and M.W. Rathke (*J. Am. Chem. Soc.*, **1977**, *99*, 4833) who found that the combination of nickel (II) bromide and *n*-butyllithium in tetrahydrofuran at -78°C leads to a Ni (0) catalyst that permits the efficient cross-coupling of β-bromostyrene with the lithium enolate of *t*-butyl acetate. In the latter case, the β,γ-unsaturated ester was obtained in 83% yield, with complete retention of olefin configuration, ruling out the possibility of radical intermediates in the reaction mechanism.

(xii) The palladium (0)-catalysed cross-coupling reaction of aryl triflates with vinylstannanes provides a pathway for accessing a wide range of functionalised styrenes under very mild, neutral, conditions (Scheme 23) (A.M. Echavarren and J.K. Stille, *J. Am. Chem. Soc.*, **1987**, *109*, 5478). Lithium chloride is an essential additive for these reactions, it allowing a catalytically-active aryl palladium chloride complex to be generated which can then participate in transmetallation and allow the catalytic cycle to continue; without it, decomposition of the Pd(0) complex takes place under the reaction conditions.

Scheme 23

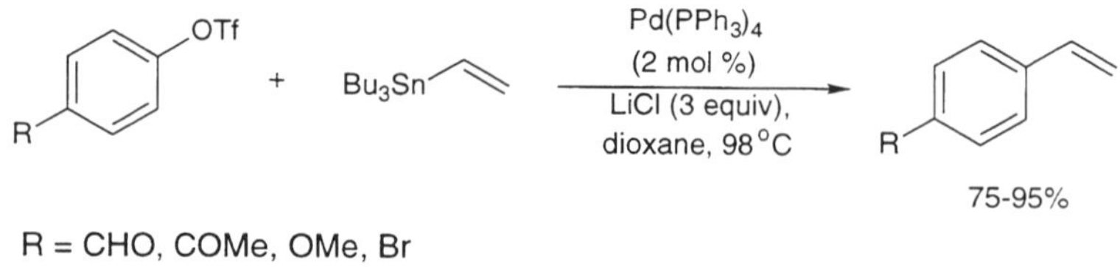

Significantly, *p*-iodophenyltriflate reacts exclusively with vinylstannanes at the carbon-iodine bond when $Pd(PPh_3)_4$ is used as the catalyst, enabling sequential cross-couplings to be achieved (Scheme 24).

Scheme 24

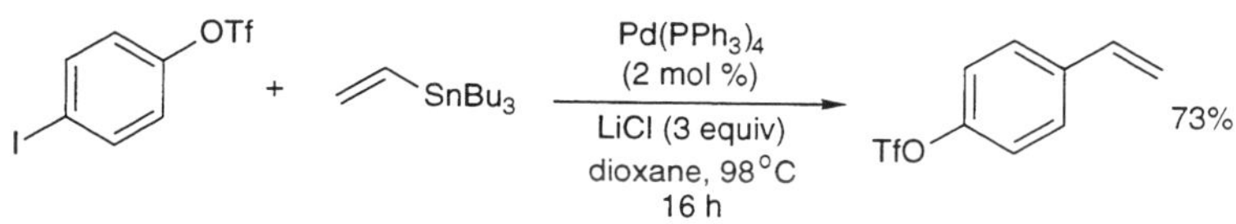

An important facet of the Stille coupling is its ability to reproducibly transfer (*E*)-alkene geometry from an (*E*)-1,2-disubstituted vinyl stannane to a styrene product (Scheme 25).

Scheme 25

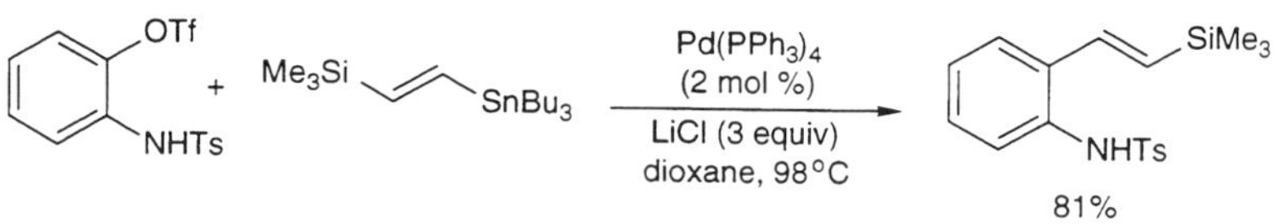

(xiii) A number of very convenient procedures now exist for the preparation of (*E*)-β-styryl iodides. One of these entails hydrozirconation of phenylacetylenes with *i*-$BuZrCp_2Cl$, generated *in situ* by treatment of Cp_2ZrCl_2 with *t*-BuMgCl (D.R.

Swanson, T. Nguyen, Y. Noda, and E. Negishi, *J. Org. Chem.*, **1991**, *56*, 2590). It is thought that hydrozirconation proceeds via

Scheme 26

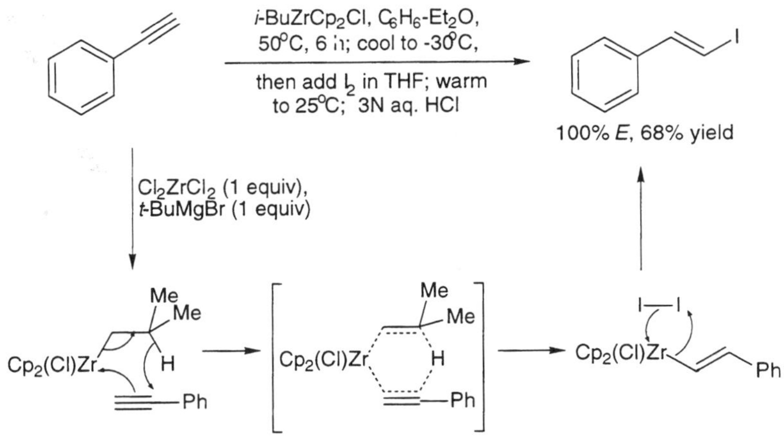

a concerted six-centred transition state in which the $Cp_2(Cl)Zr$ and β-H groups are delivered *syn* to the triple bond. The resulting (*E*)-β-styryl zirconocene chlorides are iodinated by treatment with iodine (1.5 equiv.) at -30°C and warming to room temperature (Scheme 26).

Another stereoselective route to (*E*)-β-styryl iodides is predicated on the reaction between benzaldehydes and the reagent obtained from $CrCl_2$ and iodoform in tetrahydrofuran at 0°C (Scheme 27) (K. Takai, K. Nitta, and K. Utimoto, *J. Am. Chem. Soc.*, **1986**, *108*, 7408). An alternative Cr(II) reagent, obtained by reduction of $CrCl_3$ with $LiAlH_4$, is equally effective for this reaction. Replacing iodoform with bromoform or chloroform gives access to the corresponding (*E*)-β-styryl bromides and chlorides respectively but in much lower yields (32-43%). A mechanism involving geminal carbodianion species has been invoked to rationalise formation of the (*E*)-β-styryl halide products (Scheme 27).

Scheme 27

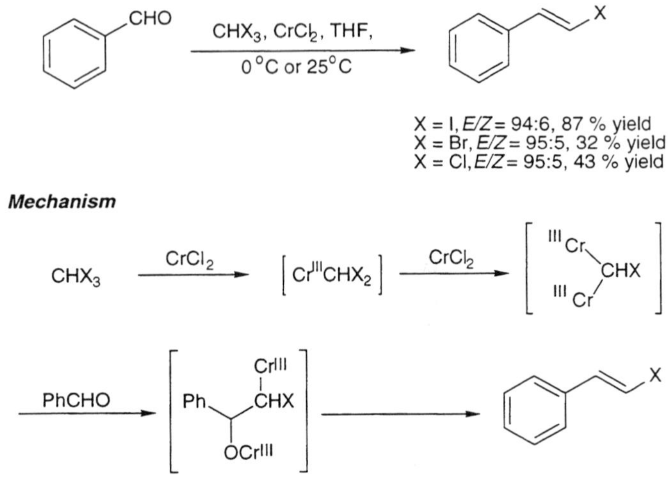

It has long been known that bromocyclooctatetraene rearranges to (*E*)-β-bromostyrene when heated at 80 °C (R. Huisgen *et al., J. Am. Chem. Soc.*, **1970**, *92*, 4102; 4104; and 4105). P.E. Eaton and D. Stossel (*J. Org. Chem.*, **1991**, *56*, 5138) have observed that when certain cubyl iodides are heated in toluene with 7 mol % of $[Rh(COD)Cl]_2$,

Scheme 28

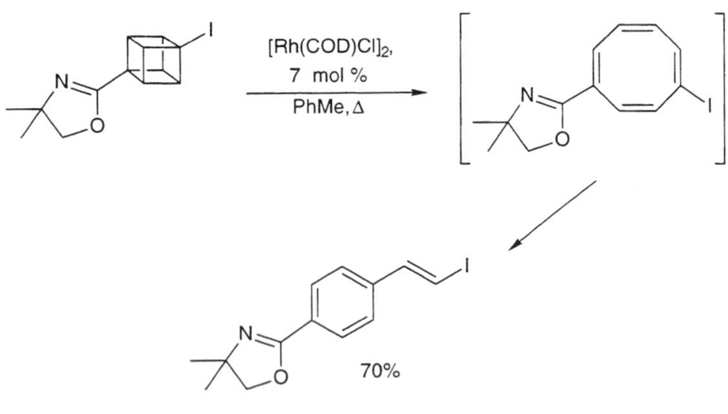

a rearrangement can occur to the tricyclooctadiene; this can then open thermally to the iodocyclooctatetraene. The latter will subsequently form substituted (*E*)-β-styryl iodides (Scheme 28).

(xiv) The reaction of benzaldehyde derivatives with zinc and 2-bis(chlorodimethylsilyl)ethane in THF at -50 ^{o}C affords symmetrical stilbenes in a single step (W.B. Motherwell *et al., Tetrahedron Lett.*, **1992**, *33*, 3899). Optimal product yields are encountered when the starting benzaldehydes bear electron-releasing groups.

(xv) β-Bis(trimethylsilyl)aminostyrene can be stereoselectively prepared by the addition of $Sn[N(SiMe_3)_2]$ to 2-phenylethanal (Scheme 29) (C. Burnell-Curty and E.J. Roskamp, *J. Org.*

Scheme 29

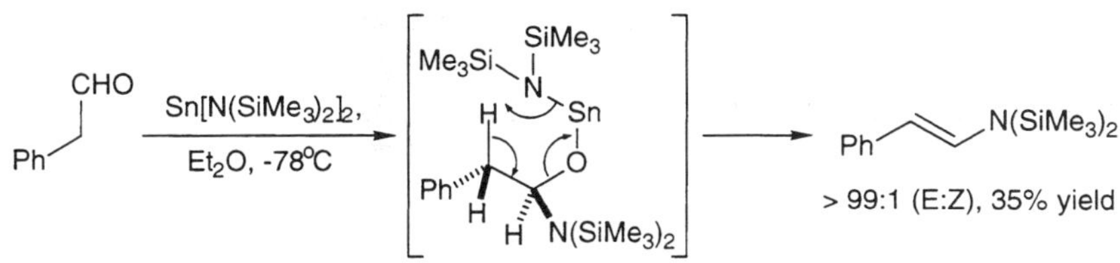

Chem., **1992**, *57*, 5063). In this reaction, coordination of the aldehyde carbonyl to the tin (II) amide is believed to initiate the transfer of a silazane ligand to the aldehyde carbonyl and generate a tin (II) alkoxy amide. This then participates in a *syn*-elimination to generate the *trans*-styryl enamine.

(xvi) Iron phthalocyanine [Fe(Pc)] has been found to catalyse the allylic amination of aromatic alkenes with phenylhydroxylamine (M. Johannsen and K.A. Jørgensen, *J. Org. Chem.*, **1994**, *59*, 214). A range of substituted styrylamine derivatives have been prepared in modest yield using this methodology. Mechanistically, the phenylhydroxylamine is believed to complex with the iron atom of the Fe(Pc) catalyst, to weaken the N-O bond. An ene-type allylic amination then occurs between the activated hydroxylamine and the alkene (Scheme 30).

Scheme 30

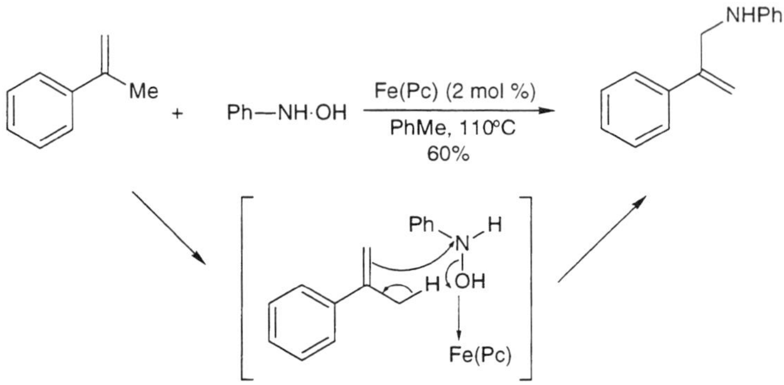

Alkene	Product	% Yield
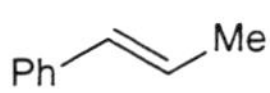	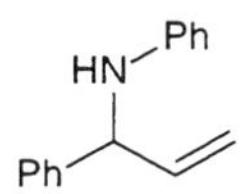	5
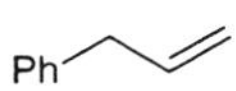	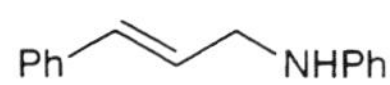	30
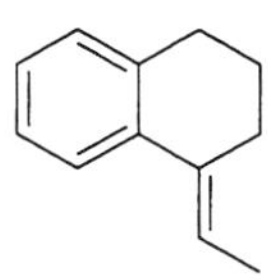	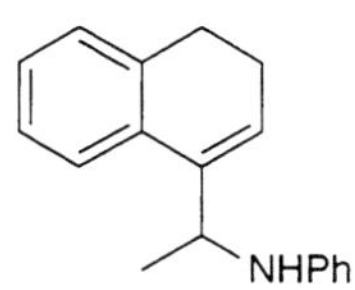	45
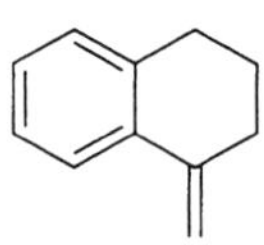	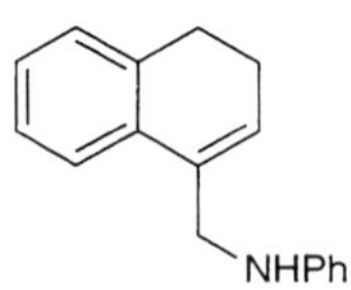	55

(b) Reactions of styrenes and derivatives

(i) The intermolecular addition of alkyl radicals to carbon monoxide produces acyl radicals that themselves add to styrene in modest yield (Scheme 31) (I. Ryu *et al., J. Org. Chem.,*

Scheme 31

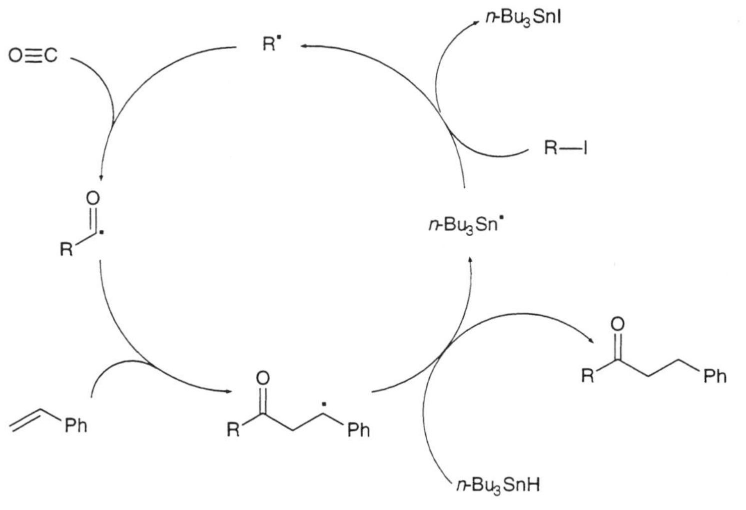

1991, *56*, 5003). Ordinarily, these double alkylation reactions proceed most efficiently when conducted at high dilution, this apparently favouring addition of the alkyl radical to the CO rather than the styrene. Two examples of this free radical mediated carbonylation reaction being applied to styrene are shown in scheme 32.

Scheme 32

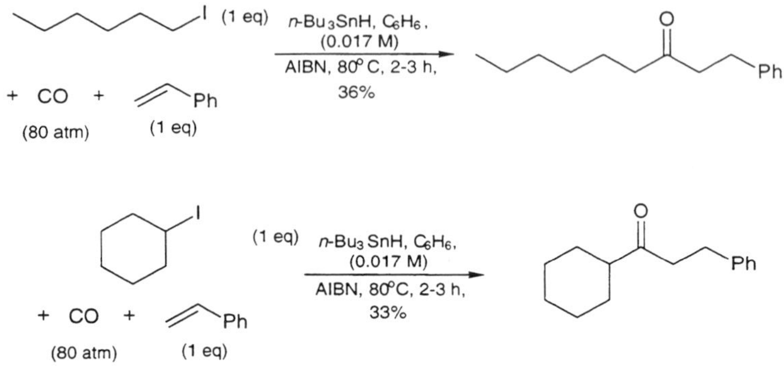

(ii) M. Brookhart *et al.* (*J. Am. Chem. Soc.*, **1983**, *105*, 6721) have devised a method for generating the chiral cationic ethylidene complex $Cp(CO)(Ph_2PR^*)Fe{=}CHCH_3^+$ *in situ.*

Scheme 33

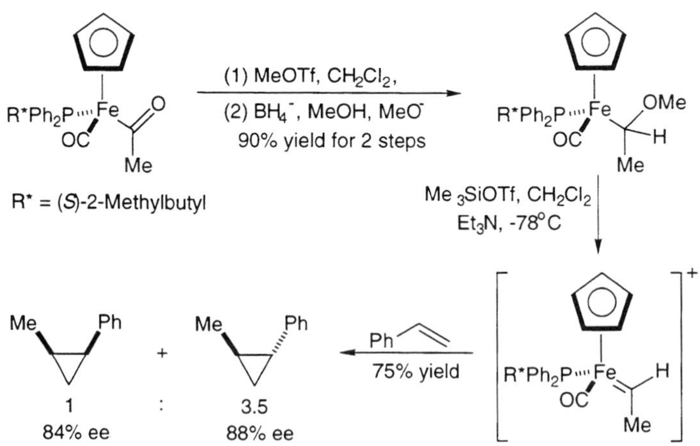

The latter undergoes efficient transfer of ethylidene to styrene to give *cis*- and *trans*-1-methyl-2-phenylcyclopropanes in 75% yield and $\geq$ 84% ee (Scheme 33).

Scheme 34

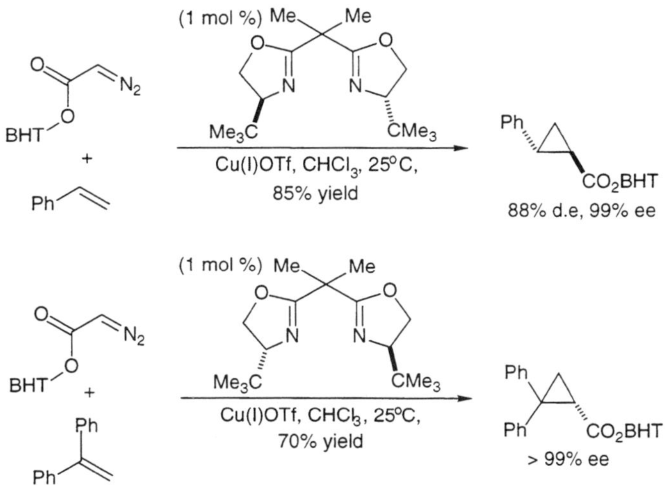

BHT = 2,6-di-*tert*-butyl-4-methylphenyl

Chiral complexes derived from bis(oxazoline) ligands and copper (I) triflate (D.A. Evans *et al., J. Am. Chem. Soc.*, **1991**, *113*, 726) have proven excellent catalysts for the enantioselective cyclopropantion of styrene and 1,1-diphenylethylene with 2,6-di-*tert*-butyl-4-methyl-phenoxy diazoacetate (Scheme 34).

T. Katsuki and coworkers (*Synlett.*, **1992**, 575; *Tetrahedron Lett.*, **1993**, *34*, 2661) have examined the utility of several chiral bis(tetrahydroquinoline)-copper (I) triflate complexes as catalysts for the asymmetric cyclopropanation of styrene with *tert*-butyl diazoacetate. One of these complexes, shown in Scheme 35, leads to the *trans*-cyclopropanated product being formed in 75% yield, with 72% d.e. and 92% ee.

Scheme 35

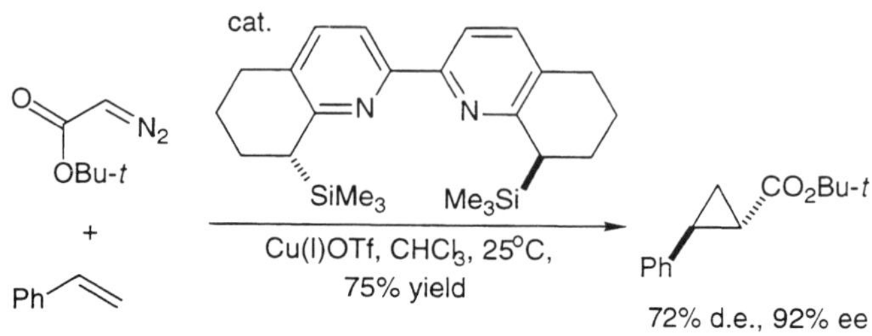

Much lower levels of stereoselectivity are encountered when (*S*)-valinol-derived bis(oxazolinyl)pyridine ligands are employed for the asymmetric cyclopropanation of styrene with *tert*-butyl diazoacetate with copper triflate (see for example, Scheme 36) (A.D Gupta, D. Bhuniya, and V.K. Singh, *Tetrahedron*, **1994**, *50*, 13725).

Scheme 36

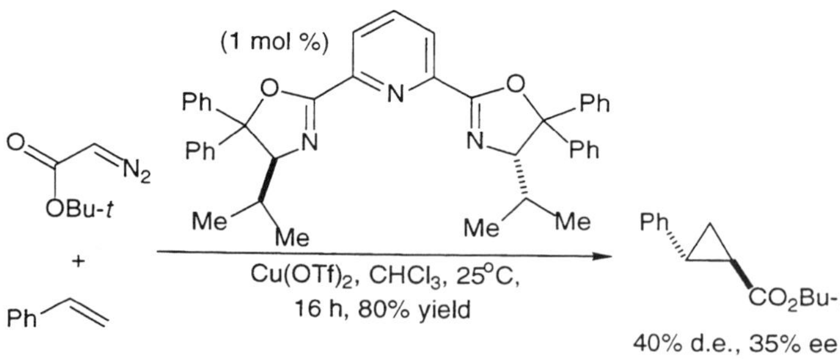

(iii) Radical cation salts such as [(4-$BrC_6H_4)_3N^{+\cdot}$ $SbCl_6^-$] have been shown to catalyse the intramolecular thermal [2+2]-cycloaddition reactions of styrenes that contain a 4-methoxy group in the phenyl ring (Scheme 37) (N.L. Bauld *et al.*, *J. Am. Chem. Soc.*, **1993**, *115*, 7653). Typically, these reactions are extremely fast, being complete within 10-15 seconds at 0°C.

Scheme 37

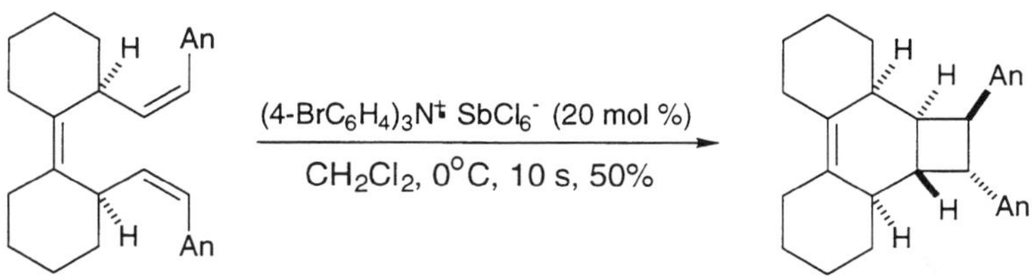

An elegant synthesis of (+)-β-cuparenone (**XVI**) by A.E. Greene *et al.* (*J. Am. Chem. Soc.*, **1987**, *109*, 4752) has demonstrated that high levels of asymmetric induction (95:5) are attainable in the [2+2]-cycloadditions of dichloroketene to chiral β-(2-phenylcyclohexyloxy)styrenes. In a rather concise route to **XVI** (Scheme 38), enol ether **XI** was stereospecifically assembled in 75% yield by base-mediated isomerisation of allylic ether **X** with potassium *tert*-butoxide in dimethylsulfoxide at 65°C. For steric reasons, enol ether **XI** was thought to adopt the s-*trans* conformation, in which the C_α-*si* face was sterically shielded by the phenyl ring of the chiral auxiliary. In accord with this thinking, dichloroketene regiospecifically added to **XI** from its more accessible C_α-*re* face to produce cyclobutanone **XII** in 92% yield. Ring expansion was then performed with diazomethane; it delivered the desired dichlorocyclopentanone **XIII**, which was liberated from the chiral auxiliary by reductive elimination with chromous perchlorate. The overall sequence to **XVI** was completed by chemoselective 1,4-conjugate addition of lithium dimethylcuprate, base-induced elimination, and nickel (0)-catalysed conjugate addition of dimethylzinc.

Scheme 38

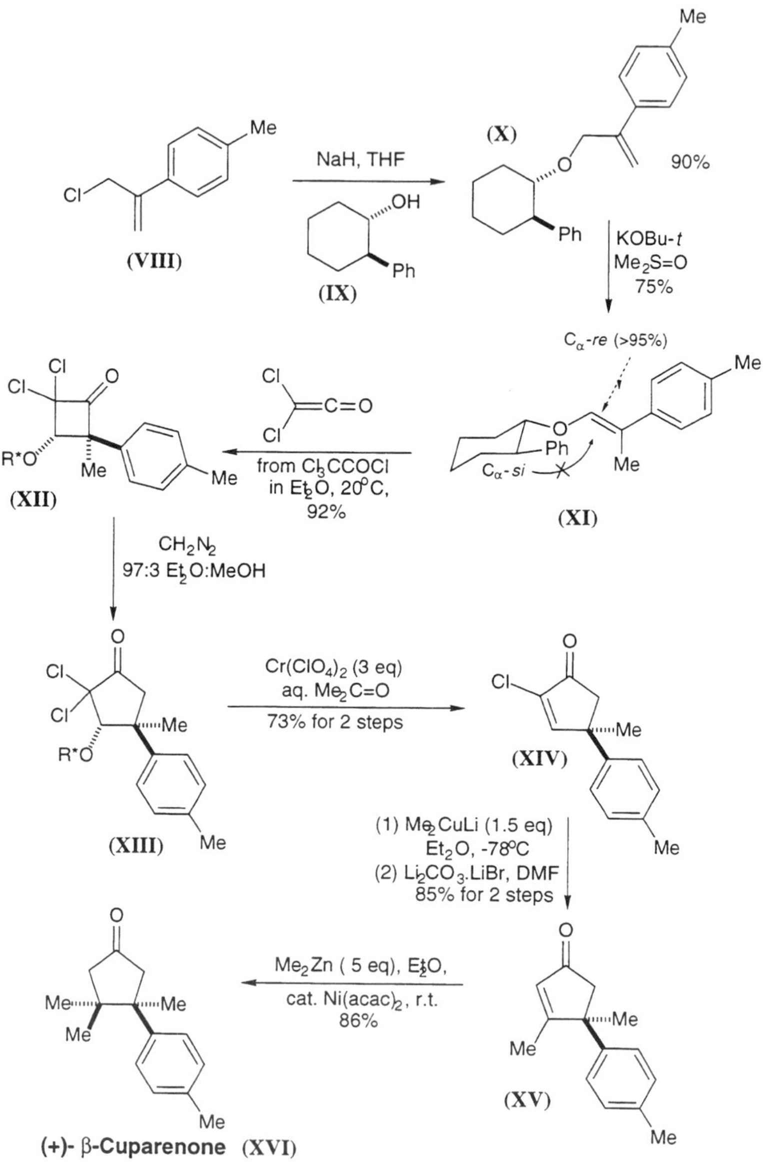

(VIII)
NaH, THF
(IX)
(X)
90%
KOBu-t
$Me_2S=O$
75%
C_α-re (>95%)
C_α-si
(XI)
from Cl_3CCOCl
in Et_2O, 20°C,
92%
(XII)
CH_2N_2
97:3 Et_2O:MeOH
(XIII)
$Cr(ClO_4)_2$ (3 eq)
aq. $Me_2C=O$
73% for 2 steps
(XIV)
(1) Me_2CuLi (1.5 eq)
Et_2O, -78°C
(2) Li_2CO_3.LiBr, DMF
85% for 2 steps
(XV)
Me_2Zn (5 eq), Et_2O,
cat. $Ni(acac)_2$, r.t.
86%
(+)- β-Cuparenone (XVI)

(iv) Styrenes are frequently unwilling participants in Diels-Alder reactions, even when partnered with reactive dienes such as 2,3-dimethyl-1,3-butadiene (K. Alder and H.F. Rickert, *Chem. Ber.,* **1938**, *70*, 379) and *trans*-1-(diethylamino)-1,3-butadiene (G. Satzinger, *Justus Liebigs Ann. Chem.,* **1969**, *728*, 64). It was striking therefore to discover that electron-rich and electron-deficient monosubstituted styrenes participate readily in [4+2]-cycloadditions with 1-(acylamino)-1,3-dienes (Scheme 39) (L.E. Overman *et al., J. Am. Chem. Soc.*, **1981**, *103*, 2816). *Endo*-cycloadducts are generally formed in excellent yield provided the reactions are conducted in the absence of oxygen and include a free radical inhibitor such as 4-*tert*-butylcatechol.

Scheme 39

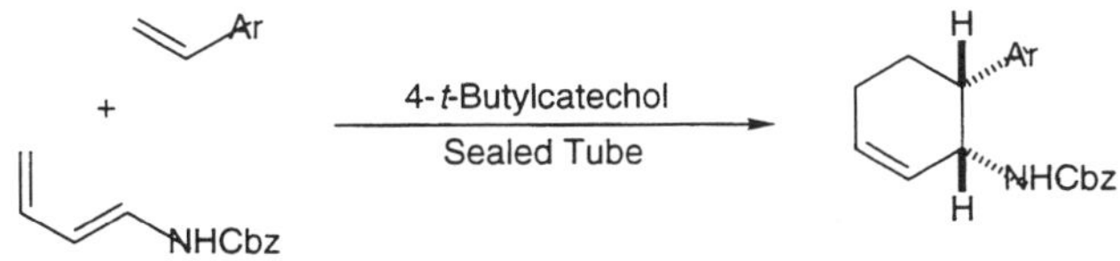

Ar	*Reaction Conditions*	*Reaction Time (h)*	*% Yield*
Ph	Styrene, 80°C	256	100
3-$NO_2C_6H_4$	Dioxane, 110°C	25	99
4-$MeOC_6H_4$	4-Methoxystyrene, 80°C	256	95
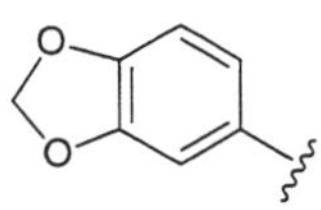	Dioxane, 140°C	144	70

(v) M. Terada and K. Mikami (*J. Chem. Soc. Chem. Comm.*, **1994**, 833) have found that the bis (*R*)-binapthol-derived titanium oxo complex shown in Scheme 40 is an efficient catalyst for the carbonyl-ene reaction between glyoxalate and α-methylstyrene. This delivers the α-hydroxyester in 98.7% ee and 88% chemical yield (Scheme 40).

Scheme 40

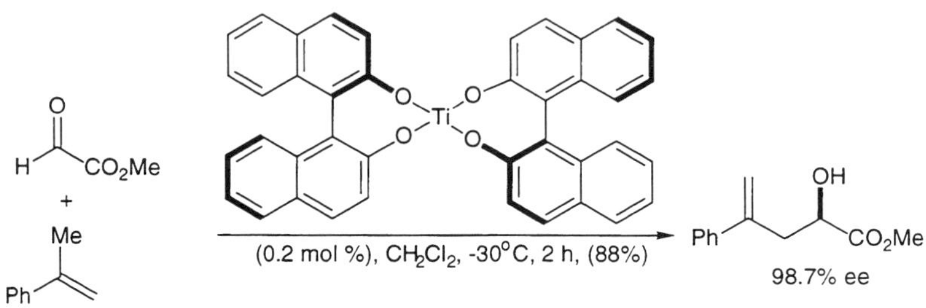

(vi) A number of carbanions add readily to the β-position of η^6-styrylchromium tricarbonyl complexes to generate stabilised benzylic anions that can be trapped with iodomethane, acid chlorides, or diphenyldisufide (Scheme 41) (M. Semmelhack, W. Seufert, and L. Keller, *J. Am. Chem. Soc.*, **1980**, *102*, 6584). It is customary to liberate the free arenes from the metal-complexed products by oxidation with excess iodine or ceric ammonium nitrate. As a rule, η^6-styryl-$Cr(CO)_3$ complexes tend to be rather unreactive towards carbanions that are highly stabilised. For example, the anion of diethylmalonate fails to add to η^6-styryl-$Cr(CO)_3$, while the lithium enolate of *tert*-butylmethyl ketone furnishes the desired β-addition product in only 10% yield. 2-Lithio-1,3-dithiane reacts with this metal complex and its α-methylstyryl counterpart, but in both cases the expected addition products are not obtained. The presence of an anion-stabilising SEt group at the benzylic position appears to expedite nucleophilic addition, but a +I substituent has a deleterious effect.

η^6-Styrylchromium tricarbonyl complexes are best prepared through the complexation of a styrene with chromium hexacarbonyl, or by a sequence involving metallation of benzenechromium tricarbonyl, addition of an aldehyde or ketone, and subsequent dehydration.

Scheme 41

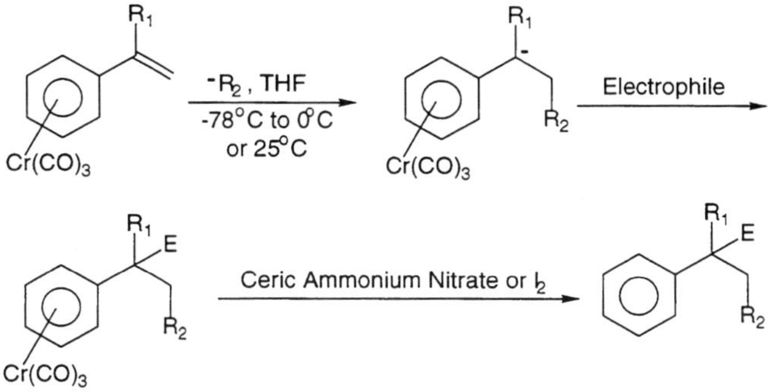

R_1	Lithiated Anion Unit, R_2	Electrophile	Product E	% Yield
H	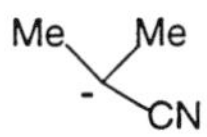	MeI	Me	82
H	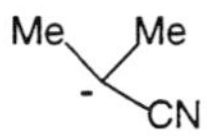	MeCOCl	COMe	71
H	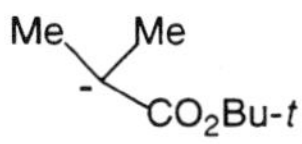	H^+	H	73
Me	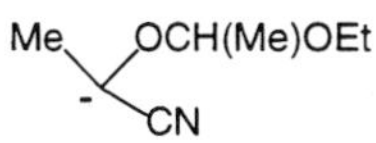	H^+	H	62
Me	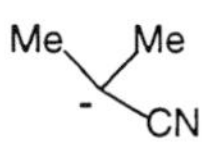	H^+	H	34
EtS	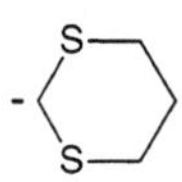	H^+	H	41
EtS	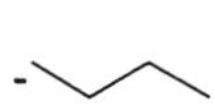	H^+	H	61

(vii) D.A. Evans, M.M. Faul, and M.T. Bilodeau (*J. Org. Chem.*, **1991**, *56*, 6744) have successfully aziridinated a range of substituted styrenes with (*N*-(*p*-toluenesulfonyl)imino)-phenyliodinane under Cu(I) or Cu(II) catalysis (Scheme 42). These nitrene insertion reactions proceed most readily when conducted in dipolar aprotic solvents such as acetonitrile or nitromethane. Solvents of lower polarity, such as dichloromethane and toluene, generally prove less satisfactory, causing extended reaction times and decomposition of the nitrene precursor.

Scheme 42

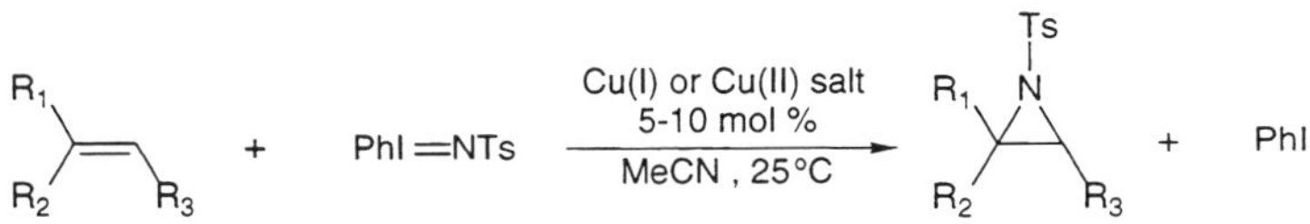

Alkene	*Catalyst*	*% Yield*
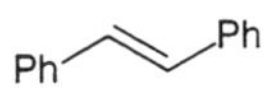	$Cu(MeCN)_4ClO_4$ $Cu(acac)_2$	81 50
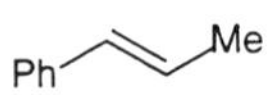	$Cu(MeCN)_4ClO_4$ $Cu(acac)_2$	81 75
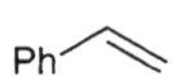	$Cu(MeCN)_4ClO_4$ $Cu(acac)_2$	89 95
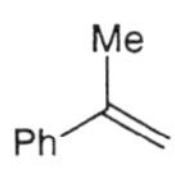	$Cu(MeCN)_4ClO_4$ $Cu(acac)_2$	89 72
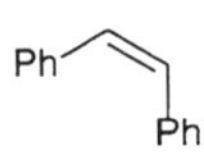	$Cu(MeCN)_4ClO_4$ $Cu(acac)_2$	80 54

Investigations with α-trimethylsilyloxystyrenes have revealed that direct α-amination is achievable with PhI=NTs under Cu(I)-catalysis. Respectable yields (*ca.* 75%) of *N*-(*p*-toluenesulfonyl)-α-amino ketones are usually obtained via internal ring-opening of the initially formed aziridine intermediate (Scheme 43).

Scheme 43

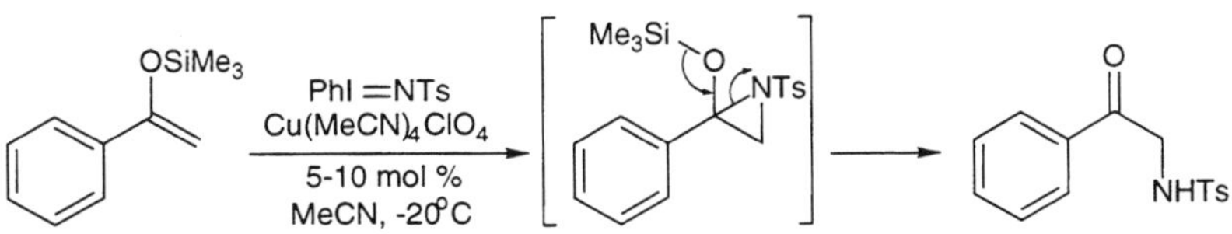

Chiral 4,4'-disubstituted bis(oxazolines) have proven useful ligands for the copper-catalysed enantioselective aziridination of mono- and (*E*)-1,2-disubstituted styrenes with PhI=NTs (D.A. Evans *et al., J. Am. Chem. Soc.*, **1993**, *115*, 5328) (Scheme 44). Ordinarily, 1,2-*trans*-disubstituted styrenes react with opposite facial-selectivity to styrene under these conditions.

Scheme 44

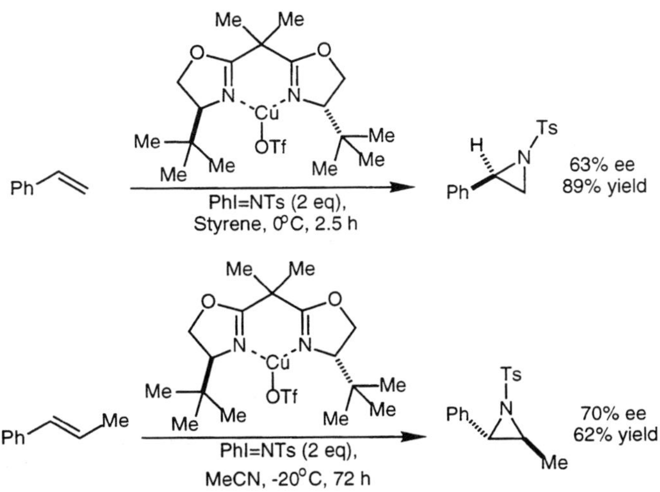

Optically active *O*-alkyl benzylidene derivatives of 1,2-diaminocyclohexane and Cu(I)OTf have been evaluated as catalysts for the asymmetric aziridination of mono- and *cis*-1,2-disubstituted styrene derivatives (Scheme 45) (Z. Li, K.R. Conser, and E.N. Jacobsen, *J. Am. Chem. Soc.,* **1993**, *115*, 5326). As with the chiral bis(oxazoline)-copper (I) triflate aziridination systems, the extent of asymmetric induction is often modest.

Scheme 45

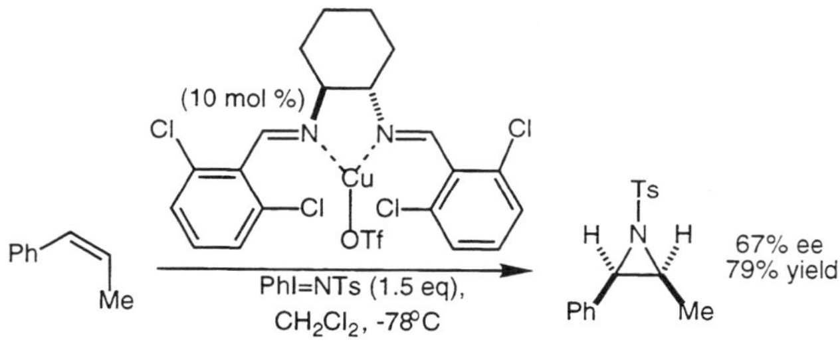

(Salen)manganese (III) complexes are another class of chiral catalysts that have been studied for the asymmetric aziridination of styrene derivatives with TsN=IPh. However, they usually lead to inferior results both in terms of chemical yield and product ee when compared with the aforementioned copper catalysts (Scheme 46) (K. Noda, N. Hosoya, R. Irie, Y. Ito, and T. Katsuki, *Synlett*, **1993**, 469).

Scheme 46

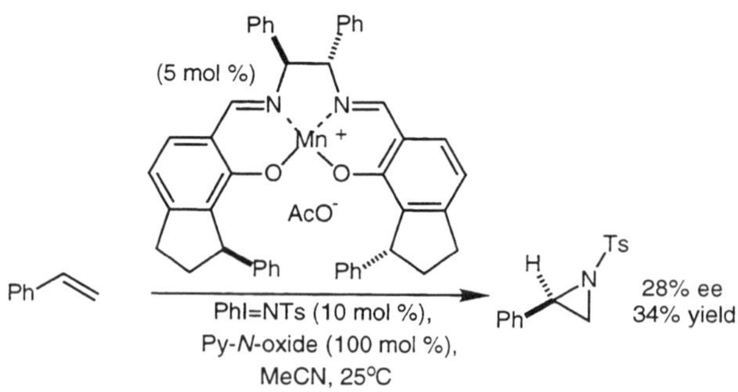

(viii) A novel method for the epoxidation of substituted styrene derivatives has been developed by N.L. Bauld *et al.* (*J. Am. Chem. Soc.*, **1993**, *115*, 7653). It is based on the reaction of a styrene with 20 mol% of tris(4-bromophenyl)aminium hexachloroantimonate and a stoichiometric amount of benzeneselenic anhydride or selenium dioxide in dichloromethane at 0°C (Scheme 47). Initially the substituted styrene is believed to donate a single electron to the electron-deficient aminium salt to create a reactive styryl radical-cation.

Scheme 47

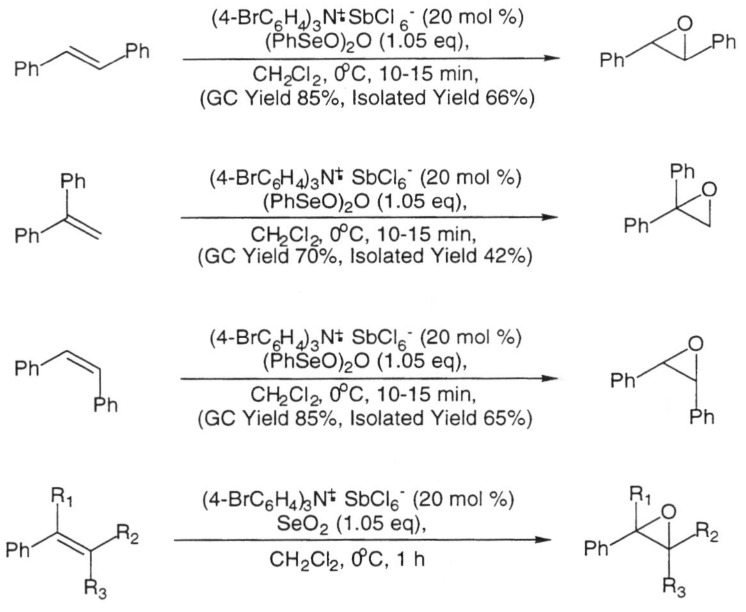

The latter is then oxidised by the selenium reagent to give a styrene oxide radical cation, which accepts an electron from another substituted styrene to continue the epoxidation cycle (Scheme 48).

Scheme 48

Mechanism of the Aminium Salt (Hole) Catalysed Oxidation of Stilbene by SeO_2

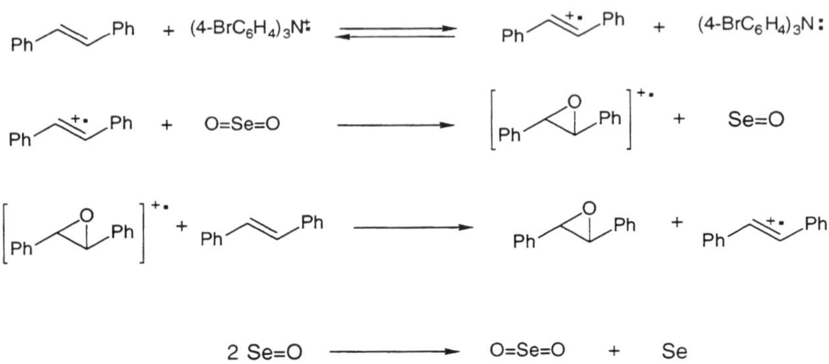

If potassium superoxide is mixed with carbon tetrachloride in acetonitrile, in the presence of 18-crown-6, an activated complex is formed that dissociates into chloride ion, dioxygen, and a CCl_3 radical. The CCl_3 radical then combines with dioxygen to create a Cl_3COO radical that epoxidises styrenes at 10 ^{o}C in somewhat capricious yield (33-83%) (M. Hirobe *et al., J. Am. Chem. Soc.,* **1986**, *108*, 539).

A range of chiral (salen) Mn (III) complexes have recently been introduced for the catalytic enantioselective epoxidation of mono- and di-substituted styrenes in the presence of a co-oxidant such as 2,4,6-trimethyl-1-iodosylbenzene (E.N. Jacobsen *et al., J. Am. Chem. Soc.*, **1990**, *112*, 2801). Although the asymmetric induction recorded with (*Z*)-β-methylstyrene was reasonably high, a significant drop in ee was encountered when α-methylstyrene, (*E*)-β-methylstyrene, and styrene itself were substrates (Scheme 49).

Scheme 49

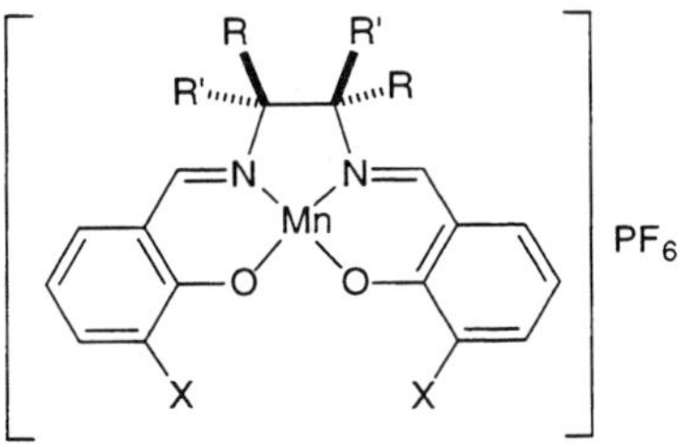

(*S,S*)-1: R = Ph, R' = H, X = H
(*R,R*)-2: R = H, R' = Ph, X = *t*-Bu

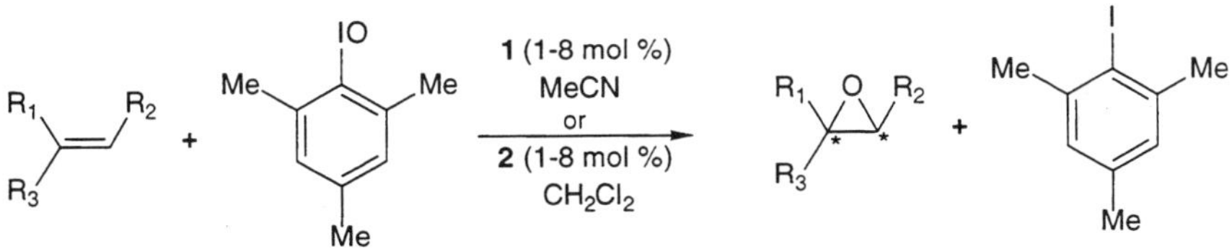

Alkene	*Chiral Catalyst*	*% Yield*	*% ee*	*Configuration*
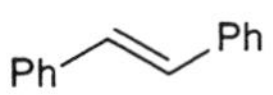	(*S,S*)-**1**	63	33	*S,S*-(-)
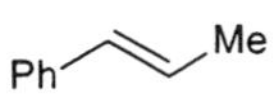	(*S,S*)-**1**	93	20	*1S,2S*-(-)
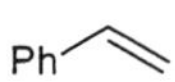	(*R,R*)-**2**	75	57	*R*-(+)
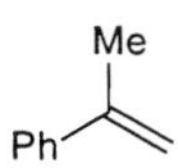	(*R,R*)-**2**	36	30	*R*-(+)
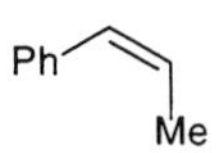	(*R,R*)-**2**	73	84	*1R,2S*-(-)

Scheme 50

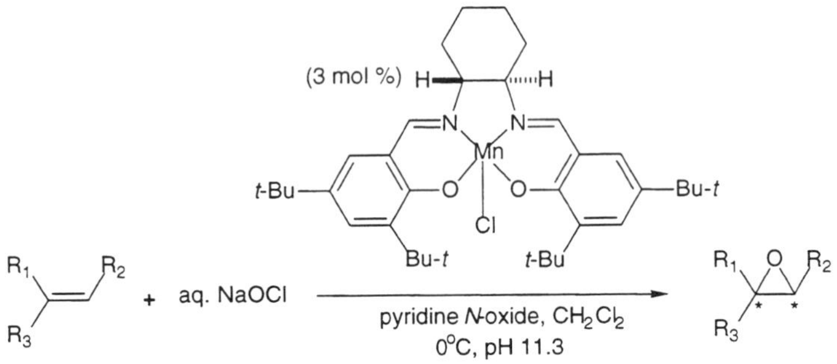

Alkene	% Yield	% ee	Configuration
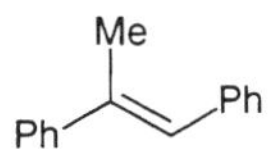	87	88	*S,S*-(-)
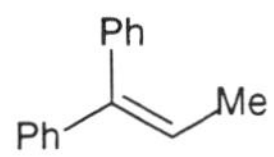	91	95	*S*-(-)
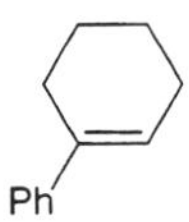	69	93	*S,S*-(-)
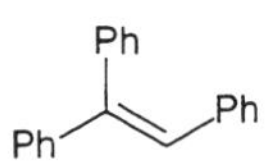	97	92	*R*-(+)

Trisubstituted styrenes are among the best substrates for the (salen) Mn(III)oxo-catalysed epoxidation reaction (Scheme 50). The highest enantioselectivities are usually recorded when the (*1R,2R*)-diaminocyclohexane-derived manganese oxo complex is employed as the catalytic oxidant with buffered bleach (pH

is employed as the catalytic oxidant with buffered bleach (pH 11.3) as the co-oxidant (B.D. Brandes and E.N. Jacobsen, *J. Org. Chem.*, **1994**, *59*, 4378). For trisubstituted olefins the observed sense of absolute asymmetric induction is opposite to that obtained with *cis*-β-methylstyrene and *trans*-stilbene. The stereochemical results are rationalised by a mechanism in which the alkene approaches the chiral (salen)Mn(oxo) complex laterally. A biradicaloid intermediate is believed to be formed in which steric interactions are minimised between the substituents on the alkene and the chiral (salen) ligand.

A recent asymmetric synthesis of the antitumour agent (+)-combretastatin D-1 by S. D. Rychnovsky and K. Hwang (*Tetrahedron Lett.*, **1994**, *35*, 8927) has featured a Jacobsen epoxidation on a (*Z*)-1,2-disubstituted styrene, embedded within a 15-membered macrolactone ring system. It utilised the (*1S,2S*)-diaminocyclohexane-derived (salen)Mn oxo complex, and yielded (+)-combretastatin D-1 in 35% ee after *O*-deacetylation (Scheme 51). The relatively low enantioselectivity encountered in this epoxidation was attributed to the

Scheme 51

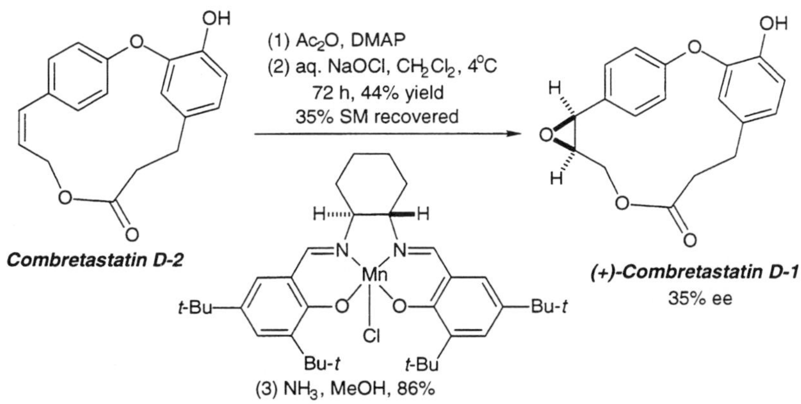

almost orthogonal conformation of the aryl ring with respect to the alkene in this substrate.

A noteworthy modification of the Jacobsen epoxidation (M. Palucki, P.J. Pospisil, W. Zhang, and E.N. Jacobsen, *J. Am. Chem. Soc.*, **1994**, *116*, 9333) has replaced the aqueous bleach co-oxidant with non-aqueous co-oxidants that permit the rapid regeneration of the chiral (salen)Mn(II)oxo complexes at -78 °C under homogenous conditions. Foremost amongst these new low-temperature stoichiometric co-oxidants is the 1:1 salt of *m*-chloroperbenzoic acid (*m*-CPBA) with *N*-methylmorpholine *N*-oxide (NMO). This simple combination allows high enantioselectivity to be attained in the epoxidation of styrene without any detrimental effect on the chemical yield (Scheme 52).

Scheme 52

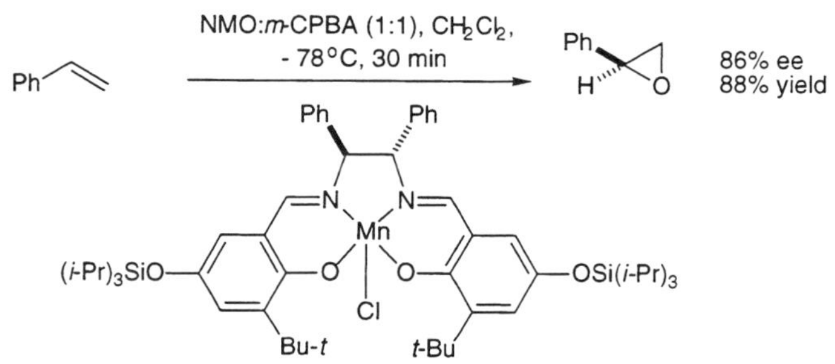

Styrene can also be epoxidised in 50-80% ee with enantiomerically pure ruthenium catalysts and iodosobenzene but the product yields are typically low (12-38%) (R.I. Kureshy *et al., Tetrahedron:Asymmetry*, **1993**, *4*, 1693).

Y. Naruta *et al.* (*Bull. Chem. Soc. Jpn.*, **1993**, *66*, 158) have found that electron-rich, binapthyl-capped, chiral iron porphyrin complexes are exceptionally good reagents for the enantioselective epoxidation of terminal styrenes bearing nitro substituents. For 3,5-dinitrostyrene, the epoxide is obtained in 96% ee, while for 2-nitrostyrene it is isolated with 89% ee. The high levels of asymmetric induction observed with these substrates are believed to be the result of favourable π-stacking interactions between the nitrobenzene ring systems and the chiral oxo-transfer complex, since much lower ee (56%) is obtained with styrene itself.

A very interesting route to (*S*)-styrene oxide (T.H. Chan, L.M. Chen, and D. Wang, *J. Chem. Soc. Chem. Comm.*, **1988**, 1280) employs a Sharpless asymmetric epoxidation on β-styryldimethylsilanol as a key step. After protodesilylation with tetraethylammonium fluoride in acetonitrile at room temperature, (*S*)-styrene oxide is isolated in good overall yield and 85-95% ee (Scheme 53).

Scheme 53

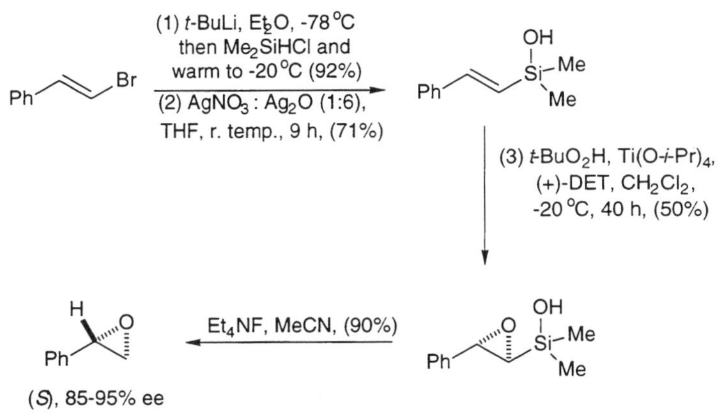

(ix) K.B. Sharpless and coworkers (*J. Org. Chem.*, **1992**, *57*, 2768; H. Kolb, M.S. VanNieuwenhze, and K.B. Sharpless, *Chem. Rev.*, **1994**, *94*, 2483) have recently introduced two new chiral *Cinchona* alkaloid-derived ligands for the highly enantioselective osmium-catalysed asymmetric dihydroxylation of substituted styrenes. The complex obtained from dihydroquinidine phthalazine [$(DHQD)_2$-PHAL] (1 mol %) and 0.2 mol % potassium osmate leads to dihydroxylation of a substituted styrene from its "top face" when the phenyl substituent is placed in the "attractive" SW quadrant, as shown in Scheme 54. The diastereomeric dihydroquinine-phthalazine complex $(DHQ)_2$-PHAL-OsO_4 favours dihydroxylation of the same styrene from its "bottom face" and produces the enantiomeric diol (Scheme 54). According to Sharpless *et al.* (*Chem. Rev.*, **1994**, *94*, 2483) the SW quadrant can be thought of as being an "enzyme-like binding-pocket" that wants to engage in attractive interactions with the aryl groups of the styrene. The co-oxidant recommended for use in the Sharpless catalytic AD reaction is the combination of potassium ferricyanide and potassium carbonate in *t*-butanol and water (1:1). Methanesulfonamide is added for most classes of styrene, except for the 1,1-disubstituted systems. Its role is to facilitate hydrolysis of the Os(VI)-glycollate ester and ensure a rapid catalytic turnover at 0 ^{o}C, the temperature at which the product ee is often maximal. The scope of the AD reaction as applied to styrenes is best appreciated after an inspection of scheme 55.

Scheme 54

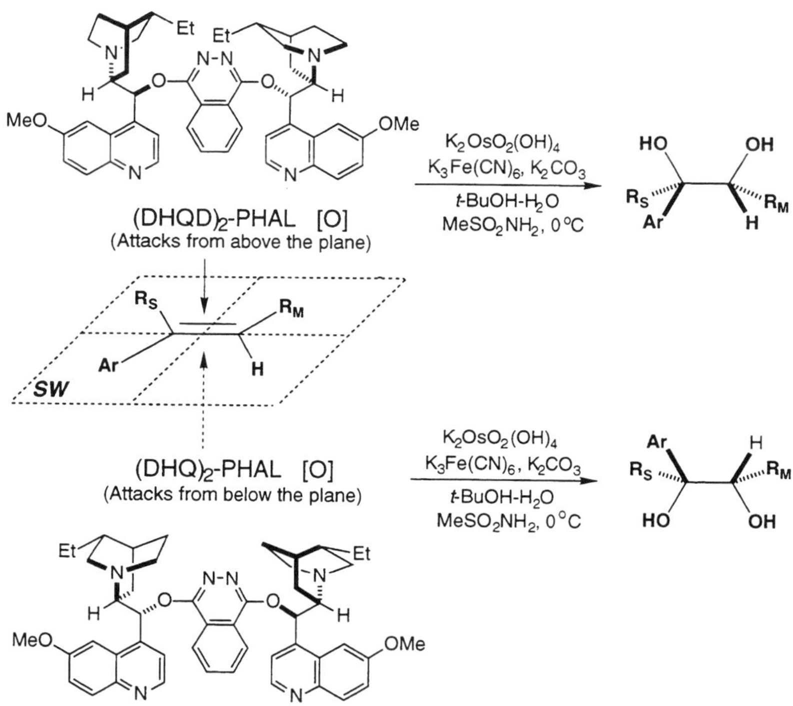

Scheme 55

Alkene	*Product*	% ee PHAL (DHQD)$_2$	% ee PHAL (DHQ)$_2$
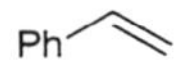	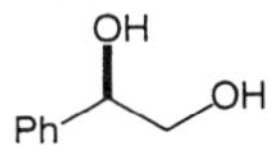	97 (*R*)	
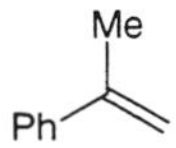	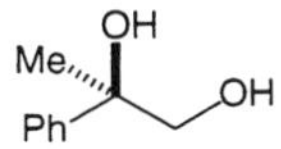	94 (*R*)	
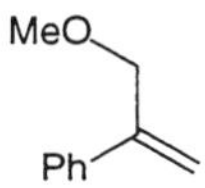	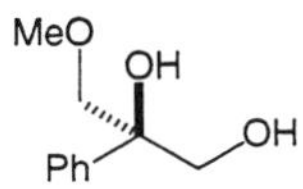	97 (*R*)	
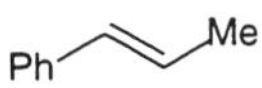	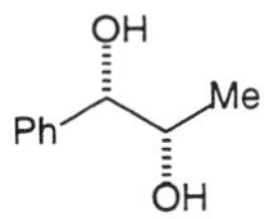		97 (*S,S*)
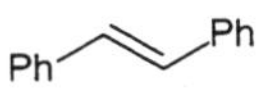	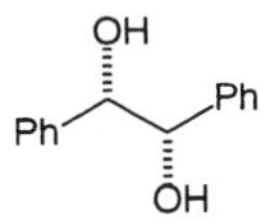		>99.5 (*S,S*)
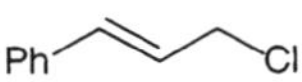	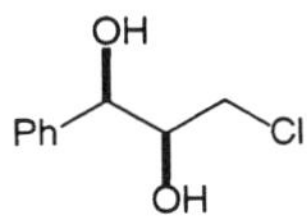	98 (*2S,3R*)	
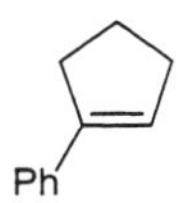	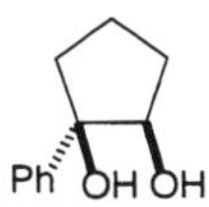	97 (*R,R*)	
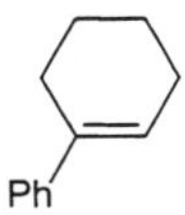	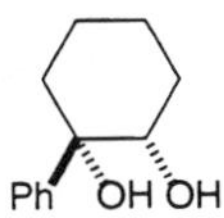		97 (*S,S*)
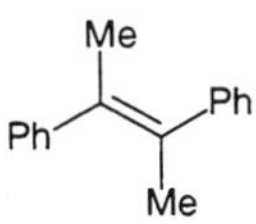	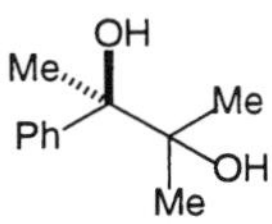	39 (*R*)	

Apart from proceeding with excellent enantioselectivity, one of the outstanding features of the AD reaction is its compatibility with readily oxidisable functionality. Thus, the AD can be

Scheme 56

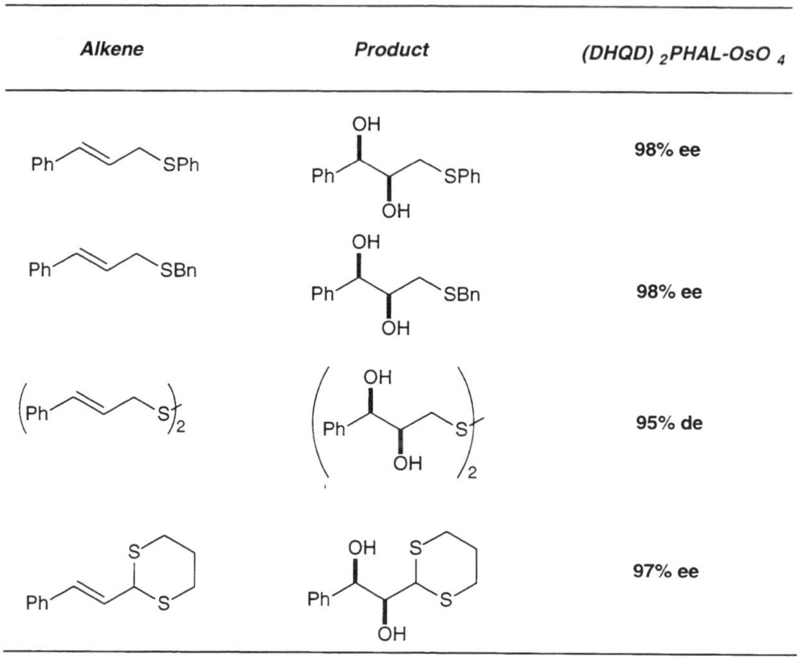

efficiently performed on (*E*)-1,2-disubstituted styrenes containing phenylthio-, 1,3-dithianyl, and even disulfide functionality without perturbation of these functional groups (Scheme 56) (P.J. Walsh, P.T. Ho, S.B. King, and K.B. Sharpless, *Tetrahedron Lett.*, **1994**, *35*, 5129).

Some rather curious results were obtained when mixtures of (*E*)- and (*Z*)-α-methoxystyrenes were examined in the AD reaction (Scheme 57). Regardless of olefin geometry in the parent enol ether, $(DHQD)_2PHAL-OsO_4$ reacted with these substrates to produce (*R*)-α-hydroxyketones in high ee, while the pseudoenantiomeric $(DHQ)_2PHAL-OsO_4$ complex afforded the corresponding (*S*)-enantiomers in comparably high ee (T.

Hashiyama, K. Morikawa, and K.B. Sharpless, *J. Org. Chem.*, **1992**, *57*, 5067).

Scheme 57

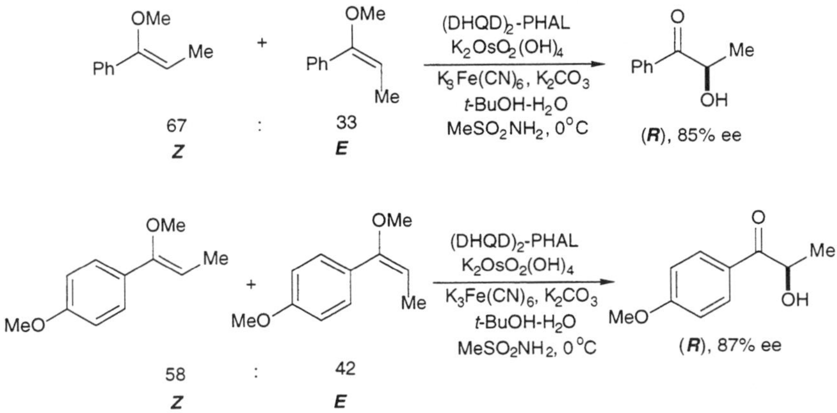

One category of styrenes that do not perform well in the osmium-catalysed AD reaction with $(DHQD)_2PHAL$ and $(DHQ)_2PHAL$ are those of the *cis*-β-substituted variety (Scheme 58). For successful AD on these substrates, 2 mol % of the

Scheme 58

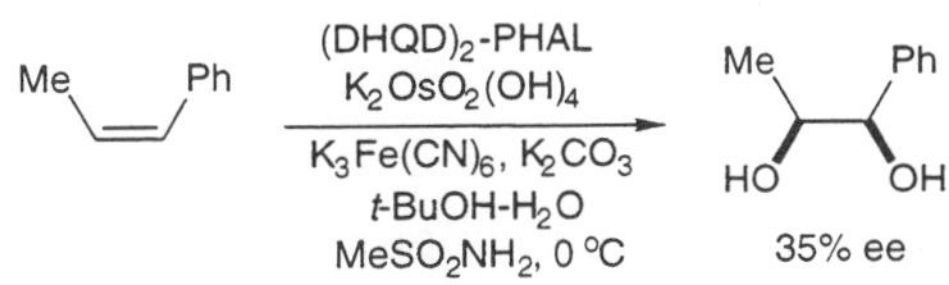

DHQD-IND or DHQ-IND ligands is required in conjunction with 0.2 mol % of OsO_4, and one equivalent of methanesulfonamide (L. Wang and K.B. Sharpless, *J. Am. Chem. Soc.*, **1992**, *114*, 7568). Reasonably high levels of enantioselectivity then result as evidenced by Scheme 59.

Scheme 59

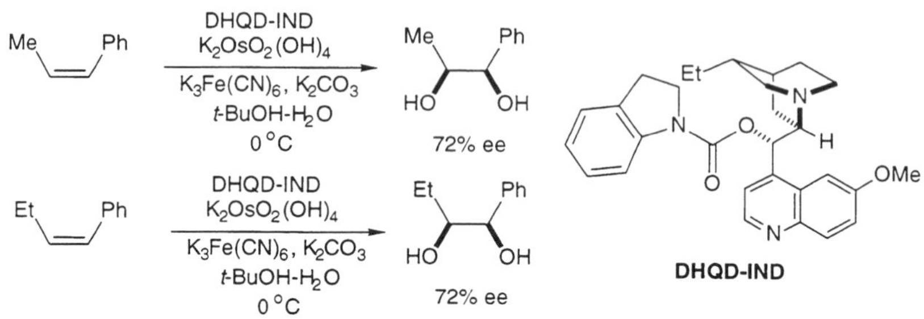

Slightly improved enantioselectivity can be realised in the AD of *cis*-β-substituted styrenes if the (*R,R*)-*trans*-1,2-diaminocyclohexane ligand shown in Scheme 60 is employed with OsO_4 under stoichiometric conditions at -90 °C in toluene (S. Hanessian *et al., J. Org. Chem.*, **1993**, *58*, 1991).

Scheme 60

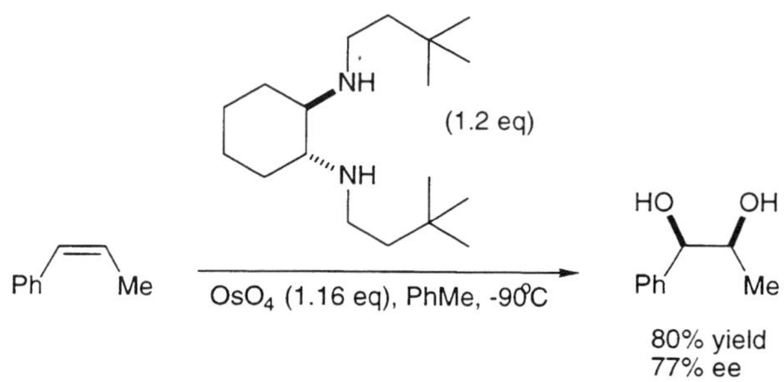

(x) A procedure for the asymmetric oxyamination of stilbene has been devised by H. Rubinstein and J.S. Svendsen (*Acta Chem. Scand.*, **1994**, *48*, 439) which utilises the chiral complex formed from interaction of trioxo(*tert*-butylimido)osmium with four equivalents of *p*-chloro-benzoyldihydroquinidine (*p*-ClBzDHQD) in dichloromethane at 25 °C. At this temperature, a 92:8 ratio

Scheme 61

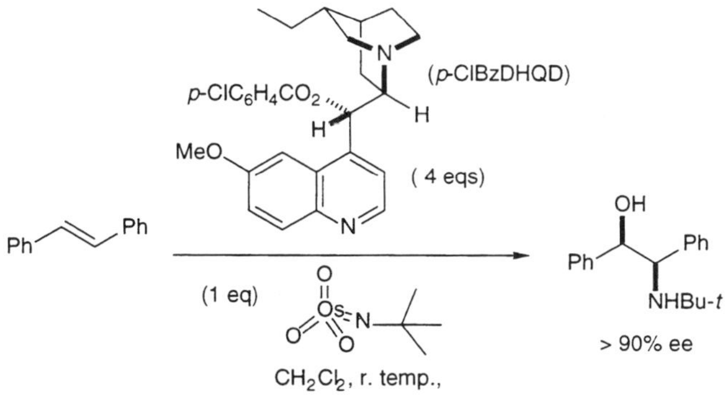

of amino alcohol:diol is obtained, with the amino alcohol having an enantiomeric excess greater than 90% (Scheme 61). Interestingly enough, the diol produced as a by-product of this reaction was virtually racemic!

(xi) 2-Bromostyrene has been oxidised microbially with *Pseudomonas putida* 39/D to afford a 4:1 mixture of (*1R*)-1-(2-bromophenyl)ethan-1,2-diol and (*1S,2R*)-4-bromo-3-ethenyl-cyclohexa-3,5-diene-1,2-diol with optical purity of ≥ 91% (Scheme 62) (K. Konigsberger and T. Hudlicky, *Tetrahedron: Asymmetry*, **1993**, *4*, 2469). Styrene derivatives possessing

Scheme 62

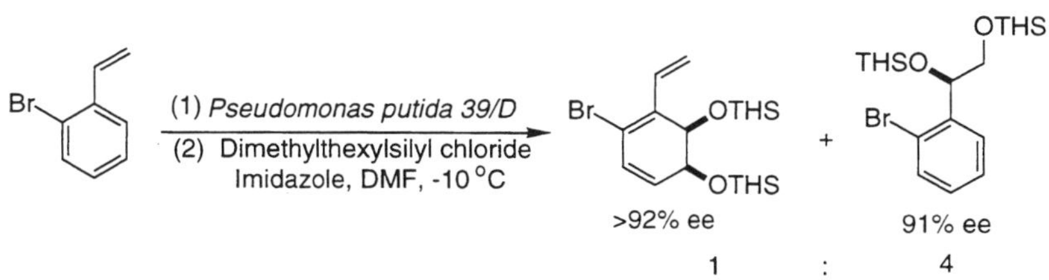

bulky substituents at the 2-position generally show a decreased tendency towards undergoing aromatic ring oxidation with *P. putida* 39/D dioxygenase, with side chain oxidation products normally predominating. Thus, when 2-chlorostyrene is oxidised, 65% of the reaction mixture is 2-chlorophenylethylene glycol, whereas when styrene is dihydroxylated only 5% of the reaction mixture is phenylethyleneglycol (T. Hudlicky, E.E. Boros, C.H. Boros, *Tetrahedron: Asymmetry*, **1993**, *4*, 1365).

(xii) Optically active 1-arylalkanols are available in up to 96% ee

Scheme 63

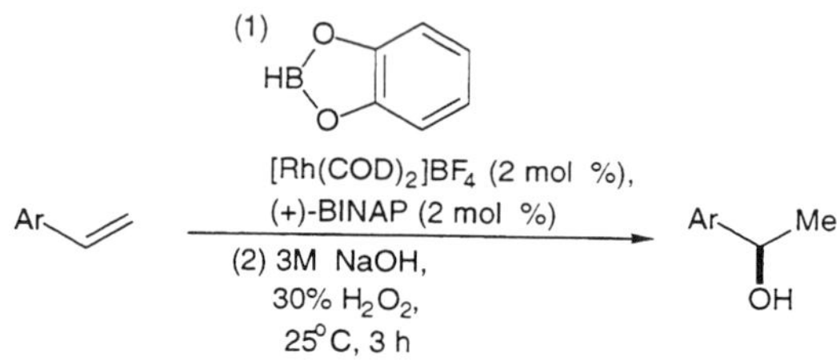

Aryl Unit	*Solvent and Temp*	*% Yield*	*% ee*
Ph	DME, -78°C	91	96
4-MeC_6H_4	DME, -78°C	77	94
4-ClC_6H_4	DME, -78°C	98	91
3-ClC_6H_4	DME, -78°C	99	85
4-$MeOC_6H_4$	DME/THF (3:1), -78°C	54	89
2-$MeOC_6H_4$	THF, -30°C	84	82

via the low temperature asymmetric hydroboration of styrenes with catecholborane in the presence of the chiral cationic complex obtained from $[Rh(COD)_2]BF_4$ and (+)-2,2'-bis(diphenylphosphino)-1,1'-binapthyl [(+)-BINAP] (Scheme 63) (T. Hayashi, Y. Matsumoto, and Y. Ito, *J. Am. Chem. Soc.*, **1989**, *111*, 3426). The regiochemistry of such metal-catalysed hydroborations is opposite to the uncatalysed reactions, being almost perfect (>99:1) for all the styrenes examined, irrespective of whether they contained electron-withdrawing or electron-releasing groups in the phenyl ring.

Scheme 64

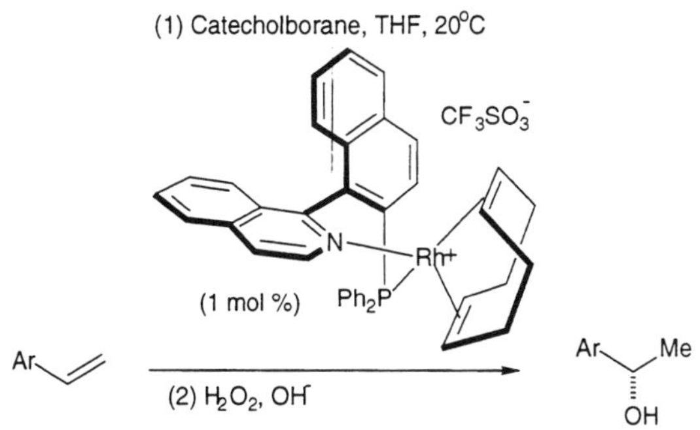

Aryl Unit	% Yield	% ee
Ph	69	88
4-ClC_6H_4	56	78
2-ClC_6H_4	59	55
4-$MeOC_6H_4$	57	94

Asymmetric hydroboration of functionalised styrenes can also be carried out at room temperature using the chiral catalyst formed from (*S*)-1-(2-diphenylphosphino-1-napthyl)isoquinoline and $[Rh(COD)(acac)-CF_3SO_3SiMe_3]$ (J.M. Brown, D.I. Hulmes and T.P. Layzell, *J. Chem. Soc. Chem. Comm.*, **1993**, 1673). After oxidative rearrangement of the styrylboronate with basic hydrogen peroxide, the 1-arylethanol is obtained in up to 94% ee and with 99% regioselectivity (Scheme 64).

(xiii) A simple and very mild procedure for the azido-phenylselenation of styrene has been reported (M. Tingolo *et al.*, *J. Org. Chem.*, **1991**, *56*, 6809); it proceeds with anti-Markovnikov regioselectivity (Scheme 65). It is thought that this regiochemical outcome results from a mechanism in which the azido anion is initially oxidised by the $PhI(OAc)_2$ to an azido radical. This then adds to the terminus of the styrene to generate a stabilised benzyl radical which is trapped by the PhSeSePh to give the observed addition product and a phenylselenyl radical. The regioisomeric product, 1-azido-1-phenyl-2-phenylseleno)ethane, is available by treating styrene with diphenyldiselenide, sodium azide, and *m*-nitrobenzenesulfonyl peroxide.

Scheme 65

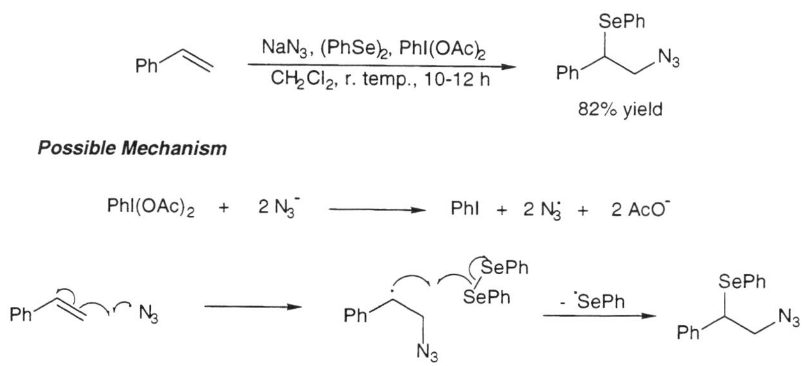

(xiv) The asymmetric hydroformylation of substituted styrenes with H_2 and CO has been studied by G. Parrinello and J.K. Stille (*J. Am. Chem. Soc.*, **1987**, *109*, 7122). With styrene itself, catalytic asymmetric hydroformylation has been achieved with $SnCl_2$ and [(-)-BPPM]$PtCl_2$ {(BPPM = (*2S,4S*)-4-(diphenyl-phosphino)-2[(diphenylphosphino)methyl]pyrollidine} in benzene

Scheme 66

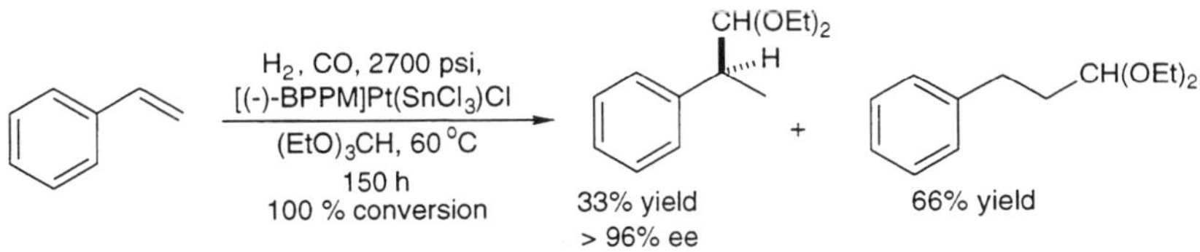

at 55-60 °C. (*2S*)-Phenylpropanal was obtained in 70-80% ee along with a roughly equal quantity of 3-phenylpropanal at 52% product conversion. Better results were obtained when hydroformylation was performed with [(-)-BPPM]Pt($SnCl_3$)Cl as catalyst in neat triethyl orthoformate at 60 °C. At 100% product conversion, the chiral acetal was obtained in approximately 33% yield, with >96% ee (Scheme 66). Hydrolysis of the acetal with pyridinium *p*-toluenesulfonate (PPTS) in acetone gave (*2S*)-phenylpropanal without racemisation.

(xv) T.V. RajanBabu and A.L. Casalnuovo (*J. Am. Chem. Soc.*, **1992**, *114*, 6265) have observed that styrenes undergo asymmetric Markovnikov addition to hydrogen cyanide when mixed with Ni (0) catalysts derived from 1,2-diol phosphinites of phenyl 4,6-*O*-benzylidene-β-D-glucopyranoside (Scheme 67). Typically, the enantioselectivities obtained are not high, but even so they are amongst the highest reported thus far.

Scheme 67

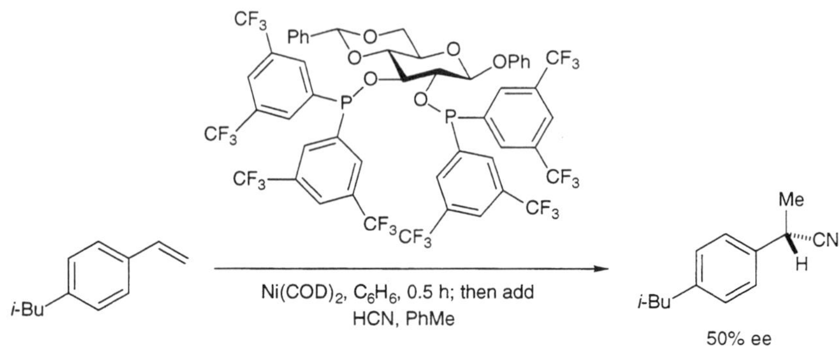

2. Arylallenes, aryldienes and aryl polyenes

(a) Synthesis

(i) A fundamentally new approach to the preparation of di-, tri-, and tetrasubstituted arylallenes takes advantage of the reaction between a titanacyclobutane and a 1,1-disubstituted allene. This produces an unsaturated titanacycle which spontaneously loses ethene to give a titanoallene. The latter combines with ketones to deliver arylallenes in reasonable yield (Scheme 68) (S.L. Buchwald and R.H. Grubbs, *J. Am. Chem. Soc.*, **1983**, *105*, 5490). Some applications of this procedure are found in Scheme 69.

Scheme 68

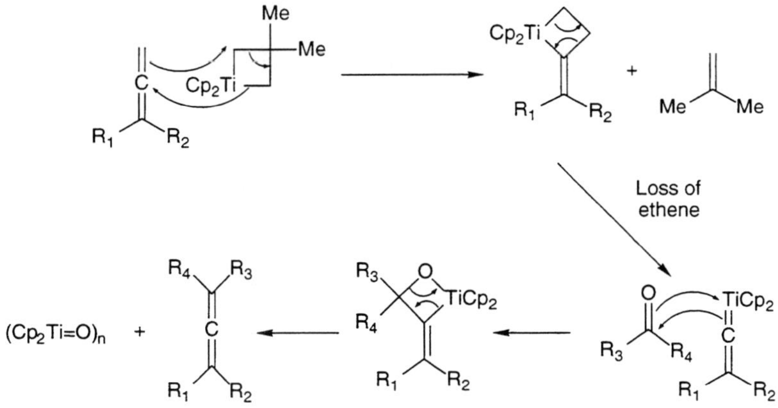

Scheme 69

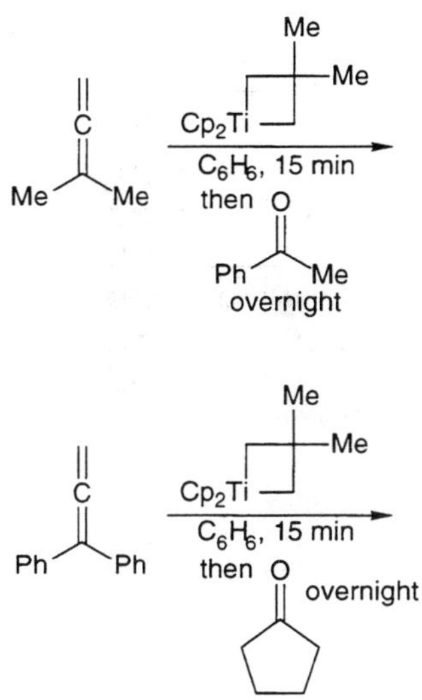

One device for obtaining terminal arylallenes is to cyclopropanate a styrene with 1,1-dibromocarbene and to then perform a halogen-metal exchange reaction with methyllithium. The latter step produces a cyclopropyllithium which undergoes an eliminative rearrangement to give the arylallene (Scheme 70) (J. Woning, F.A.T. Lijten, and W.H. Laarhoven, *J. Org. Chem.*, **1991**, *56*, 2427).

Scheme 70

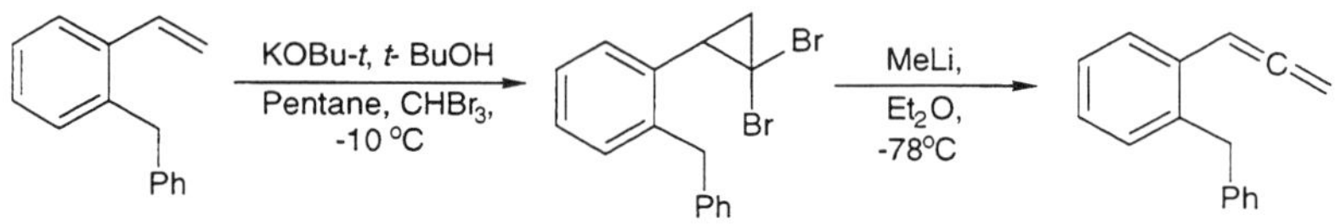

α-Lithio(γ-trimethylsilyloxy)arylallenes are sufficiently nucleophilic to react with alkyl iodides and form α-substituted (γ-trimethylsilyloxy)-arylallenes (Scheme 71) (I. Kuwajima and M. Kato, *Tetrahedron Lett.*, **1980**, *21*, 623).

Scheme 71

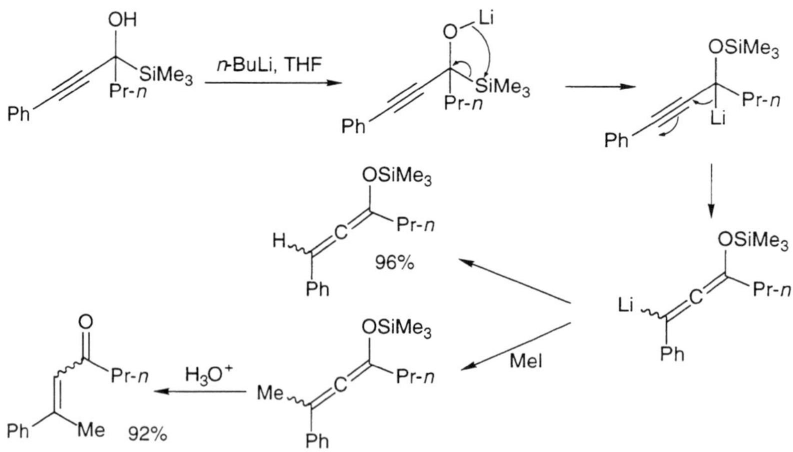

(ii) Vinyl catecholboronate esters readily cross-couple with β-styrylhalides to furnish conjugated aryldienes with retention of configuration in both coupling partners if treated with $(Ph_3P)_4Pd$ (Scheme 72) (N. Miyaura, K. Yamada, H. Suginome, and A. Suzuki, *J. Am. Chem. Soc.*, **1985**, *107*, 972).

Scheme 72

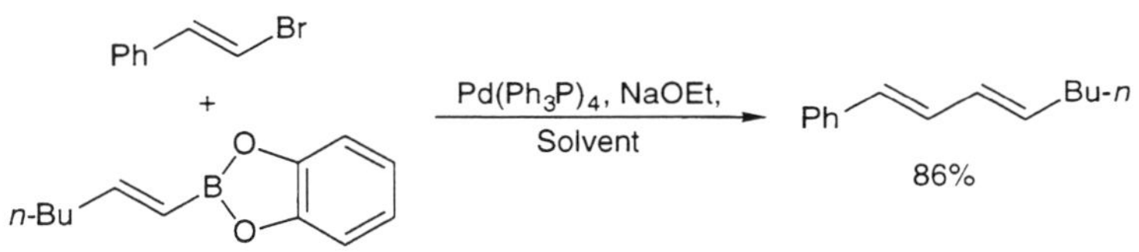

R.M. Moriarty and W.R. Epa (*Tetrahedron Lett.*, **1992**, *33*, 4095) have demonstrated that β-styryl iodonium salts readily participate in palladium (0)-catalysed cross-coupling reactions with vinyl stannanes (Stille couplings) (Scheme 73). Typically, such reactions are complete within five minutes at room temperature and furnish conjugated aryldienes without loss of parent alkene geometry.

Scheme 73

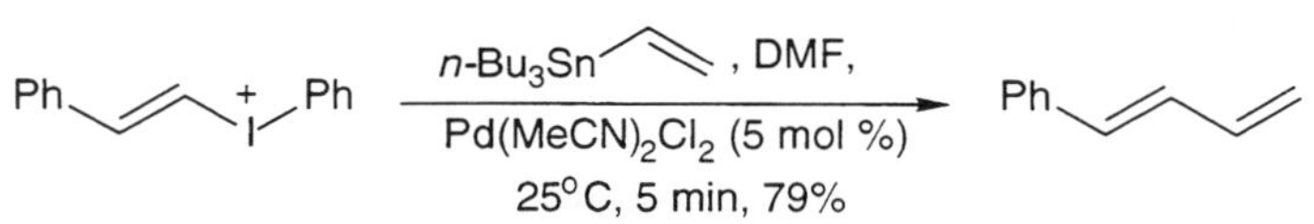

β-(Styryl)iodopalladium (II) derivatives, available from the oxidative addition of palladium (0) complexes to β-styryl iodides, readily undergo *cis*-addition to disubstituted alkynes to create aryldienylpalladium intermediates that give 1,2,4-trisubstituted 1,3-aryldienes after protonation (Scheme 74) (S. Cacchi *et al.,Tetrahedron Lett.*, **1989**, *30*, 3465).

Scheme 74

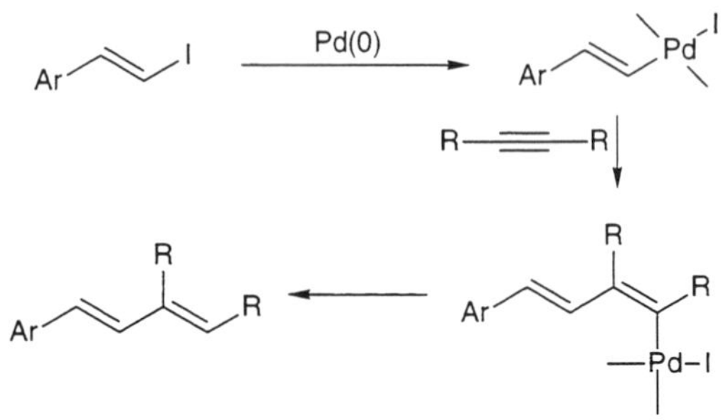

B.M. Trost and J. Dumas (*Tetrahedron Lett.*, **1993**, *34*, 19) have applied this reaction to β-styryl bromides and 1,6-enynes possessing a terminal acetylene. They discovered that tandem intermolecular carbapalladation/Heck cyclisation occurs, usually via a 6-*endo*-trig pathway (Scheme 75). Interestingly, 1,6-enynes containing a disubstituted acetylene preferred to cyclise via the 5-*endo*-trig carbapalladation mode (B.M. Trost, W. Pfrengle, H. Urabe, and J. Dumas, *J. Am. Chem. Soc.*, **1992**, *114*, 1923).

Scheme 75

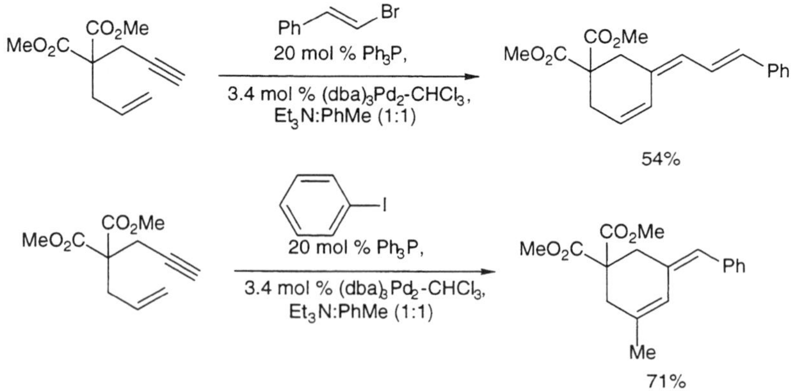

Scheme 75 (continued)

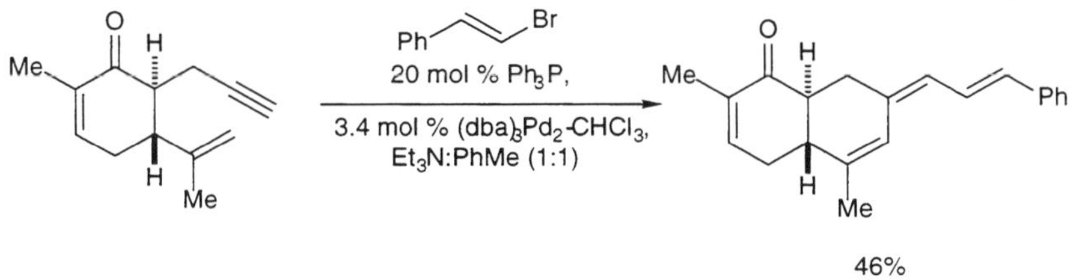

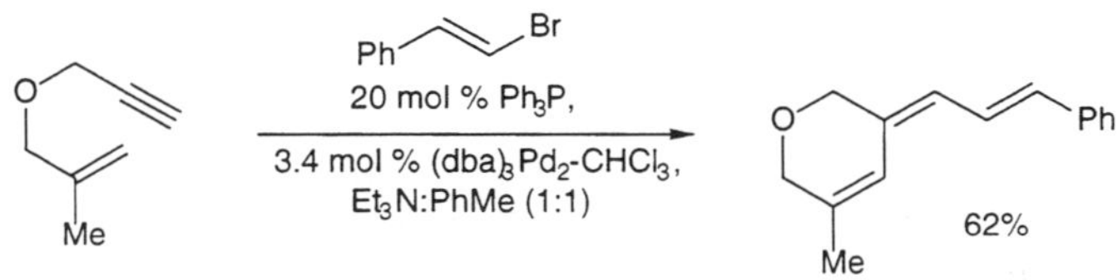

1,7-Enynes apparently favour the 6-*exo*-trig cyclisation pathway under these conditions and produce novel bicyclic conjugated aryl dienes from a single acyclic precursor (Scheme 76).

Scheme 76

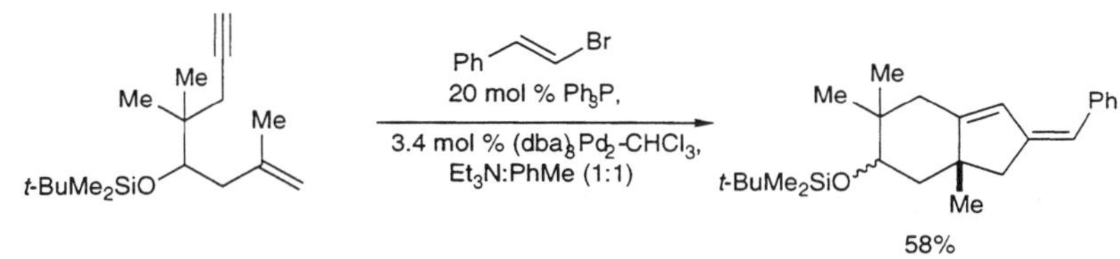

When (*E*)-cinnamyl acetate is reacted with sodium bromide and triphenylphosphine in the presence of 5 mol% $Pd(PPh_3)_4$, it is directly transformed into an (*E*)-cinnamyl phosphonium bromide. The latter can be converted to the reactive ylide *in situ* by deprotonation with *n*-butyllithium; this ylide reacts with aldehydes to give aryl-1,3-dienes in good yield (Scheme 77) (H. Kinoshita *et al., Bull. Chem. Soc. Jpn.*, **1984**, *57*, 3013).

Scheme 77

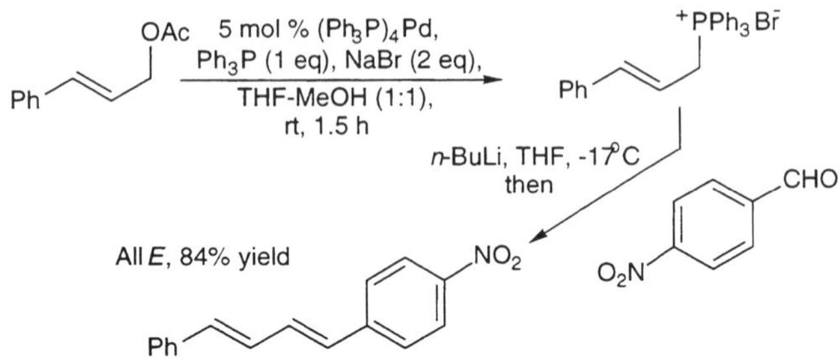

A strategy for obtaining symmetrical 2,3-aryl-1,3-butadienes is to effect a double S_N2'-type displacement on but-2-yne 1,4-diol dimethyl ether with an arylmagnesium bromide and cuprous bromide (Scheme 78) (Y. Ishino, I. Nishiguchi, F. Takihira, and T. Hirashima, *Tetrahedron Lett.*, **1980**, *21*, 1527).

Scheme 78

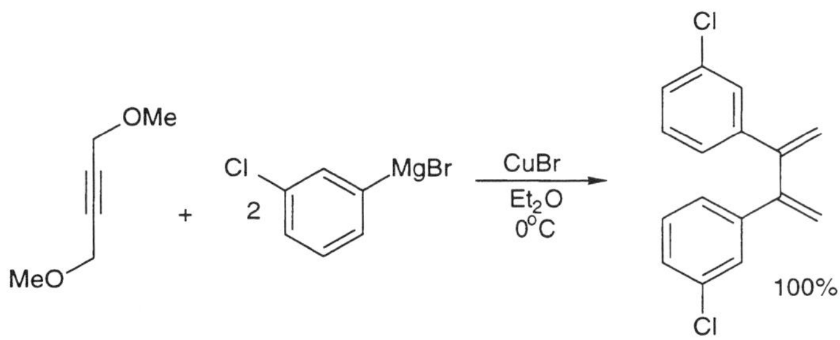

Synthetically useful (*E*)-1-phenyl-2-methoxy-1,3-butadiene is accessible by the acid-catalysed isomerisation of 1-(phenylmethyl)-1-methoxyallene with pyridinium *p*-toluenesulfonate in dichloromethane (Scheme 79) (A. Kucerovy, K. Neuenschwander, and S.M. Weinreb, *Synth. Comm.*, **1983**, *13*, 875).

Scheme 79

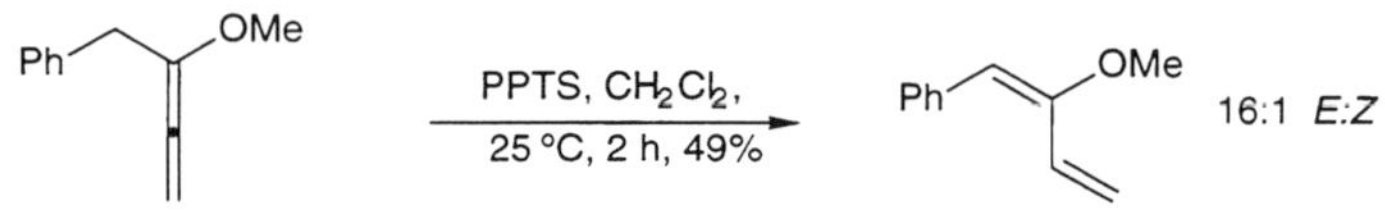

1,5-Aryl dienes are available in good yield by the regiospecific high-pressure cross-coupling of cinnamyl halides with allylstannanes (Scheme 80) (Y. Yamamoto, K. Maruyama, and K. Matsumoto, *J. Chem. Soc. Chem. Comm.,* **1984**, 548).

Scheme 80

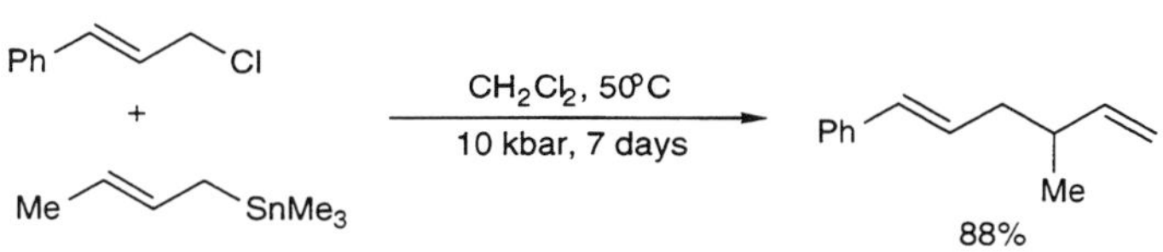

Symmetrical "skipped" aryldienes have been prepared, with retention of starting olefin configuration, from (*E*)-β-trimethylsilylstyrenes, a dialkoxyacetal, and the Lewis acid catalyst $MoCl_5$ (Scheme 81) (T. Hirao *et al., Tetrahedron Lett.*, **1981**, *22*, 3633).

Scheme 81

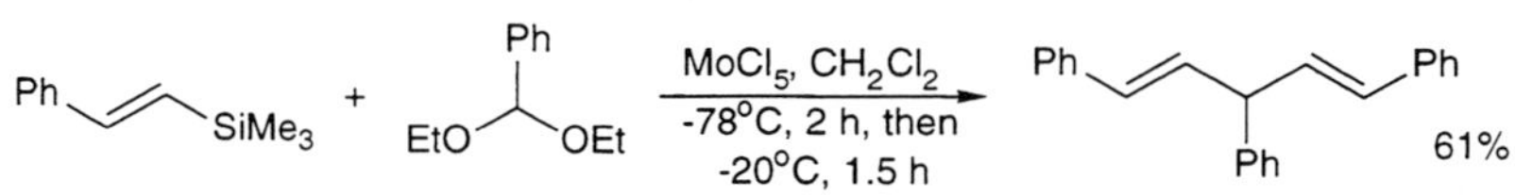

2-Aryl-1,4-pentadienes have been synthesised by the reaction of benzylic cyclic 1,3-dithioketals with allylmagnesium halides and nickel (0) catalysts (Scheme 82) (T.-Y. Luh *et al., J. Org. Chem.,* **1991**, *56*, 4035).

Scheme 82

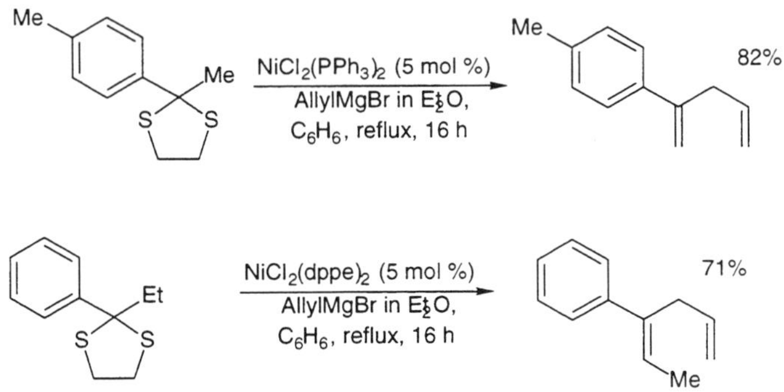

One gateway to (*E,Z,E*)-aryltrienes is through the Heck reaction of *cis*-vinyl-1,2-dibromides with styrenes (Scheme 83) (A. de Meijere and F.E. Meyer, *Angew. Chem. Int. Edn. Engl.*, **1994**, *33*, 2379; A. Lansky, O. Reiser, and A. de Meijere, *Synlett*, **1990**, 405).

Scheme 83

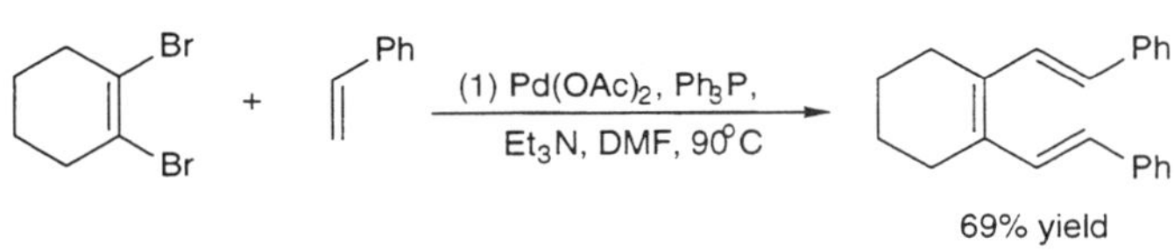

Use of 2-chlorovinyl nonaflate derivatives allows two successive Heck couplings to be performed with differing alkenes, one of which may be a styrene (Scheme 84). It is noteworthy that operating at high pressures facilitates the oxidative addition of the Pd (0) species to the dienyl chloride; ordinarily, such additions occur reluctantly under the standard Heck conditions.

Scheme 84

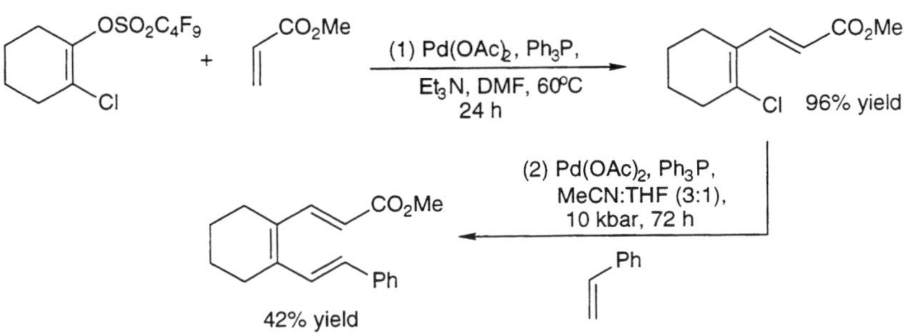

(b) Reactions

(i) The selective asymmetric dihydroxylation of a wide range of conjugated aromatic 1,3-dienes has been described by H. Becker, M.A. Soler, and K.B. Sharpless (*Tetrahedron*, **1995**, *51*, 1345; see also: D. Xu, G.A. Crispino, and K.B. Sharpless, *J. Am. Chem. Soc.*, **1992**, *114*, 7570; H.C.Kolb, M.S. VanNieuwenhze, and K.B. Sharpless, *Chem. Rev.*, **1994**, *94*, 2483). It appears that if a single aryl ring is in conjugation with the 1,3-diene, then oxidation will occur preferentially at the distal double bond so as to minimise disruption of conjugation (Scheme 85). Somewhat surprisingly, the cyclohexadiene and cyclopentadiene systems shown in Scheme 85 turned out to be excellent substrates for the AD process, despite the fact that both the alkenes undergoing attack were of the *cis*-1,2-disubstituted variety, normally a poor substrate class for the AD reaction (*vide infra*).

Scheme 85

Alkene	*Product*		*(DHQD)$_2$PHAL % ee*
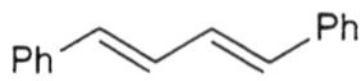	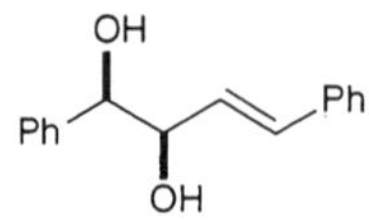		≥ 99
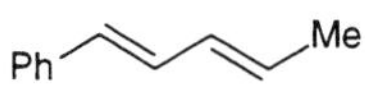	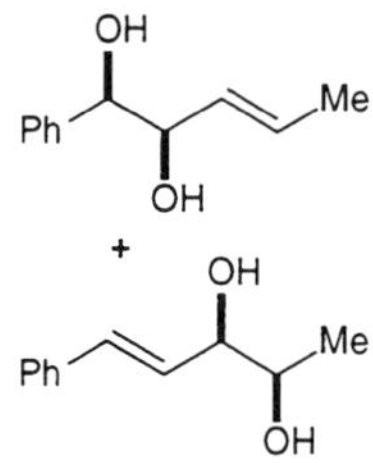	1 : 4	92
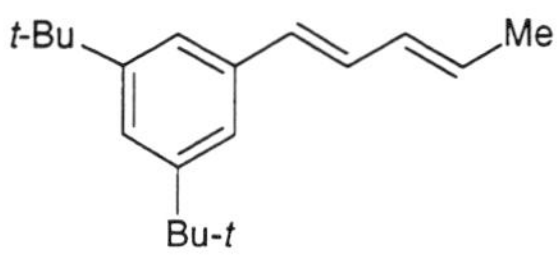	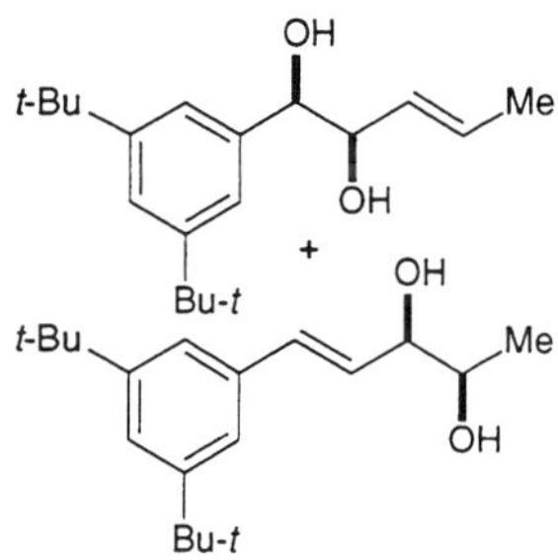	1 : 15	92
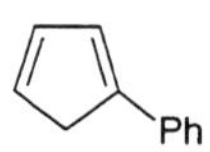	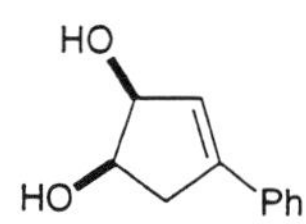		99
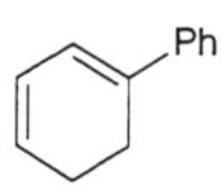	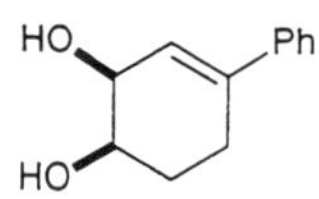		91
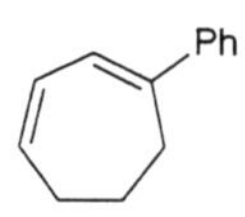	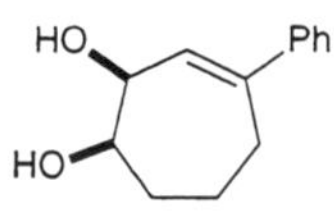		41

(ii) Selective epoxidation of the central double bond in an (*E,Z,E*)-aryltriene has been achieved in 89% yield by treatment with *meta*-chloroperbenzoic acid in dichloromethane (Scheme 86) (A. de Meijere and F.E Meyer, *Angew. Chem. Int. Edn. Engl.*, **1994**, *33*, 2379).

Scheme 86

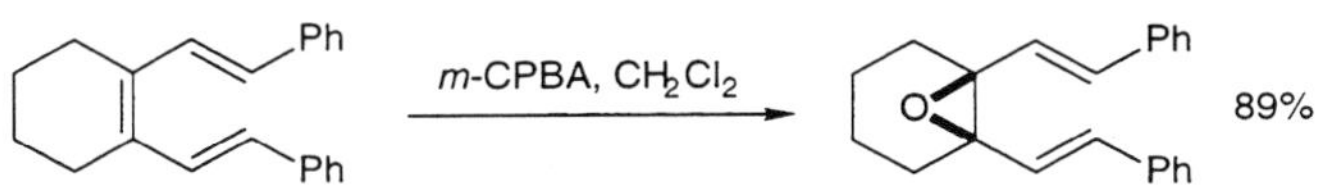

(iii) B.M. Trost *et al.* (*J. Am. Chem. Soc.*, **1992**, *114*, 1923) have disclosed a strategy for the construction of ring annelated 1-phenyl-1,3,5-hexatrienes. These are fairly reactive species

Scheme 87

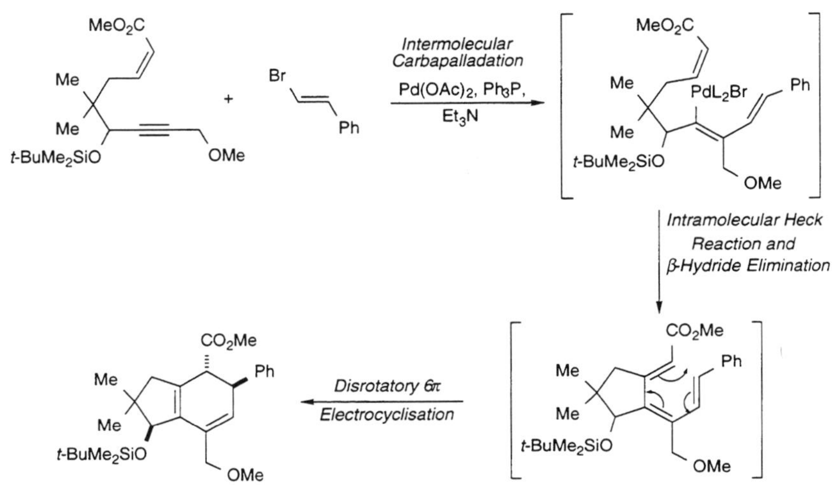

that participate readily in disrotatory 6π electrocyclic ring closure under the reaction conditions to produce annelated 1,3-cyclohexadienes (Scheme 87).

3. Arylalkenols and related structures

(a) Preparative methods

(i) A high yielding procedure for obtaining (*E*)-arylalkenols involves reduction of (*E*)-α,β-unsaturated esters with diisobutylaluminium hydride at low temperature (-78°C). The method shows excellent chemoselectivity for 1,2- rather than 1,4-reduction; an example of this tactic being applied can be found in Scheme 88 (S. Hanessian and R. Leger, *Synlett*, **1992**, 402).

Scheme 88

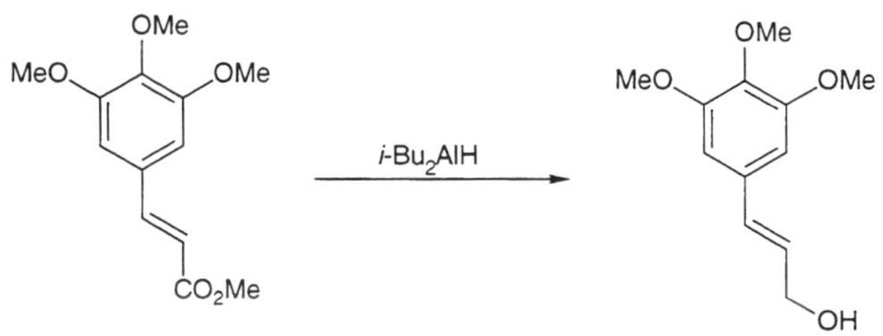

(ii) Triethoxysilane activated by KF is a particularly effective reagent for the selective 1,2-reduction of cinnamaldehyde derivatives to cinnamyl alcohols (Scheme 89) (R.J.P. Corriu *et al., J. Chem. Soc. Chem. Comm.*, **1981**, 121). The method is compatible with a wide range of functionality that includes

Scheme 89

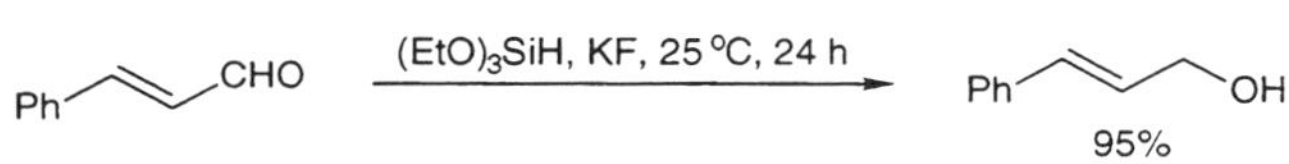

bromo, nitro, amido, ester or terminal alkene groups. Since aldehydes are more reactive than ketones towards the $(EtO)_3SiH/KF$ system, this suggests that selective reductions may be possible in molecules containing both structural units.

5-10 Mol % of tetra-*n*-butylammonium fluoride in HMPA or DMPU is sufficient to catalyse the markedly stereocontrolled 1,2-

reduction of chiral styryl ketones with dimethylphenylsilane (Scheme 90) (M. Fujita and T. Hiyama, *J. Am. Chem. Soc.*

Scheme 90

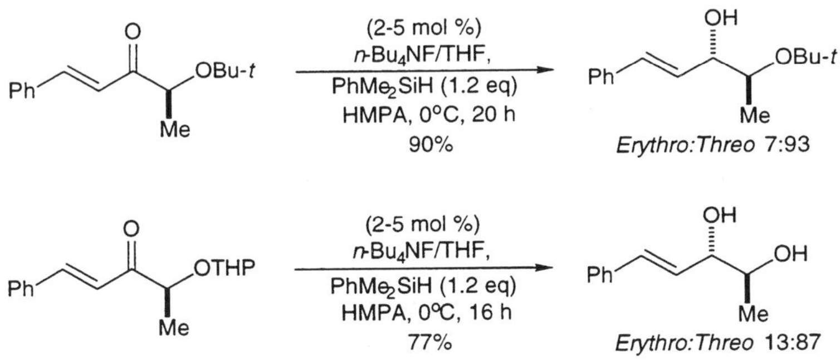

1984, *106*, 4629). The noteworthy stereocontrol observed in these reductions is presumably attributable to the steric bulk of the $[n\text{-}Bu_4N]^+$ $[PhMe_2SiFH]^-$ reagent which favours attack on the carbonyl via a Felkin-Ahn transition state.

(iii) E.J. Corey and R.K. Bakshi (*Tetrahedron Lett.*, **1990**, *31*, 611) have reported a chiral oxazaborolidine derivative of L-proline which catalyses the asymmetric 1,2-reduction of styryl ketones with catecholborane in toluene at -78 °C (Scheme 91).

Scheme 91

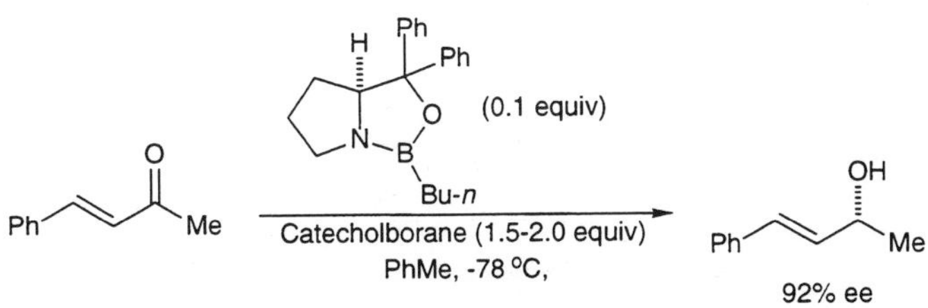

Apart from imparting excellent levels of enantiocontrol, the method appears to be highly chemoselective, it being tolerant of olefinic and amide functionality that may be present within a substrate.

(iv) K. Yamakawa and coworkers (*J. Org. Chem.*, **1989**, *54*, 3130) have devised a route to optically active cinnamyl alcohols that is based on alkylation of the anion derived from (-)-chloro-

Scheme 92

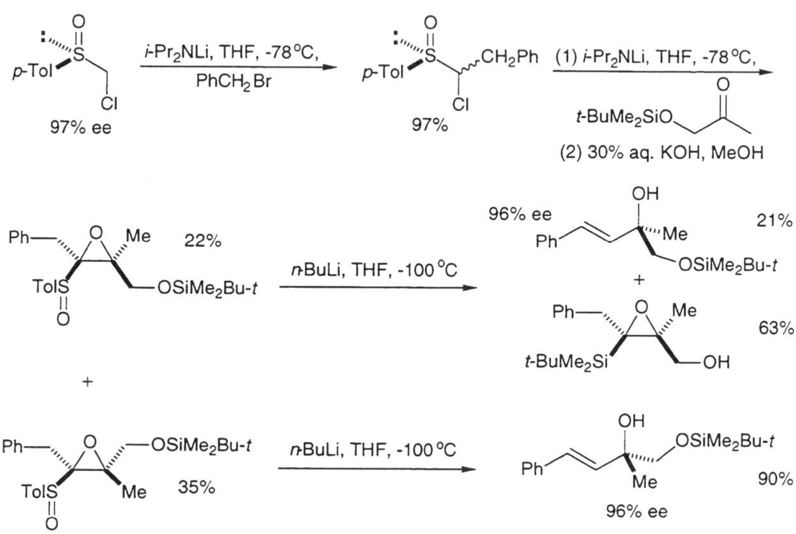

methyl *p*-tolyl sulfoxide with benzyl bromide, deprotonation of the resulting mixture of chloroalkyl *p*-sulfoxides with LDA, and addition of this anion to a ketone. In the case of acetol *tert*-butyldimethylsilyl ether, the crude chlorohydrins were readily cyclised with base to give two α,β-epoxy sulfoxides in 22 and 35% yield respectively. After treatment with *n*-butyllithium (3.0 equiv) at low temperature, both epoxides underwent elimination to the corresponding (*E*)-cinnamyl alcohol derivatives, but in yields that were dependant on the starting epoxide stereochemistry (Scheme 92).

(v) Allylic transposition of aryl vinyl alcohols provides a convenient pathway to (*E*)-cinnamyl alcohol derivatives in good yield (Scheme 93) (B. Goument, L. Duhamel, and R. Mauge, *Tetrahedron*, **1994**, *50*, 171).

Scheme 93

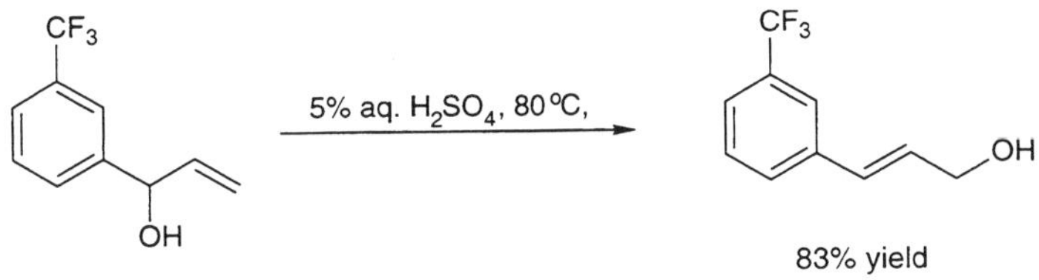

(b) Reactions

(i) A.B. Charette and coworkers (*J. Am. Chem. Soc.*, **1991**, *113*, 8166) have utilised partially-protected monosaccharides with a free hydroxyl group at C(2) to mediate highly enantioselective Simmons-Smith cyclopropanations on β-cinnamyl glycosides using diethylzinc and diiodomethane (Scheme 94). A novel ring contraction process was employed to liberate the optically pure cyclopropylmethyl alcohols from the parent chiral auxiliary. Although the chemical and optical yields of the cyclopropanated products were indeed impressive, the method was disadvantaged by the fact that the chiral auxiliary was not reusable.

Scheme 94

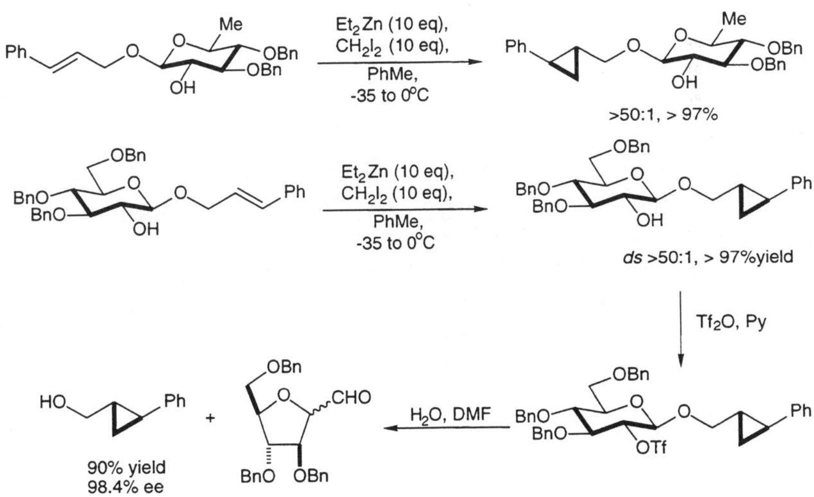

Scheme 95

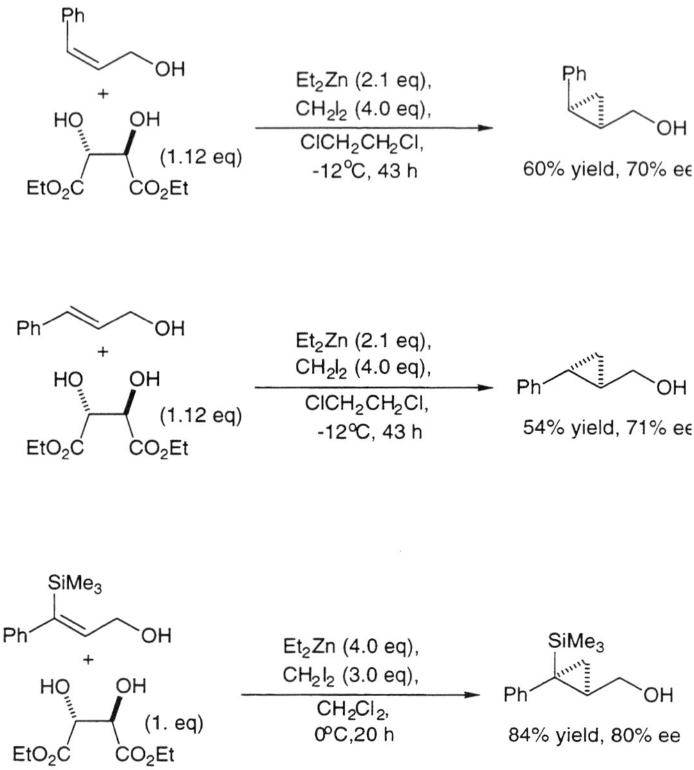

Addition of (+)-(*R,R*)-diethyl tartrate (1.1 equiv) to a solution of a cinnamyl alcohol, diethylzinc, and diiodomethane in dichloroethane at 0 °C leads to cyclopropylmethyl alcohols in 70-80% ee (Scheme 95) (Y. Ukaji, M. Nishimura, and T. Fujisawa, *Chem. Lett.,* **1992**, 61; Y. Ukaji, K. Sada, and K. Inomata, *Chem. Lett.,* **1993**, 1227).

Considerably higher enantioselectivity (93% ee) is realisable in the asymmetric Simmons-Smith cyclopropanation of (*E*)-3-phenyl-2-propen-1-ol, if conducted with the *n*-butyldioxaborolane derived from (*R,R*)-(+)-*N,N,N',N'*-tetramethyltartaric acid diamide (Scheme 96) (A.B. Charette and H. Juteau, *J. Am. Chem. Soc.*, **1994**, *116*, 2651). A commendable feature of this methodology is that it operates at room temperature, a facet that may be advantageous for industrial applications.

Scheme 96

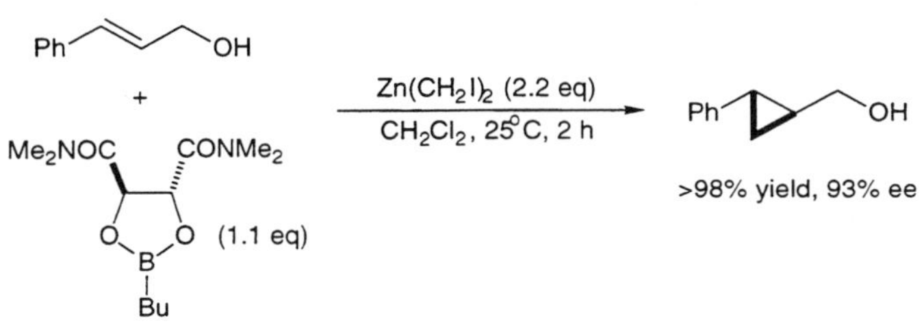

S. Kobayashi and coworkers (*Tetrahedron Lett.*, **1992**, *33*, 2575) have discovered that zinc carbenoids generated *in situ* from diethylzinc, diiodomethane, and a catalytic quantity of the C_2-symmetric chiral disulfonamide shown in Scheme 97 will readily cyclopropanate (*E*)-3-phenyl-2-propen-1-ol in 75% ee and 92% chemical yield.

Scheme 97

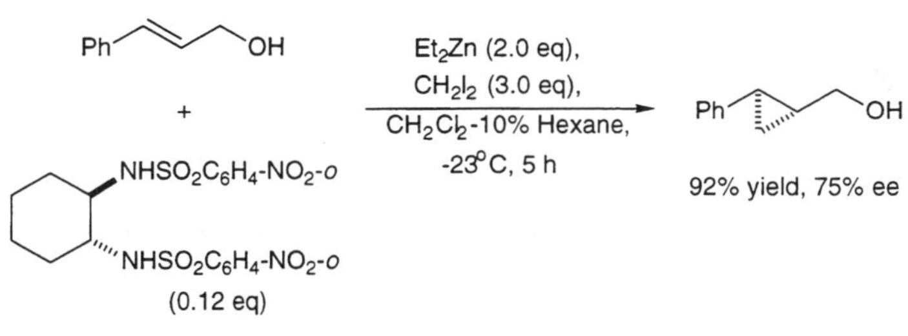

The double Simmons-Smith reaction of 5-phenylpenta-2,4-dien-1-ol proceeds through a racemic monocyclopropanated allylic alcohol that reacts further with the methylene zinc carbenoid to generate an *anti*-bicyclopropane with selectivity at the 5:1 level (Scheme 98) (A.G.M. Barrett and G.J. Tustin, *J. Chem. Soc. Chem. Comm.*, **1995**, 355).

Scheme 98

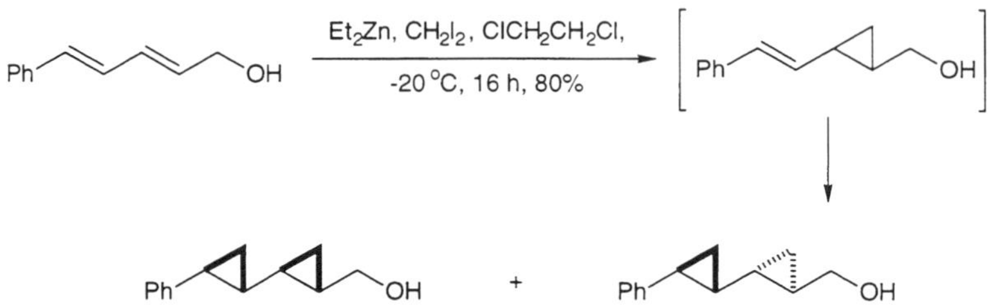

1 : 5

(ii) The Sharpless asymmetric epoxidation (AE) of allylic alcohols (T. Katsuki and K.B. Sharpless; *J. Am. Chem. Soc.*, **1980**, *102*, 5974; M.G. Finn and K.B. Sharpless, in *Asymmetric Synthesis*, Ed. J.D. Morrison, Academic Press, **1985**, *Vol. 5*, Chapter 7, 193) is of special importance for the construction of synthetically useful 2,3-epoxy alcohols in high optical yield. In this method, allylic alcohols are oxidised with *tert*-butyl hydroperoxide in dichloromethane at -20 °C in the presence of either L-(+)- or D-(-)-diethyl tartrate (DET) and titanium tetraisopropoxide. A face-selection rule has been proposed for predicting the absolute stereochemistry of the epoxides produced in the AE (Scheme 99). If the allylic alcohol is drawn in an horizontal plane, such that the hydroxymethyl group is located at the top left hand side, the D-(-)-DET-derived oxidant will epoxidise the olefin from above this horizontal plane, while the oxidant obtained from L-(+)-DET will epoxidise from below the plane.

Scheme 99

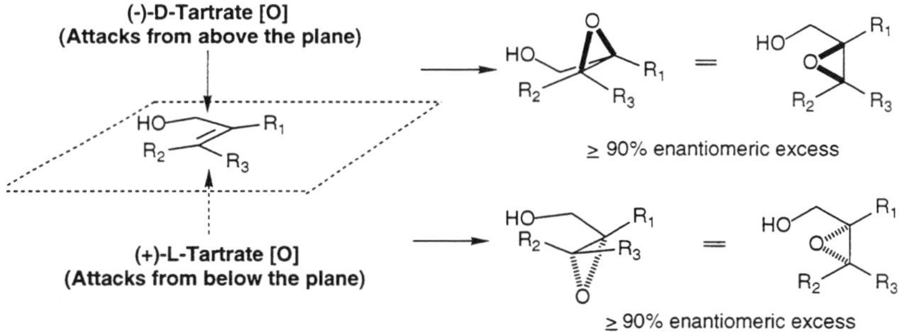

The Sharpless AE has been successfully applied on a range of cinnamyl alcohol substrates with optimal results accruing under the catalytic AE conditions. These utilise 5-10 mol % of titanium tetraisopropoxide and the tartrate ester in conjunction with 3A or 4A molecular sieves (Scheme 100) (K.B. Sharpless *et al., J. Am. Chem. Soc.*, **1987**, *109*, 5765). The catalytic conditions are especially recommended for cinnamyl alcohol derivatives that contain electron-donating groups (e.g. 4-methoxycinnamyl alcohol). This is due to the known proclivity with which these product epoxy alcohols undergo epoxide ring-opening *in situ* under the stoichiometric AE conditions.

As a rule, (*Z*)-1,2-disubstituted allyl alcohols react considerably more slowly than their (*E*)-1,2-disubstituted counterparts in the AE. Indeed, when (*E*)-3-phenyl-2-propenol is subjected to the catalytic AE conditions, epoxidation is complete within three hours at -20 ^{o}C (K.B. Sharpless *et al., J. Am. Chem. Soc.*, **1987**, *109*, 5765). However, when the corresponding (*Z*)-isomer is epoxidised under the stoichiometric conditions, the reaction takes 5-6 days at -30°C to reach completion, and furnishes the epoxy alcohol in diminished ee (76-80%) and reduced chemical yield (61-65%) (A.E. Greene *et al., J. Org. Chem.*, **1986**, *51*, 46) (Scheme 100). According to Sharpless, the most reactive category of cinnamyl alcohol substrates in the catalytic AE are those of the trisubstituted class. However, the catalytic AE epoxidises these systems in slightly lower ee than does the stoichiometric process. For example, when α-phenylcinnamyl alcohol is subjected to the catalytic AE, the 2,3-

epoxy alcohol is isolated in 91% ee whereas under the stoichiometric conditions the product is of > 98% ee (K.B. Sharpless *et al., J. Am. Chem. Soc.*, **1987**, *109*, 5765).

Scheme 100

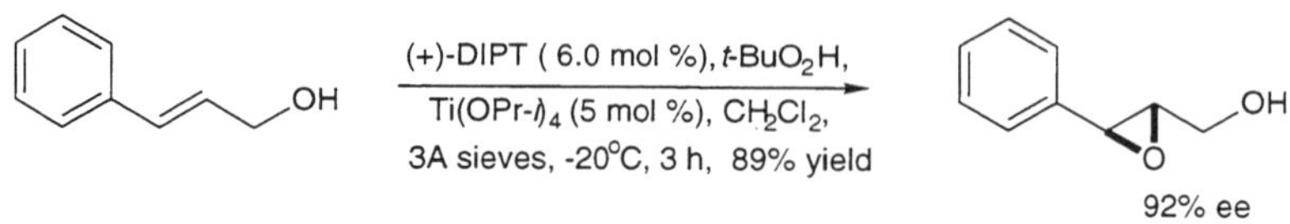

K.B. Sharpless *et al., J. Am. Chem. Soc*, **1987** , *109*, 5765.

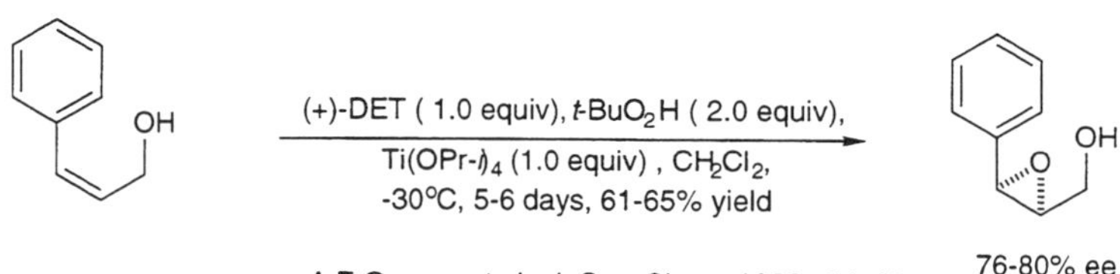

A.E Greene *et al., J. Org. Chem*, **1986** , *51*, 46.

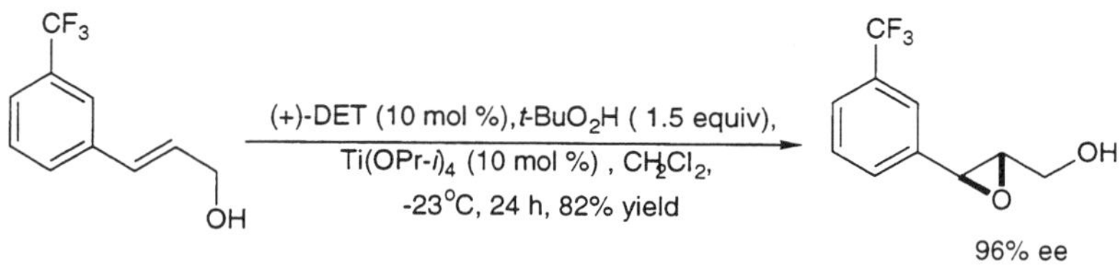

B. Goument, L. Duhamel, and R. Mauge, *Tetrahedron*, **1994** , *50*, 171.

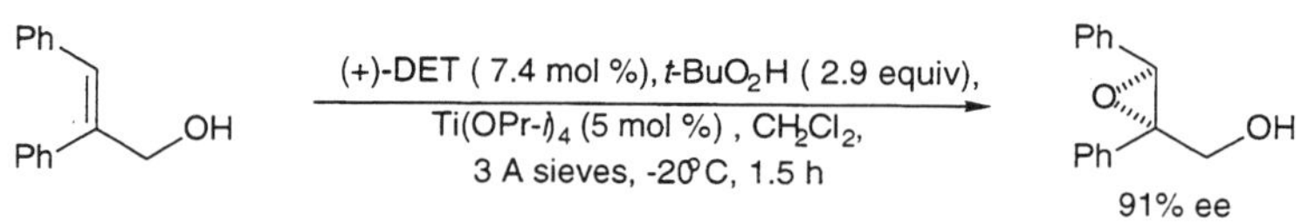

K.B. Sharpless *et al., J. Am. Chem. Soc*, **1987** , *109*, 5765.

The mechanism of the AE has been the subject of much debate (see for example: P.R. Carlier and K.B. Sharpless, *J. Org. Chem.*, **1989**, *54*, 4016; C.J. Burns, C.A. Martin, and K.B. Sharpless, *J. Org. Chem.*, **1989**, *54*, 2826; E.J. Corey, *J. Org.*

Scheme 101

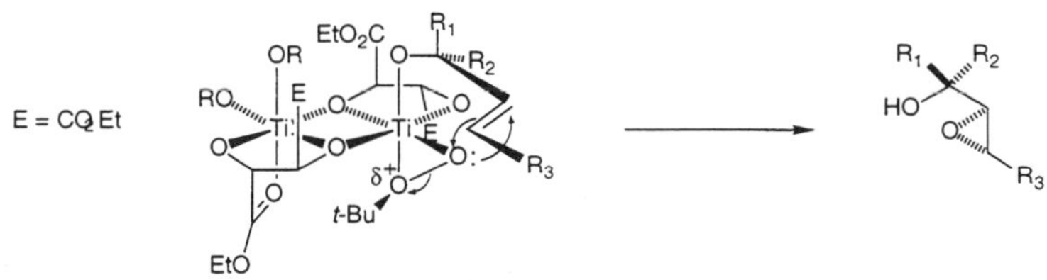

Chem., **1990**, *55*, 1693; S.S. Woodard, M.G. Finn, and K.B. Sharpless, *J. Am. Chem. Soc.*, **1991**, *113*, 106; M.G. Finn and K.B. Sharpless, *ibid.*, **1991**, *113*, 113; P.G. Potvin and S. Bianchet, *J. Org. Chem.*, **1992**, *57*, 6629; Y.-D. Wu and D.K.W. Lai, *ibid.*, **1995**, *60*, 673). Spectroscopic, osmometric, and extensive kinetic studies all support the idea that the active epoxidation catalyst is dimeric in nature. In the case of the (+)-DET oxidant, Sharpless has proposed that the hydroperoxide becomes engaged in axial-equatorial bidentate coordination to the underside of one of the titanium atoms in the dimeric complex, and that the allylic alcohol also coordinates axially to the same titanium atom from its topside (Scheme 101). Once this has happened, the activated *tert*-butyl hydroperoxide delivers its oxygen to the underside of the double bond and produces the observed product (B.H. McKee, T.H. Kalantar, and K.B. Sharpless, *J. Org. Chem.*, **1991**, *56*, 6966).

K.B. Sharpless *et al.* (*Pure & Appl. Chem.*, **1983**, *55*, 589) have successfully utilised the AE for the efficient kinetic resolution of racemic cinnamyl alcohol derivatives (Scheme 102). In the case of (±)-(*E*)-5-phenyl-4-penten-3-ol, the (*3R*)-allylic alcohol was kinetically resolved in an ee greater than 99%.

Scheme 102

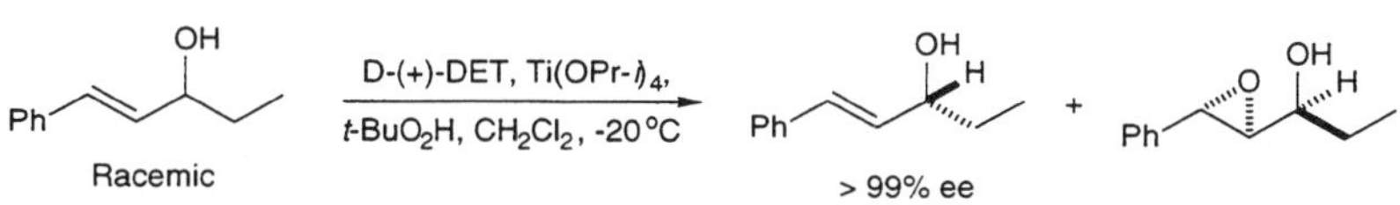

The observed stereoselectivity with the D-(+)-DET oxidant and (*E*)-5-phenyl-4-penten-3-ol can be rationalised after inspecting the loaded catalyst shown in Scheme 101. For the (*3R*)-allylic alcohol, the ethyl substituent (R_2 = Et) has to be positioned on the underside of the alkene during the epoxidation which leads to unfavourable steric clashes in the transition state. For the (*3S*)-enantiomer, however, no such unfavourable steric interactions exist between the ethyl group (R_1= Et) and the chiral catalyst. As a result, epoxidation proceeds much more rapidly from the underside of the (*3S*)-enantiomer than its optical antipode, allowing a kinetic resolution of the original racemic mixture to be achieved. By applying similar logic to other allylic alcohols, the course of an AE kinetic resolution can be conveniently predicted.

Tandem AE and enantiospecific ring expansion of 2-alkoxyaryl-substituted cyclopropylideneethanols leads to cyclobutanones with a quaternary chiral centre α- to the carbonyl

Scheme 103

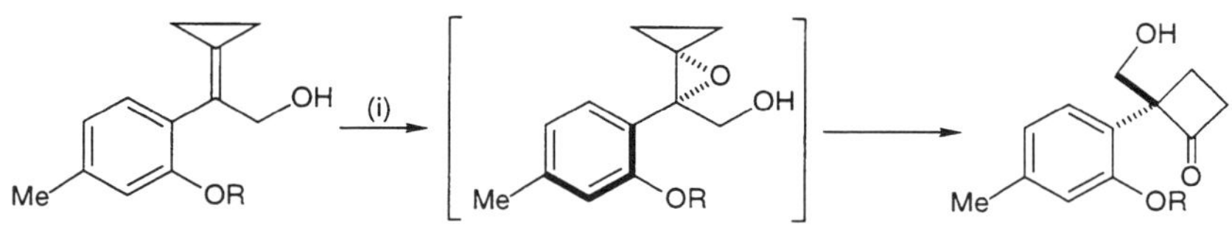

(i) Ti(OPr-*i*)$_4$, *t*-BuO$_2$H, (+)-DIPT, CH_2Cl_2, 3Å MS, -50°C, 48-72 h.

R = MOM, 61% Yield, 55% ee
R = Ts, 65% Yield, 81% ee
R = $SiMe_2$Bu-*t*, 98% Yield, 95% ee

group. The enantioselectivity of these sequences depends quite markedly on the nature of the protecting group used for the phenoxy, with highest product ee being recorded when a *tert*-butyldimethylsilyl group is employed (Scheme 103) (H. Nemoto, M. Nagamochi, H. Ishibashi, and K. Fukumoto, *J. Org. Chem.*, **1994**, *59*, 74).

Scheme 104

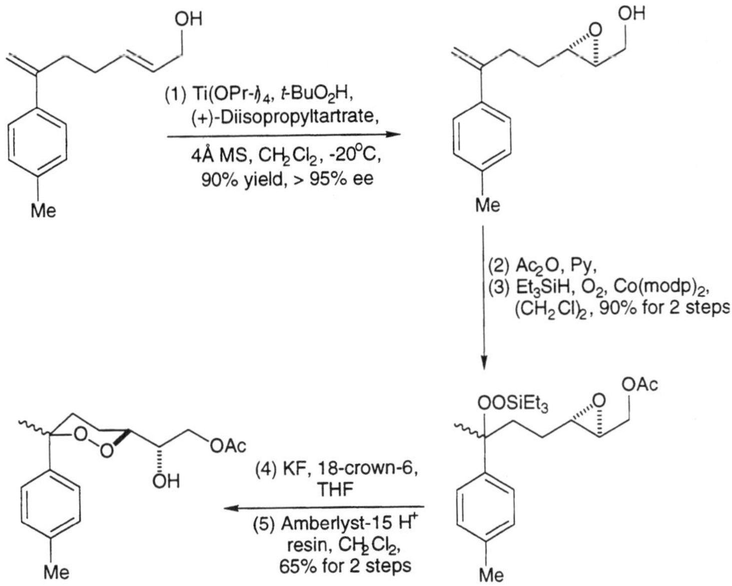

An enantioselective total synthesis of all four stereoisomers of the naturally-occurring antimalarial peroxy-sesquiterpene Yingzhaosu C has been described by X.-X. Xu and H.-Q. Dong (*Tetrahedron Lett.*, **1994**, *35*, 9429). Part of the synthesis of two of these diastereoisomers is outlined in Scheme 104; a Sharpless AE was the key asymmetry-inducing step.

(iii) High degrees of stereoselection (97:3) are manifest in the homogenous hydrogenation of (*2S*)-3-phenylbut-3-en-2-ol with the complex 1,4-bis(diphenylphosphino)butane bicyclo[2,2,1]-heptadiene rhodium tetrafluoroborate in dichloromethane (Scheme 105) (J.M. Brown and R.G. Naik, *J. Chem. Soc. Chem. Comm.*, **1982**, 348). Unfortunately, this reaction suffers from a competing isomerisation of the starting alkene to 3-phenylbutan-

2-one. At room temperature, 60% of the starting alkene reacts via this pathway, whereas at 0 °C only 23% of the starting material is consumed in this manner. The homologue (*2S*)-4-phenylpent-4-en-2-ol is hydrogenated by the same complex in THF with lower levels of stereoselection (88:12) and with an opposite sense of asymmetric induction (Scheme 105).

Scheme 105

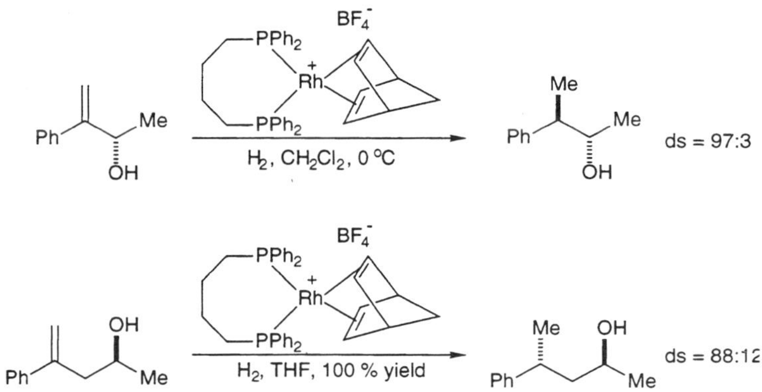

(iv) (*E*)-1-Phenyl-propene is formed in 87% yield when cinnamyl alcohol is deoxygenated with the hydridopentacyanaocobaltate anion using β-cyclodextrin as a phase transfer reagent (Scheme 106) (J.-T. Lee and H. Alper, *Tetrahedron Lett.*, **1990**, *31*, 4104). Reactions of this type are normally stereospecific with none of the *Z*-alkene being formed with this or related substrates.

Scheme 106

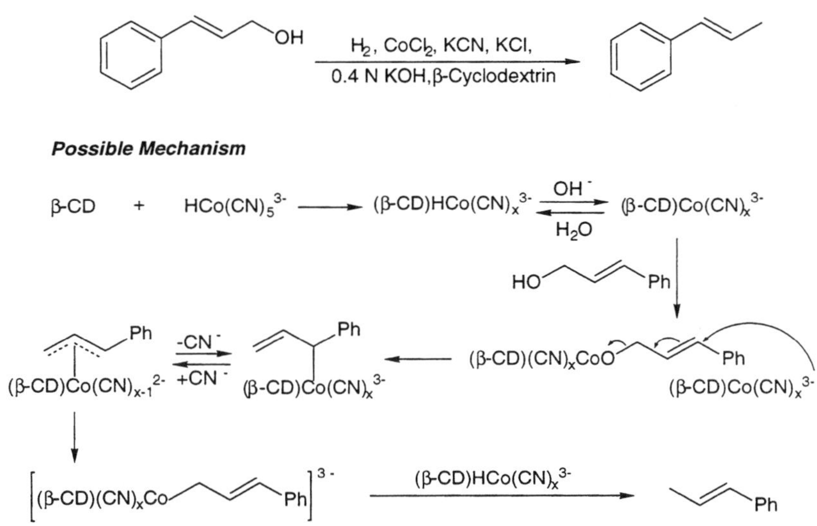

(v) "Hole" catalysed intramolecular Diels-Alder reactions may be performed on (*E*)-4-methoxycinnamyl ethers that are tethered to a reactive diene unit. Remarkably, these cycloadditions are complete within 60 seconds at 0 °C and excellent yields of product are isolated (Scheme 107) (N.L. Bauld *et al., J. Am. Chem. Soc.*, **1993**, *115*, 7653). Intramolecular [2+2] cycloadditions will also occur readily when two (*E*)-4-methoxycinnamyl ethers are interlinked and treated with a catalytic amount of a radical cation salt such as $[(4\text{-BrC}_6H_4)_3N^{+\cdot} SbCl_6^-]$ in dichloromethane at 0°C (Scheme 108) (N.L. Bauld *et al., loc cit.*).

Scheme 107

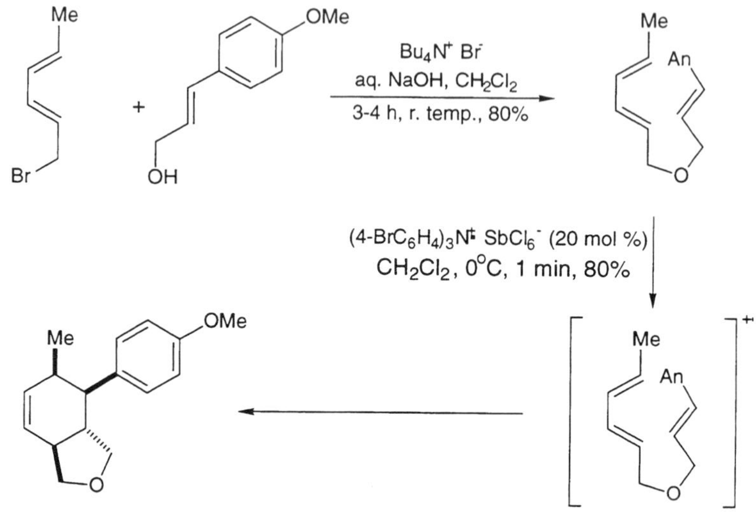

Scheme 108

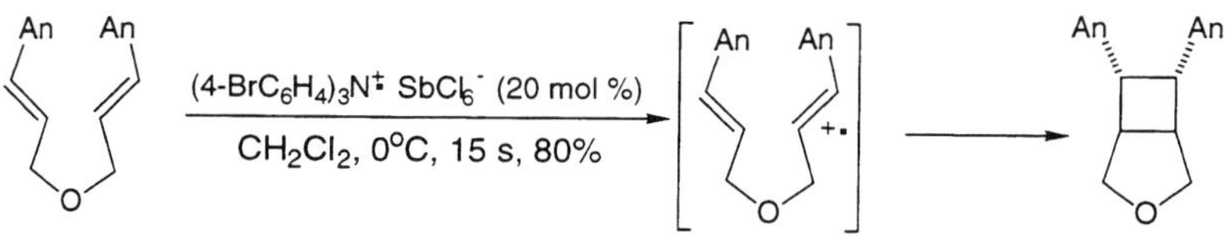

(vi) During their total synthesis of the lignan antibiotic (±)-burseran, S. Hanessian and R. Léger (*Synlett*, **1992**, 402) relied on the addition of a trimethyltin radical to an allylic 3,4,5-trimethoxycinnamyl ether to assemble a 3-trimethylstannylmethyltetrahydrofuran (Scheme 109). The carbon-tin bond was then oxidatively cleaved with ceric ammonium nitrate to give an aldehyde that was equilibrated to the *trans*-isomer with >23:1 selectivity. The methylene dioxyphenyl group present in the natural product was installed by addition of 3,4-methylenedioxyphenyllithium. The resulting alcohols were then separated, *O*-acetylated, and hydrogenolysed to provide (±)-burberan with high stereocontrol.

Scheme 109

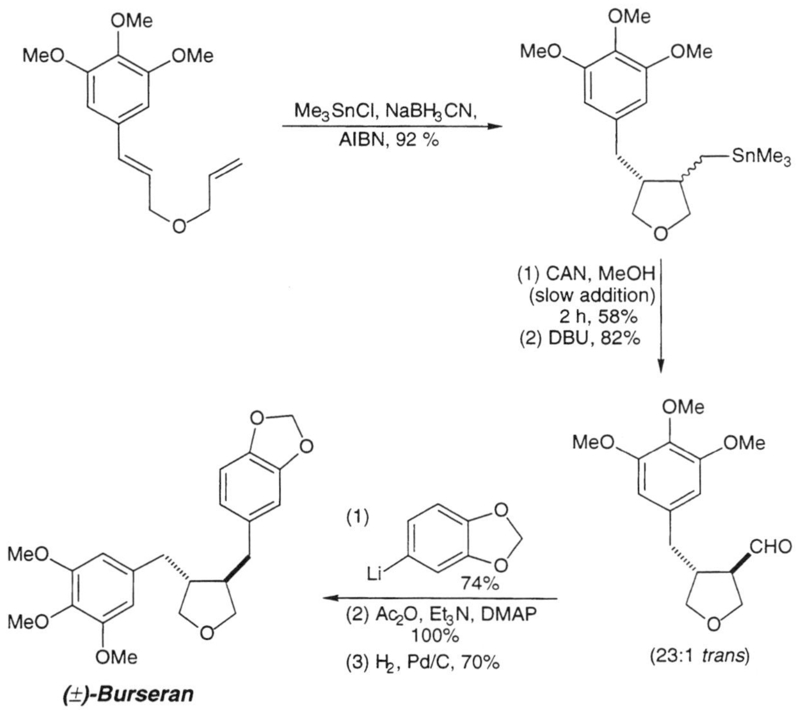

(vii) The chiral Lewis acid methylaluminium (*R*)-(+)-3,3'-bis(triphenylsilyl-1,1'-binapthalene-2,2'-diolate brings about the asymmetric Claisen rearrangement of (*Z*)-and (*E*)-cinnamyl vinyl ethers to optically active 3-substituted 1-(trimethylsilyl)-4-penten-1-ones in good to modest ee (Scheme 110) (K. Maruoka and H. Yamamoto, *Synlett*, **1991**, 793). Somewhat surprisingly, the absolute configuration of the product acylsilanes is independant of alkene geometry in the starting cinnamyl ether substrate (Scheme 110) (H. Maruoka, H. Banno, H. Yamamoto, *J. Am. Chem. Soc.,* **1990**, 112, 7791; *Tetrahedron: Asymmetry*, **1991**, *2*, 647).

Scheme 110

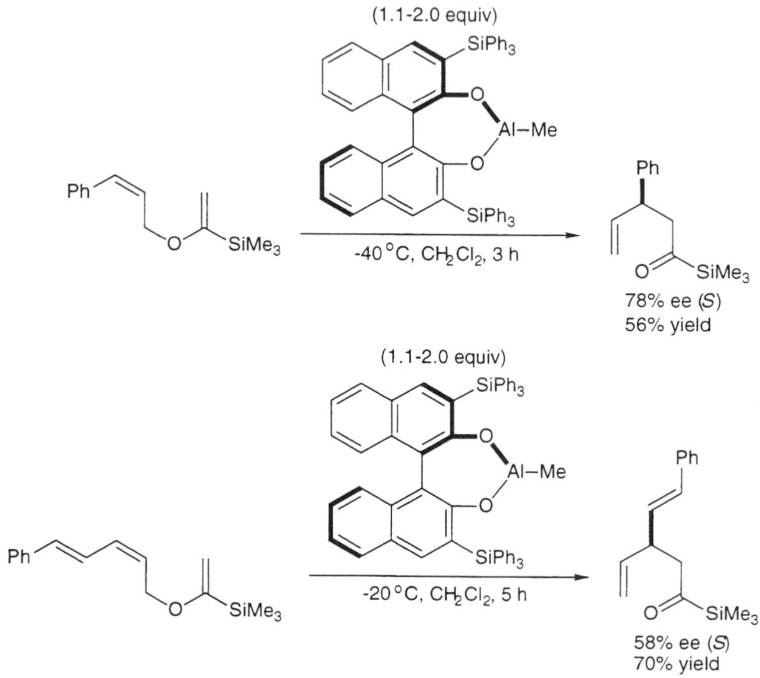

J. Kallmerten and T.J. Gould (*J. Org. Chem.*, **1986**, *51*, 1152) have demonstrated that (*E*)-cinnamyl alcohol glycolate esters will undergo diastereoselective Ireland ester enolate Claisen rearrangements with 6.1-4.2:1 selectivity when a remote and removable chiral auxiliary is bound to the glycolate hydroxyl.

(viii) S. Lamothe and T.H. Chan (*Tetrahedron Lett.*, **1991**, *32*, 1847) have reported that good regio- and stereocontrol can be realised in the reactions of chiral α–silylcinnamyl carbanions with primary alkyl iodides in toluene. Normally, (*E*)-allylsilanes

Scheme 111

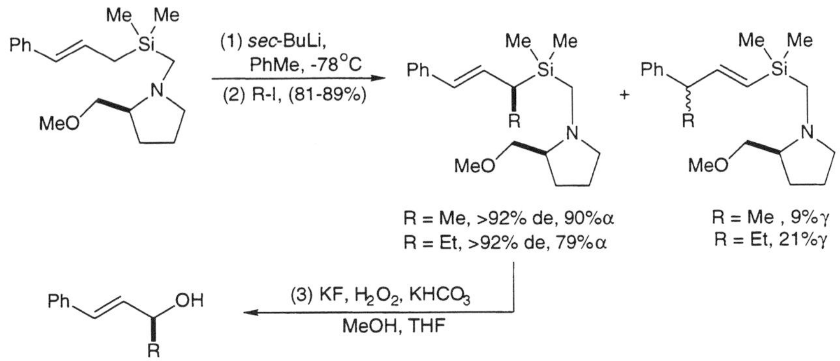

predominate in these alkylations; product regiochemistry is determined primarily by the size of the alkylating agent and the choice of reaction solvent (Scheme 111).

4. Arylalkenals

(a) Preparative methods

(i) 3-Methylcinnamaldehydes are available as *E/Z* mixtures from the reaction of 1,2-dien-4-ols with aryl halides under Pd(0) catalysis with 1,2-bis(diphenylphosphino)ethane (dppe) as a ligand for palladium. Conjugated aryldienals may also be prepared if a β-styryl bromide replaces the aryl iodide component (Scheme 112) (I. Shimizu, T. Sugiura, and J. Tsuji, *J. Org. Chem.*, **1985**, *50*, 537).

Scheme 112

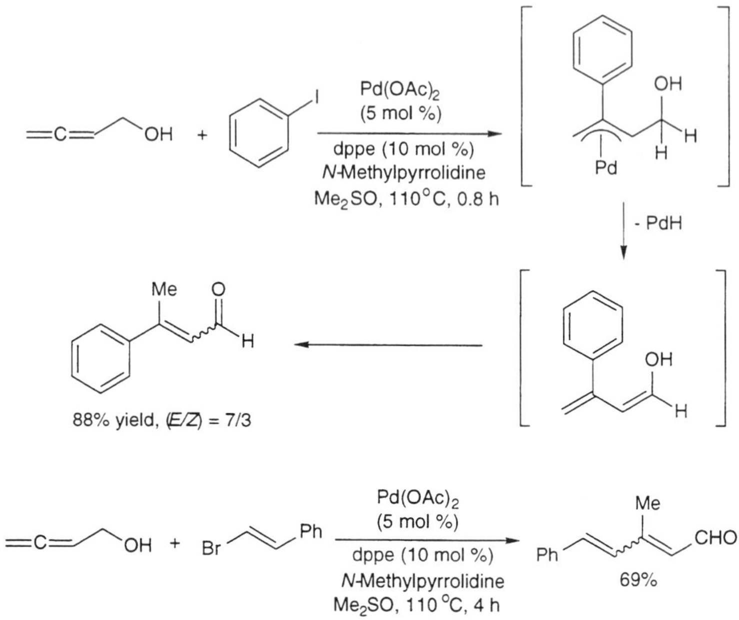

(ii) Another route to (*E*)-cinnamaldehyde derivatives capitalises on the Heck arylation reaction between a *4H*-1,3-dioxine and an aryl triflate or iodide. Initially this affords a phenyldioxine in reasonable yield which is converted to the (*E*)-cinnamaldehyde derivative by a retro Diels-Alder reaction in boiling toluene (Scheme 113) (T. Sakamoto, Y. Kondo, and H. Yamanaka, *Tetrahedron Lett.*, **1992**, *33*, 6845).

Scheme 113

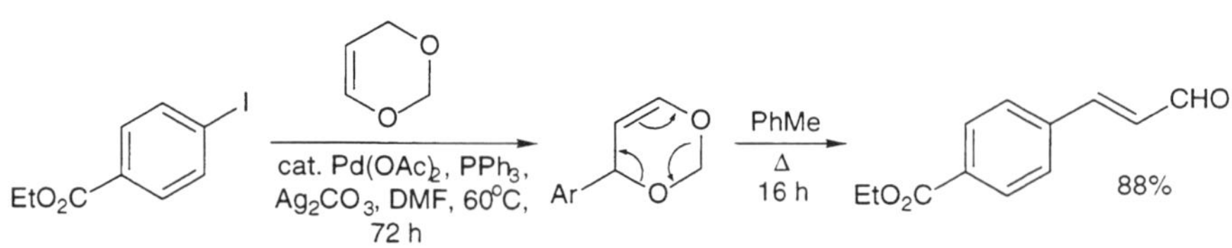

(b) Reactions

(i) Cinnamaldehyde will take part in a *syn*-selective asymmetric aldol reaction with 1-trimethylsiloxy-1-ethylthio-2-*tert*-butyldimethylsiloxyethene when mixed with a chiral tin (II)

Scheme 114

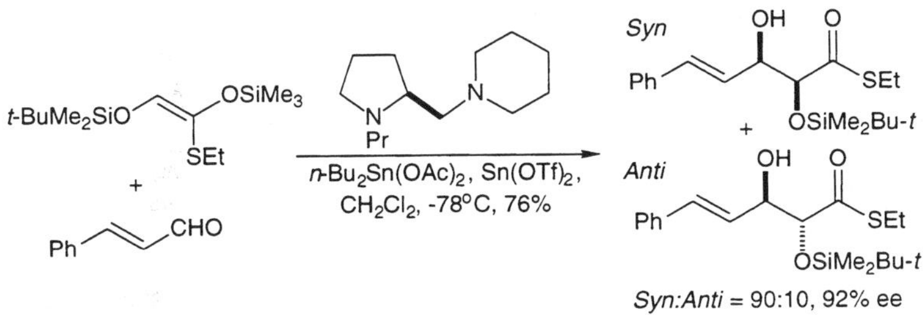

Lewis acid derived from tin (II) triflate, *N*-propyl-(*2S*)-[(1-piperidin-1-yl)methyl]pyrrolidine and dibutyltin acetate (Scheme 114) (T. Mukaiyama, I. Shiina, and S. Kobayashi, *Chem. Lett.*, **1991**, 1901). The high *syn* selectivity is thought to originate from

Scheme 115

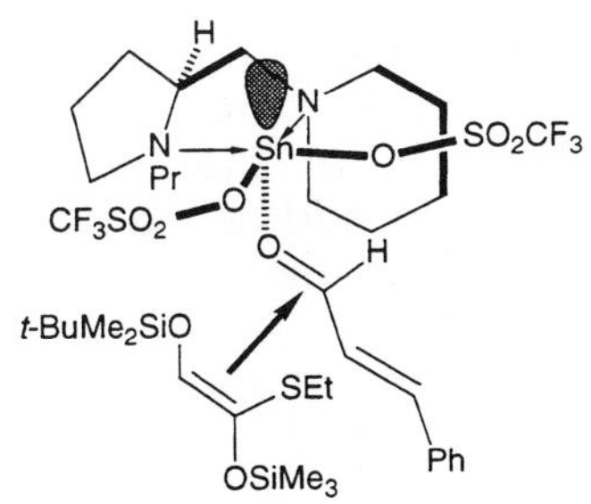

a rigid bicyclic complex being formed between the chiral diamine and the $Sn(OTf)_2$; this then coordinates to the aldehyde as shown in Scheme 115. In this chelate, the aldehyde hydrogen adopts an *endo* orientation to prevent the bulky carbonyl substituent from interacting with the ligands of the complex. The *O*-silyl ketenethioacetal is then assumed to approach the activated aldehyde from its "western" underside.

Scheme 116

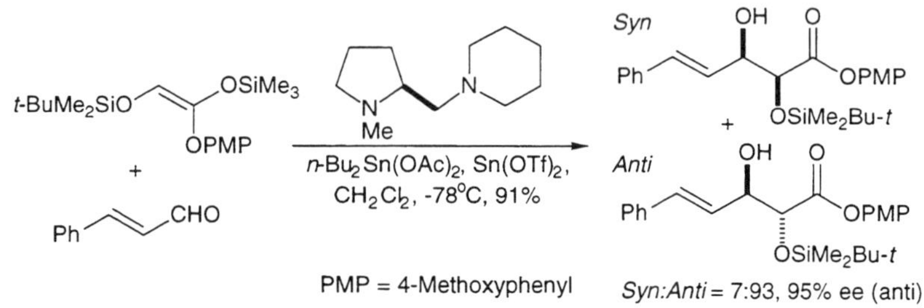

The stereochemistry of this aldol reaction can be modulated to favour formation of the *anti*-product if the thioethyl component of the *O*-silyl thioketene acetal is replaced with a *p*-methoxyphenyloxy group (S. Kobayashi and T. Kawasuji, *Tetrahedron Lett.*, **1994**, *35*, 3329). This leads to a 97:3 ratio of *anti:syn* aldol adducts with cinnamaldehyde (Scheme 116).

R.O. Duthaler *et al.* (*Angew. Chem. Int. Edn. Engl.*, **1989**, *28*, 495) have recently used bis(1,2:5,6-di-*O*-isopropylidene-α-D-glucofuranose)-cyclopentadienyltitanium chloride to transmetallate the lithium enolate obtained from *tert*-butyl acetate. The resulting chiral titanium enolate shows an overwhelming preference to add to the *re*-face of the carbonyl group in cinnamaldehyde to furnish the aldol addition product in very high ee (Scheme 117).

Undoubtedly, the catalytic variant of the externally mediated asymmetric aldol reaction represents its pinnacle in terms of synthetic efficiency. A number of research groups have described chiral complexes that will catalyse the asymmetric aldol reaction between cinnamaldehyde and achiral *O*-silyl

Scheme 117

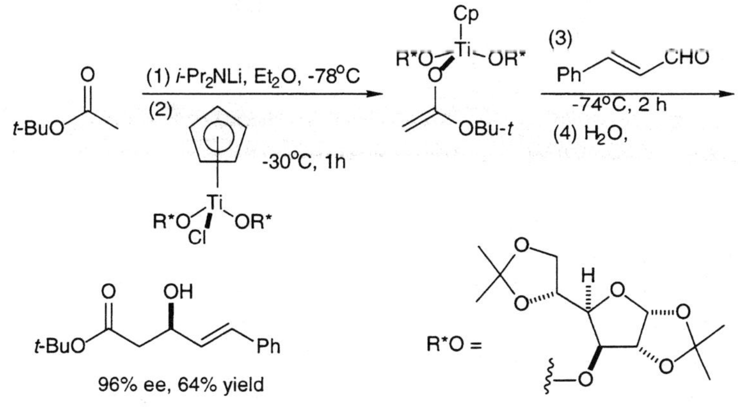

ketene acetals. For example, S. Kiyooka, Y. Kaneko, and K. Kume (*Tetrahedron Lett.*, **1992**, *33*, 4927) have made a chiral borane reagent from (*S*)-valine *p*-nitrobenzenesulfonamide and BH_3-THF complex, and employed it *in situ* to catalyse the addition of 1-trimethylsilyloxy-1-phenoxy-2-methyl-2-propene to cinnamaldehyde at -78 °C (Scheme 118). The use of nitroethane proved critical for efficient propagation of the catalytic cycle in this reaction, other solvents leading to diminished product yields and longer reaction times.

Scheme 118

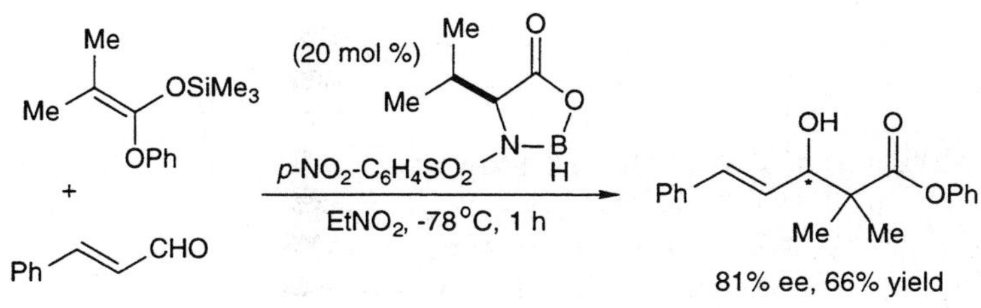

A chiral titanium-binapthyl complex has recently been introduced by E.M. Carreira, R.A. Singer, and W. Lee (*J. Am.

Chem. Soc., **1994**, *116*, 8837) for use in the asymmetric "acetate" aldol reaction. When employed at concentrations as low as 2 mol %, it externally mediates the addition of 1-trimethylsilyloxy-1-methoxy-ethene to cinnamaldehyde to give the (*R*)-aldol addition product in superb ee (Scheme 119).

Scheme 119

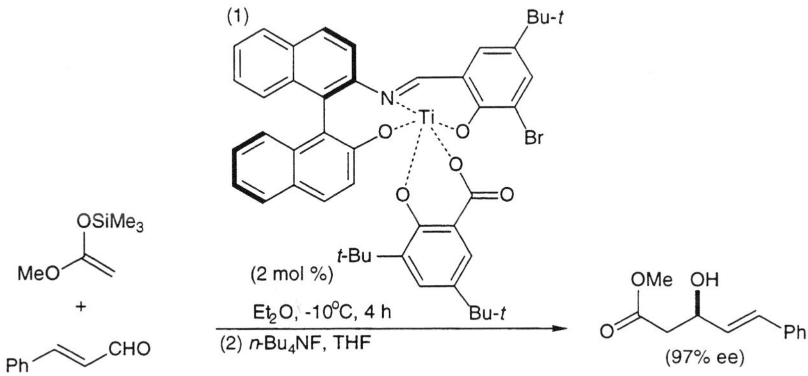

(ii) (*E*)-Cinnamaldehyde participates in enantioselective hetero-Diels-Alder additions with siloxy dienes, if catalysed by 10 mol % of the chiral Lewis acid methylaluminium (*R*)-(+)-3,3'-bis[tri(3,5-xylyl)silyl-1,1'-binapthalene-2,2'-diolate] (Scheme 120) (H. Yamamoto *et al., J. Am. Chem. Soc.*, **1988**, *110*, 310). A nonpolar solvent such as toluene is beneficial for these cycloadditions since it leads to enhanced enantiofacial selectivity. Donor solvents are unsatisfactory, however, since they retard the rate of cycloaddition quite significantly. In the reaction shown in Scheme 120, only a small amount of the *trans* adduct was formed, the *cis* product being obtained in 93% yield and 96% ee.

Scheme 120

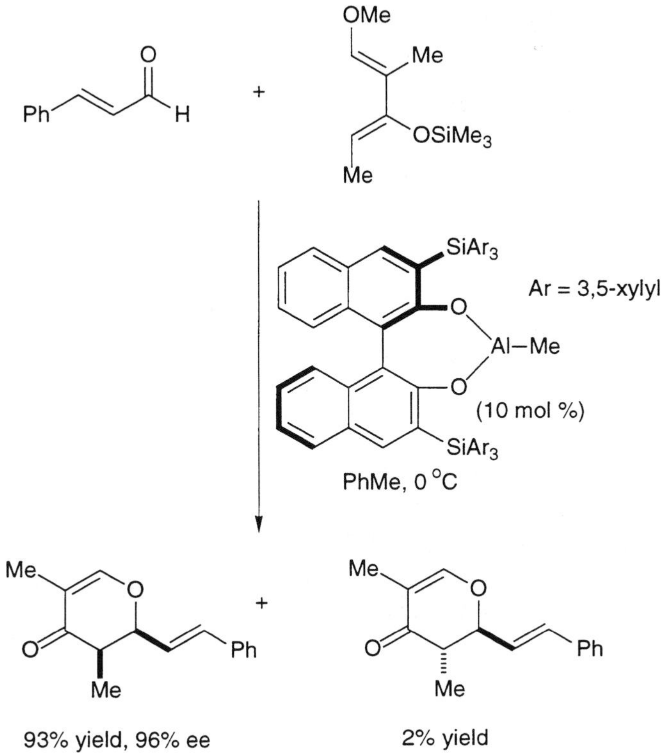

(iii) Incubation of α-methylcinnamaldehyde with fermenting baker's yeast and glucose affords crystalline, optically pure, (*3R,4S*)-2-methyl-1-phenylpent-1-ene-3,4-diol in 30 % yield. Typically, 83 g of α-methylcinnamaldehyde is added to 2.5 kg of yeast suspended in 10 L of water containing 2 kg of D-glucose and 100 g of Na_2HPO_4, and the mixture stirred at 30°C for 12 h

(Scheme 121) (C. Fuganti, P. Grasselli, and S. Servi, *J. Chem. Soc. Perkin Trans. 1*, **1983**, 241; C. Fuganti and P. Grasselli, *Chem. Ind.,* **1977**, 983). The (*3R,4S*)-2-methyl-1-phenylpent-1-ene-3,4-diol thus obtained is an extremely versatile chiral building block for the asymmetric synthesis of natural products (see for example: C. Fuganti *et al.*, *J. Org. Chem.*, **1984**, *49*, 543; *Tetrahedron Lett.*, **1983**, *24*, 3753; *ibid.*, **1982**, *23*, 4143). Its utility is underscored by a synthesis of (-)-frontalin (Scheme 121).

Scheme 121

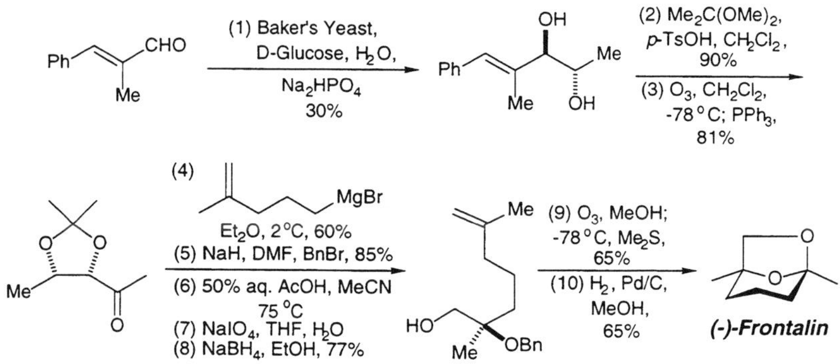

5. Arylalkenones, chalcones, etc.

(a) Preparative Methods

(i) Under palladium (0) catalysis, zinc salts of enol ether anions readily cross-couple with stereodefined β-styryl bromides to give arylalkenones after acid hydrolysis; parent alkene geometry is retained in the products (Scheme 122) (C.E. Russell and L.S. Hegedus, *J. Am. Chem. Soc.*, **1983**, *105*, 943). The

Scheme 122

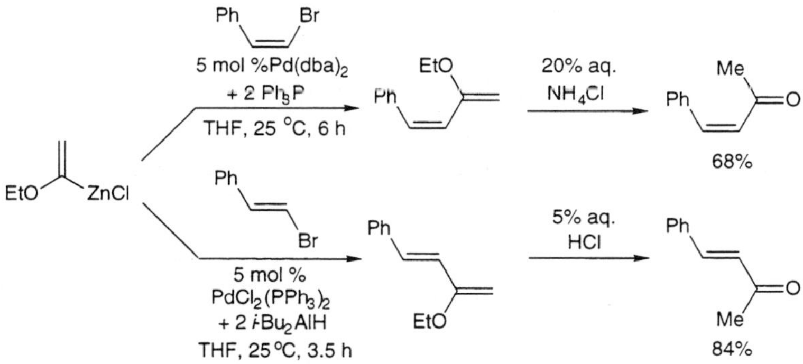

procedure is tolerant of a variety of functional groups in the aromatic ring (e.g. NO_2, NH_2, CO_2Me, $SiMe_3$).

(ii) γ–Iodo-γ-alkoxy-arylallenes are arylated in modest yield with aryl zinc chlorides when catalytic quantities (5 mol %) of palladium (0) complexes are additives (Scheme 123). A more

Scheme 123

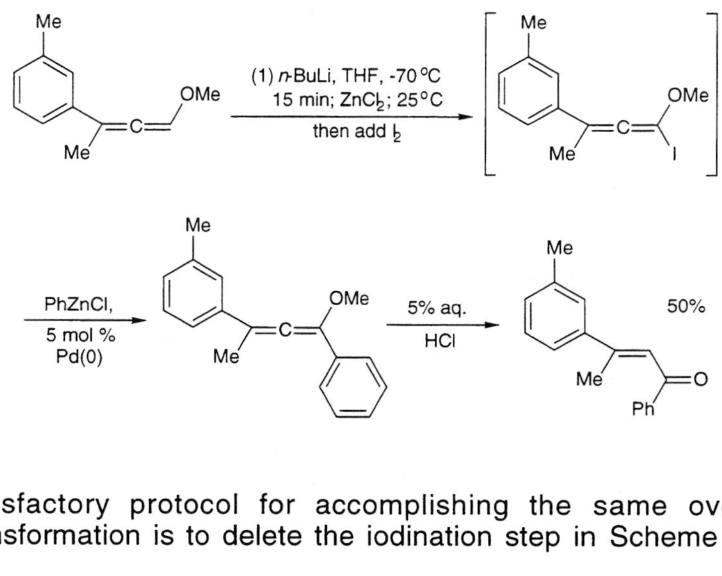

satisfactory protocol for accomplishing the same overall transformation is to delete the iodination step in Scheme 123 and effect transmetallation of the allenyllithium species with zinc chloride. This generates an allenylzinc reagent that more readily cross-couples with an aryl iodide under Pd(0) catalysis (C.E. Russell and L.S. Hegedus, *J. Am. Chem. Soc.*, **1983**, *105*, 943).

(iii) A concise genesis of unsymmetrical arylalkenones is the palladium (0)-catalysed carbonylative cross-coupling of a vinyl iodide with a tetraalkylstannane under a carbon monoxide atmosphere. It is customary to carry out such reactions at 40-50 °C and 50 psig pressure (Scheme 124) (J.K. Stille *et al., J. Am. Chem. Soc.*, **1984**, *106*, 6417). A distinct advantage of this process over other methods lies in its successful prosecution at neutral pH, an asset that enables acid- and base-sensitive functionality to be preserved within the reactants. It has been observed that if either of the coupling partners possesses (*E*)-alkene geometry, this is usually transferred to the product. However, if one of the starting components has (*Z*)-geometry, isomerisation often occurs, with the result that an all (*E*)-product is usually isolated. The rate of Pd(0)-catalysed carbonylative cross-coupling is affected quite substantially by substituent size in the vinyltin partner, with larger groups generally slowing the transmetallation step in the catalytic cycle.

Scheme 124

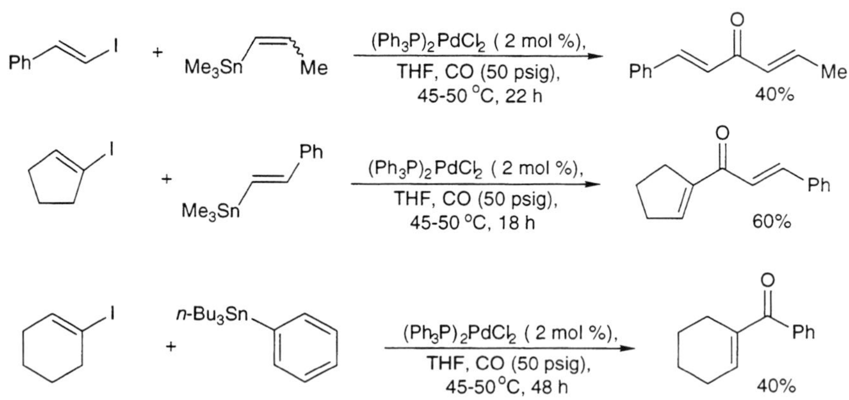

(iv) Arylalkenones may be prepared in high yield by palladium (0)-catalysed carbonylative coupling of vinyl triflates with arylstannanes in THF at 55 °C and 15 p.s.i (Scheme 125) (G.T. Crisp, W.J. Scott, and J.K. Stille, *J. Am. Chem. Soc.,* **1984**, *106*, 7500).

Scheme 125

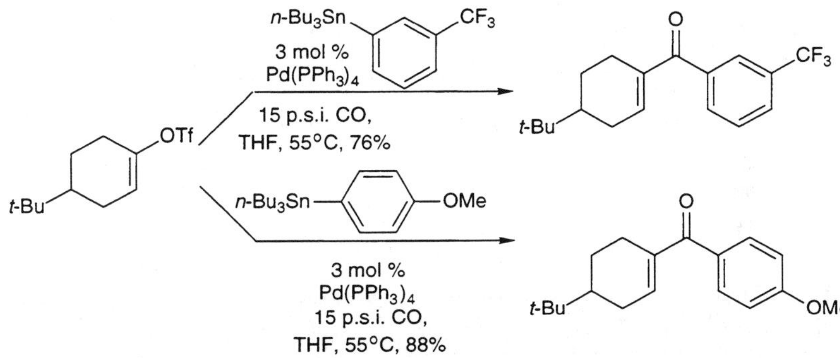

Lithium chloride is an essential ingredient for these reactions since it enables transmetallation to proceed between the organotin reagent and the vinylpalladium (II) triflate; the chloride ion simply exchanges with the triflate ligand in this intermediate, and generates a vinylpalladium (II) chloride complex that more readily engages in transmetallation. Importantly, when a (*Z*)-vinylstannane is employed as one of coupling partners, its alkene geometry is maintained in the product. Another attractive aspect of this procedure lies in the ease with which stereodefined enol triflates can be readily prepared from ketone enolates by trapping with *N*-phenyltriflimide.

(v) R.A. Kjonaas (*J. Org. Chem.*, **1986**, *51*, 3708) has studied the ketovinylation of various arylthallium derivatives with methyl vinyl ketone under lithium tetrachloropalladate catalysis.

Scheme 126

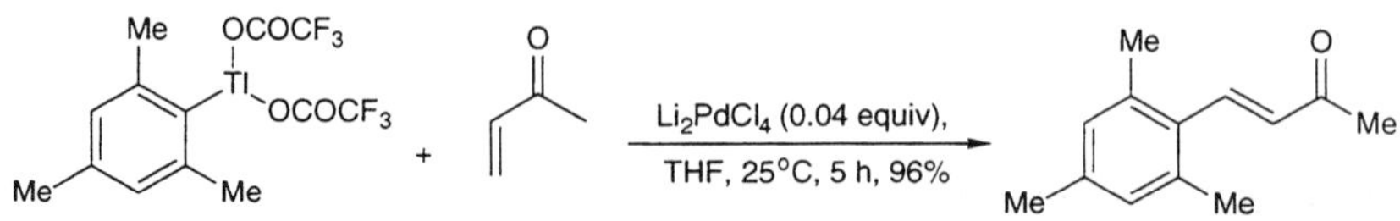

Excellent yields of β-arylation products are obtained even when three contiguously substituted aryl carbons are present in the target structure (Scheme 126). The reaction mechanism is believed to operate via transmellation, insertion, and reductive elimination, with the Pd(0) being reoxidised by the Tl(III) and then recycled.

(b) Reactions of Arylalkenenones, Chalcones etc.

(i) E. Wada, H. Yasuoka, and S. Kanemasa (*Chem. Lett.*, **1994**, 1637) have found that arylalkenones containing a terminal β-ketosulfonyl group will readily engage in chiral Lewis acid-

Scheme 127

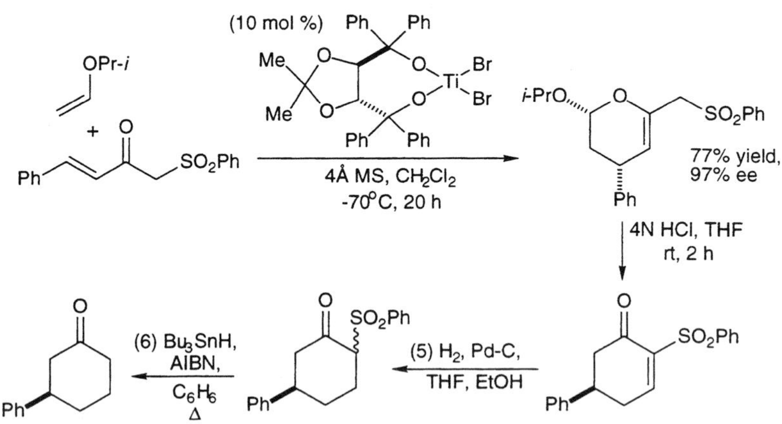

catalysed inverse-demand hetero Diels-Alder reactions with vinyl ethers (Scheme 127). High *endo*-selectivity is seen for these cycloadditions, with enantioselectivity improving as the size of the alkoxy group is increased in the dienophile. By employing this tactic, (*3R*)-3-phenyl-cyclohexanone was synthesised in 97% ee. An acid-mediated intramolecular aldol condensation/elimination, and a novel reductive β-keto desulfonylation with tributyltin hydride (A.B. Smith, III, K.J. Hale, and J.A. McCauley, Jr., *Tetrahedron Lett.*, **1989**, *30*, 5579) formed key steps in this sequence.

(ii) When (2*S*)-1-methyl-2-hydroxymethylpyrrolidine is combined with cuprous bromide and methyl magnesium bromide in THF at -30 °C, a chiral donor is fashioned that readily installs a methyl group at the β-position of 1,3-diphenyl-2-propen-1-one in 88% chemical yield and 68% optical yield (Scheme 128) (T. Imamoto and T. Mukaiyama, *Chem. Lett.*, **1980**, 45).

Scheme 128

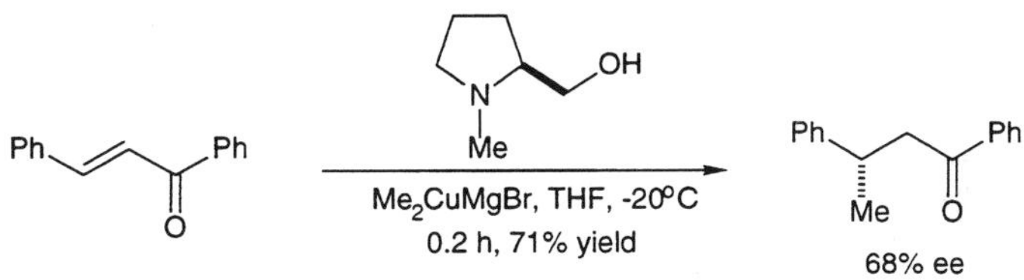

The related ligand, (*2S*)-methoxymethyl-(*4S*)-*tert*-butylthio-*N*-pivaloyl pyrrolidine, also forms a chiral complex with lithium dimethylcuprate. The latter conjugatively adds to 1,3-diphenyl-2-propen-1-one (Scheme 129) with opposite enantioselectivity to that encountered in Scheme 128 (F. Leyendecker and D. Laucher, *Tetrahedron Lett.*, **1983**, *24*, 3517).

Scheme 129

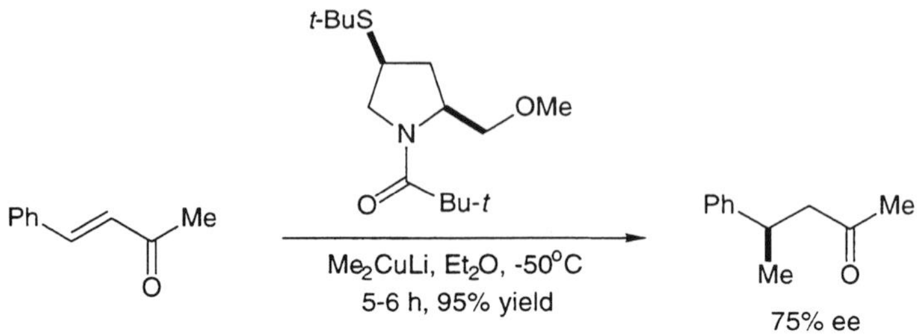

Modest levels of asymmetric induction have been noted in the tin (II) triflate promoted 1,4-addition of methyl dithioacetate to arylalkenones (Scheme 130) (T. Yura, N. Iwasawa, and T. Mukaiyama, *Chem. Lett.*, **1988**, 1021).

Scheme 130

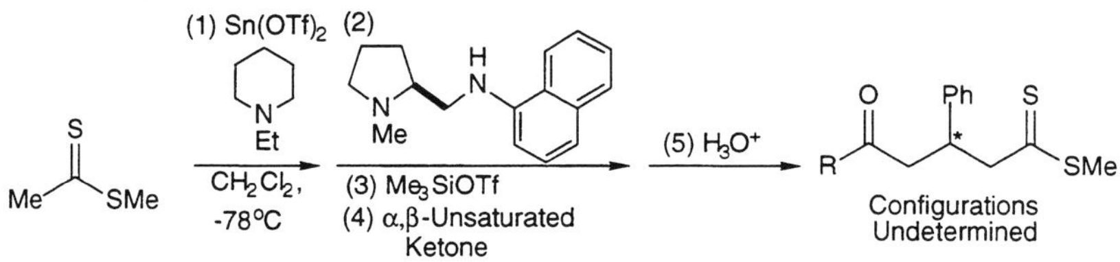

α,β -Unsaturated Ketone	*R*	*% Yield*	*% ee*
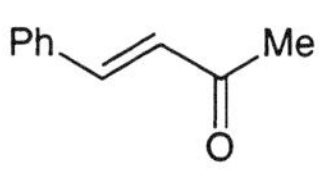	Me	82	70
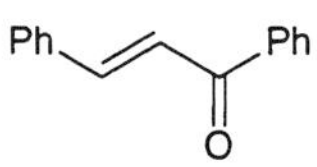	Ph	75	40
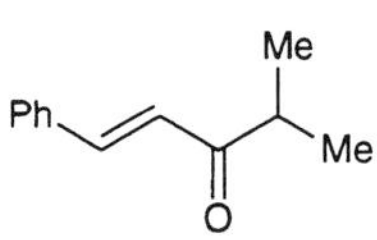	*i*-Pr	79	60

A modification of this method, which makes it catalytic in the chiral tin complex, employs an *S*-trimethylsilyl ketene thioacetal as the nucleophilic component (Scheme 131) (T. Mukaiyama *et al., Chem. Lett.*, **1988**, 1025).

Scheme 131

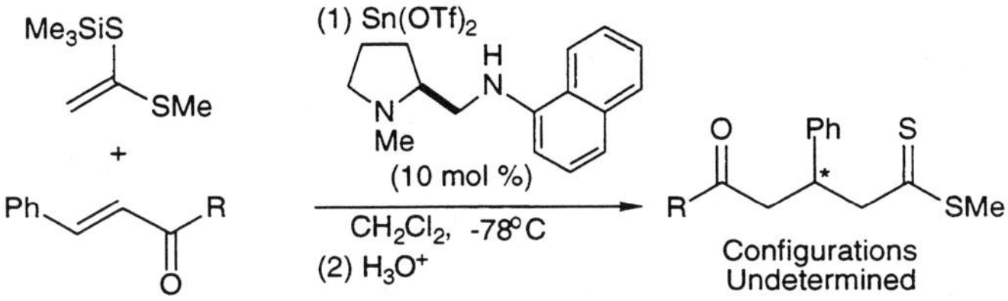

R	*% Yield*	*% ee*
Me	80	70
Ph	79	40

When executing these reactions it is usually advisable to add the trimethylsilyl enethiolate slowly to a solution of the arylalkenone and the chiral tin (II) catalyst in dichloromethane at -78 °C. This enables adducts to be isolated in up to 70% ee. The catalytic cycle that is considered to operate in these reactions is revealed in Scheme 132 .

Scheme 132

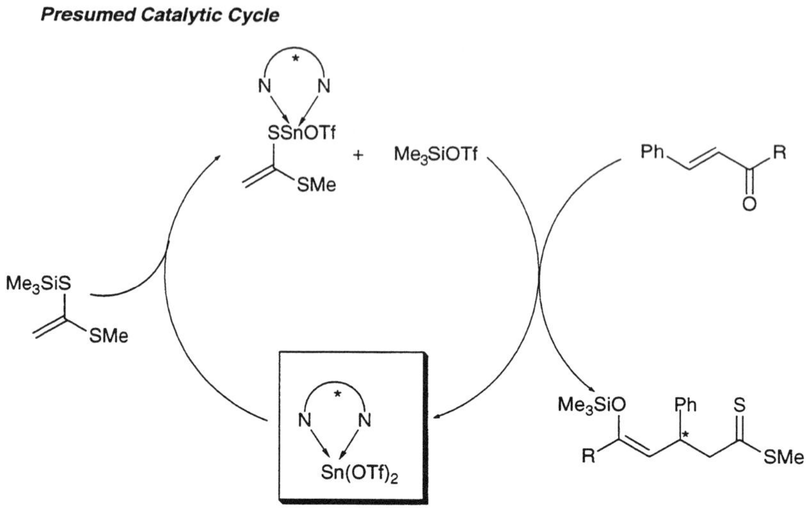

A catalytic quantity of nickel (II) chloride in association with (2*S*)-1-pentyl-2-anilinomethylpyrrolidine is an effective promoter for the enantioselective 1,4-addition of diethylzinc to chalcones (Scheme 133) (M. Asami *et al., Chem. Lett.*, **1994**, 297). Other nickel (II) species such as $Ni(acac)_2$ or $NiBr_2$ may also function in this capacity but tend to impart lower levels of asymmetric induction.

Scheme 133

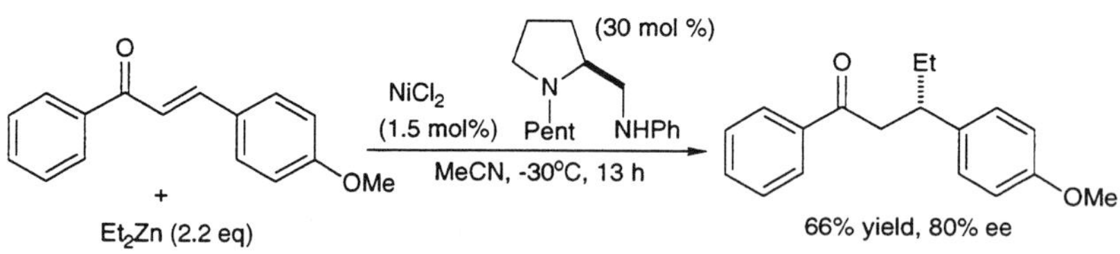

Useful levels of enantiocontrol (60-80% ee) have been recorded when 25 mol % of (*1R,2S*)-1-phenyl-2-(1-piperidinyl)propan-1-ol is adopted as a catalyst for the conjugate addition of diethylzinc to arylalkenones (Scheme 134). A stoichiometric amount of this chiral ligand does, however, lead to considerably improved optical purities in the products obtained (81-94% ee) (K. Soai, M. Okudo, and M. Okamoto, *Tetrahedron Lett.*, **1991**, *32*, 95).

Scheme 134

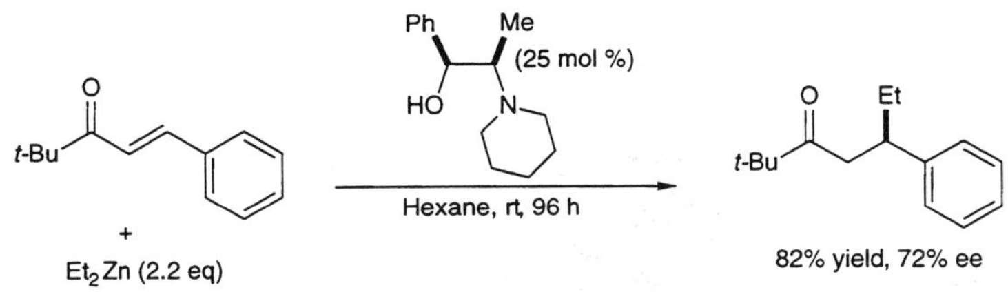

A range of alkylboranes undergo transmetallation with dimethylzinc in hexane at 0°C to give alkylzincs that conjugatively add to arylalkenones in the presence of copper (I) cyanide and lithium chloride (Scheme 135) (F. Langer, J. Waas, and P. Knochel, *Tetrahedron Lett.*, **1993**, *34*, 5261).

Scheme 135

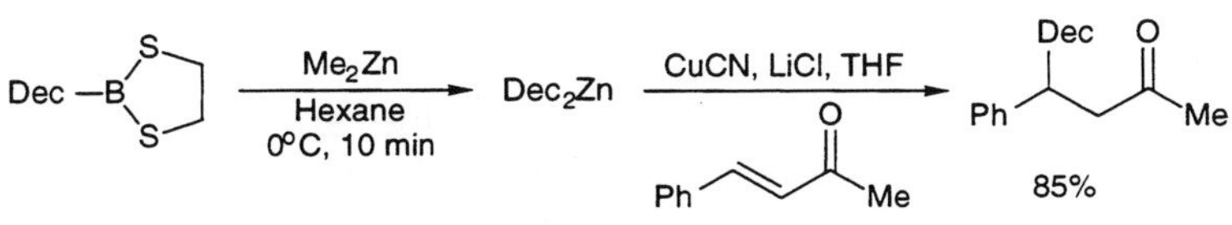

T. Hayashi, Y. Matsumoto, and Y. Ito (*J. Am. Chem. Soc.*, **1988**, *110*, 5579) have devised a convenient process for obtaining *anti*-aldol adducts (Scheme 136) that involves: (a) asymmetric addition of a disilane to an arylalkenone under

Scheme 136

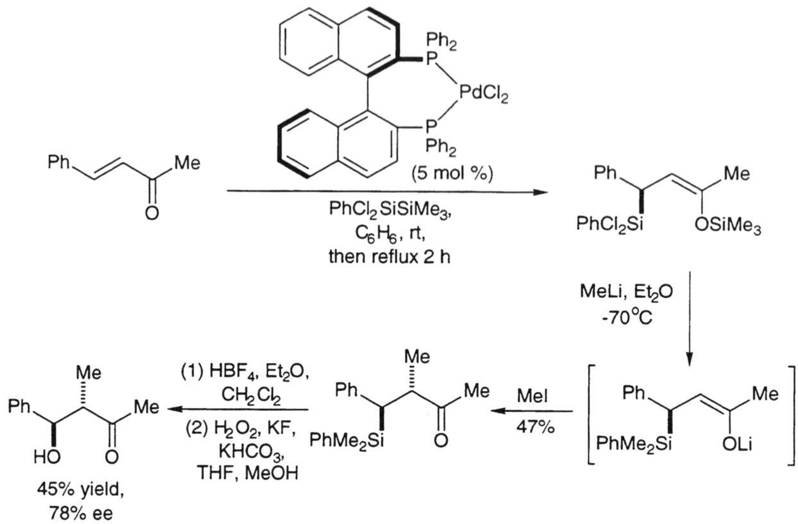

$PdCl_2$[(+)-BINAP] catalysis; (b) generation of a chiral lithium enolate from the intermediary *O*-silyl enol ether by treatment with methyllithium; (c) trapping of this enolate with a reactive alkylating agent and; (d) oxidative desilylation with retention of stereochemistry to install the hydroxy group β- to the ketone.

(iii) Certain poly-L-α-amino acids have been shown to catalyse the asymmetric epoxidation of chalcone with basic hydrogen peroxide in carbon tetrachloride and water. Poly-L-alanine and poly-L-isoleucine are the best chiral additives in this regard, each affording epoxy chalcone in approximately 95% ee and good chemical yield (Scheme 137) (S. Colonna and S. Julia *et al., Tetrahedron*, **1983**, *39*, 1635). Substituted chalcones may also be gainfully utilised as substrates but many react with lower levels of enantioselectivity (Scheme 137) (D. Ferreira *et al., Tetrahedron Lett.*, **1987**, *28*, 4857).

Scheme 137

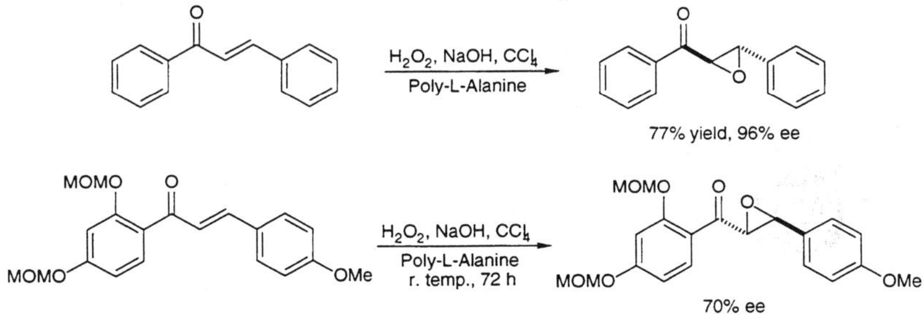

6. Cinnamic acids, cinnamate esters, cinnamylamides and related compounds

(a) Preparative methods

(i) A recent publication by A.-S. Carlstrom and T. Frejd (*J. Org. Chem.*, **1991**, *56*, 1289) has described the successful stepwise coupling of bromo(iodo)benzenes with different 2-amidoacrylates in the presence of catalytic palladium (II) acetate, tetra-*n*-butylammonium chloride, and sodium bicarbonate in DMF (Scheme 138). In the case of alkenes which polymerise readily under the normal Heck reaction conditions, it is sometimes advantageous to incorporate a free radical scavenger such as hydroquinone to suppress this side reaction and maximise the yield.

Scheme 138

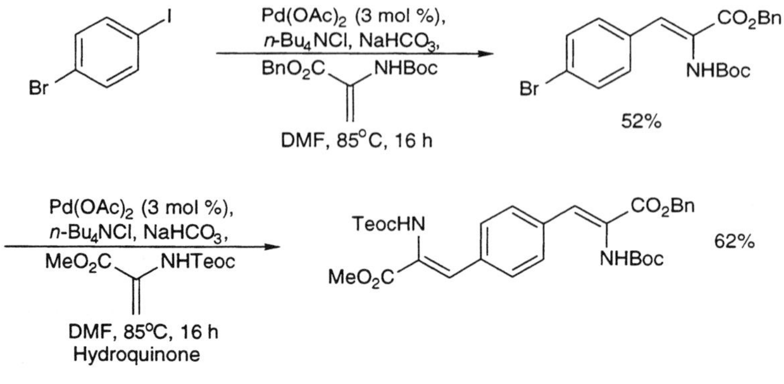

A procedure is now available for conducting the Heck reaction in an aqueous medium; it employs aryl diazonium salts and acrylate derivatives (S. Sengupta and S. Bhattacharya, *J. Chem. Soc. Perkin. Trans. 1*, **1993**, 1943). More specifically, an aniline is diazotised with $NaNO_2$ and 42% aq. HBF_4 at 0 °C for 1 hour, whereupon an acrylate and 2 mol % of $Pd(OAc)_2$ is added to the reaction mixture; the mixture is then heated at 50-60 °C for 1 hour (Scheme 139). According to these workers, aryl

Scheme 139

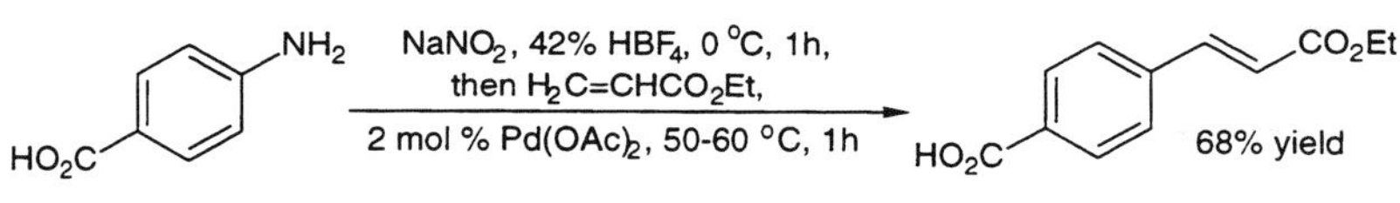

diazonium salts are considerably more reactive than the corresponding aryl bromides in the Heck reaction for preparing cinnamic acid derivatives.

Polymer-supported aryl iodides undergo Heck reactions with acrylates to produce cinnamic acid esters in good yield (Scheme 140) (K.-L. Yu, M.S. Deshpande, and D.M. Vyas, *Tetrahedron Lett.*, **1994**, *35*, 8919).

Scheme 140

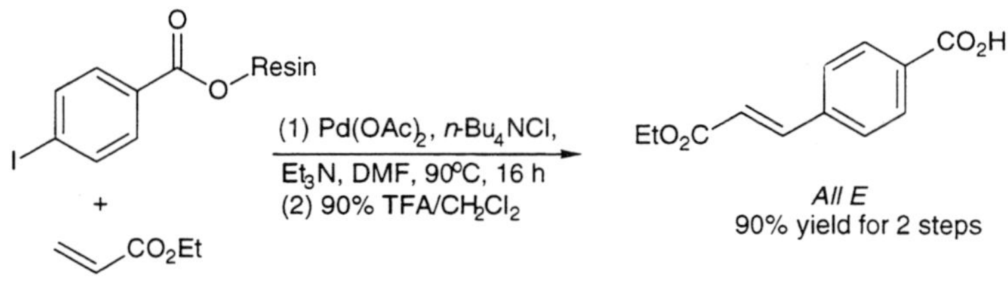

For quite some time chloroarenes were not considered viable substrates for the Heck reaction due to their inability to participate in oxidative addition reactions with palladium (0) under the standard reaction conditions. It was an important breakthrough, therefore, when J.J. Bozell and C.E. Vogt (*J. Am. Chem. Soc.*, **1988**, *110*, 2655) discovered that the combination of nickel (II) bromide and sodium iodide considerably facilitates the oxidative addition of $[Pd_2(dba)_3]$ to substituted chlorobenzenes, since this has considerably extended the arsenal of starting substrates that can now be considered for cinnamic acid assembly (Scheme 141). From an industrial perspective this finding is particularly significant, since chloroarenes are very cheap chemical feedstocks.

Scheme 141

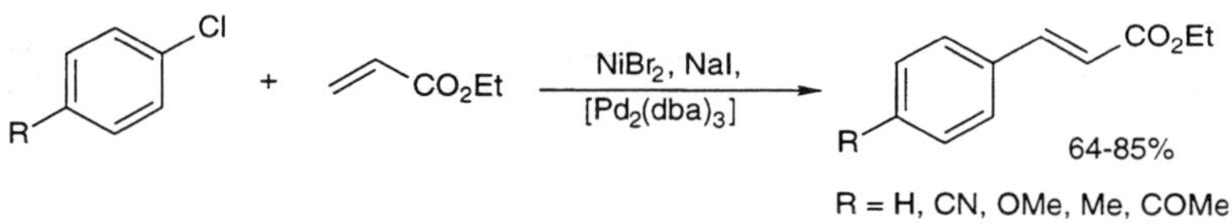

(ii) A.M. Echavarren and J.K. Stille (*J. Am. Chem. Soc.*, **1987**, *109*, 5478) have successfully cross-coupled benzyl (*E*)-3-(tri-*n*-butylstannylpropenoate) with 4-nitrophenyl triflate using 2 mol % of $(Ph_3P)_4Pd$ in dioxane at 98 °C (Scheme 142). The corresponding (*Z*)-organostannane proved to be a reluctant cross-coupling partner under identical reaction conditions, affording the isomerised (*E*)-cinnamate in a meagre 5% yield after 66 h. However, a change of catalyst to $Cl_2Pd(PPh_3)_2$, and use of DMF as the reaction solvent led to a considerably improved reaction rate (5 h) and a significant increase in chemical yield (47%). Again, olefin isomerisation was problematic under these conditions, with a 2:1 mixture of (*Z*)- and (*E*)-cinnamates being the final outcome.

Scheme 142

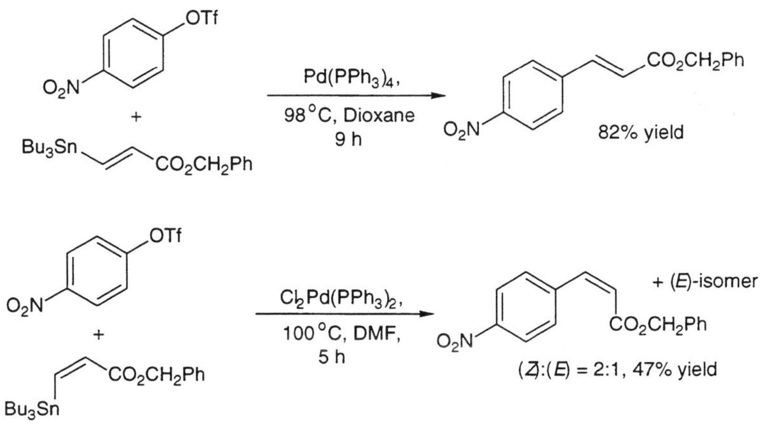

(iii) During a total synthesis of the naturally-occurring HIV-1 inhibitor thiangazole, a novel base-mediated cyclocondensation reaction was attempted between cinnamonitrile and (*2R*)-2-methyl-cysteine methyl ester hydrochloride. It produced a rather unusual chiral 4-methylthiazoline ester in 40% yield (Scheme 143) (R.J. Boyce, G.C. Mulqueen, and G. Pattenden, *Tetrahedron Lett.*, **1994**, *35*, 5705).

Scheme 143

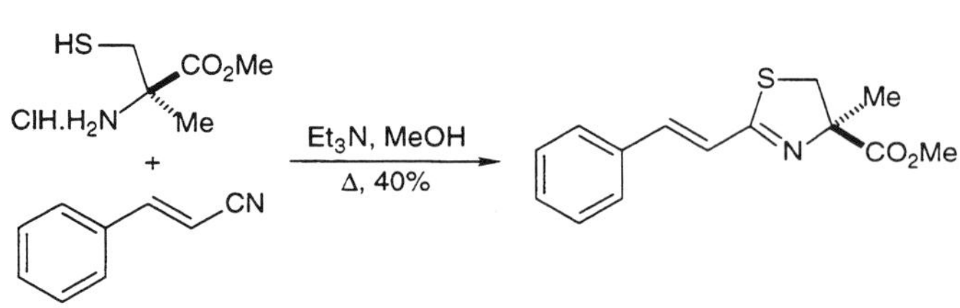

(b) Reactions

(i) Both cinnamic acid and its sodium salt can be reduced to cinnamaldehyde in 99% and 93% yield respectively when exposed to thexylbromoborane-dimethyl sufide (Scheme 144) (J.S. Cha, J.E. Kim, and K.W. Lee, *J. Org. Chem.*, **1987**, *52*,

Scheme 144

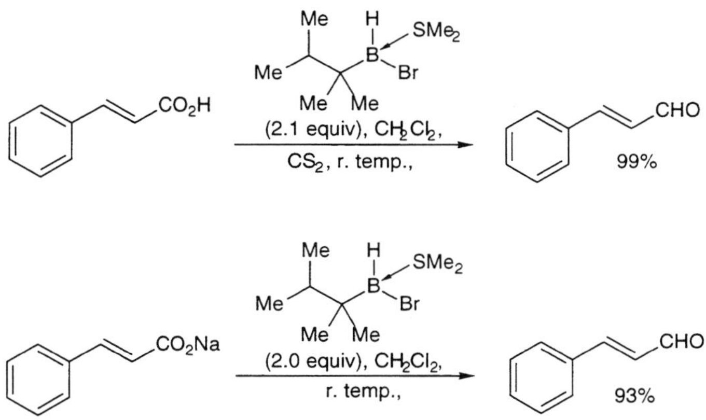

5030). Reductions with this reagent are very chemoselective, proceeding without attack on the double bond. Chloride, bromide, and nitro groups are also left undisturbed, but nitriles are reduced to imines that hydrolyse to aldehydes upon aqueous work-up.

(ii) J.M. Brown and A.P. James (*J. Chem. Soc. Chem. Comm.*, **1987**, 181) have demonstrated that the homogenous hydrogenation of 3-phenylitaconate esters proceeds with high diastereoselectivity when catalysed by chiral cationic rhodium-phosphine complexes. Efficient kinetic resolution of racemic mixtures is possible using such catalysts, one enantiomer being hydrogenated much more rapidly than the other (Scheme 145).

Scheme 145

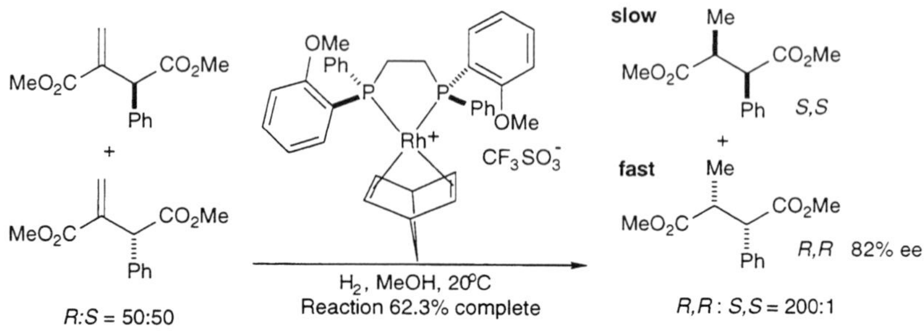

Chiral BINAP-ruthenium (II) carboxylate complexes serve as efficient catalysts for the asymmetric high-pressure hydrogenation of α-phenyl acrylic acid (Scheme 146) (R. Noyori *et al.*, *J. Org. Chem.*, **1987**, *52*, 3174).

Scheme 146

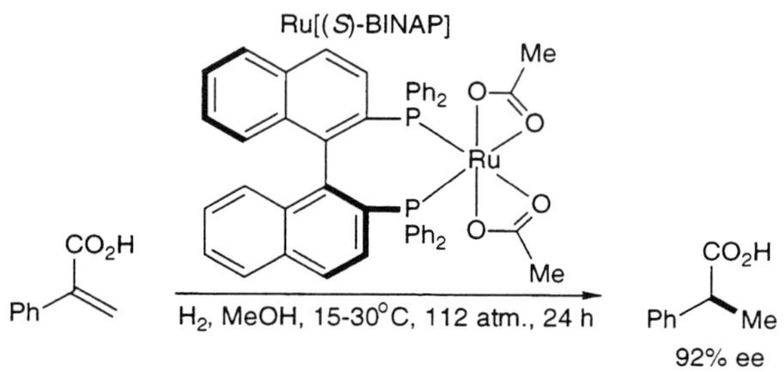

(iii) (*N*-(*p*-Toluenesulfonyl)imino)phenyliodinane in conjunction with copper (I) triflate and a chiral bis(oxazoline) ligand is a useful catalytic system for the asymmetric aziridination of phenyl cinnamate (Scheme 147) (D.A. Evans *et al.*, *J. Am. Chem. Soc.*, **1993**, *115*, 5328).

Scheme 147

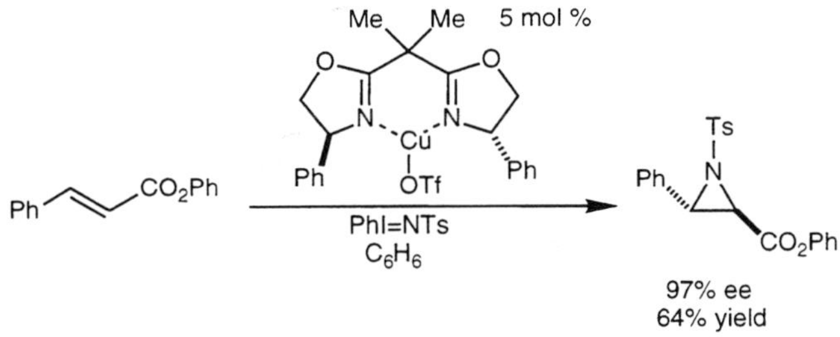

(iv) *N*-Acetoxyamino-2-isobutylquinazolin-4-one may be used to selectively hydraziridinate a wide range of cinnamates at -20 °C in dry methylene chloride; the yields are quite decent (58-72%) (M. Deshmukh, P. Chavan, and D. Kharade, *Monatsh. Chem.*, **1994**, *125*, 743).

(v) An *anti*-selective tandem conjugate addition/electrophilic α-hydroxylation results when (*E*)-*tert*-butyl cinnamate is treated with lithium (*R*)-(α-methylbenzyl)benzylamide in THF at -78 °C for 1.5 h, and (+)-(camphorsulfonyl)oxaziridine is then added (Scheme 148) (M.E. Bunnage, S.G. Davies, and C.J. Goodwin, *J. Chem. Soc. Perkin Trans 1*, **1993**, 1375).

Scheme 148

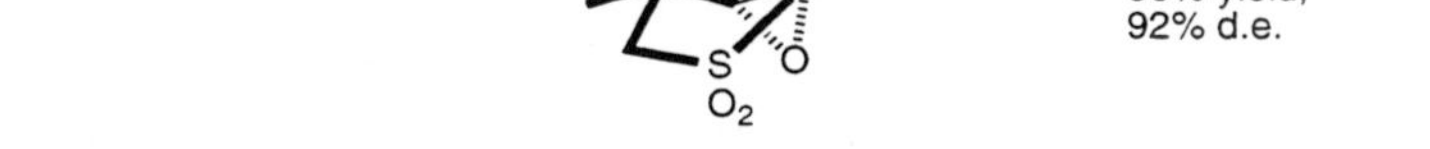

S.G. Davies *et al.* (*Tetrahedron: Asymmetry*, **1994**, *5*, 35) have also discovered that bromomagnesium (*R*)-(α-methylbenzyl)benzylamide will undergo conjugate addition to

(*E*)-*tert*-butyl cinnamate, and that the resulting enolate can be methylated with excellent *syn* diastereoselectivity (19:1, 90% d.e.) (Scheme 149). The sense of asymmetric induction is identical for both the lithium and magnesium amides in the

Scheme 149

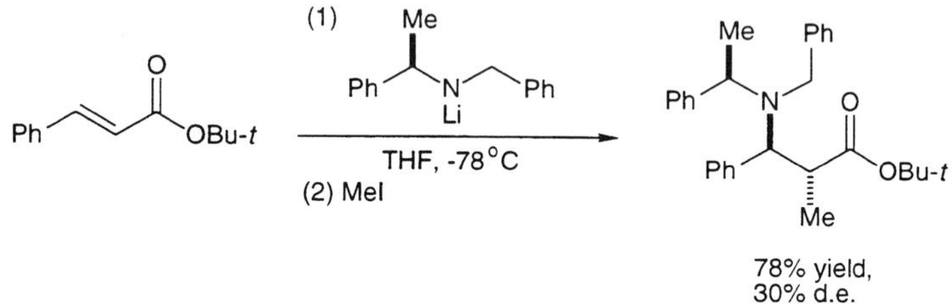

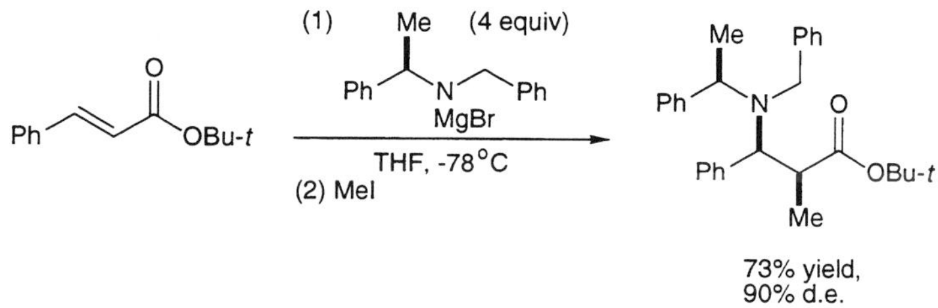

initial Michael addition step, but the differing enolates show opposite facial preferences in their subsequent reactions with methyl iodide (see: S.G. Davies *et al., J. Chem. Soc. Chem. Comm.*, **1993**, 1153).

One approach to β-aryl-δ-keto esters involves conjugate addition of a SAMP-hydrazone carbanion to an (*E*)-cinnamic acid ester (Scheme 150). Such reactions usually proceed with

Scheme 150

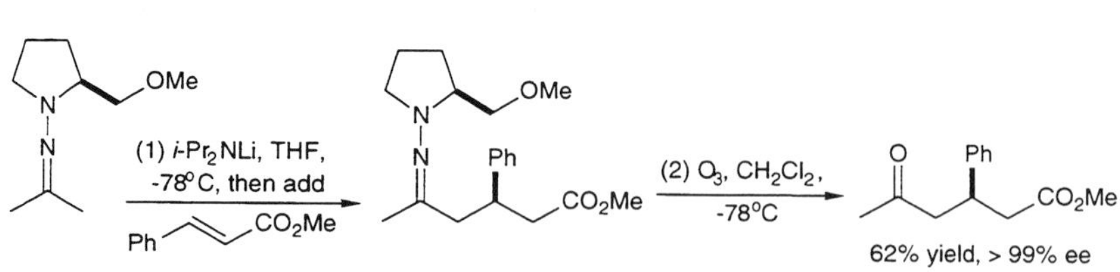

excellent levels of enantioselectivity, furnishing almost optically pure products after removal of the chiral auxiliary by ozonolysis (D. Enders and K. Papadopoulos, *Tetrahedron Lett.*, **1983**, *24*, 4967).

Scheme 151

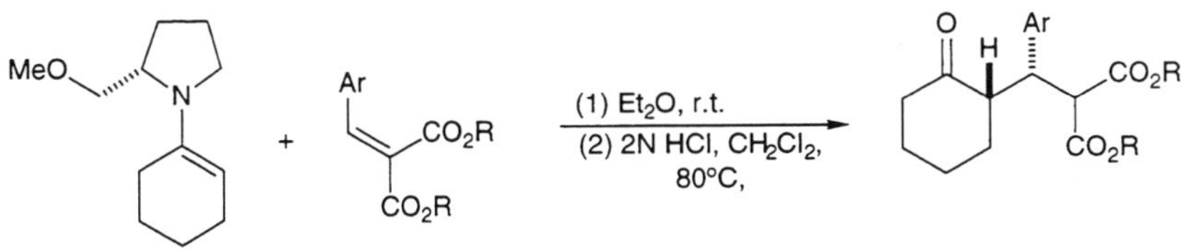

Ar	*R*	*% yield*	*% de*	*% ee*
Ph	Et	70	>95	95
Ph	Me	76	>95	92
4-ClC_6H_4	Me	53	88	80
4-$NO_2C_6H_4$	Me	35	88	83
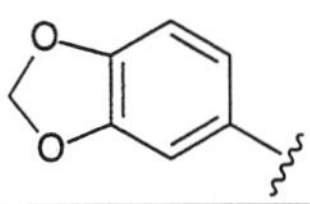	Me	55	95	82

Asymmetric Michael addition of (*2S*)-1-(1-cyclohexen-1-yl)-2-(methoxymethyl)pyrrolidine to α-carboxyalkylcinnamates also furnishes δ-keto-β-aryl ester derivatives in > 88% diastereomeric purity and > 80% ee (Scheme 151) (S.J. Blarer and D. Seebach, *Chem. Ber.*, **1983**, *116*, 2250).

Organocuprate reagents will add 1,4- to (*E*)-cinnamates bearing a chiral imidazoline group at the *ortho*-position. The high diastereoselectivity obtained in these additions is attributed to steric hindrance from the chiral auxiliary (Scheme 152) (A. Alexakis *et al.*, *Tetrahedron Lett.*, **1988**, *29*, 4411).

Scheme 152

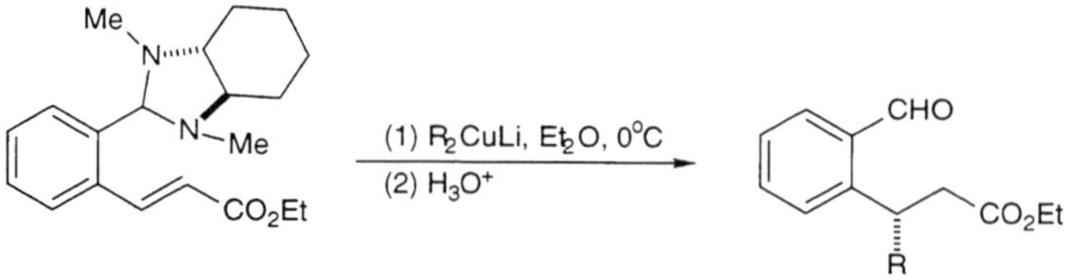

Organocuprate Reagent	R Group	% yield	% ee
Me_2CuLi	Me	85	94
Bu_2CuLi	Bu	90	95
Ph_2CuLi	Ph	84	96
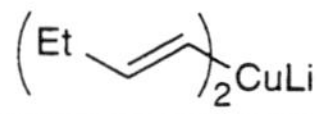	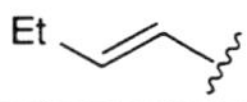	80	90

Asymmetric conjugate additions to achiral (*E*)-cinnamates have also been performed with lithiated thioacetals and various external chiral chelating ligands. However, none of these

Scheme 153

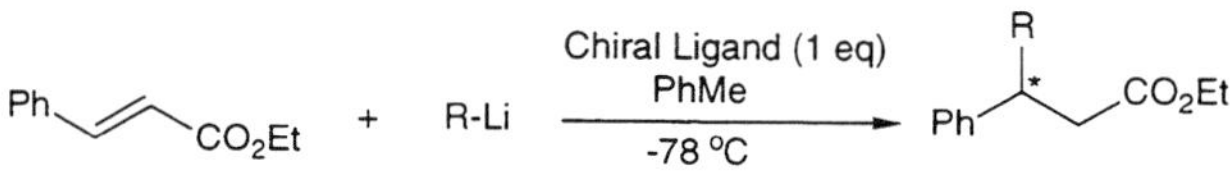

R-Li	Chiral Ligand	% Yield	% ee	Config'n
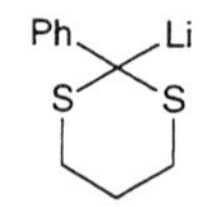	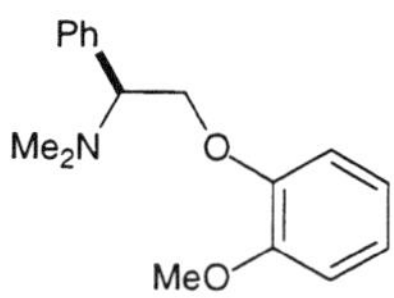	22	53	***R***
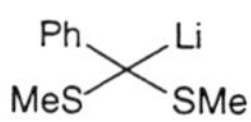		53	36	***R***
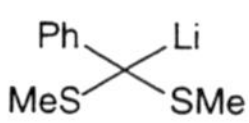	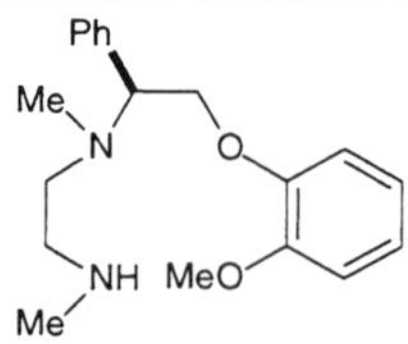	81	43	***S***

reactions have as yet displayed high levels of enantioselectivity (Scheme 153) (K. Koga *et al., Chem. Lett.*, **1985**, 329).

V.J. Hruby *et al.* (*Tetrahedron Lett.*, **1993**, *34*, 2565) have devised a stereocontrolled route to novel tyrosine analogues that is based on the 1,4-addition of MeMgBr/CuBr-Me_2S to chiral *N*-cinnamylated oxazolidinones followed by *in situ* bromination with *N*-bromosuccinimide. The resulting bromides are then subjected to S_N2 displacement with tetramethylguanidinium azide, the auxiliary removed with LiOH/H_2O_2, and the azido group reduced by hydrogenation over palladium on charcoal (Scheme 154).

Scheme 154

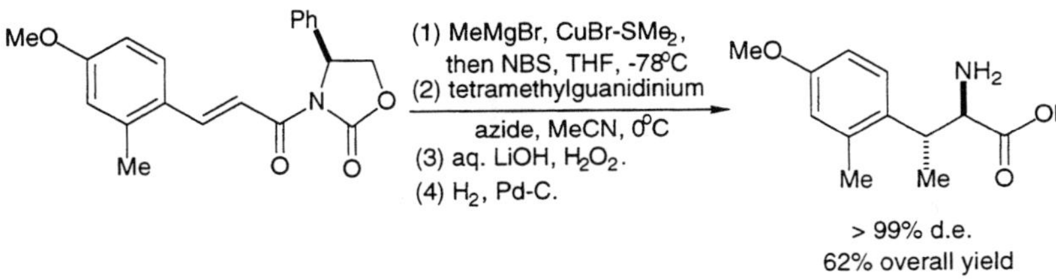

(vi) (*Z*)-Ethyl cinnamate has been directly epoxidised in 97% ee and 56% yield with a Mn^{III}(salen) complex (6 mol %) and aqueous sodium hypochlorite (Scheme 155) (L. Deng and E.N. Jacobsen, *J. Org. Chem.*, **1992**, *57*, 4320).

Scheme 155

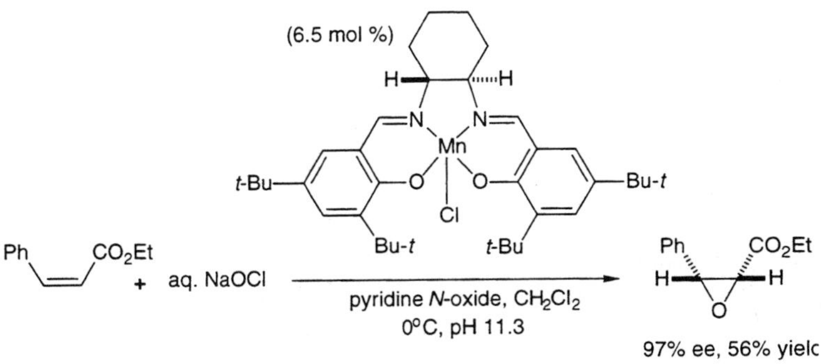

(vii) The osmium-catalysed asymmetric dihydroxylation of 2-aryl acrylates with the $(DHQD)_2$PHAL ligand yields diols with good to excellent optical purity (Scheme 156) (Z.-M. Wang and

K.B. Sharpless, *Synlett*, **1993**, 603). The $(DHQD)_2PHAL$ ligand likewise affords *(2S,3R)*-diols in > 96% ee when used for the AD on (*E*)-ethyl cinnamate and (*E*)-*N*-methoxy-*N*-methyl cinnamide at room temperature with 1 mol % of OsO_4 (Scheme 156) (K.B. Sharpless *et al., J. Org. Chem.*, **1992**, *57*, 2768).

Scheme 156

Alkene	*Product*	*$(DHQD)_2PHAL$-OsO_4 % ee*	*Yield*
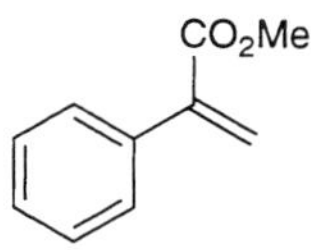	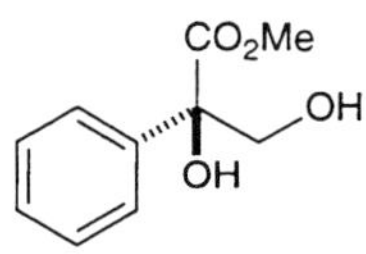	**88**	**86**
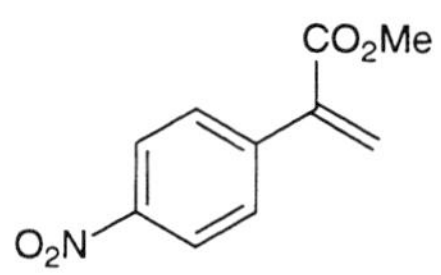	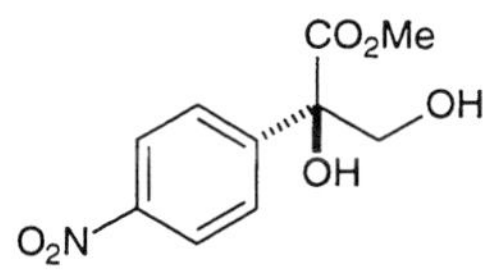	**93**	**91**
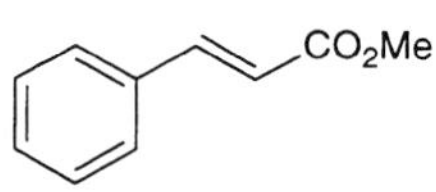	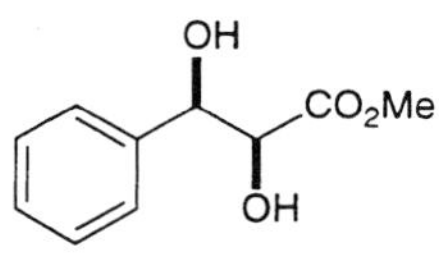	**97**	**>80**
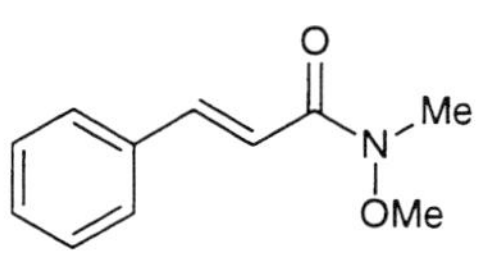	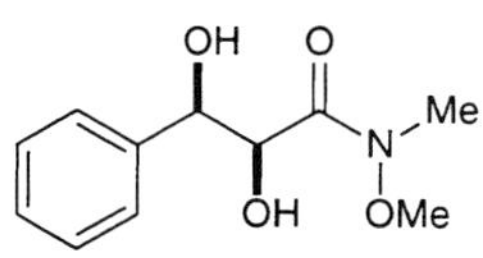	**96**	**92**

It is generally best to perform the asymmetric dihydroxylation of (*Z*)-ethyl cinnamate with the DHQD-IND ligand; this furnishes the *(2R,3R)*-diol in 78% ee. Use of the isopropyl ester raises the ee very slightly to 80% (L. Wang and K.B. Sharpless, *J. Am. Chem. Soc.*, **1992**, *114*, 7568).

(viii) Chiral *exo*- and *endo*-3-amino-2-triisopropylsilyloxy-bornanes have proven effective chiral auxiliaries for the diastereofacially-controlled cyclopropanation of (*E*)-cinnamic acid (Scheme 157) with Et_2Zn/CH_2I_2 in dichloromethane at room

Scheme 157

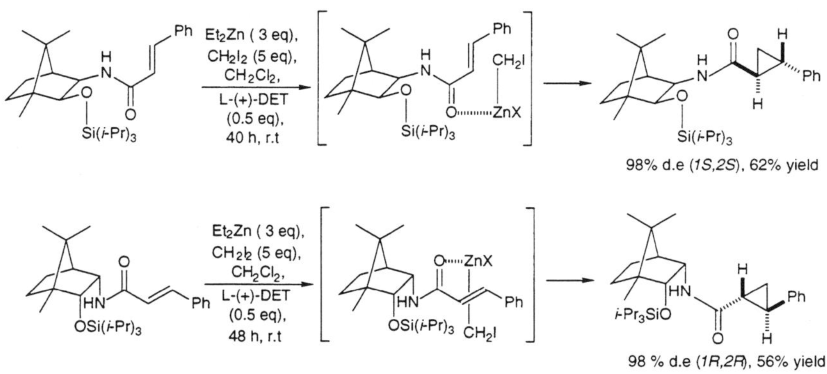

temperature. Curiously, if the 0.5 equiv of (L)-(+)- or D-(-)-diethyl tartrate is not added to the reaction mixture, the chemical yield of cyclopropanated product is very poor (16-20%), although stereoselectivity is still high (98% d.e); the mechanistic origins of this ligand effect await clarification (K. Tanaka *et al.*, *Tetrahedron: Asymmetry*, **1994**, *5*, 1175).

(ix) The intermolecular [2+2] photocycloadditions of various cinnamic acid derivatives to a substituted chromone have been examined in aqueous media in the presence of surfactants. Some of the results are displayed in Scheme 158 (H.C. Hailes, R.A. Raphael, and J. Staunton, *Tetrahedron Lett.*, **1993**, *34*, 5313).

Scheme 158

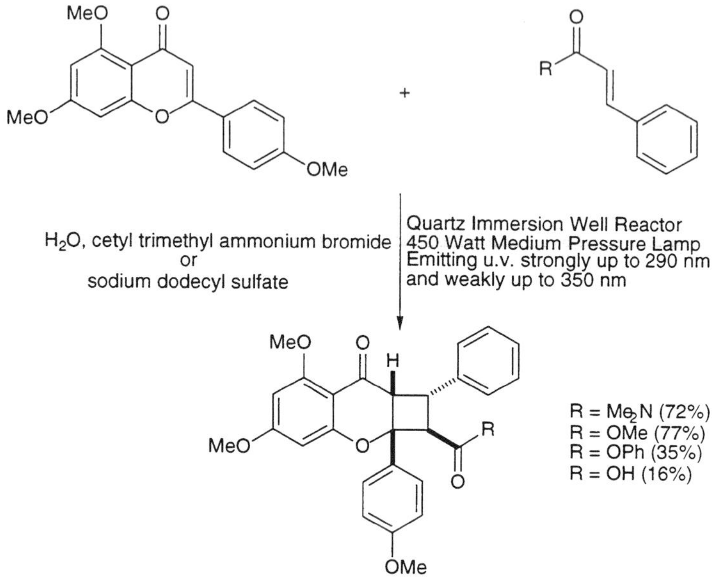

7. Arylalkynes

(a) Preparation

(i) J.S. Schumm, D.L. Pearson, and J.M. Tour (*Angew. Chem. Int. Edn. Engl.*, **1994**, *33*, 1360) have described a general procedure for converting an aryl alkyl ketone to an arylalkynylsilane. It involves enolisation of the ketone with LDA, trapping of the enolate as its enol phosphate, base-induced elimination to give the terminal arylalkynyl anion, and quenching with trimethylsilyl chloride (Scheme 159).

Scheme 159

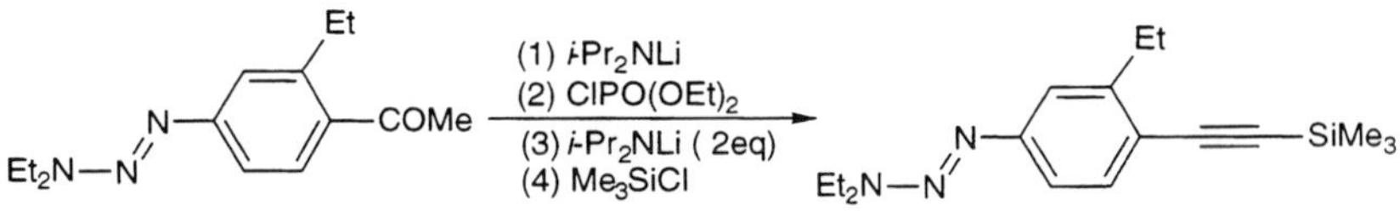

(ii) Arylalkynes are also obtainable from the cross-coupling of terminal alkynes with aryl iodides mediated by Pd(0) complexes (5 mol %), copper (I) iodide (10 mol %), and triphenylphosphine (20 mol %) in diisopropylamine and THF (Scheme 160) (J.S. Schumm, D.L. Pearson, and J.M. Tour, *Angew. Chem. Int. Edn. Engl.*, **1994**, *33*, 1360).

Scheme 160

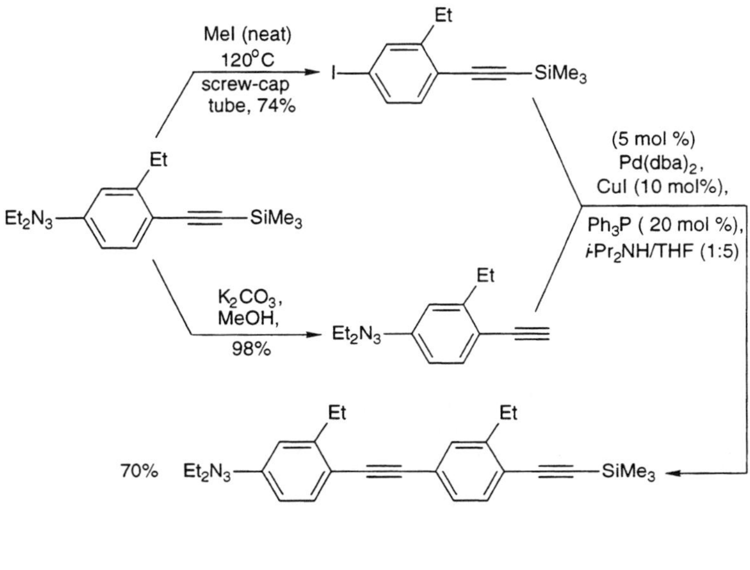

A synthesis of hexa(trimethylsilylethynyl)benzene from hexabromobenzene relied on this method (Scheme 161) (K.P.C Vollhardt *et al., Angew. Chem. Int. Edn. Engl.*, **1986**, *25*, 268).

Scheme 161

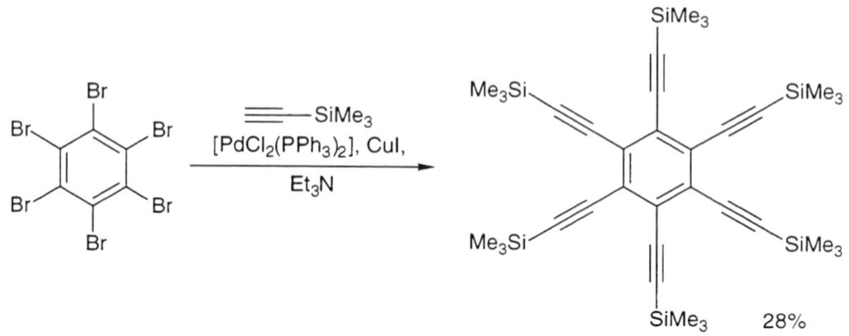

It has also been used by K.S. Feldman *et al.* (*J. Am. Chem. Soc.*, **1994**, *116*, 9019) for the preparation of 1,4-bis(3,3-diethynylpenta-1,4-diynyl)benzene from 3,3-bis[(trimethylsilyl)-ethynyl]-5-(trimethyl-silyl)pent-1,4-diyne (Scheme 162).

Scheme 162

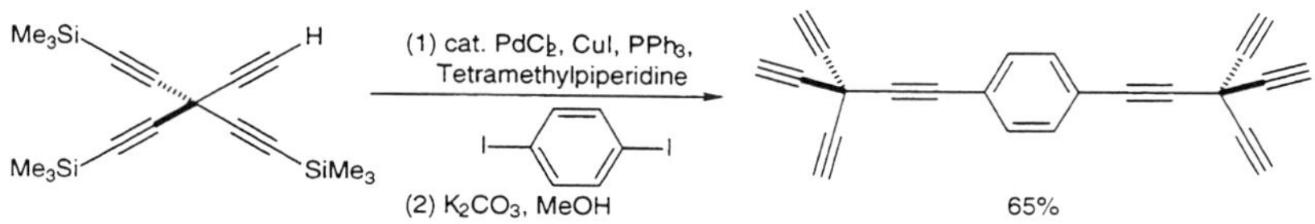

(iii) Terminal aryl acetylenes can be prepared through the palladium (0)-catalysed cross-coupling reaction of an aryl halide with trimethylsilylacetylene (Scheme 163) (K.S.Y. Lau *et al., J. Org. Chem.*, **1981**, *46*, 2280). Initially, this leads to an aryl silyl acetylene which is converted to the ethynyl arene by methanolysis. The method is recommended for the preparation

Scheme 163

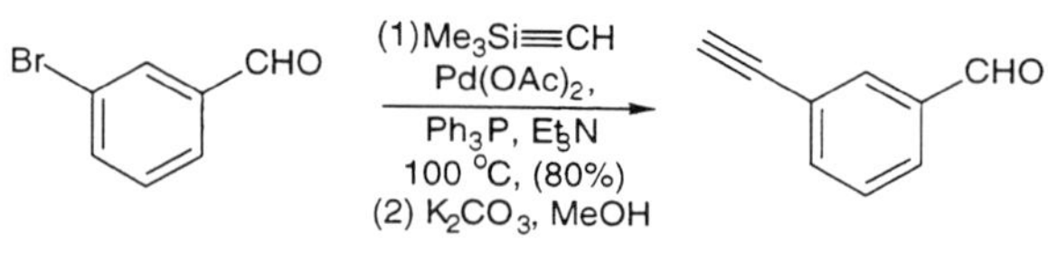

of sterically hindered arylacetylenes (M.E. Garst and B.J. McBride, *J. Org. Chem.*, **1989**, *54*, 249).

(iv) A catalytic quantity of $(Ph_3P)_4Pd$ in hexane at 70 °C is sufficient to bring about the union of arylalkynylstannanes with perfluoroalkyliodides to produce arylacetylenes containing a perfluoroalkyl group (Scheme 164) (S. Matsubara, M. Mitani, and K. Utimoto, *Tetrahedron Lett.*, **1987**, *28*, 5857).

Scheme 164

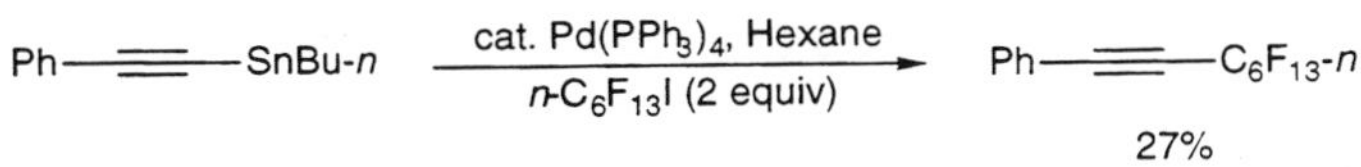

(v) Aryltriflates have been converted to arylalkynes in good yield by palladium (0)-catalysed cross-coupling with alkynyl-

Scheme 165

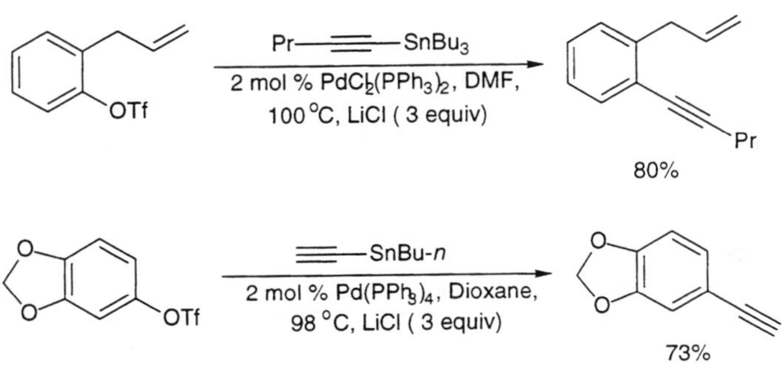

stannanes (Scheme 165) (A.M. Echavarren and J.K. Stille, *J. Am. Chem. Soc.*, **1987**, *109*, 5478).

(vi) A.M. Gilbert and W.D. Wulff (*J. Am. Chem. Soc.*, **1994**, *116*, 7449) have reported that arene-$Cr(CO)_3$ triflate complexes will undergo Stille couplings with substituted stannylacetylenes at around 65 °C in THF containing 2 mol % of $Pd(PPh_3)_4$ (Scheme 166).

Scheme 166

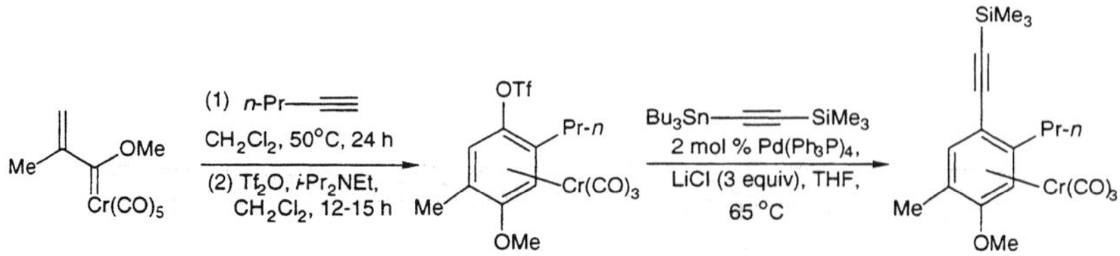

(vii) β,β-Dibromostyrenes undergo dehydrobromination to give aryl bromoacetylenes when treated with 5N or 10N NaOH and poly(ethyleneglycol) (PEG-400) as a phase transfer catalyst (Scheme 167) (P. Li and H. Alper, *J. Org. Chem.*, **1986**, *51*, 4354).

Scheme 167

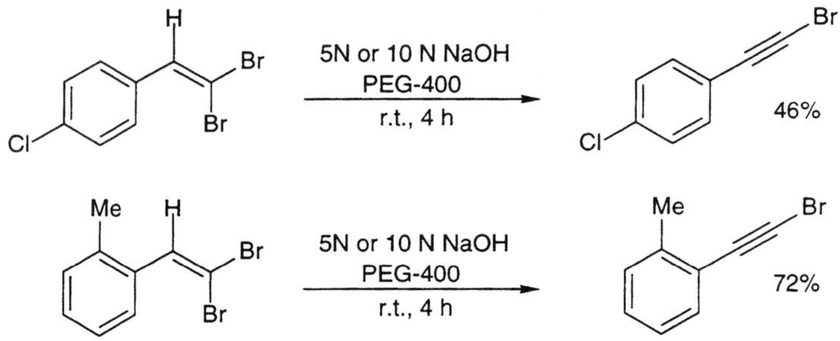

(viii) A facile synthesis of aryl acetylenes is founded on the low temperature addition of the anion derived from dimethyl (diazomethyl)-phosphonate to arylaldehydes and arylketones. This produces a transient betaine which eliminates phosphate to give a diazoethene. The latter is unstable at -78 °C and spontaneously eliminates nitrogen to generate a vinylidene carbene that rearranges to the acetylene (Scheme 168) (J.C.

Scheme 168

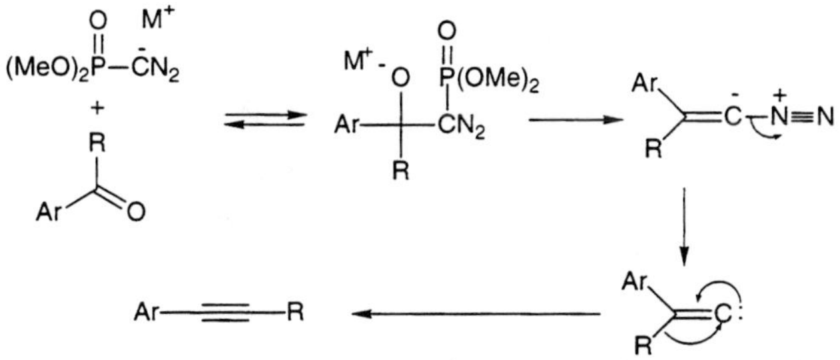

Gilbert and U. Weerasooriya, *J. Org. Chem.*, **1982**, *47*, 1837; see also: E.W. Colvin and B.J. Hamill, *J. Chem. Soc. Perkin Trans. 1*, **1977**, 869). Some examples of this process are found in Scheme 169.

Scheme 169

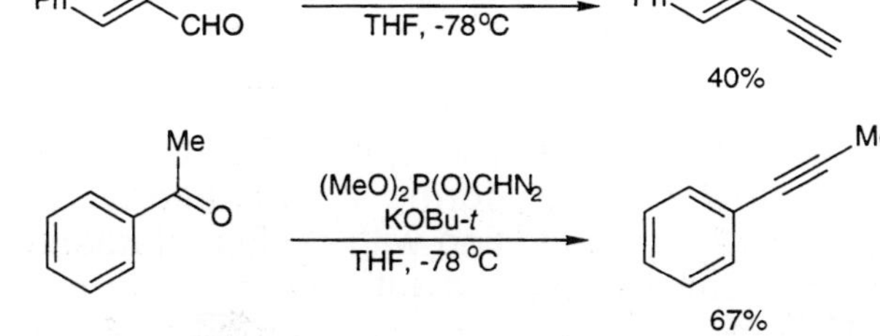

(ix) Symmetrical aryldiynes are produced when carbonylation is attempted on β,β-dibromostyrenes with $Pd(diphos)_2$ as a catalyst in benzene (Scheme 170) (P. Li and H. Alper, *J. Org. Chem.*, **1986**, *51*, 4354).

Scheme 170

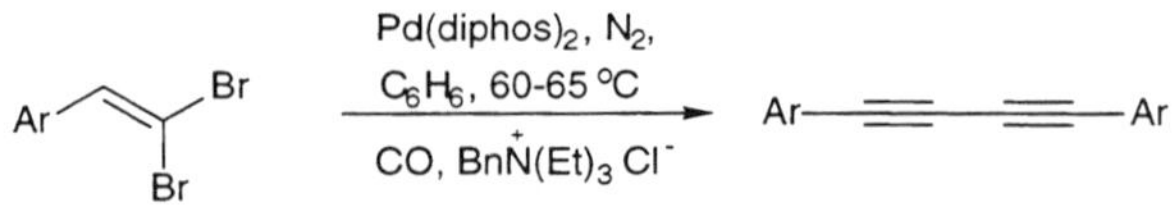

(b) Reactions

(i) Cyclic *N*-(acyloxy)iminium cations tethered to an arylalkyne willingly participate in asynchronous intramolecular [4+2] cyclocondensation reactions (Scheme 171). Such reactions are useful for the assembly of bicyclic α-amino ketone derivatives (M.J. Fisher and L.E. Overman, *J. Org. Chem.*, **1990**, *55*, 1447).

Scheme 171

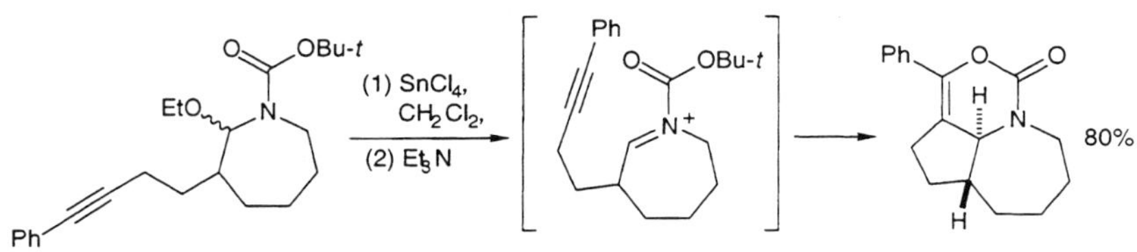

(ii) Upon treatment with Lewis acid at -55 °C, 1,1-dimethyl-3-phenylprop-2-ynyl chloride forms an arylallenyl cation that engages in a [4+2]-cycloaddition with cyclopentadiene to give a 1:1 adduct that can be trapped with chloride ion in 65% yield (Scheme 172); a 2:1 adduct is also formed in 6% yield. The latter is believed to arise from a [$_{\pi}2s+_{\pi}2a$]-cycloaddition between the arylallenyl cation and the cyclopentadiene. This leads to a [4,5]-bicyclic allylic cation that can engage in [3+4]-cycloaddition with cyclopentadiene. A 1,2-alkyl shift and π-cyclisation is then believed to occur to produce a carbocation that is finally

Scheme 172

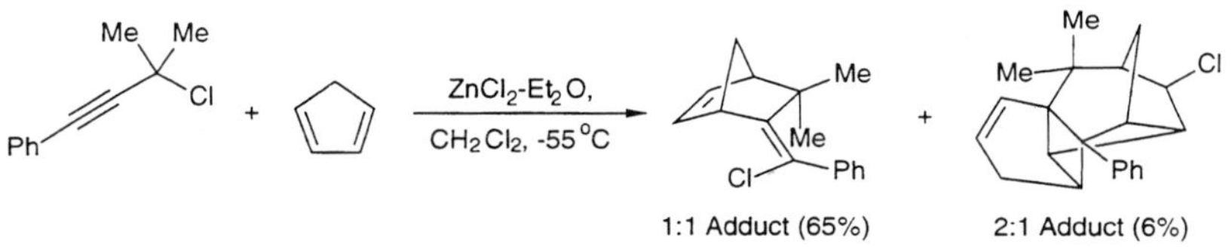

quenched with chloride ion (Scheme 173) (H. Mayr, E. Wilhelm, and C. Kaliba, *J. Chem. Soc. Chem. Comm.*, **1981**, 683).

Scheme 173

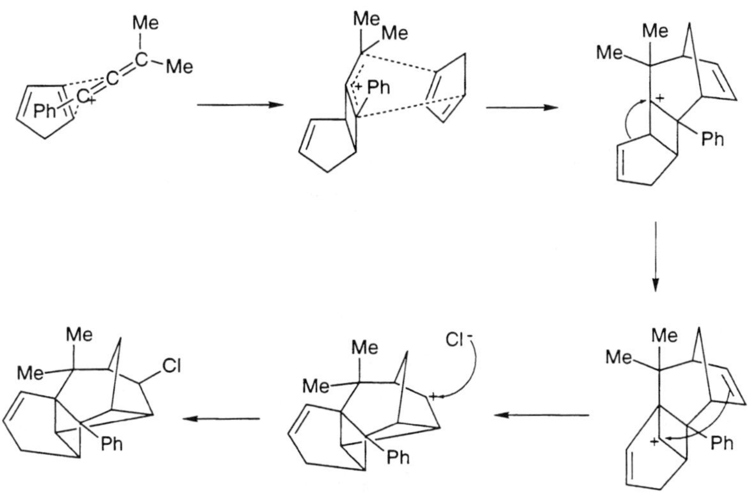

(iii) Phenylacetylene adds directly to aldehydes and ketones to give phenylacetylenic alcohols in good yield when promoted by tin (II) triflate and tertiary amines. Arylalkynylsilanes are also accessible in high yield via this procedure if the aldehyde or ketone is replaced by a trialkylsilyl chloride (Scheme 174) (M. Yamaguchi, A. Hayashi, and T. Minami, *J. Org. Chem.*, **1991**, *56*, 4091).

Scheme 174

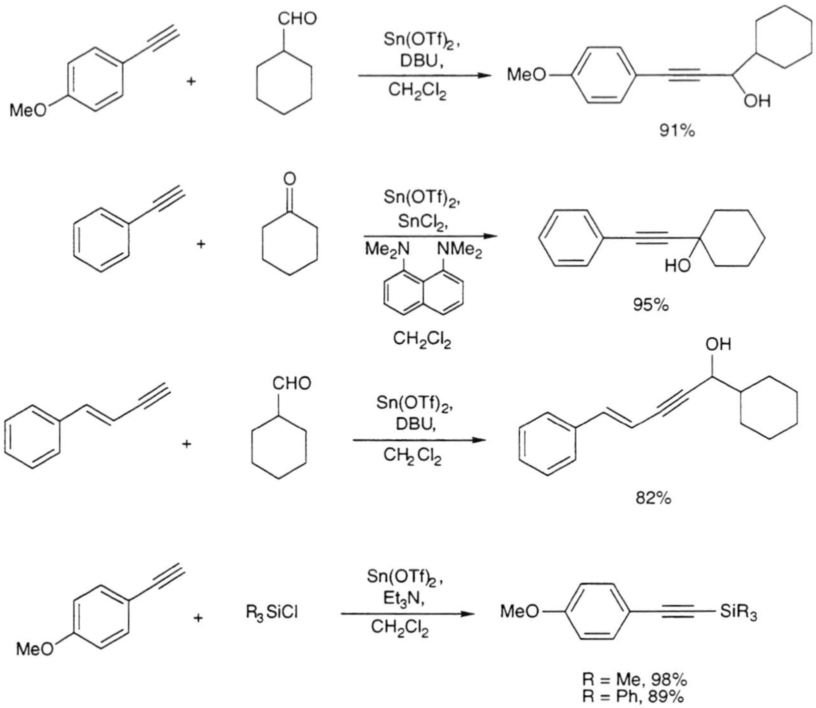

(iv) E.J. Corey and K.A. Cimprich (*J. Am. Chem. Soc.*, **1994**, *116*, 3151) have disclosed a highly enantioselective method for the arylalkynylation of aldehydes with (phenylethynyl)tri-*n*-butyl-

Scheme 175

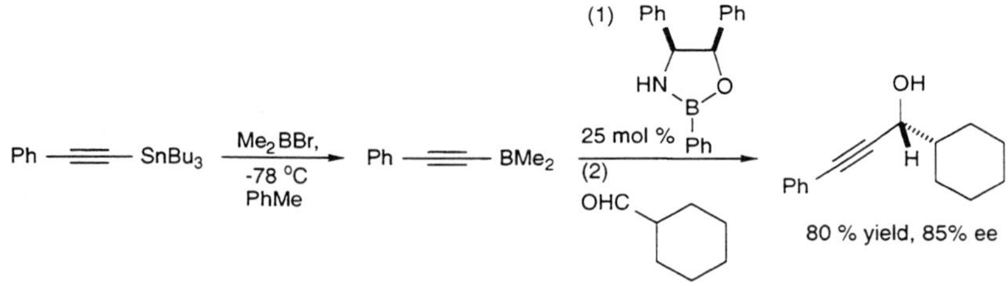

tin, which utilises an external chiral oxazaborolidine catalyst at low temperature (Scheme 175). For all the aldehydes examined, the product yields were >71%, while the optical purities were in the 85-97% range.

(v) In the presence of catalytic quantities of $(Ph_3P)_4Pd$, silylstannanes combine with terminal arylacetylenes in a highly regio- and stereoselective manner. In most cases, the products arise from *cis* addition of the organometallic to the triple bond, the tin moiety preferring to domicile itself at the carbon atom

Scheme 176

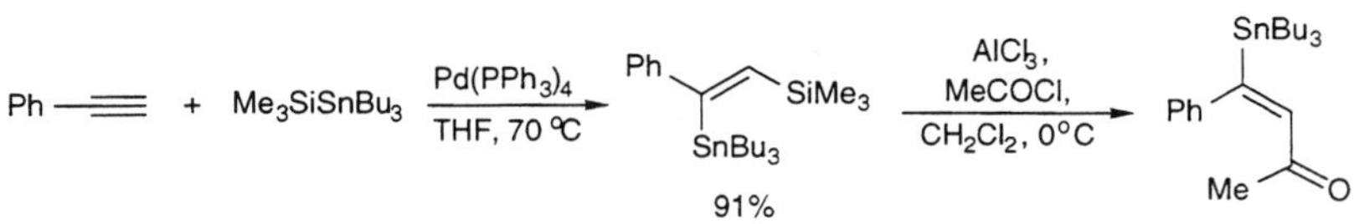

attached to the aryl unit (Scheme 176) (B.L. Chenard and C.M. Van Zyl, *J. Org. Chem.*, **1986**, *51*, 3561). The resulting adducts undergo some remarkably chemoselective reactions. For instance, the Lewis acid-catalysed acylation depicted in Scheme 176 proceeded in 30-50% yield.

(vi) In a completely stereospecific synthesis of (*Z*)-tamoxifen (Scheme 177), R.B. Miller and M.I. Al-Hassan (*J. Org. Chem.*, **1985**, *50*, 2121) have regiospecifically *cis*-carbometallated phenyl(trimethylsilyl)acetylene with diethylaluminium chloride and titanocene dichloride in dichloromethane at room temperature. This generated a β-styrylmetal species that was brominated with *N*-bromosuccinimide at -78 °C to produce an (*E*)-vinylic bromide. The bromine group was stereospecifically replaced by reaction with phenylzinc chloride and catalytic $Pd(PPh_3)_4$. The trimethylsilyl moiety was then substituted with a bromide by treatment with bromine and sodium methoxide at -78 °C. After Pd(0) cross-coupling with (*p*-methoxyphenyl)zinc chloride, and *O*-demethylation, (*Z*)-tamoxifen was prepared in a further three steps.

Scheme 177

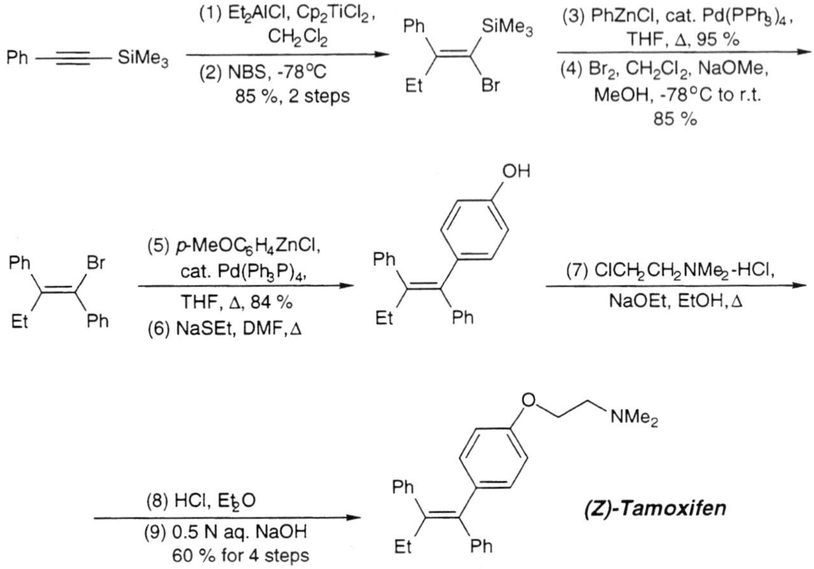

(vii) Mixtures of trimethylaluminium and titanium tetrachloride in dichloromethane will *cis*-carbometallate aryl homopropargyl alcohols with high regiocontrol at -45 °C (Scheme 178) (D.W. Thompson *et al.*, *J. Org. Chem.*, **1985**, *50*, 2124).

Scheme 178

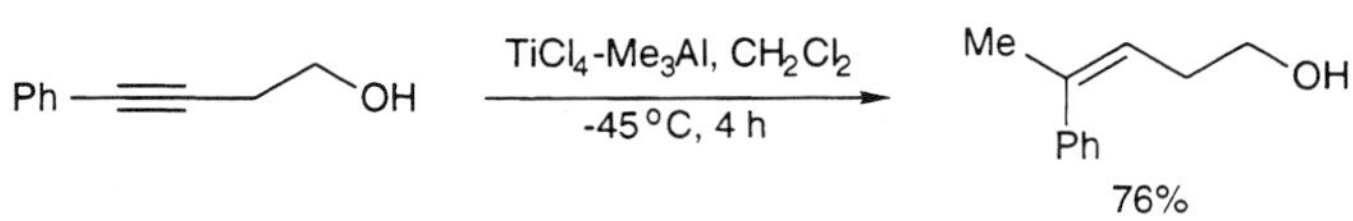

(viii) 10 Mol % of titanocene dichloride is all that is needed to catalyse the hydromagnesiation of disubstituted arylacetylenes with *iso*-butylmagnesium bromide (F. Sato, H. Ishikawa, and M. Sato, *Tetrahedron Lett.*, **1981**, *22*, 85; F. Sato *et al.*, *J. Chem.*

Scheme 179

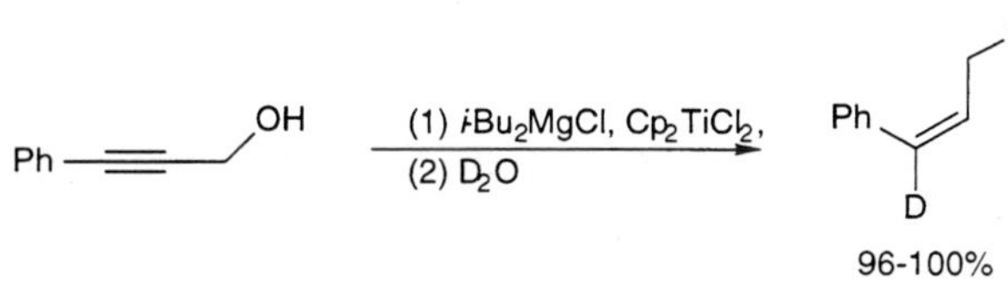

Soc. Chem. Comm., **1981**, 718). The resulting α-styryl Grignard reagents readily combine with a host of electrophiles to afford stereodefined tri- and disubstituted styrenes (Scheme 179).

(ix) A high yielding example of a lithiated aryl(trimethylsilyl)-acetylene undergoing a stereoselective intramolecular carbolithiation has been published by G. Wu, F.E. Cederbaum, and E. Negishi, (*Tetrahedron Lett.*, **1990**, *31*, 493) (Scheme 180).

Scheme 180

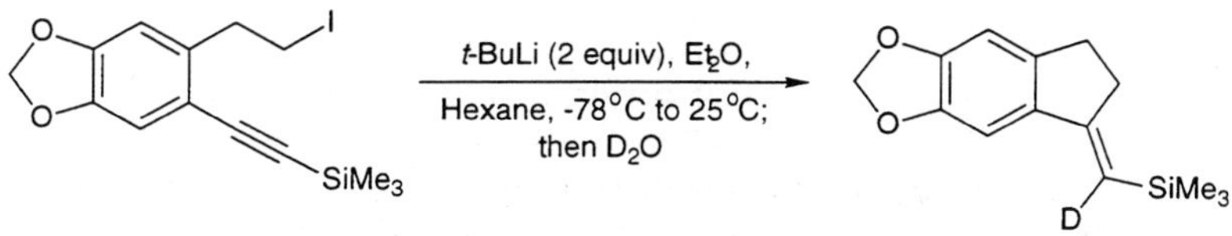

(x) The palladium-catalysed intramolecular bis-silylation of arylacetylenes by appropriately tethered disilanes has been described by Y. Ito *et al.* (*Tetrahedron*, **1993**, *49*, 3933). The

Scheme 181

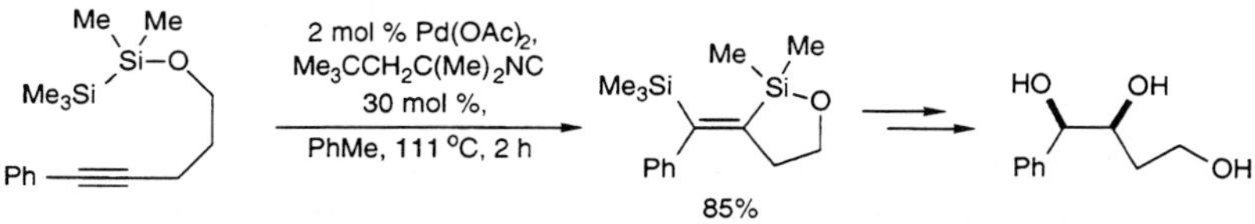

example shown in Scheme 181 proceeded exclusively via a *cis*-addition pathway. Bis-silylated adducts of this type yield 1,2,4-triols with excellent *syn*-selectivity after hydrogenation and oxidation.

(xi) It has recently been demonstrated that substituted alkynes which are connected to alkyl titanocene chlorides will undergo intramolecular insertion into the Ti-C bond if the latter is activated with a Lewis acid such as $EtAlCl_2$ or Me_2AlCl. For the phenyl-substituted alkyne shown in Scheme 182, the four-membered exocyclic trisubstituted alkene was produced via such a process in 84% yield (A.E. Harms and J.R. Stille, *Tetrahedron Lett.*, **1992**, *33*, 6565).

Scheme 182

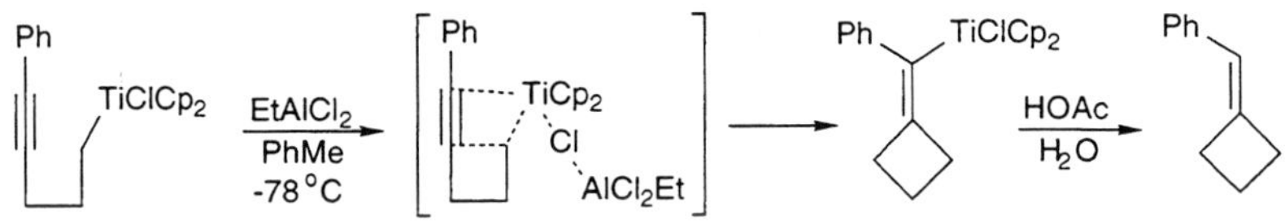

(xii) A highly stereo- and regio-selective cross-coupling reaction occurs when diphenylacetylene and ethylene are combined with Cp_2ZrBu_2 in THF at room temperature; a zirconacyclopentene intermediate is believed to form under the reaction conditions (Scheme 183) (T. Takahashi *et al., Chem. Lett.*, **1993**, 1001).

Scheme 183

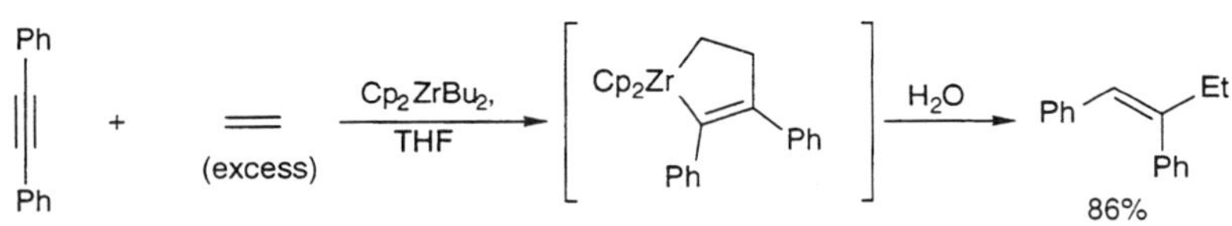

(xiii) Vinyl and thioamide cyclopropanes which are attached to *o*-phenoxy(trimethylsilyl)acetylenes have proven useful precursors of cyclopenta[*b*]benzofuran ring systems via radical

Scheme 184

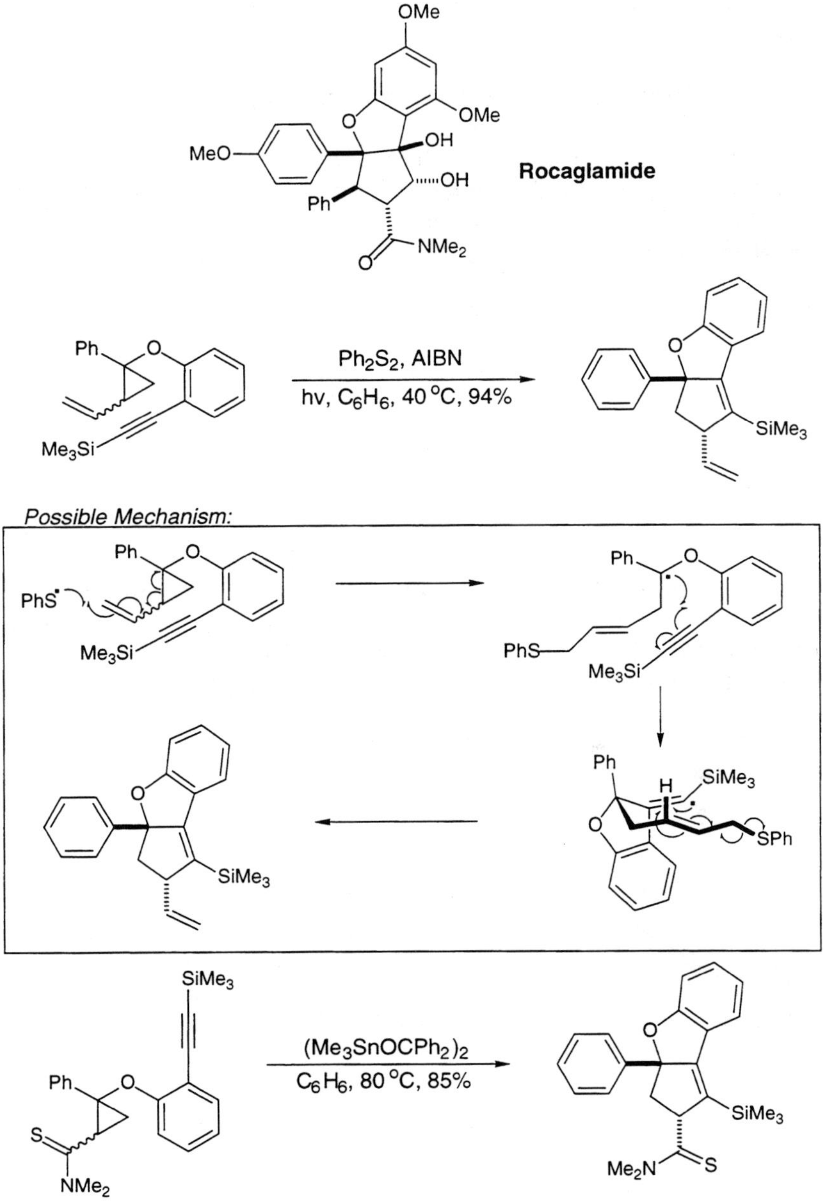
Rocaglamide
Ph2S2, AIBN
hv, C6H6, 40 °C, 94%
Possible Mechanism:
(Me3SnOCPh2)2
C6H6, 80 °C, 85%

mediated intramolecular [3-atom+2-atom] addition (Scheme 184) (K.S. Feldman and C.J. Burns, *J. Org. Chem.*, **1991**, *56*, 4601). Such a ring system can be found in the powerful antileukaemic agent rocaglamide.

(xiv) Phenylacetylene participates in a facile [2+2+2] cycloaddition with norbornadiene when treated with azeotropically predried $Co(acac)_3$ in benzene at room temperature; the product deltacyclene is isolated in 75-80% yield (Scheme 185) (M. Lautens and C.M. Crudden, *Organometallics*, **1989**, *8*, 2733).

Scheme 185

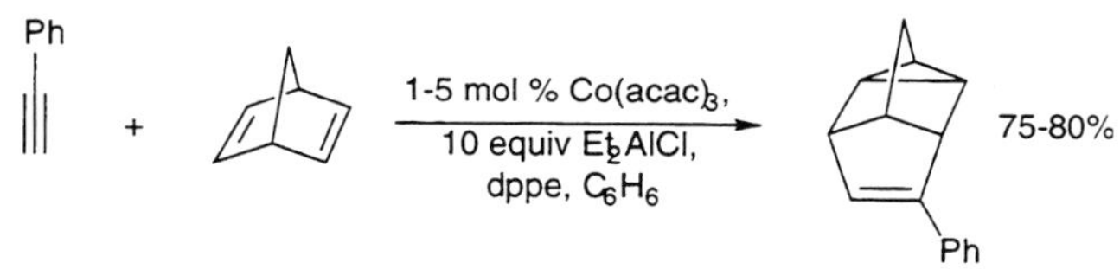

(xv) The thermally-induced decomposition of the aziridinylimine shown in Scheme 186 proceeds smoothly in benzene to afford a benzobicyclo[3,1,0]hexene as the sole product in 72% yield (A. Padwa, Y. Gareau, and S.L. Xu, *Tetrahedron Lett.*, **1991**, *32*, 983). It is thought that an initially

Scheme 186

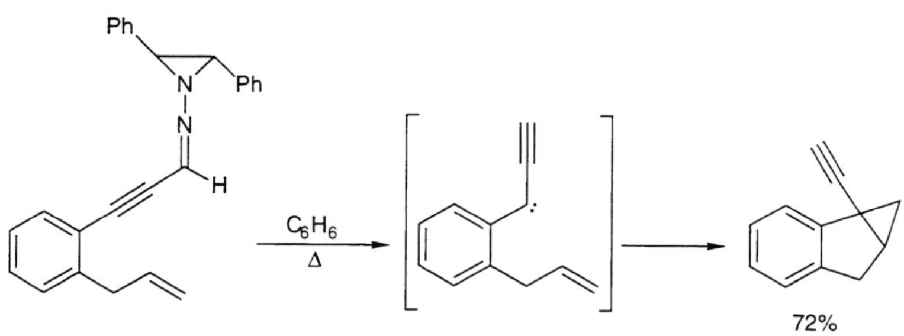

generated carbene undergoes rearrangement prior to intramolecular insertion into the *ortho*-propenyl side chain.

Scheme 187

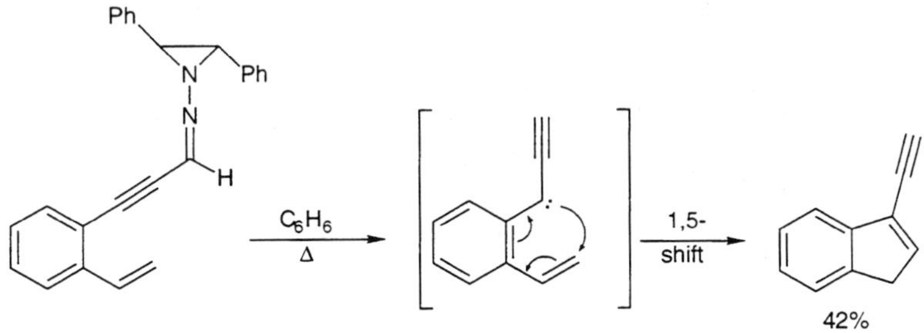

When a similar decomposition was attempted on a related substrate (Scheme 187) an electrocyclisation/1,5-H shift occurred to give an alkynylindene.

(xvi) N. Iwasawa and T. Matsuo (*Chem. Lett.*, **1993**, 997) have reported that 1-(1-phenylethynyl)cyclopropanol rearranges to 3-phenyl-2-cyclopentenone when heated with 10 mol % of dicobalt octacarbonyl and 20 mol % of $P(OPh)_3$ in dimethoxyethane (Scheme 188).

Scheme 188

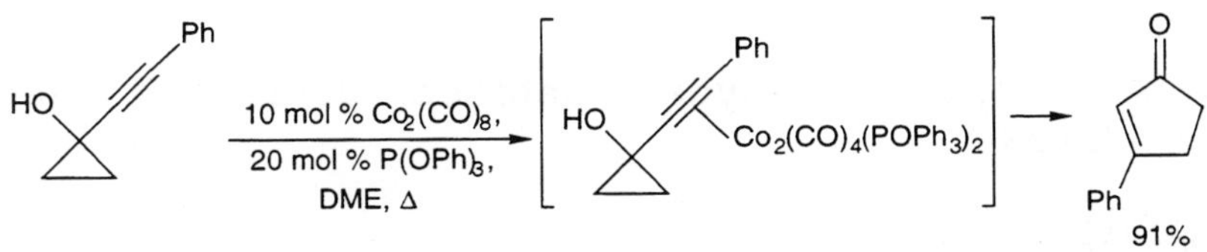

(xvii) C. M. Marson *et al.* (*J. Org. Chem.*, **1993**, *58*, 5944) have observed that two equivalents of tin (IV) chloride in dichloromethane at 0 °C is sufficient to bring about the highly stereoselective semipinacol rearrangement of arylacetylenic 2,3-epoxy alcohols. This novel reaction is one of the few general methods available for constructing an aryl quaternary centre next to an alcohol (Scheme 189).

Scheme 189

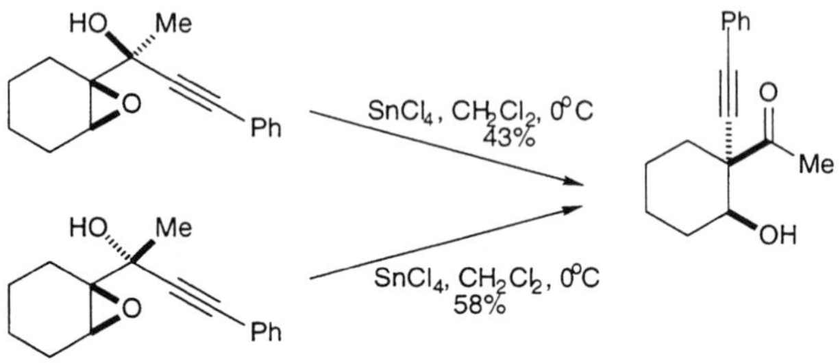

(xviii) Terminal and disubstituted arylacetylenes react smoothly with diphenyldiselenide and ammonium peroxydisulfate in methanol to give α-ketoacetals and α-ketoketals respectively in good yield (Scheme 190) (M. Tiecco *et al., J. Org. Chem.*, **1991**, *56*, 4529).

Scheme 190

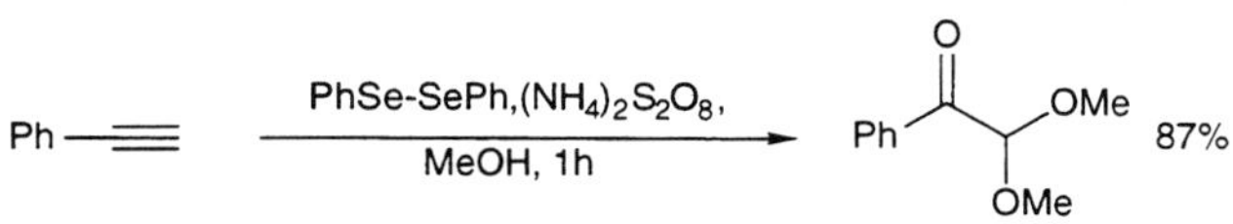

8. Arylalkynyl aldehydes, ketones, acids and their derivatives

(a) Preparation

(i) Phenylacetylene has been converted to a range of arylacetylenic ketones through reaction with aryl and vinyl halides under a carbon monoxide atmosphere in the presence of an amine and $PdCl_2$[1,1'-bis(diphenylphosphino)ferrocene] [$PdCl_2$(dppf)] (Scheme 191) (T. Kobayashi and M. Tanaka, *J. Chem. Soc. Chem. Comm.*, **1981**, 333).

Scheme 191

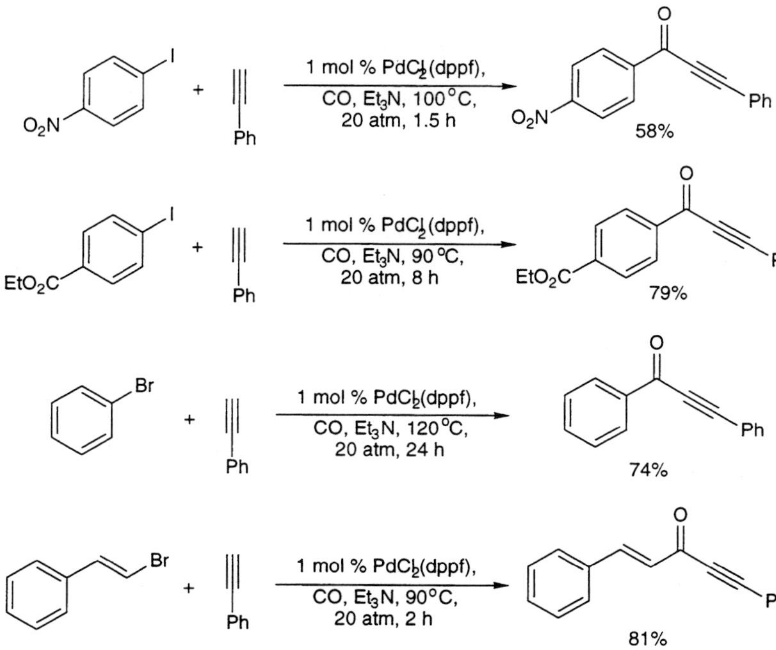

(ii) The coupling of cuprous phenylacetylide with acryloyl chlorides enables aryl enynones to be prepared in high yield (Scheme 192) (G.A. Kraus and M.J. Taschner, *J. Am. Chem. Soc.*, **1980**, *102*, 1974).

Scheme 192

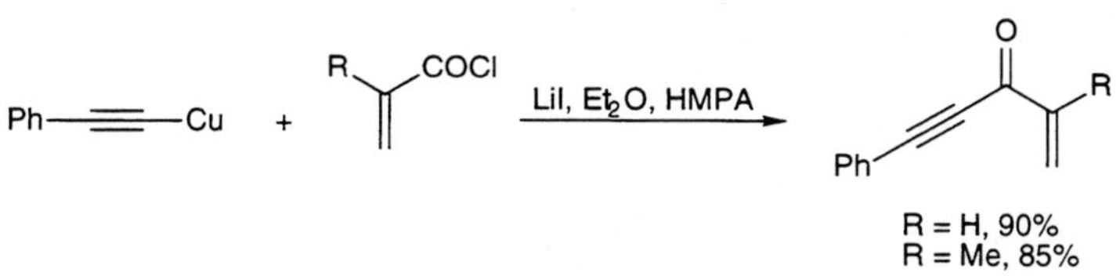

(b) Reactions

(i) "Timed" Diels-Alder reactions can be successfully executed on non-symmetrical bis dienes and aryl enynones. In this approach to polycyclic ring assembly, an intermolecular Diels-Alder reaction is initially performed between the more reactive enone component of the enynone and the more electron-rich diene partner. An intramolecular Diels-Alder reaction then occurs between the arylalkynone and the remaining less reactive diene. A good illustration of the "timed" Diels-Alder strategy being applied can be seen in scheme 193 (G.A. Kraus and M.J. Taschner, *J. Am. Chem. Soc.*, **1980**, *102*, 1974).

Scheme 193

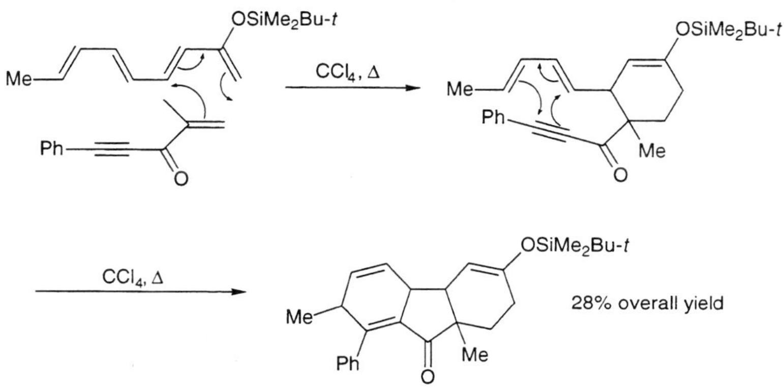

(ii) A very interesting modification of the Evans asymmetric aldol reaction adds tin (II) enolates obtained from chiral *N*-acyl oxazolidinones to aldehydes (A.S. Kende, K. Kawamura, and M.J. Orwat, *Tetrahedron Lett.*, **1989**, *30*, 5821). When applied on the conjugated aldehyde shown in scheme 194, the oxazinedione was obtained as the exclusive product in 76% yield. Difficulties were encountered in the hydrolytic removal of the auxiliary with aqueous lithium hydroxide; attack also occurred at the urethane carbonyl.

Scheme 194

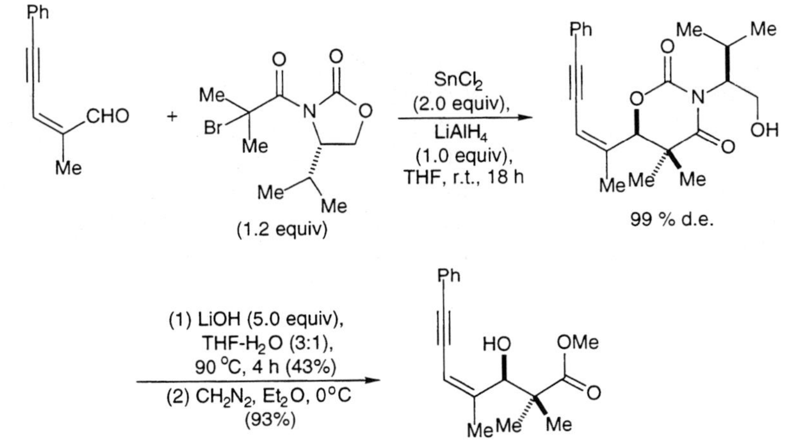

(iii) Arylalkynyl esters react with phenyl chromium carbene complexes, as shown in Scheme 195, to give substituted hydronapthoquinones in good yield. It is often best to isolate the product phenol after acetylation *in situ* with acetic anhydride and triethylamine (A. Yamashita and A. Toy, *Tetrahedron Lett.*, **1986**, *27*, 3471).

Scheme 195

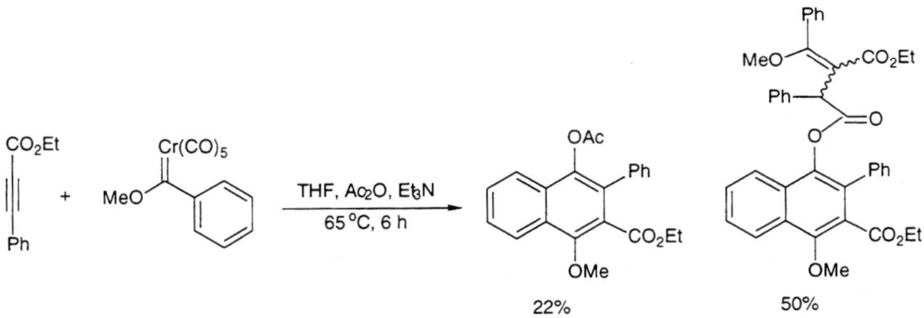

9. Aryl enediynes

Preparation

S. Tsuboi *et al.* (*J. Org. Chem.*, **1993**, *58*, 5952) have reported a highly stereoselective coupling between ethyl (*Z*)-2,3-dibromopropenoate and excess phenylacetylene in the presence of catalytic quantities of $Pd(PPh_3)_4$ and copper (I) iodide. The product, ethyl (*E*)-2-(phenylethynyl)-5-phenyl-2-penten-4-ynoate, was isolated as the sole product in 97% yield, via the monovinyl bromide (Scheme 196).

Scheme 196

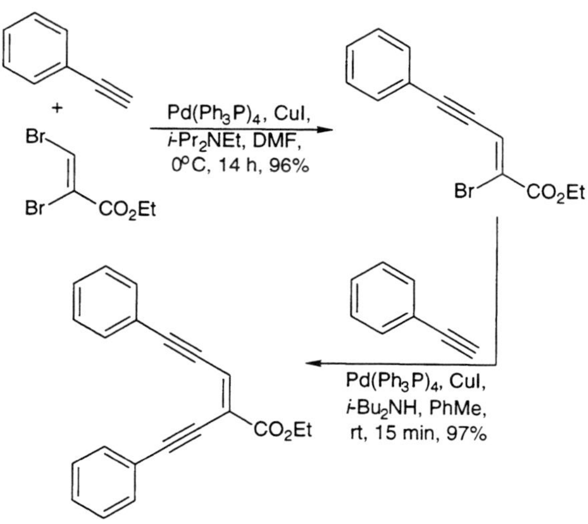

Second Supplements to the 2nd Edition of Rodd's Chemistry of Carbon Compounds, Vol.III D/E/F (Partial), edited by M. Sainsbury

Chapter 20

ARYL BENZENES AND THEIR DERIVATIVES

J.A.H. MacBride

1. Introduction

Interest in the synthesis and properties of molecular systems containing two or more directly linked benzene rings has continued, and perhaps intensified, since publication of the previous Supplement to Rodd's C.C.C. on this topic (J.A.H. MacBride, Supplement to 2nd Edn. Part III F, Chapter 20) but the literature shows a distinct change of emphasis towards work with potential application, such as the preparation and study of solid-state materials, and away from natural product synthesis for its own sake. Indeed, the majority of natural product studies have been directed towards compounds with promising biological activity like gilvocarcin, a group of anti-tumour antibiotics (see Section 3(c) (iii)). In all this work formation of the aryl-aryl bond remains a key synthetic step.

The properties of biphenyls and their oligomers and polymers which have stimulated much work aimed at the development of new materials are firstly electronic conjugation between adjacent rings, giving the possibility of electrical conductivity and related electronic properties. Secondly, the thermal, chemical, and photochemical stability of polyphenylenes gives promise of materials resistant to high temperatures and other aggressive conditions; and thirdly the steric character of the *para-para*-biphenylene unit, combining relatively rigid linearity with some degree of conformational freedom, leading to liquid-crystalline properties in many of its derivatives. Rigidity combined with limited conformational freedom also underlies the incorporation of biphenyl and oligophenyl groups, often *meta*-linked, into macrocyclic and related compounds, including crown-ethers, prepared in studies of host-guest interactions (Section 4).

Improved methods for formation of the aryl-aryl bond have centred on organometallic intermediates and catalysts, particularly those of nickel and palladium in low oxidation states, and Grignard or aryllithium reagents. Several such reactions noted at a preliminary stage in the previous Supplement have been developed into powerful procedures for synthesis and polymerisation. These have to a noticeable extent superseded the classical Ullmann coupling of aryl halides with copper, and to a lesser extent its developments using pre-formed arylcopper reagents to effect specific cross-coupling of dissimilarly substituted rings (Section 3a (i)). Specific cross-coupling remains a difficulty with some recent methods, including that of halides mediated by nickel(0), whereas with the more complex Meyers coupling (Section 3(d)) complete regio-specificity, and in suitable examples a high degree of enantiospecificity, can be achieved.

For polymerisation to proceed to useful chain lengths the coupling reaction must continue very close to completion (e.g. 95% reaction gives an average degree of polymerisation, DP, of only twenty monomer units) and be virtually free of side reactions which interrupt chain extension. Recent advances in aryl-aryl coupling using nickel or palladium catalysts can meet these criteria (Section 3).

The layout of this chapter is similar to that of the previous Supplement: The introductory paragraphs are divided into general and environmental aspects (Sections 1 and 2). Section 3 treats developments in arylbenzene synthesis (with particular emphasis on linked phenylene systems) including examples of applications to natural product synthesis and extensions to terphenyls and higher oligomers. Section 4 presents a brief structural and synthetic survey of phenylene oligomers applied in host-guest chemistry and Section 5 reports studies of the chemical and physico-chemical properties of biphenyls. New Sections 6 and 7 describe materials based on bi- and polyphenylenes, and the synthesis and properties of biphenylene, respectively.

Scheme and structure numbers are preceded by the Section number in which they (first) occur and then numbered independantly within each Section.

General synthetic aspects of biphenyl chemistry have been reviewed by: M. Sainsbury, *Tetrahedron*, 1980, **36**, 3327; D.W. Knight in *Comprehensive Organic Synthesis*, B.M. Trost Ed., Pergamon 1991; and with particular emphasis on synthetic strategy by G. Bringmann, R. Walter and R. Weirich, *Angew Chem. Int. Edn. Engl.* 1990, **29**, 977. More specialised reviews are referenced in the appropriate Sections.

2. Environmental Aspects of Biphenyls

In the two decades or so since the use of polychlorinated biphenyls (PCBs) as dielectric fluids in electrical capacitors, as transformer coolants, and in agriculture has been generally discontinued by international agreement very large quantities of these materials have been destroyed by burning, much of it at sea.

Despite the stability of these compounds in the environment, the hoped for reduction in their concentration seems to have been achieved, and populations of predatory birds, whose fertility suffered because of bioconcentration of PCBs along the food chain from algae through fish, are largely restored; simultaneous control of chlorinated insecticides has doubtless played a large part in this improvement.

Some use of polybrominated biphenyls (PBBs) mainly as agricultural fungicides continues and a major incident of human food contamination in the USA, when a large amount of PBB was accidentally added to animal feed used for meat production, resulted in a plethora of medical claims for compensation.

Disposal of PCBs by burning has itself caused concern since incomplete combustion of these compounds, usually during electrical fires, has been found to lead to readily detectable amounts of chlorinated dibenzodioxins (Scheme 2.1), dibenzofurans (2.2) and, remarkably and often predominantly, biphenylenes (2.3) (E.R. Barnhart, *et al.*, *Anal. Chem.*, 1987, **59**, 2248 and references therein). Some twelve compounds in the oxygenated series ("the dirty dozen") are highly toxic, causing acute as well as longer-term genetic effects. Synthetic difficulties have severely limited biological study of the chlorinated biphenylenes (see Section 7), but there is some indication that they

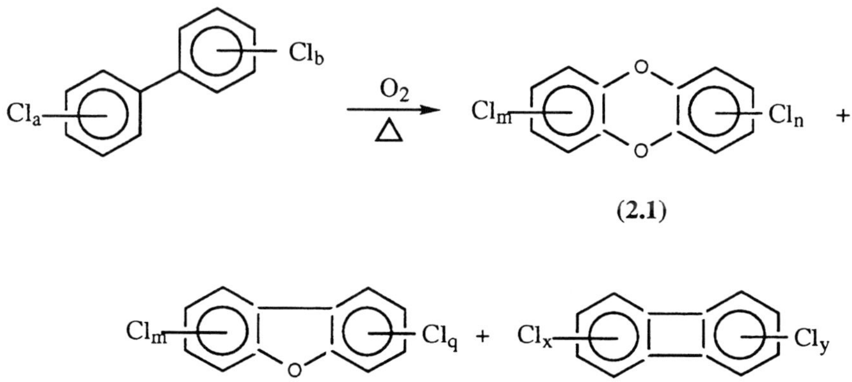

(2.1)

(2.2) (2.3)

Scheme 2.1

have properties related to those of their oxyanalogues (A. Poland, E. Glover and A.S. Kende, *J. Biol. Chem.* 1976, **251**, 4936).

Despite these anxieties it is thought that a large part of the PCB stocks have been destroyed by incineration under controlled conditions, *i.e.* injection into oxygen-rich hydrocarbon flames (giving H_2O, CO_2 and HCl), without increasing back-ground levels of the toxic by-products, which probably arise principally from municipal incineration of domestic waste.

3. Synthesis of Arylbenzenes

(a) Coupling Reactions Mediated by Arylcopper Derivatives

(i) Coupling of aryl halides by metallic copper (the 'classical' Ullmann reaction)

Although most authors presenting new or modified aryl-aryl coupling methods describe the Ullmann reaction as "unsatisfactory" it remains, either under the original solvent-less conditions or with a donor solvent such as dimethylformamide (DMF), the method of choice for many biaryl syntheses. For example the iodide (3.1) gives the biphenyl diester (3.2) in 91% yield (M.A. Rizzacasa and M.V. Sargeant, *Australian J. Chem.*, 1988, **41**, 1087). Similarly high yields are obtained in the preparation of symmetrical biphenyls with four substituents *ortho* to the biphenyl bond, including the di-aldehyde

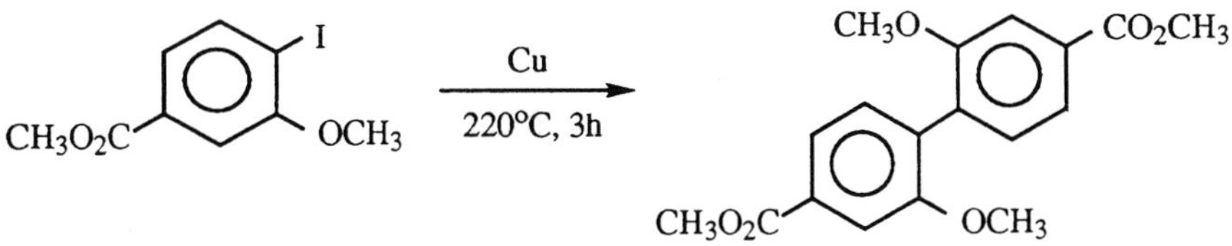

(3.1) **(3.2)** 91%

(3.4), a substituent group incompatible with more reactive organometallic intermediates (see following Sections) using a solvent-less centrifugation technique, described in earlier work, with aryl bromides (E. Brown, R. Dahl and J.-P. Robin, *Tetrahedron*, 1982, **38**, 2569; see also previous Supplement). Good yields, 30-70%, of unsymmetrically substituted biphenyls such as the diester (3.7, Scheme 3.1) can be similarly prepared by condensing an aryl iodide (3.5) with a bromide (3.6).

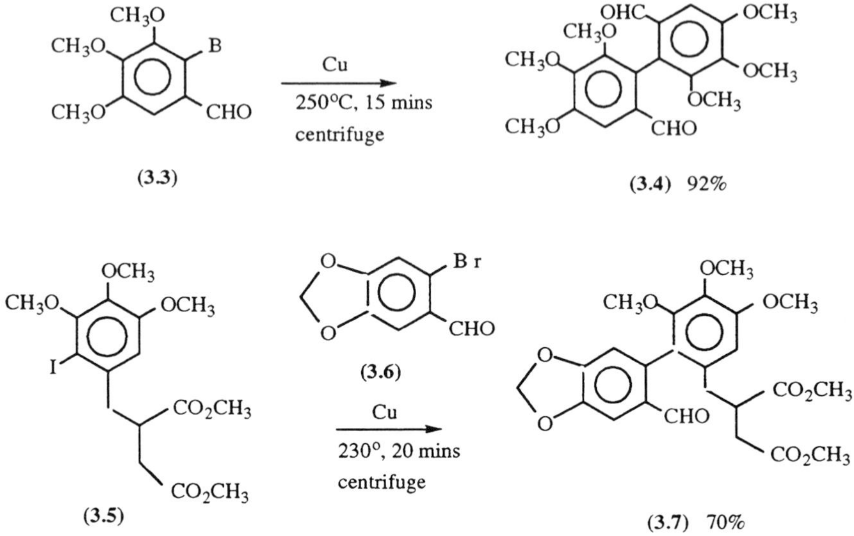

Scheme 3.1

Intramolecular coupling to give bridged biphenyls is also effective under 'classical' conditions: the dibromide (3.8, Scheme 3.2) gives the phenanthrene (3.9) from which effusol (3.10), isolated from marsh grasses, has been synthesised (C.F. Caruhalo, M.V. Sargeant, and E. Stanojevic, *Australian J. Chem.*, 1984, **37**, 2111; for cyclisation of an optically resolved binaphthyl dibromide with almost complete stereoselectivity using copper in boiling DMF, see S. Miyano *et. al.*, *J. Chem. Soc. Chem. Commun.*, 1980, 1233).

It is noteworthy that these efficient examples of the Ullmann coupling are between electron-rich benzene rings, contrary to the often quoted generalisation that the reaction works best when strongly electron withdrawing substituents, particularly nitro-groups, are present. In each case at least one oxygen function is *ortho* to the biphenyl bond, so that chelation of the arylcopper intermediate, highlighted in previous work (see previous Supplement p. 303), may be important.

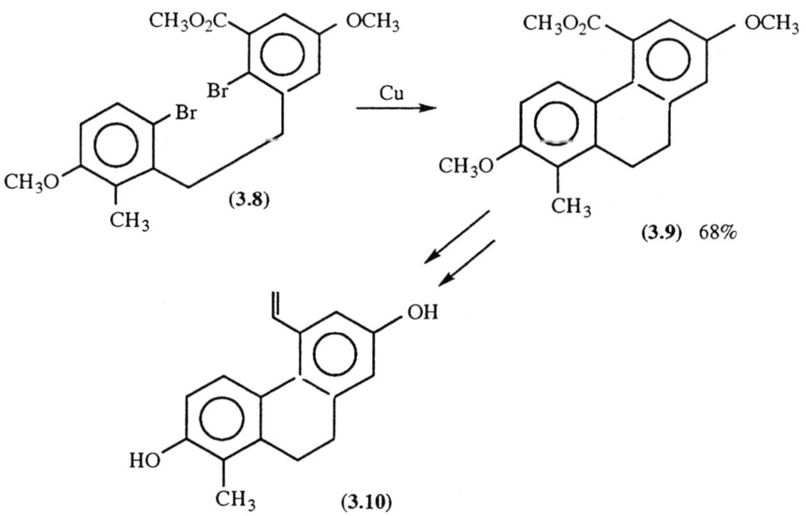

Scheme 3.2

Ultrasonic irradiation increases the rate of coupling of 2-iodonitrobenzene by copper to give 2,2'-dinitrobiphenyl in DMF at *ca.* 60°C by a factor of about 50 with yields up to 81%. The effect is attributed partly to reduction in particle size, but more particularly to rapid removal of reaction products from the metal surface (J. Lindley, J.P. Lorimer and T.J. Mason, *Ultrasonics*, 1986, **24**, 293).

In connection with the mechanism of the Ullmann (and similar) reactions, earlier authors disagreed on whether copper(3) intermediates are of low enough energy to be involved (see previous Supplement pp. 302-304), so that it is interesting to note that stable compounds of copper in this oxidation state have since been isolated without the need for severe oxidising conditions (see for example L. Eberson, L. Jonsson, and L.-G. Wistrand, *Tetrahedron*, 1982, **38**, 1087; F.C. Anson, *et al.*, *J. Amer. Chem. Soc.*, 1987, **109**, 2974; M.A. Willert-Pavada, D.J. Burton and N.C. Baenziger, *J. Chem. Soc., Chem. Commun.*, 1989, 1633).

(ii) *Coupling of arylcopper intermediates generated via aryllithiums*

This more recent modification of the Ullmann system, which is usually conducted in the presence of an electron pair donor catalyst such as pyridine, which may be the solvent, is applicable to unsymmetrically substituted biphenyls. Applied to symmetric examples it has the advantage that both the arylcopper and the aryliodide with which it can be coupled can be prepared from the same aryllithium, as in the preparation of 2,2',6,6'-tetramethoxybiphenyl (3.11, Scheme 3.3) (G. Lindsten, O. Wennerstrom, and R. Isaksson, *J. Org. Chem.*, 1987, **52**, 547). Demethylation of the product with boron tribromide gave the corresponding tetrahydroxybiphenyl, which was required for the synthesis of biphenyl bis-crown ethers (see Section 4).

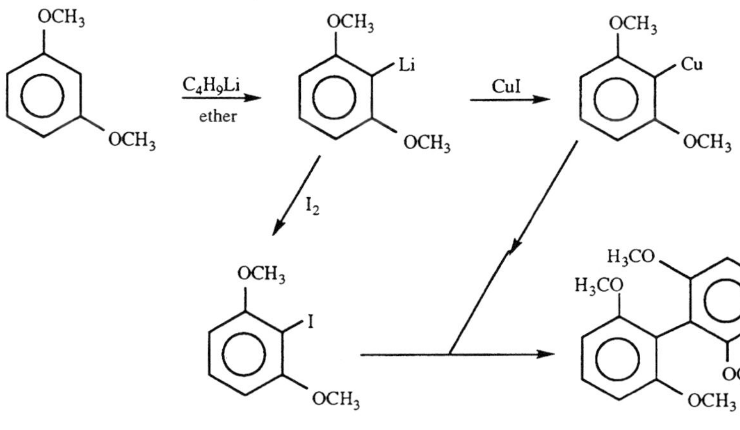

(3.11)

Scheme 3.3

(iii) *Coupling by oxidation of diarylcuprates*

Interaction of arylcopper reagents, generated by treatment of copper halides with aryllithium or Grignard reagents, with a second molecule of an aryllithium gives a lithium diarylcuprate (Gilman reagent), Equation 1:

$$CuX + ArLi \longrightarrow CuAr + LiX$$

$$CuAr + Ar'Li \longrightarrow ArAr'CuLi \qquad \textbf{Equation 1}$$

Decomposition of such diarylcuprates under oxidising conditions at low temperature has previously been shown to give biaryls (see previous Supplement, p. 304; also G.M. Whitesides *et al.*, *J. Amer. Chem. Soc.*, 1967, **89**, 5302), although those containing two different organic groups e.g. phenyl and butyl gave mixtures of coupled products (S.H. Bertz and C.P. Gibson, *J. Amer. Chem. Soc.*, 1986, **108**, 8286).

When copper(1) cyanide is treated with two molecular proportions of an aryllithium a "higher order" (H.O.) dilithium cuprate is formed, Equation 2:

$$2ArLi + CuCN \longrightarrow Ar_2CuCNLi_2 \qquad \textbf{Equation 2}$$

Oxidation of these compounds with *o*-dinitrobenzene, or more conveniently with molecular oxygen in the presence of N,N,N'N'-tetramethylethylenediamine (TMEDA) gives symmetrical biaryls in 80-90% yield (B.H. Lipshutz, K. Siegmann and E. Garcia, *J. Amer. Chem. Soc.*, 1991, **113**, 8161; for a more general account of the development of this work see *idem*, *Tetrahedron*, 1992, 2599). These authors found that the H.O. cuprate formed by mixing different aryllithiums (3- and 4-methoxyphenyl) with copper(1) cyanide in equimolar proportions at -75°C gave the statistical mixture of the three isomeric biphenyls (3,3'-, 3,4'- and 4,4'-biphenyl in 1:2:1 ratio) on oxidation, but that when the cuprate formed from one of these lithium reagents and copper(1) cyanide was treated with an equal molecular proportion of the other lithium reagent at -125°C, and the product was oxidised at this temperature, the unsymmetrical biaryl was formed in 93% yield. Several similar experiments with phenyllithium substituted by methoxy, methyl, chloro-, fluoro- and trifluoromethyl groups, including an example with one substituent *ortho* to the coupling position in each phenyl group (methyl and methoxy), gave yields of 78-90% of unsymmetrical product, e.g. Scheme 3.4.

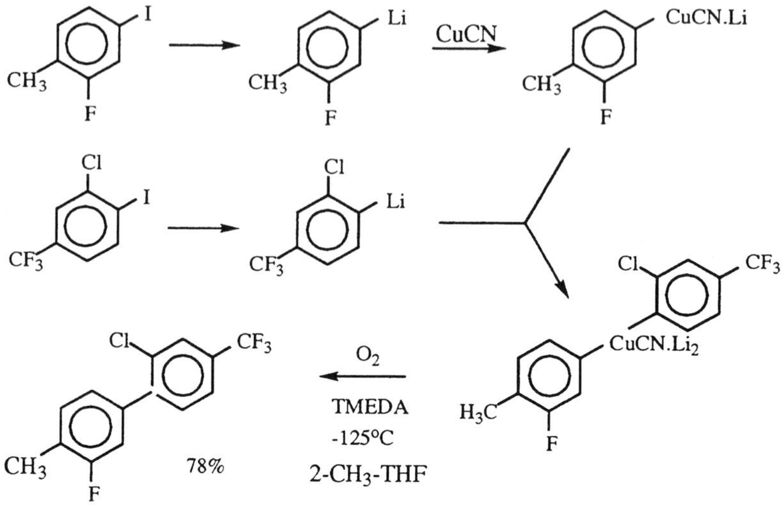

Scheme 3.4

Stirring the reaction mixtures at -125°C was facilitated by using 2-methyltetrahydrofuran as solvent.

When the mixed H.O. cyano-cuprate prepared at -125°C was allowed to reach temperatures up to -80°C and then oxidised (at either temperature) mixtures of biaryls were formed, the yield of unsymmetrical product falling in proportion to the temperature.

The authors suggest an explanation for these results on the basis that "the combination of temperature and sequence of ligand addition to copper appears to allow for maintenance of the initial positioning of the aryl groups in the organometallic complex"; this complex is termed a "kinetic cuprate".

A simpler explanation is that the aryl groups exchange between copper atoms in the H.O. cuprate species at the higher but not the lower temperatures (and that oxidative coupling is, as the authors imply, unimolecular).

Dissociation to the lower order cuprate (ArCuCN.Li) plus aryllithium, which has been detected in the NMR spectrum at -80°C of the related dilithium triphenylcuprate $(C_6H_5)_3CuLi_2$ prepared from copper halides (S.H. Bertz and G. Dabbagh, *J. Amer. Chem. Soc.*, 1988, **100**, 3668) with random recombination is a likely mechanism for aryl scrambling in the H.O. cuprate intermediates at -75°C.

The X-ray crystal structures of several oligomeric lithium phenylcuprate species, crystallised from dimethylsulphide which forms an additional ligand, have been determined (M.M. Olmstead and P.P. Power, *J. Amer. Chem. Soc.*, 1990, **112**, 8008).

(b) Aryl-Aryl Coupling Catalysed or Mediated by Nickel Compounds

(i) Coupling between two aryl halide molecules

(A) Using catalytic amounts of nickel with a reducing metal

Earlier work by Kende *et al* ., on the coupling of aryl halides, usually bromides or iodides, using complexes of Ni(0) generated *in situ* from Ni(2) compounds and excess zinc (see RCCC Vol. III, 2nd Supplement to Part F p. 305 *et seq.*) has been developed to provide an efficient bi-aryl synthesis (I. Colon and D.R. Kelsey, *J. Org. Chem.*, 1986, **51**, 2627). Thus use of a catalytic amount (1 mol) of Ni(2) halide (usually chloride) with an excess (30 mol) of zinc powder (magnesium or manganese are also effective) with triphenyl phosphine (7.6 mol) in a dipolar aprotic solvent (dimethylacetamide or dimethylformamide) leads to rapid formation of symmetrical biaryls from aryl chlorides (20 mol) at 50-80°C. Yields are near quantitative with suitably substituted chlorides (e.g. 4-cyano or 4-acetyl [4-ethanoyl]) but a side-reaction leading to unsubstituted biphenyl by-products, e.g. biphenyl and 4-aminobiphenyl from 4-chloroaniline, is due to aryl group exchange with the triphenylphosphine ligand and can be effectively suppressed by addition of 2,2'-bipyridyl which also accelerates coupling. Bromide and iodide ions, e.g. from added sodium bromide, also catalyse biaryl formation. Nitro or acidic, e.g. hydroxy, substituents are precluded the latter leading, like added water, to reductive de-halogenation, a minor side reaction in any case, particularly when aryl bromides or iodides are coupled.

Reaction under these conditions, *i.e.* with a catalytic proportion of nickel in the presence of excess reducing metal, is distinct from aryl halide coupling with a stoichiometric amount of a pre-formed Ni(0) complex, which is much less effective for aryl chlorides than for bromides or iodides (see (B) below).

Unsymmetrical biaryls are not formed by simultaneous coupling of two differently substituted chlorides, but small amounts of cross-coupled products were observed when a second chloride was added after the first had reacted completely; this result is attributed to residual relatively stable arylnickel species (7.6% in the case of 4-chloroanisole) which are decomposed to the dechlorinated arene during normal work-up.

Although the precise mechanism of aryl-aryl coupling in this system is not established the considerable number of previous papers on the organometallic chemistry of nickel, well referenced by Colon and Kelsey (*loc. cit.*), leave little doubt that (a) arylnickel(2) derivatives are formed initially by oxidative addition of aryl halide to a nickel(0) species (step 1 below); (b) biaryls result from reductive elimination (e.g. step 5 below) from diarylnickel complexes, whose oxidation state remains a particular point in question, and that free aryl radicals are not involved; and (c) more than one mechanism operates, depending upon solvent polarity, the availability of unreacted halide, the particular halogen(s) and ligands employed and the continuing presence of oxidising or reducing agents.

On the basis of their kinetic and related observations Colon and Kelsey (*loc. cit.*) have proposed the mechanistic sequence, Scheme 3.5, in which, following the proposal of Tsou and Kochi (T.T. Tsou and J.K. Kochi, *J. Amer. Chem. Soc.*, 1979, **101**, 7547), the immediate biaryl precursor (step 5) is the paramagnetic Ni(3) complex $Ar_2Ni(3)ClL_2$ formed by oxidative addition of aryl chloride to a Ni(1) derivative (step 4) arising initially from Ni(2) by step 3 and re-cycling via step 6 or via steps 7 and 8:

$$Ni(2)Cl_2 + 3L + Zn \rightarrow Ni(0)L_3 + ZnCl_2 \qquad \text{(step 1)}$$

$$Ni(0)L_3 + ArCl \rightarrow ArNi(2)ClL_2 + L \qquad \text{(step 2)}$$

$$ArNi(2)ClL_2 + \tfrac{1}{2} Zn + L \rightarrow ArNi(1)L_3 + \tfrac{1}{2} ZnCl_2 \qquad \text{(step 3)}$$

$$ArNi(1)L_3 + ArCl \rightarrow Ar_2Ni(3)ClL_2 + L \qquad \text{(step 4)}$$

$$Ar_2Ni(3)ClL_2 + L \rightarrow Ar_2 + Ni(1)ClL_3 \qquad \text{(step 5)}$$

$$Ni(1)ClL_3 + \tfrac{1}{2} Zn \rightarrow Ni(0)L_3 + \tfrac{1}{2} ZnCl_2 \qquad \text{(step 6)}$$

$$Ni(1)ClL_3 + ArCl \rightarrow ArNi(3)Cl_2L + 2L \qquad \text{(step 7)}$$

$$ArNi(3)Cl_2L + Zn + 2L \rightarrow ArNi(1)L_3 + ZnCl_2 \qquad \text{(step 8)}$$

Scheme 3.5, $L = (C_6H_5)_3P$

It is suggested that step 3 is rate determining, consistent with faster reaction observed in the presence of a greater excess of zinc, and that bromide and iodide catalysis may be due to easier electron transfer from the metal surface to $ArNi(2)ClL_2$ *via* a halide bridge between Ni and Zn; 2,2'-bipyridyl may behave similarly since it is known to increase the rate of reduction of nickel(2) chloride by zinc.

An alternative mechanism with rate-determining elimination of Ar_2 from a square-planar Ni(2) complex (3.12), in which the aryl groups are likely to be *trans* on the basis of earlier studies of analogous methyl derivatives (D. G. Morrell and J.K. Kochi, *J. Amer. Chem. Soc.*, 1975, **97**, 7262), could also explain catalysis by heavier halide ions *via* nickelate adducts (3.13) in which the aryl groups are brought closer together in the pentacoordinate geometry. Similarly 2,2'-bipyridyl could facilitate biaryl elimination by ligand exchange to give the intermediate (3.14) in which the aryl groups are forced to adopt adjacent coordination sites.

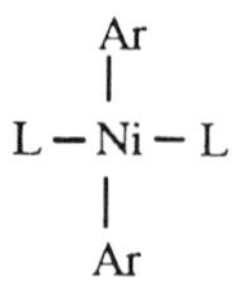

(3.12)

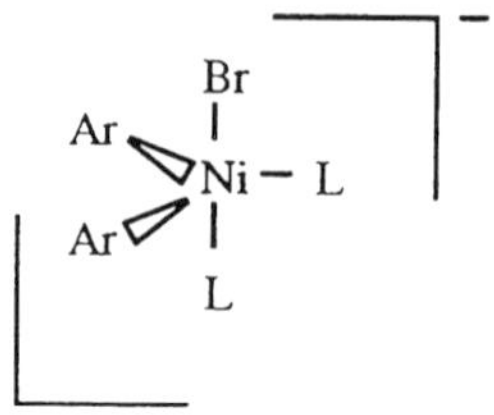

(3.13)

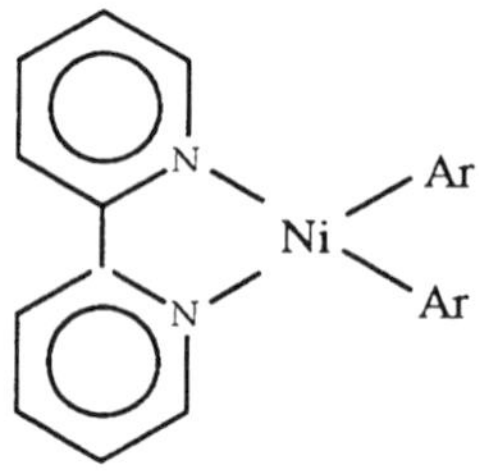

(3.14)

Coupling has also been demonstrated using electrochemical reduction in place of the reducing metal in an otherwise similar reagent system (G. Schiavou, G. Bontempelli, and B. Carcin, *J. Chem. Soc. Dalton Trans.* 1981, 1074).

A curious note records that an unspecified, presumably aryl, bromide gave similar coupling products under the conditions of Kende *et al.*, (2nd Supplement p. 306) in the presence or absence of nickel reactant (A. Stanger and A. Shazar, *J. Organometal. Chem.* 1993, **458**, 233, footnote 3).

(B) Coupling between two aryl halide molecules using stoichiometric amounts of pre-formed nickel(0) complexes

Semmelhack and his coworkers have followed up their pioneer work (see previous Supplement to RCCC, p. 305) on this biaryl synthesis (and the analogous coupling of alkenyl halides to give dienes) with a comparison of the isolated yields of biaryl from variously substituted aryl halides using the stoichiometric amount of nickel(0) complex (*i.e.* 0.5 mol) corresponding with equation 3 (M.F. Semmelhack *et al.*, *J. Amer. Chem. Soc.* 1981, **103**, 6460):

$$2ArX + Ni(0)L_4 \rightarrow Ar_2 + Ni(2)X_2L_2 + 2L \qquad \textbf{Equation 3}$$

Nickel(0) biscyclo-octadiene is the preferred reagent for intermolecular coupling although nickel(0) tetrakistriphenylphosphine is better for intramolecular cyclisation, as in the synthesis of alnusone (previous Supplement, p. 312). While bromides are the most effective halides, aryl iodides can also be coupled satisfactorily but chlorobenzene gave a significantly lower yield of biphenyl.

Reaction of aryl bromides containing either electron withdrawing or releasing substituents at the 2- or 4-positions at 35-60°C in dimethylformamide (DMF) gives symmetrical biaryls, usually in *ca.* 80% yield. Thus nitrile and carbonyl substituents, including aldehyde, ketone or ester groups, methyl and methoxy functions can be present; free amino-groups can also be tolerated (unlike Ullmann coupling) but the yield of benzidine from 4-bromoaniline is reduced to 54%. Acidic groups, hydroxyl and carboxyl, and their salts (sodium in the experiments) are precluded, as well as nitro. Methyl groups at both *ortho*-positions prevented reaction, presumably by steric hindrance.

Cyclisation of *ortho*-polymethylene bridged bisaryl iodides (3.16) to medium and large ring biaryls (3.17) using $Ni(0)[(C_6H_5)_3P]_4$, without high dilution, gives remarkably high yields, e.g. 76% for the eight (3.17, n = 4) and 58% for the fourteen membered ring (3.18) similarly formed in 0.1 M solution. The lowest yield, 38%, for the ten-membered ring (3.17 ,n = 6) contrasts with the minimum rate of cyclisation, and therefore minimum yield since the rate of competing linear oligomerisation is essentially independant of chain length, often observed for formation of cycles of eight atoms (see, e.g., Rodd's C.C.C., Supplement to 2nd Edn. Vol IV, Part K, p. 437).

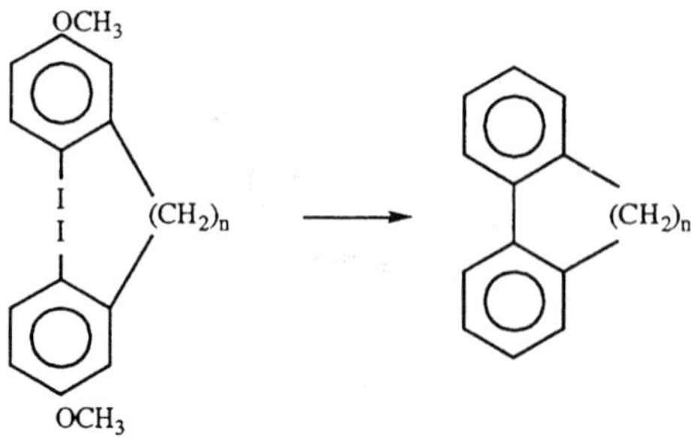

(3.16)

(3.17)

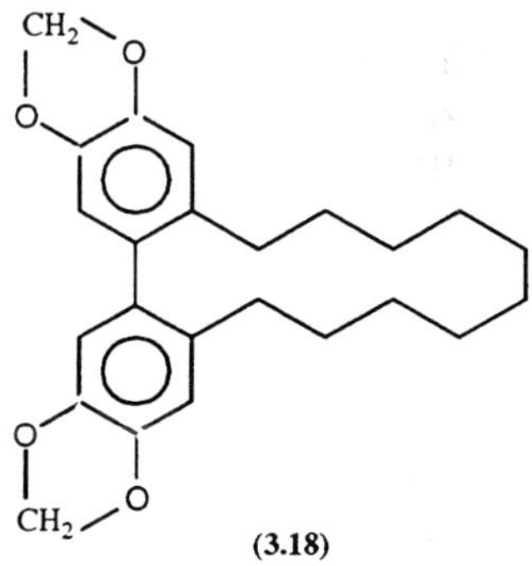

(3.18)

Semmelhack and coworkers (*loc. cit.*) investigated the mechanism of nickel(0) mediated coupling with the particular intention of devising conditions for specific cross-coupling of differently substituted halides, and following related work (see previous Section) inferred that complexed four coordinate arylnickel(2) halides ($ArNiHalL_2$) are important intermediates. Bistriphenylphosphine phenylnickel(2) bromide (3.19), prepared by addition of bromobenzene to tetrakistriphenylphosphinenickel(0) under the conditions of M. Hidai *et. al.* (*J. Organometal. Chem.*, 1971, **30**, 279), formed biphenyl slowly in dimethylformamide (DMF) solution at 25°C; 99% was obtained after 12h at 50°C. The methylenedioxy iodide (3.20) similarly gave the biphenyl (3.21) in 70% yield after 12h at 25°C.

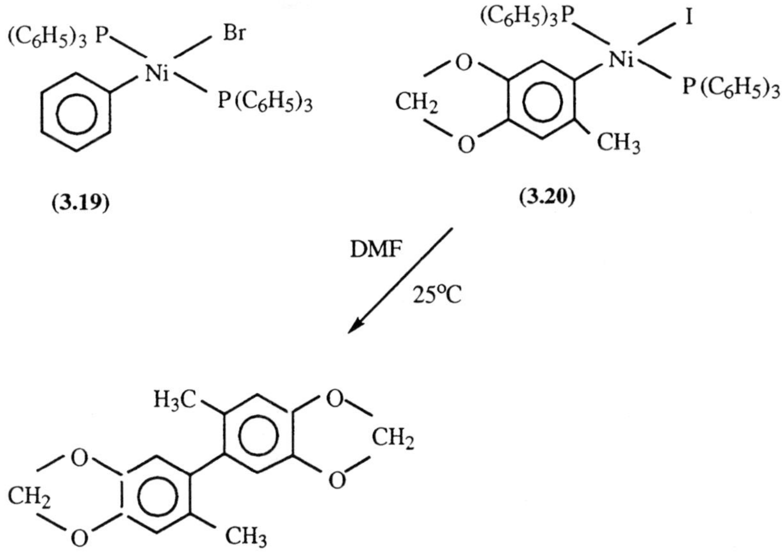

(3.19) (3.20) (3.21)

Addition of iodobenzene to the phenyl-compound (3.19) in DMF gave biphenyl a little faster but in only 69% yield, while o-iodotoluene gave similar proportions of the three possible biaryls with slight preference for symmetric coupling. None of a series of similar experiments gave more than 50% of cross-coupled product, the proportion expected if the three coupling reactions are statistically equivalent. More specific cross coupling has been described by Tsou and Kochi (*loc. cit.*) in analogous experiments using triethylphosphine as ligand.

A recent kinetic study of the coupling of aryl halides by biscyclo-octadienenickel(0) (Ni(COD)$_2$) (T. Yamamoto, S. Wakabayashi and K. Osakada, *J. Organometal. Chem.* 1992, **428**, 223) uses 2,2'-bipyridyl as auxiliary ligand (like Colon and Kelsey (*loc. cit.*, Section 3b(i)A), whose paper is omitted from the bibliographic survey) and DMF solvent, equation 4.

$$Ni(COD)_2 + 2ArX + bipy \rightarrow Ar_2 + NiX_2\ bipy \qquad \textbf{Equation 4}$$

Under these conditions chlorides react only slowly and most kinetic data were obtained for bromides; iodides reacted too fast for accurate rate measurement. The reaction of bromobenzene proved to be second order in $Ni(COD)_2$ and the rate almost independent of bromobenzene concentration. Measurements between 323 and 343 K gave activation parameters: $\Delta G^{\ddagger}_{333}$ 90 kJ Mole^{-1}; $\Delta H^{\ddagger}_{333}$, 60 kJ mole^{-1}; and $\Delta S^{\ddagger}$ 0.9 kJ mole^{-1} K^{-1}. Electron donating substituents, *p*-methoxy and *p*-methyl, increased the rate constant and an electron withdrawing one, *p*-acetyl (*p*-ethanoyl) reduced it compared with bromobenzene, but the claimed linearity of the Hammett plot, and its ρ value, remain in doubt owing to discrepancy between the tabulated and log-plotted values of K.

On the basis of previous work and the colour changes during coupling, the reaction sequence, Scheme 3.6, is proposed:

$$Ni(0)(COD)_2 + bipy \longrightarrow Ni(0)(COD)bipy + COD \qquad \text{step 1}$$

$$Ni(0)(COD)\ bipy + C_6H_5Br \longrightarrow bipy\ Ni(2)\begin{smallmatrix} C_6H_5 \\ Br \end{smallmatrix} \qquad \text{step 2}$$

(3.22)

$$2\ bipy\ Ni(2)\begin{smallmatrix} C_6H_5 \\ Br \end{smallmatrix} \longrightarrow bipy\ Ni(2)Br_2 + bipy\ Ni(2)\begin{smallmatrix} C_6H_5 \\ C_6H_5 \end{smallmatrix} \qquad \text{step 3}$$

$$bipy\ Ni(2)\begin{smallmatrix} C_6H_5 \\ C_6H_5 \end{smallmatrix} \longrightarrow (C_6H_5)_2 + bipy\ Ni(0) \qquad \text{step 4}$$

Scheme 3.6

The kinetic results then require step 3 to be rate-determining. The intermediate (3.22) formed in the proposed step 2 can be isolated as a red solid when equimolar amounts of $Ni(COD)_2$ and 2,2'-bipyridyl are treated with excess bromobenzene at room temperature. Complex (3.22) gives biphenyl in 80% yield on dissolution in DMF at room temperature, but remarkably the presence of an equimolar proportion of bromobenzene prevents this reaction, in contrast to Semmelhack's result with the analogous bis-triphenylphosphine complex (3.19) and iodobenzene; it is not clear how this result can be reconciled with the observed null effect of bromobenzene on the rate.

Treatment of the complex (3.22) with 4-bromotoluene in DMF at 60°C gives biphenyl in 61% yield but no methyl- or dimethylbiphenyl was detected (by G.C.). Phenyl group exchange between complex (3.22) and C_6D_5Br was observed in toluene at 70°C, and decomposition of the complex (3.22) in the presence of 4-bromotoluene under these conditions gave rather low yields of the three biaryls with the dimethyl compound predominating and little cross-coupling.

Yamamoto *et al.* propose that the disproportionation step 3 proceeds through an intermediate such as (3.23), or (3.24) in which aryl groups bridge two nickel centres; participation of bridged structures during transfer of aryl groups between nickel atoms has previously been discussed by Tsou and Kochi (*loc. cit.*). The ionic structure (3.24) has the advantage of explaining the pronounced effect of polar solvents, and the authors relate the effect of different halogens and of aryl-4-substituents to weakening of the Ni-X (Cl stronger than I) and Ni-ArR (R = $COCH_3$ stronger than R = OCH_3) bonds leading to faster formation of the intermediate (3.23) or (3.24).

The effect of aryl group substituents might alternatively be considered on the basis that if the transition state for the rate-determining step of coupling is taken to have aryl-bridged character similar to the proposed intermediates (3.23) and (3.24), whose 4-coordinate carbon must be electron deficient, the system would be stabilised (lower E_a and faster reaction) by electron donating substituents.

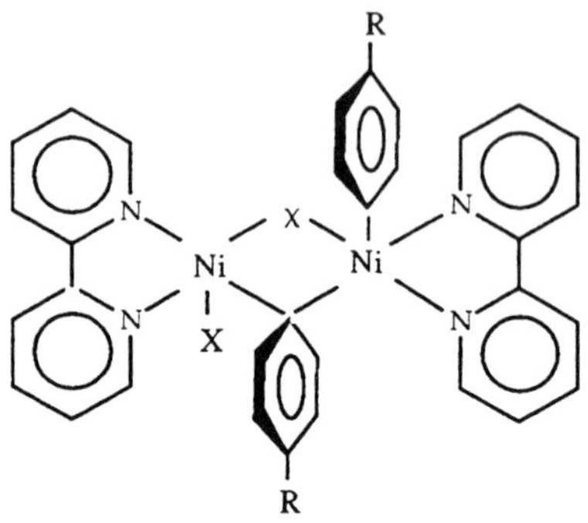

(3.23)

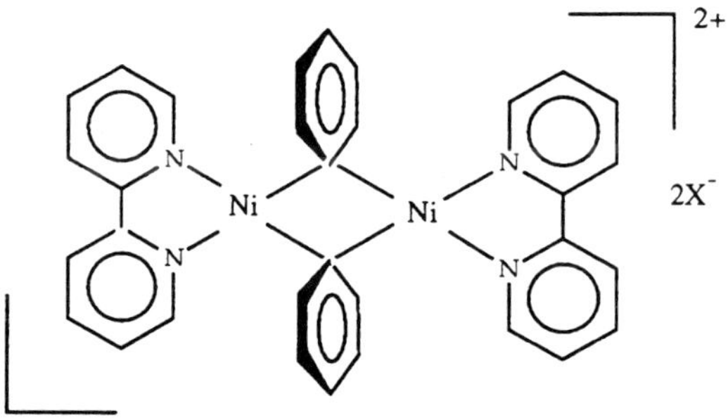

(3.24)

The mechanisms of nickel-mediated coupling reactions have been reviewed by P.W. Jolly in *Comprehensive Organometallic Chemistry*, G. Wilkinson, F.G.A. Stone and E.W. Abel Eds., Pergamon, New York 1982, Vol. 8, p. 713; A. Yamamoto *et al.*, *Pure Appl. Chem.*, 1984, **56**, 1621; K. Tamao and M. Kumada in *Chemistry of the Metal-Carbon Bond*, F.R. Hartley and S. Patai Eds., Wiley, Chichester 1987, Vol. 4, p. 819; and with particular reference to aryl chlorides by U.U. Grushin and H. Alper, *Chem. Rev.*, 1994, **94**, 1047.

(ii) Coupling of Aryl trifluoromethylsulphonates with Complexes of Ni(0)

In a reagent system analogous to that used to couple aryl halides, a nickel catalyst generated *in situ* from nickel(2) chloride, zinc, triphenylphosphine and sodium iodide in dimethylformamide (DMF) couples aryl trifluoromethanesulphonates (triflates) in yields of 65-95% (measured by gas chromatography). Methoxy, methyl and methoxycarbonyl substituents may be present at the 2-, 3- or 4-positions but two *ortho* substituents, e.g. with 2, 4, 6-trimethylphenyltriflate, prevent reaction. The reaction rate, at 60°C, is increased by up to a factor of two by ultrasonic irradiation (J. Yamashita *et al*, *Chem Lett.*, 1986, 407).

(iii) Nickel(0) catalysed condensation of arylzinc chlorides with aryl iodides

This reaction, often referred to as the Negishi method, is also catalysed by complexes of palladium(0) and is discussed with comparison of the two metal catalysts under Section 3c.

A notable nickel catalysed example is the synthesis of the natural leavorotatory enantiomer of the lignan steganone (3.28) (E.R. Larson and R.A. Raphael, *J. Chem. Soc. Perkin 1*, 1982, 521). A previous synthesis of the racemic compound by the same group, which used photochemical cyclisation to form the biphenyl bond (see previous Supplement, p. 313) proved suitable only for small quantities but pharmacological assessment of this and related lignans required larger samples.

Scheme 3.7 outlines the revised synthesis in which connection of the biaryl bond between the zinc reagent (3.25) and the iodide (3.26) precedes construction of the bridging side-chain, which had been needed *before* photochemical cyclisation, using the imine-protected aldehyde function of the biphenyl (3.27, R = CH = NC_6H_{11}). The chelating *ortho*-imino group is necessary for coupling since no biaryl was formed on corresponding treatment of the carbomethoxy iodide (3.26, R = CO_2CH_3). The Ni(0) catalyst is prepared by treating nickel(2) acetylacetonate with di-isobutylaluminium hydride in the presence of excess triphenyl phosphine in THF.

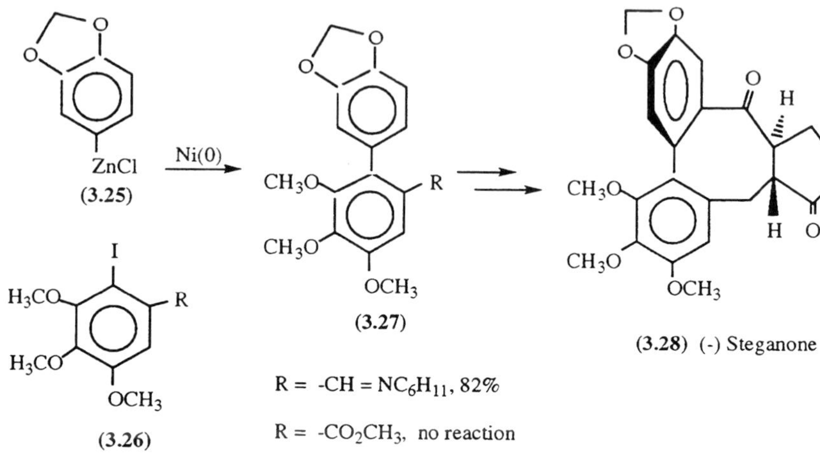

R = $-CH=NC_6H_{11}$, 82%

R = $-CO_2CH_3$, no reaction

Scheme 3.7

(iv) Coupling of Aryl Grignard Reagents with Aryl Bromides

The nickel(0) catalysed coupling of aryl Grignard reagents with aryl bromides has been applied by Yamamoto *et al.*, to the polymerisation of 4-bromophenylmagnesium bromide to poly-*para*phenylene (Y. Yamamoto, Y. Hayashi and A. Yamamoto, *Bull. Chem. Soc. Japan*, 1978, **51**, 2091; see Section 6(b) i).

(v) Coupling of Aryl Iodides or Bromides Using Activated Nickel Powder

Metallic nickel, precipitated from solutions of nickel(2) halides in 1,2-dimethoxyethane by lithium in the presence of catalytic amounts of naphthalene, couples aryl iodides at 80°C to give biphenyls in 74-100% yield

(H. Matsumoto, S. Inaba, and R.D. Rieke, *J. Org. Chem.* 1983, **48**, 840). Aryl bromides react more slowly, giving generally lower yields, and the reagent prepared from nickel(2) iodide is more effective, probably owing to initial halogen exchange. Coupling is prevented by a single *ortho*-substituent and only reductive dehalogenation occurs, but 4-methoxy, 4-chloro and 4-cyano-halobenzenes give the corresponding biphenyls.

Pentafluoro-halobenzenes give decafluorobiphenyl, quantitatively from the iodo-compound at 80°C, but this product is not formed at room temperature when addition of triphenylphosphine gives bis(triphenylphosphine)bis(pentafluorophenyl)nickel(2) ($[(C_6H_5)_3P]_2[C_6F_5]_2Ni$) in 45% yield. When triphenylphosphine is present at the start of this experiment bis(triphenylphosphine)pentafluorophenylnickel(2) iodide ($[(C_6H_5)_3P]_2C_6F_5NiI$) can be isolated in 15% yield.

These intermediates trapped by triphenylphosphine are consistent with a mechanism in which the solvent-coordinated monoarylnickel(2) halide disproportionates to the diarylnickel(2) which undergoes reductive elimination of biaryl in a sequence analogous to Scheme 3.6.

(c) Aryl-aryl Coupling Catalysed by Palladium or its Compounds

(i) Palladium(0) Catalysed Coupling of Aryl Iodides or Bromides with Aryl Grignard Reagents

A systematic assessment of this method has been made with a view to its application to natural product synthesis (D.A. Widdowson and Y.-Z. Zhang, *Tetrahedron*, 1986, **42**, 2111). Using arylmagnesium bromides in THF, aryl iodides react at room temperature but bromides require reflux (65°C), e.g. Scheme 3.8. Although electron withdrawing substituents in the halide are known to enhance reactivity, methoxy groups, common in natural biphenyls, do not inhibit reaction from the 3-, 4- or 5-positions and are often beneficial. A limitation is that coupling is prevented by more than two substituents adjacent to the potential biphenyl bond and by two *ortho*-substituents on the halide; an *ortho*-substituent on each reactant reduces reactivity. Grignard reagents, e.g. mesityl (3.29), with two *ortho*-substituents react satisfactorily but give improved yields by addition of hexamethylphosphoramide (HMPA) and higher catalyst concentration.

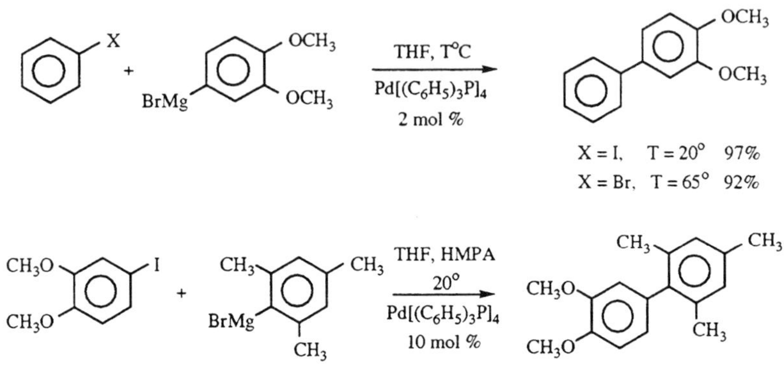

Scheme 3.8

The mechanism proposed for this coupling requires reductive elimination of two aryl groups from a palladium complex as the product-forming step, Scheme 3.9. Oxidative addition of the aryl halide to the Pd(0) catalyst (3.30) gives the arylpalladium(2) halide complex (3.31) and the organometallic reactant (Grignard) substitutes halogen by aryl, metathesis step, forming (3.32). It is known, however, that these reactions lead to *trans* diaryl square planar complexes of Pd(2), but reductive elimination of the aryl groups, as biaryl, requires (by orbital symmetry as well as chemical common sense) that they occupy adjacent, e.g. *cis* square planar, coordination sites (see A. Gillie and J.K. Stille, *J. Amer. Chem. Soc.*, 1980, **102**, 4933 and references therein). These authors have shown that the *trans* complex (3.32) can be isomerised to *cis* (3.34) by polar coordinating solvents, probably by facilitating displacement and

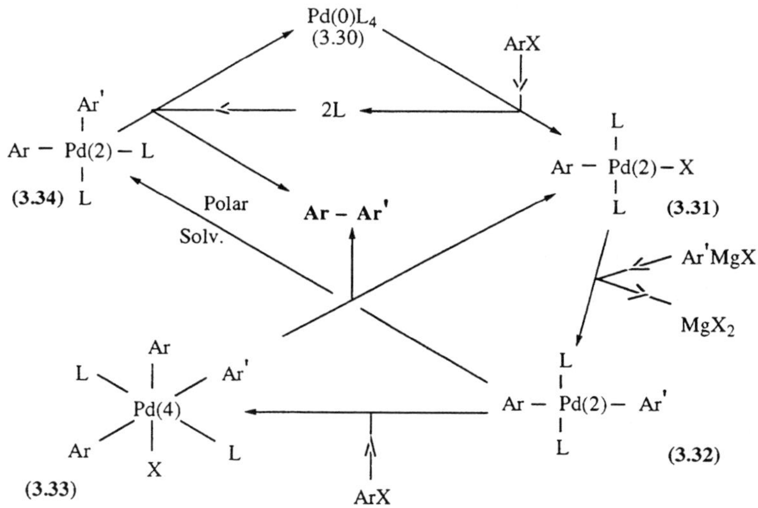

Scheme 3.9 $L = (C_6H_5)_3P$

re-coordination of a phosphine ligand since added phosphine suppresses product formation in some examples. Elimination may then occur from the *cis*-complex (3.34), but these authors favour an alternative reaction to bring the aryl groups into adjacent positions: a second oxidative addition of aryl halide to the *trans*-complex (3.32) giving the octahedral palladium(4) derivative (3.33), whence elimination of biaryl returns the catalyst to the palladium(2) state (3.31). This suggestion does not appear to explain the usually high specificity for cross-coupling products, however, since elimination of both halide-derived aryl groups (Ar in 3.33) to a symmetric product may be similarly probable (unless these specifically occupy *trans*-octahedral sites?).

Widdowson (*loc. cit.*) has rationalised the inhibiting effect of *ortho*-substituents upon palladium catalysed Grignard-to-halide coupling on the basis of this general mechanism, taking the metathesis step (with ArMgX) as rate determining. It is, however, likely that the corresponding reaction with arylzinc halides (Section 3c ii) has a similar mechanism. The zinc reaction can sustain three (or four?) *ortho*-substituents and the magnesium example only two, so that the metal of the organometallic reagent may be more important than these mechanistic proposals infer.

(ii) Palladium(0) catalysed coupling of aryl iodides or bromides with arylzinc chlorides

This coupling and the related one with a nickel(0) catalyst (see also Section 3B iii) have come to be known as the Negishi method, e.g. Scheme 3.10.

Experimental details, a list of examples and a discussion of its scope and limitations are available (E. Negishi, T. Takahashi, and A.O. King, *Organic Syntheses*, 1988, **66**, 67).

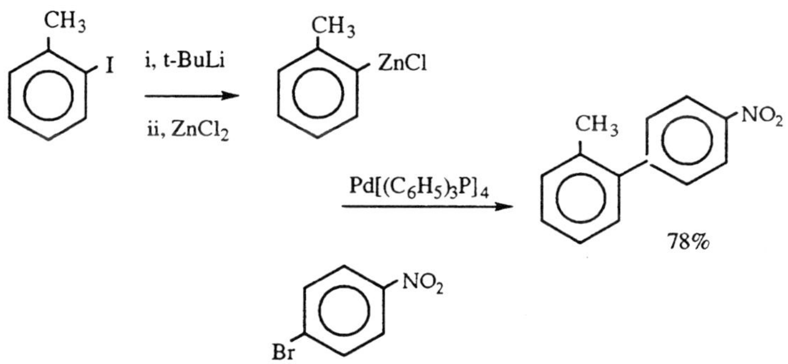

Scheme 3.10

Comparison of nickel and palladium catalysts shows that in general nickel is more reactive, allowing coupling of bromides without activating electron withdrawing groups such as nitro, but the nickel catalyst is incompatible with nitro and to a lesser extent ethynyl, substituents. The bromopyridines provide an example showing how bromide reactivity in coupling with phenylzinc chloride using Pd(0) parallels its ease of nucleophilic displacement: 2-bromopyridine reacts quantitatively but 3-bromopyridine gives no coupled product.

An example with three methyl substituents *ortho* to the biphenyl bond, from mesitylzinc chloride and 2-iodotoluene, was formed in 93% yield using nickel and 88% with palladium catalysts. No example with two substituents *ortho* to the reacting halide is presented.

(iii) Palladium(0) or (2) catalysed condensation between aryl bromides and arylboronic acids

Palladium(0) tetrakistriphenylphosphine catalyses the coupling of aryl bromides (or iodides) with arylboronic acids to give unsymmetrical (or symmetrical) biaryls in high yield (N. Miyaura, T. Yang and A. Suzuki, *Synth. Comm.*, 1981, **11**, 513).

This method has the advantage over those using more reactive organometallic intermediates (e.g. of Li, Mg, Zn, or Cu) that dry conditions are not needed; indeed the reaction is usually conducted in a vigorously stirred two-phase system of benzene and 2M aqueous sodium or potassium carbonate (Suzuki conditions). Where an aqueous reagent is unsuitable, the coupling can be accomplished in DMF at 100°C in the presence of 2 mol. % of palladium(2) acetate, triphenylphosphine (or tri-*o*-tolylphosphine) and triethylamine (W.J. Thompson and J. Gaudino, *J. Org. Chem.*, 1984, **49**, 5237).

Arylboronic acids (3.36, Scheme 3.11) are available from reaction between aryl Grignard or lithium reagents and trialkyl (usually trimethyl) borates, the intermediate dialkyl arylboronate (3.35) being hydrolysed to the boronic acid, without isolation, by aqueous work-up. Alternatively the aryllithium can be treated with trimethylsilyl chloride to form the aryltrimethylsilane (3.37) which with boron tribromide gives the dibromoborane (3.38) leading to the boronic acid (3.36) by similar hydrolytic work-up. (W. Haubold *et al. J. Organometal. Chem.*, 1986, **315**, 1; M.J. Sharp, W. Cheng and V. Snieckus, *Tetrahedron Lett.*, 1987, **28**, 5093). Aryltrimethylsilanes have also been prepared by treating unactivated aryl halides, e.g. 4-halotoluenes, with hexamethyldisilane and potassium methoxide in hexamethylphosphoramide (HMPT) at 25°C (M.A. Shippey and P.B. Dervan, *J. Org. Chem.*, 1977, **42**, 2654).

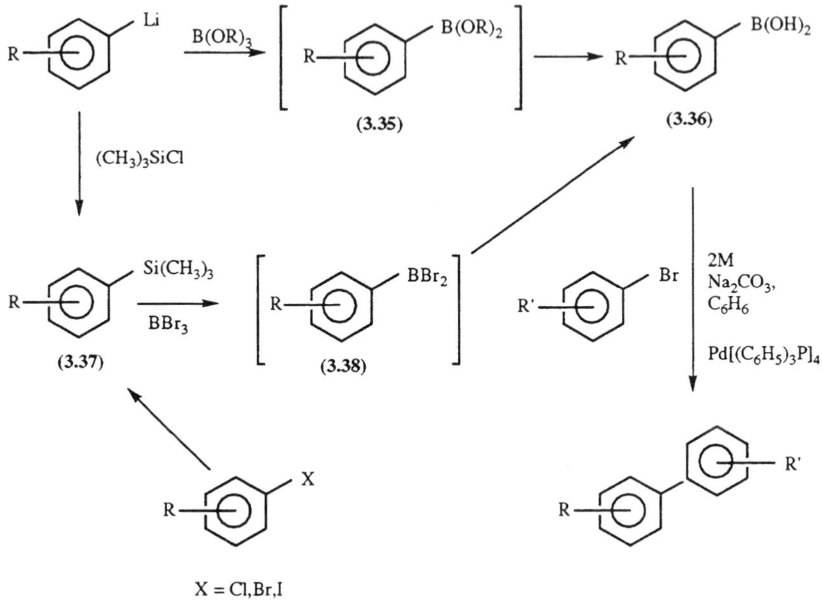

Scheme 3.11

A second advantage of this coupling reaction is that it can be used in the presence of nitro-groups (which are incompatible with lithium or Grignard reagents) attached to either the aryl bromide or arylboronic acid component (R.B. Millar and S. Dugar, *Organometallics*, 1984, **3**, 1261). The latter intermediates are formed by nitration of arylboronic acids; e.g. phenylboronic acid gives (mainly) the 3-nitro-derivative with nitric acid (d. 1.5 g. cm^{-3}) below 0°C but the 2-nitro-isomer by similar nitration in acetic (ethanoic) anhydride (W. Seaman and J.R. Johnson, *J. Amer. Chem. Soc.*, 1931, **53**, 711).

Suzuki coupling is not precluded by three substituents *ortho* to the biphenyl bond, although mesitylboronic acid reacted more slowly (R.B. Millar, *loc. cit.*). Early workers used a 10% molar excess of arylboronic acid over the aryl bromide (and 3 mol % Pd catalyst) which was destroyed before isolation of the product by oxidation (to the phenol) with hydrogen peroxide, but successful polymerisation of 4-bromophenylboronic acid (see Section 6(b)) shows that the excess is not essential.

A modification of this method, also due to Suzuki, employs esters of arylboronic acids which can be coupled efficiently to aryl bromides or iodides in the presence of thallium(1) carbonate (1.5 mol.) and palladium(0) tetrakistriphenylphosphine in benzene at reflux temperature (non-aqueous system), or using aqueous thallium(1) hydroxide (3 mol. 0.5 M) with bis(diphenylphosphino)ethane (dppe) palladium(2) chloride in benzene at 50°C. The former reagent coupled the dibutyl ester of 4-nitrophenylboronic acid to 4-bromoacetophenone in 90% isolated yield; the latter gave 4-methoxybiphenyl (82% by g.c.) from the cyclic phenylboronate ester of propan-1,3-diol and 4-iodoanisole (M. Sato, N. Miyaura, and A. Suzuki, *Chem. Lett.* 1989, 1405).

Snieckus has combined the technique of *ortho*-lithiation directed by dialkylamido (-$CONR_2$), carbamate (-$OCONR_2$), methoxymethoxy (-OCH_2OCH_3), and similar groups with boronic acid formation by the trimethylborate or intermediate silylation methods to form substituted biphenyls, Scheme 3.12 (the products described in several papers are catalogued: V. Snieckus, *Chem. Rev.* 1990, **90**, 879, Table 36). By repeating the sequence, Scheme 3.13 *meta*-terphenyls (e.g. 3.39) can be prepared:

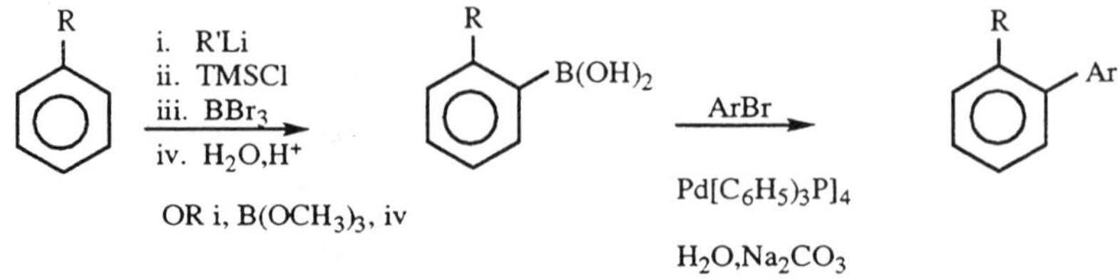

R = $CONR_2$, $OCONR_2$, OCH_2OCH_3, NHCO.OBu[t]

Scheme 3.12

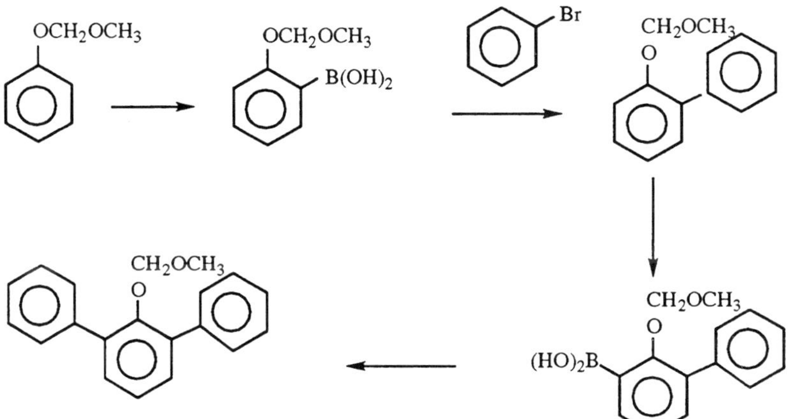

(3.39)

Scheme 3.13

This synthesis can be adapted to form *para*-terphenyls (3.42, R = H, and quaterphenyls e.g. R = C_6H_5), Scheme 3.14, by making use of the sec.-butyllithium promoted rearrangement of the phenylcarbamate group. The carboxamido moiety of (3.40) migrates from oxygen to the *ortho* ring position, an anionic analogue of the Fries rearrangement. After methylation of the phenolic group which is generated, the carboxamide function of (3.41) serves to direct a further *ortho*-lithiation, and hence boronation and aryl coupling, to the position *para* to the initial biphenyl bond (W. Cheng and V. Snieckus, *Tetrahedron Lett.*, 1987, 28, 5097).

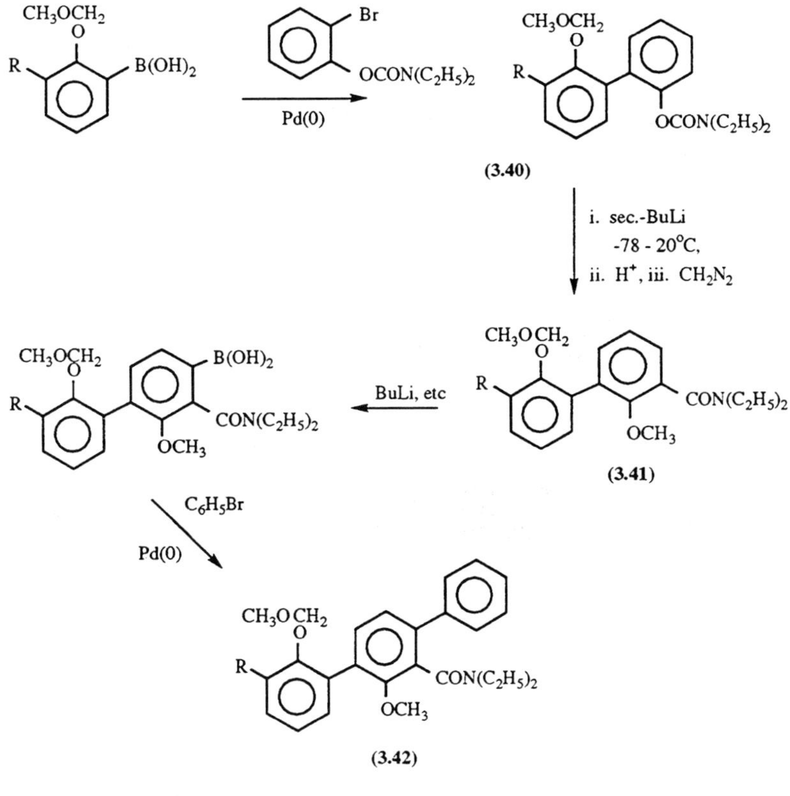

Scheme 3.14

The boronic acid coupling has been applied to synthesis of aglycones in the gilvocarcin series of anti-tumor antibiotics (M.E. Jung and Y.H. Jung, *Tetrahedron Lett.* 1988, 2517; see also Sections 3(d) and 3(m) using an aryl iodide (3.43) under Suzuki conditions, Scheme 3.15.

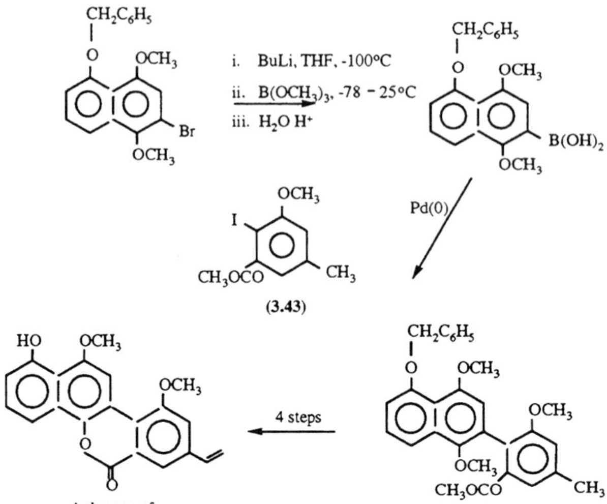

(3.43)

Aglycone of Gilvocarcin M

(3.44)

Scheme 3.15

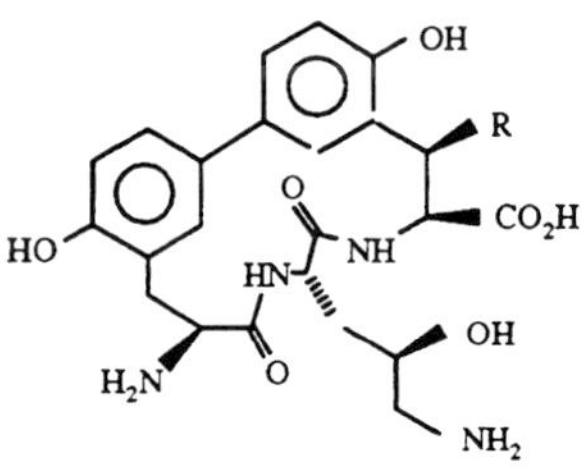

(3.45) R = H or OH

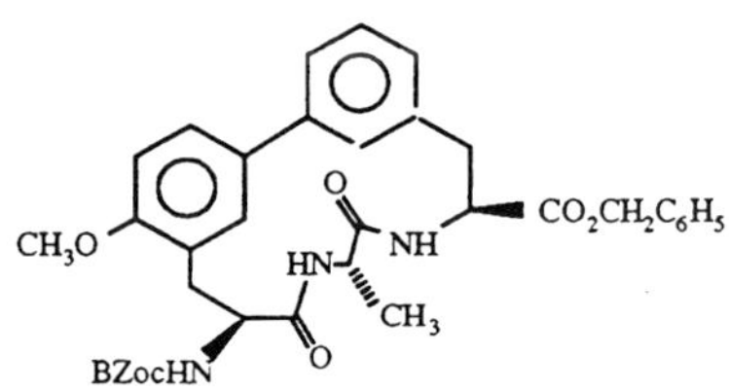

(3.46)

BZoc = CO.O.$CH_2C_6H_5$

Antibiotics of the vancomycin group incorporate the biphenyl system (review: D.H. Williams, Acc. Chem. Res., 1984, 364), as does the simpler cyclic tripeptide antibiotic WS-43708A (3.45) (R. Kannan and D.H. Williams, *J. Org. Chem.*, 1987, **52**, 5435 and references therein). A model for this ring system (3.46) and the dityrosine derivatives (3.47) have been synthesised using boronic acid coupling, Scheme 3.16 (U. Schmidt *et al. Angew. Chem. Int. Edn. Eng.*, 1989, **28**, 929).

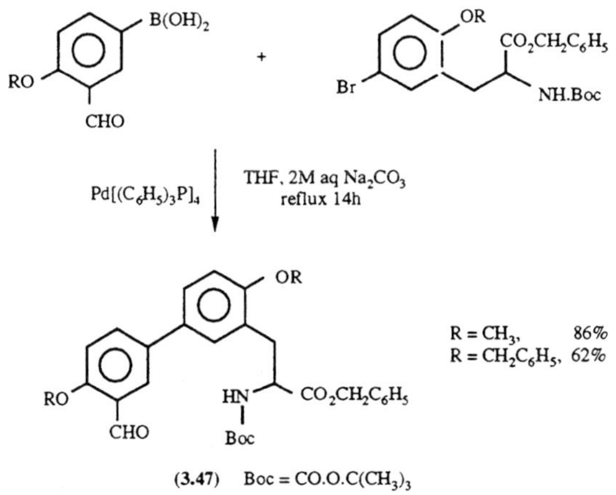

Scheme 3.16

(iv) Palladium catalysed coupling of aryltrialkylstannanes with aryl trifluoromethane sulphonates

Unsymmetric biaryls are obtained, often in excellent yield, when aryltrialkylstannanes (e.g. 3.48, Scheme 3.17) condense with aryl trifluoromethane sulphonates (triflates) (e.g. 3.49) in the presence of lithium chloride (3 mol) using either palladium(0)tetrakistriphenylphosphine (2 mol %) in boiling dioxan, or bistriphenylphosphinepalladium(2) chloride in DMF at 100° (A.M. Echwarren and J.K. Stille, *J. Amer. Chem. Soc.*, 1987, **109**, 5478). Examples with 4-nitro (3.50), 4-acetyl (3.49) and, remarkably, 3-hydroxy (3.52) substituents on the triflate component and alkoxy groups in the 3- or 4-positions of the arylstannane have been demonstrated in 53-85% yield.

Symmetric coupling of aryl triflates is effected in near quantitative isolated yield when they are treated with a half molecular proportion of hexamethylditin under the same conditions, Scheme 3.17. Aryltrimethylstannanes are evidently intermediates, which react with a second molecule of the triflate, since these stannanes are the major products using an equimolar proportion of the ditin reagent (Echwarren and Stille, *loc. cit.*).

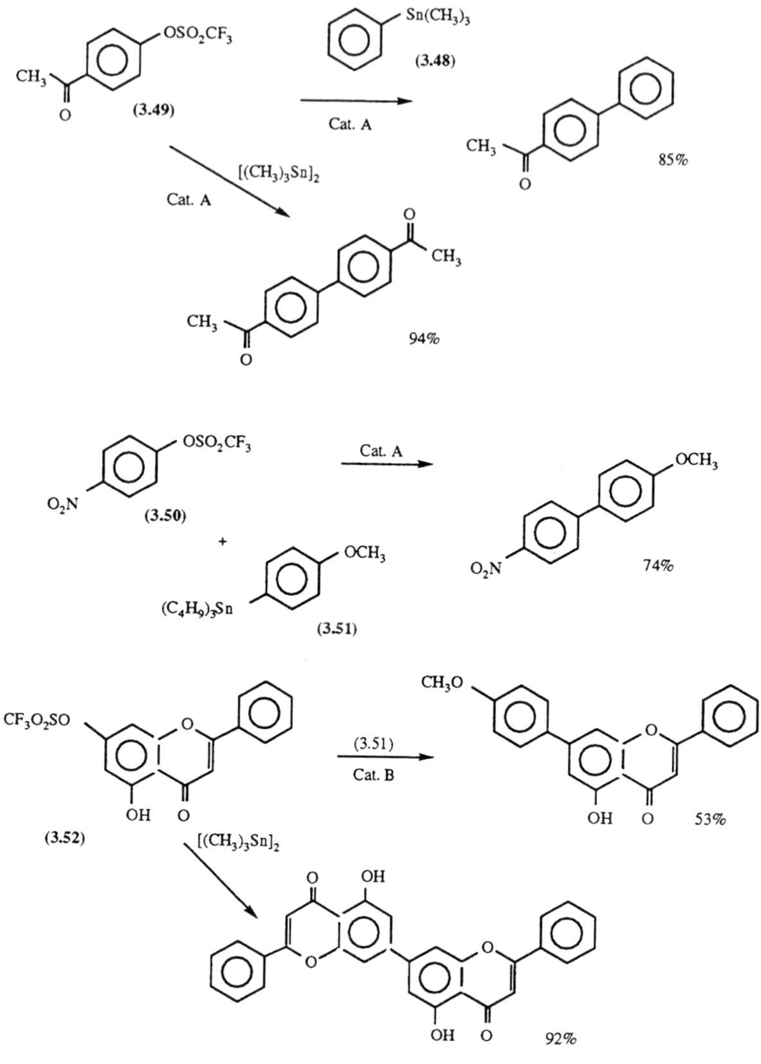

Cat. A = $Pd[(C_6H_5)_3P]_4$, dioxan, 100°
Cat. B = $Pd[C_6H_5)_3P]_2Cl_2$, DMF, 100°

Scheme 3.17

The aryltrialkylstannanes required for unsymmetrical coupling are available by treating the corresponding aryllithium with trialkyltin chlorides, and the technique has been combined with that of directed *ortho*-lithiation (V. Snieckus, *loc. cit.* Section 3(c) iii).

Aryl triflates are obtained by treating phenols with trifluoromethane sulphonic anhydride in the presence of pyridine (Stille, *loc. cit.*). Since this initial phenolic group activates the benzene ring (etc.) to a wide range of electrophilic substitution reactions and is then removed in the coupling step, this method has particular synthetic potential.

Successful couplings, symmetrical and unsymmetrical, of an example, the flavone (3.52), bearing a free phenolic hydroxyl group is unusual in biaryl coupling reactions involving organometallic intermediates (but see para (VI) below and Section 3(j) for phenol oxidative coupling). In this example the dihydroxyflavone (chrysin) specifically gave the 7- trifluoromethane sulphonyl ester (3.52) on treatment with pyridine and one molecular proportion of the sulphonic anhydride.

Saá and coworkers have made a systematic study of the application of this coupling reaction (and more particularly its vinylstannane analogue) to hindered, electron-rich substrates and found that in general these require larger amounts, 10-15 mol. % of palladium catalyst ($PdCl_2[(C_6H_5)_3P]_2$) and additional triphenylphosphine, in DMF. Two *ortho*-methoxy substituents are tolerated on the arylstannane but only one on the aryl triflate, although this may be otherwise pentasubstituted. The limiting example, 2,2',6-trimethoxybiphenyl was isolated in 33% yield after 24h at 140°C (J.M. Saá, G. Martorell and A. Garcia Raso, *J. Org. Chem.*, 1992, **57**, 678).

(v) Reductive coupling of aryl halides with aqueous alkali and palladium on carbon

Observation of biphenyl by-products during reduction of nitrohalobenzenes with palladium on carbon in an aqueous system lead to a procedure for symmetric coupling of bromo or chlorobenzenes, which may bear amino, acetyl (ethanoyl), fluoro, or methyl substituents. Thus refluxing the halobenzene with aqueous sodium hydroxide, sodium formate, 3% palladium-carbon, a surfactant and xylene (for starting materials melting above 100°C) gives the biphenyl after filtration or steam distillation in 35-65% yield (P. Bamfield and P.M. Quan, *Synthesis*, 1978, 537).

This is not the first observation of aryl halide coupling with palladium metal and an alkaline reducing agent: an early report (M. Busch *et al.*, *J. Prakt. Chem.*, 1936, **146**, 1; see also C. Weygand, *Organic Preparations*, Interscience Publishers, New York, 1945, p. 354, *et seq.*) describes preparation of biphenyl and its 4,4'-diamino and 4,4'-dicarboxylic acid derivatives in 40-80% yield by heating the bromobenzenes with palladium on calcium carbonate in an alkaline solution of hydrazine in ethanol, methanol or dioxan to reflux or autoclave temperatures. *Meta*- and *para*-dibromobenzenes gave phenylene oligomers with molecular weights thought to be over 1000.

(vi) Palladium Catalysed Coupling of Arylalkyldifluorosilanes with Aryl Iodides

Arylalkyldifluorosilanes (or diaryldifluorosilanes), available by fluorinating the corresponding dichlorosilanes with antimony trifluoride, couple with aryl iodides in the presence of η^3-allylpalladium chloride (5 mol %) and potassium fluoride (2 mol.) in dimethylformamide at 70-100°C. The dibenzylideneacetone and triphenylphosphine complexes of Pd(0) or Pd(2)Cl_2

$$ArSiRF_2 + Ar'I \xrightarrow[\text{KF, DMF}\;\; 70\text{-}100^\circ C]{\eta^3\text{-}C_3H_5PdCl)_2} Ar\text{-}Ar'$$

were not effective catalysts. Yields from 52-89% of biphenyls were isolated using iodides substituted at the 3 or 4-positions with alkoxy, acetyl, cyano, carboxymethyl, acetoxymethyl and, remarkably, free hydroxyl groups exemplified by 4-iodophenol and 3-iodobenzylalcohol. Silanes carried 3- or 4-methyl, trifluoromethyl, methoxy or chloro-substituents, but the one *ortho*-substituted example, 2-iodoanisole, gave only 45% of 2-methoxy-4-methylbiphenyl (Y. Halanaka, S. Fukushima and T. Hiyama, *Chem. Lett.* 1989, 1711).

Symmetric coupling of 4-ethoxyiodobenzene (to 4,4'-diethoxybiphenyl) is a major side reaction with phenylethyldifluorosilane at 100°C but 4-ethoxybiphenyl is formed in 81% yield at 70°C.

(d) Displacement of Oxazoline Activated Alkoxy Groups by Aryl Grignard or Lithium Reagents

While aryl ethers are generally inert to Grignard and similar organometallic reagents, and indeed often serve to mask acidic hydroxy functions, Meyers and coworkers discovered that nucleophilic displacement of methoxy groups by the aryl anion moiety of Grignard and lithium reagents to form biaryls occurs when an oxazoline group is placed *ortho*- to the methoxy (A.I. Meyers, R. Gabel, and E.D. Mihehich, *J. Org. Chem.*, 1978, **43**, 1372).

The necessary oxazoline group is usually formed from the corresponding carboxylic acid, *via* its acid chloride, which may be regenerated by acid or alkaline hydrolysis after coupling, Scheme 3.18 (for a review of further transformations of the oxazoline group and its applications in synthesis see M. Reuman and A.I. Meyers, *Tetrahedron*, 1985, **41**, 837-860).

The activating effect of the oxazoline group is thought to result from initial chelation of the reagent metal atom by methoxy and nitrogen, Scheme 3.19, leading by nucleophilic addition of the aryl anion to a relatively stable Meisenheimer complex (3.54) which can re-aromatise to the biaryl (3.55) with elimination of methoxide assisted by the metal.

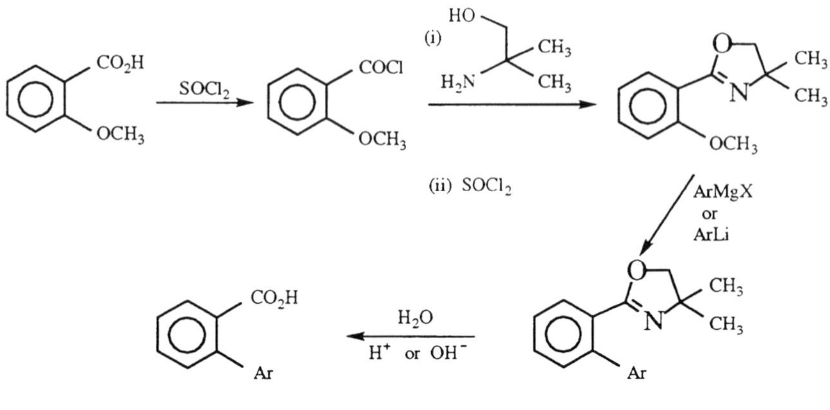

Scheme 3.18

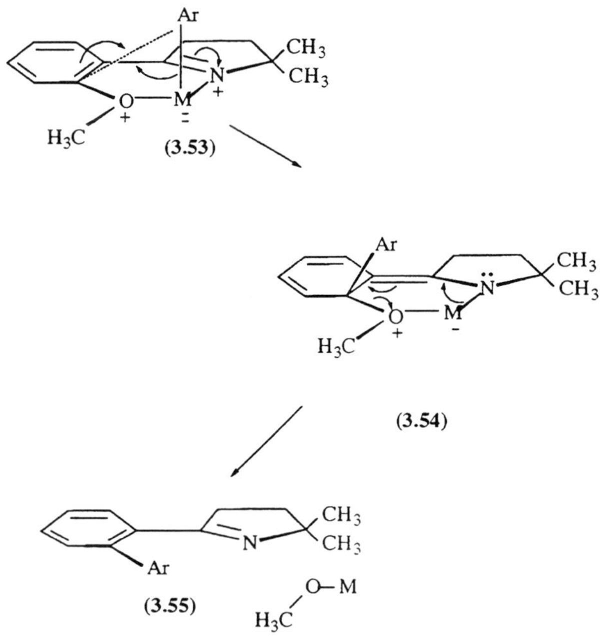

Scheme 3.19

When the two positions *ortho* to oxazoline are occupied by methoxy both can be arylated, giving *meta*-terphenyls, but only in a single reaction with two (or more) molecular proportions of reagent; once isolated the mono-aryl methoxy derivative does not react further, showing that the second arylation involves attack on an intermediate such as (3.54).

The standard procedure involves addition of pre-formed Grignard or lithium reagent to the methoxy-oxazoline, but the terphenyl (3.57) is formed in low yield when the dimethoxy-oxazoline (3.56) is refluxed in THF with

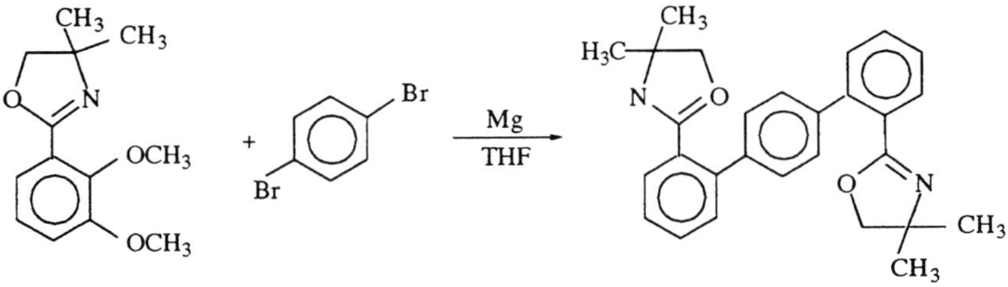

(3.56) (3.57)

Scheme 3.20

para-dibromobenzene and magnesium turnings, presumably by stepwise metallation of the two bromine positions *in situ*.

Meyers biaryl coupling is a key step in the synthesis of defulgogilvocarcin (3.44, R = H), the aglycone of the C-glycoside gilvocarcin V (3.44, R = 3.58) an antibiotic with anti-tumour activity apparently associated with its ability to induce bacteriophage after photochemical activation, a property retained by the aglycone (3.44, R = H) (A.D. Patten, N.H. Nguyen, and S.J. Danishefsky, *J. Org. Chem.*, 1988, **53**, 1003).

The Grignard reagent (3.59, X = MgBr), formed from the bromide (3.59, X = Br) by successive treatment with n-butyllithium and magnesium bromide, condenses with the dimethoxyoxazoline (3.60) to give the coupled product (3.61), Scheme 3.21. Simultaneous acid hydrolysis of the oxazoline and methoxymethyl groups leads spontaneously to the lactone (3.44). (For an alternative synthesis of the defulgocarcin skeleton see Section 3(c) iii, and 3(m)).

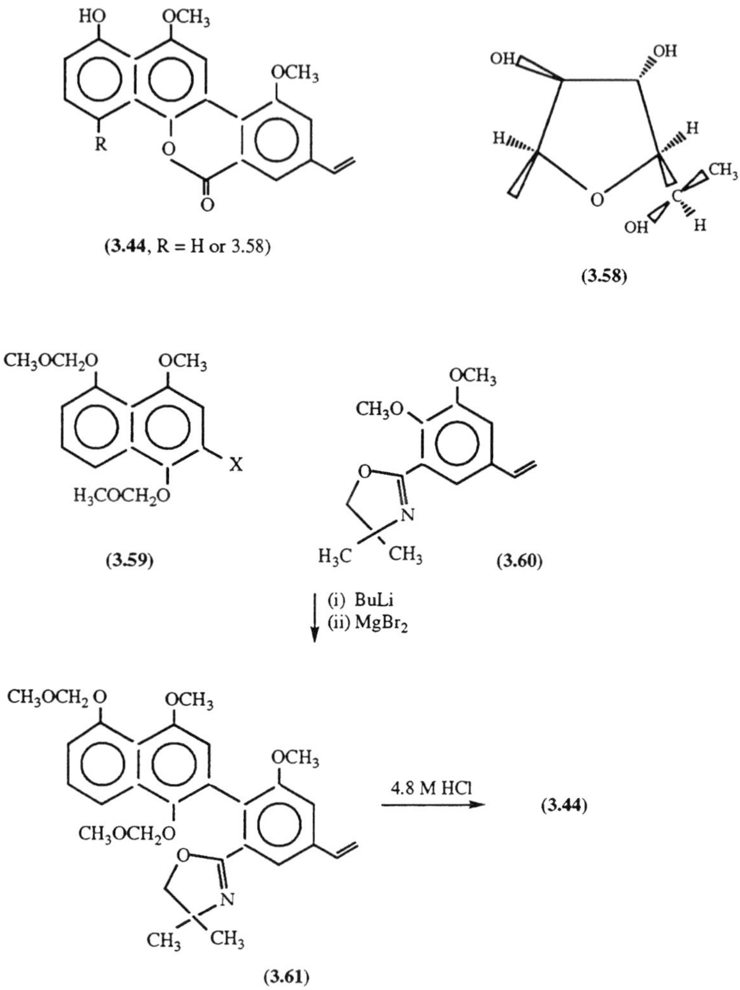

Scheme 3.21

Asymmetric induction is observed with a chiral oxazoline activating group, formed for example by condensation of the resolved hydroxy-amine (3.63), Scheme 3.22, with the methoxybenzamide (3.62). Reaction with an *ortho*-substituted phenylmagnesium bromide (3.65) then gives a preponderance of one diastereomer (arising by hindered rotation about the biphenyl bond) of the biphenyl (3.66) varying from 24:1 (for $R^1 = CH_3$, $R^2 = OCH_3$) to nil (i.e. 1:1 for $R^1 = R^2 = OCH_3$) (A.I. Meyers and R.J. Himmelsbach, *J. Amer. Chem. Soc.*, 1985, **107**, 682; for asymmetric induction in the corresponding coupling of naphthalene derivatives see A.I. Meyers and K.A. Lutomski, *J. Amer. Chem. Soc.*, 1982, **104**, 879, and by use of chiral alkoxy leaving groups, J.M. Wilson and D.J. Cram, *J. Org. Chem.*, 1984, **49**, 4930).

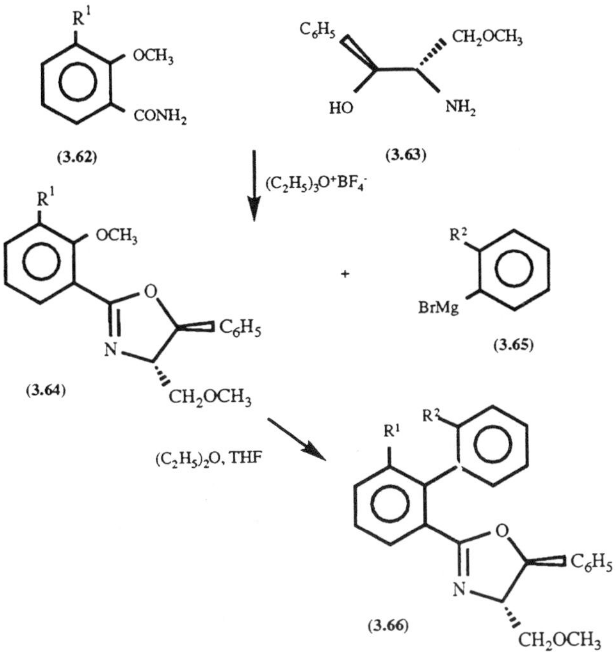

Scheme 3.22

Asymmetric induction is observed with a chiral oxazoline activating group, formed for example by condensation of the resolved hydroxy-amine (3.63), Scheme 3.22, with the methoxybenzamide (3.62). Reaction with an *ortho*-substituted phenylmagnesium bromide (3.65) then gives a preponderance of one diastereomer (arising by hindered rotation about the biphenyl bond) of the biphenyl (3.66) varying from 24:1 (for R^1 = CH_3, R^2 = OCH_3) to nil (i.e. 1:1 for R^1 = R^2 = OCH_3) (A.I. Meyers and R.J. Himmelsbach, *J. Amer. Chem. Soc.*, 1985, **107**, 682; for asymmetric induction in the corresponding coupling of naphthalene derivatives see A.I. Meyers and K.A. Lutomski, *J. Amer. Chem. Soc.*, 1982, **104**, 879, and by use of chiral alkoxy leaving groups, J.M. Wilson and D.J. Cram, *J. Org. Chem.*, 1984, **49**, 4930).

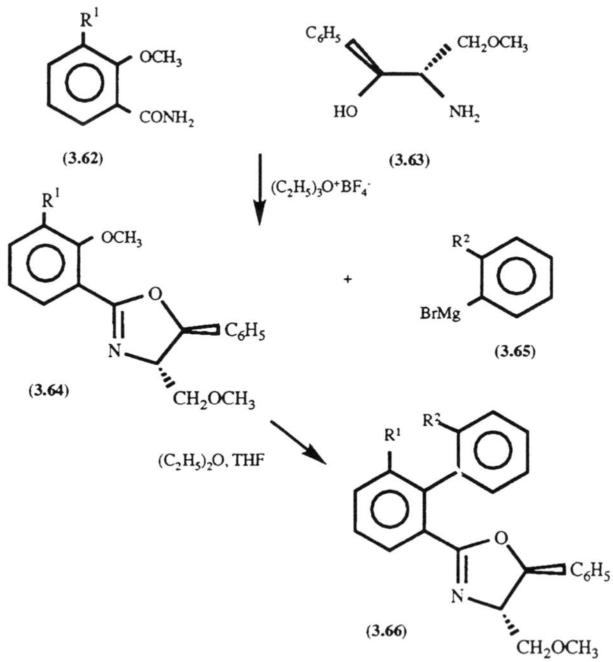

Scheme 3.22

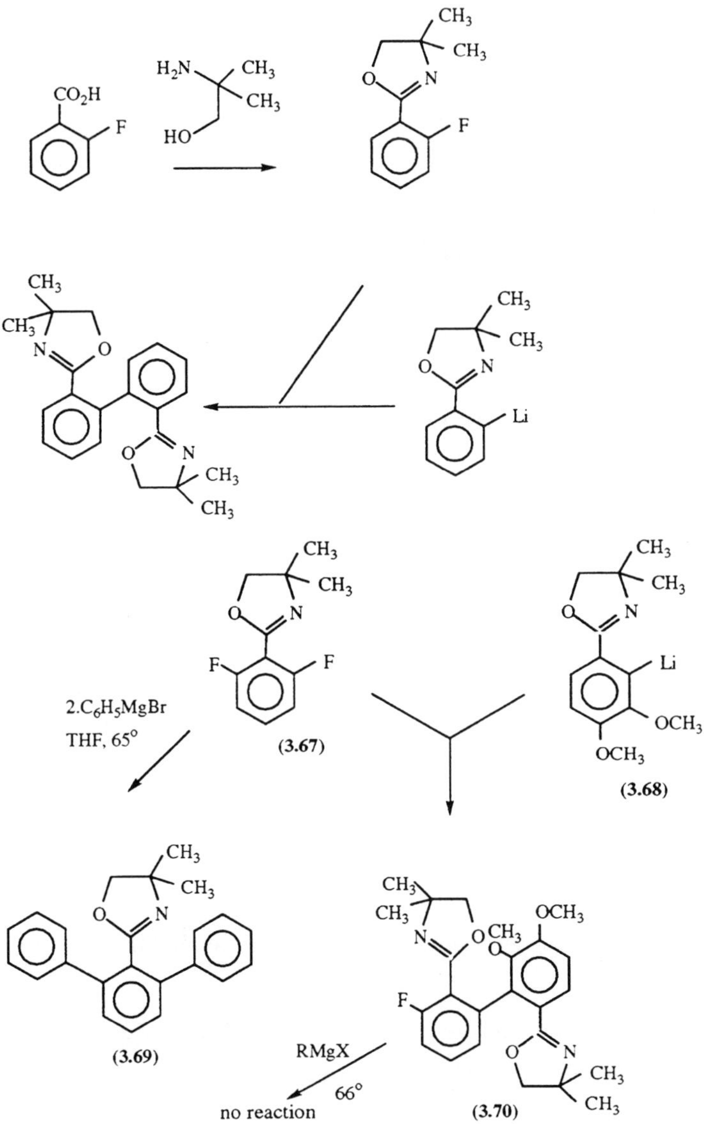
CO2H
F
H2N
CH3
CH3
HO
CH3
CH3
O
N
F
CH3
CH3
N
O
O
N
CH3
CH3
CH3
CH3
O
N
Li
CH3
CH3
O
N
F
F
(3.67)
CH3
CH3
O
N
Li
OCH3
OCH3
(3.68)
2.C6H5MgBr
THF, 65°
CH3
CH3
O
N
(3.69)
CH3
CH3
N
O
OCH3
CH3
O
F
O
N
CH3
CH3
(3.70)
RMgX
66°
no reaction

Scheme 3.23

Replacement of the Grignard reagent (3.65) by the corresponding lithio-derivative gives coupling without diastereoselectivity, while introduction of a second *ortho*-substituent into this reagent gives poor yields, also unselectively. Hydrolysis of the oxazoline group of the biphenyl (3.66) to carboxyl by hot acid or alkali caused racemisation (which was also studied kinetically in toluene at 110°C), but chiral biphenyl aldehydes with enantiomeric excess up to 64% (by NMR) were obtained from some examples by a four-step sequence at room temperature.

(e) Displacement of fluorine activated by ortho-oxazoline or oxazole groups by aryllithium or Grignard reagents

An *ortho*-oxazoline group activates fluorine to nucleophilic displacement by aryl (and other) lithium or Grignard reagents in a reaction analogous to the displacement of methoxy in the Meyers coupling, Section 3(d). Biaryls are formed in high yield, Scheme 3.23 (A.I. Meyers and B.E. Williams, *Tetrahedron Letters*, 1978, 223) and at higher temperature the difluoro-compound (3.67) gives the *meta*-terphenyl (3.69) with two molecules of phenylmagnesium bromide, but only one fluorine could be displaced using the dimethoxy analogue (3.68) with two substituents *ortho* to lithium; indeed, the resulting monofluorobiphenyl (3.70) failed to react with simple alkyl Grignard reagents.

Similar reactions have been demonstrated when oxazole groups replace oxazoline, e.g. Scheme 3.24 (D.J. Cram, J.A. Bryant and K.M. Doxsee, *Chem. Lett.*, 1987, 19); again, *meta*-terphenyls (e.g. 3.72) are available from difluoro-derivatives (3.71).

Oxidation of the oxazole group in the coupled products with singlet oxygen, using tetraphenylporphyrin as singlet sensitiser during irradiation, followed by hydrolysis with lithium hydroxide (and acidification) gives the corresponding carboxylic acid, while amides e.g. (3.73) are more efficiently obtained from diphenyloxazoles by oxidation with pyridinium chlorochromate and similar hydrolysis (D.J. Cram *et al., loc. cit.*).

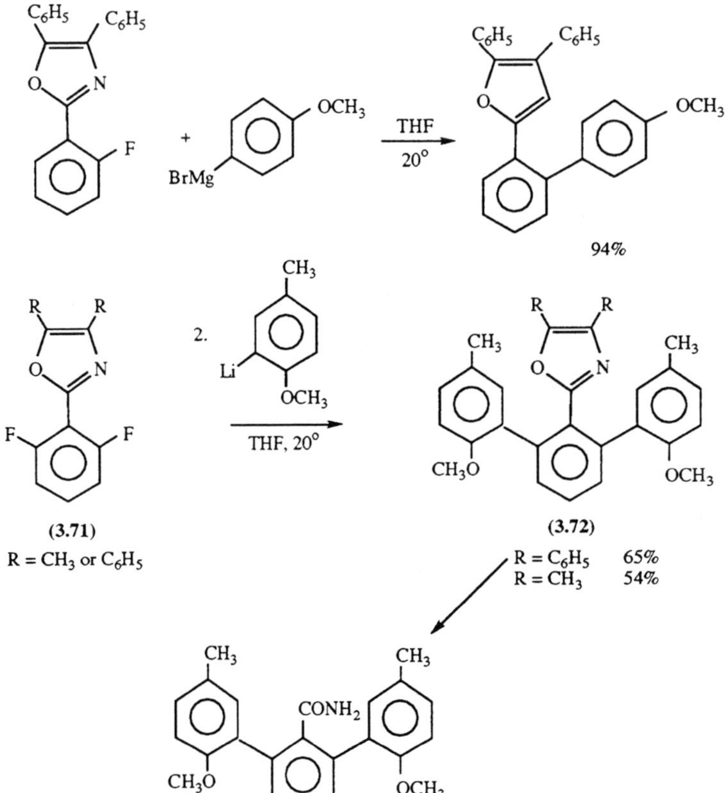

(3.71)
R = CH_3 or C_6H_5

(3.72)
R = C_6H_5 65%
R = CH_3 54%

(3.73)
R = C_6H_5 69%

Scheme 3.24

4-Fluorophenyloxazole does not couple under these conditions, showing that chelation of an organometallic intermediate is almost certainly necessary to fluorine displacement, and that the oxazole group is not acting primarily by stabilising an anionic intermediate (Meisenheimer complex) in an 'ordinary' $S_N(Ar)$ reaction, when a *para*-substituent (nitro, etc.) is normally more effective than an *ortho* one.

(f) Reaction of Grignard Reagents with Di- and Tetra-halobenzenes

Biphenyls are formed, usually in 60-70% yield, when 2-bromoiodobenzene is slowly added to two molecular proportions of aryl Grignard reagent in THF at room temperature (H. Hart, K. Harada, and C-J. F. Du, *J. Org. Chem.* 1985, **50**, 3104).

The reaction is thought to proceed by initial metal-iodine exchange, elimination of magnesium bromide to give an intermediate benzyne (3.76, Scheme 3.25) which adds a second molecule of Grignard to form the biaryl-2-magnesium bromide (3.75). Quenching with acid or with an electrophile (bromine or iodine were used) gives the biaryl or its derivative (3.74). 1,2-di-iodobenzene reacts similarly, but not 1,2-dibromobenzene.

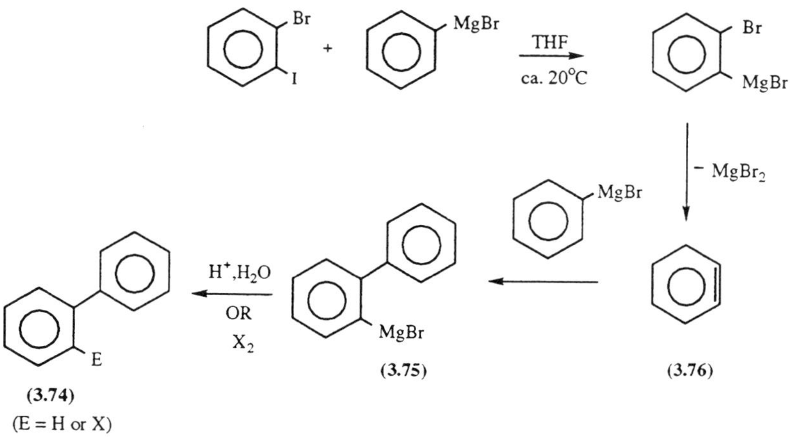

Scheme 3.25

Para-terphenyls, with less than 1% *meta*-terphenyl, are similarly formed from 1,4-dibromo-2,5-di-iodobenzene (3.77, Scheme 3.26) in about 40% yield using four proportions of Grignard reagent; biphenyl-4-magnesium bromide reacts in an elegant one step synthesis of paraquinquephenyl (3.80 E = H, R = C_6H_5). Hart *et al.* (*loc. cit.*) favour a sequence in which the di-Grignard (3.78) undergoes two elimination-addition steps. This mechanism requires regio-specificity at the second addition step giving the *para*-adduct (3.79) (as would repetition of Scheme 3.25 for the second coupling). In this case

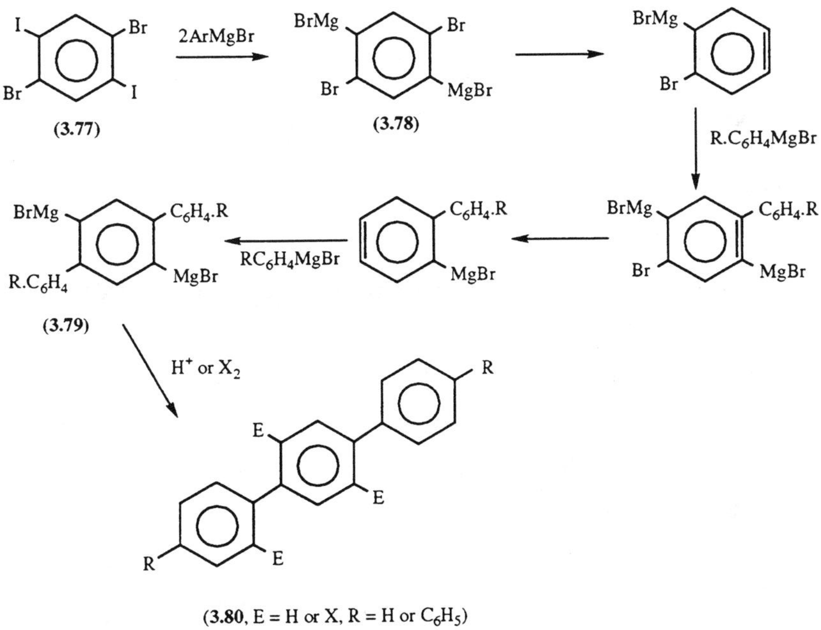

Scheme 3.26

1,5-dibromo-2,4-di-iodobenzene (3.81) should also give only *para*terphenyl, formed in 37% yield, but *meta*terphenyl, 6%, was also isolated, Scheme 3.27.

This might be attributable to a minor alternative reaction or to a small degree of concerted substitution character in the formal elimination-addition sequence.

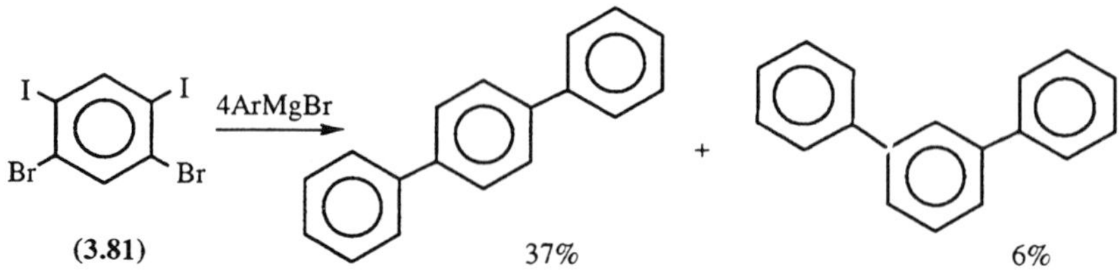

(3.81) 37% 6%

Scheme 3.27

It is probably relevant that a 5% yield of biphenyl is reported from an *uncatalysed* reaction (cf. Section 3(c) i) between iodobenzene and phenylmagnesium bromide (A. Sekiya and M. Ishikawa, *J. Organometal. Chem.*, 1976, **118**, 349).

(g) Coupling of Aryl Halides using Lithium with Ultrasonic Irradiation

This modern variation of the classical Wurtz-Fittig biaryl synthesis gives biphenyl, 3,3'- or 4,4'-dimethylbiphenyl in 36-73% yield using lithium wire, tetrahydrofuran (THF) as solvent and 10-12h at room temperature in an ultrasonic cleaning bath (B.H. Han and P. Boudjouk, *Tetrahedron Lett.* 1981, 2757). Interestingly, chloro-, bromo- and iodobenzenes gave almost the same yield, 70-73%, in similar reaction times. A detailed description of this reaction, with 4h reaction time, suitable for a student experiment is available (T.D. Lash and D. Berry, *J. Chem. Educ.*, 1985, **62**, 85).

(h) *Coupling of Nitrobenzenes with Phenolate Anions*

Nitrobenzenes substituted by a leaving group at the 2- or 4-position (3.82) react with the anion of 2,6-di-tert-butylphenol (3.81) in dimethylsulphoxide (DMSO) to give biphenyls by an SN(Ar) reaction, Scheme 3.28.

Coupling with displacement of hydrogen occurs under similar conditions when two electron withdrawing groups, NO_2 or CN, are present, usually at positions *ortho* and *para* to the reacting position (3.83). 2,6-Dimethyl- and unsubstituted phenol give lower yields (G.P. Stahly, *J. Org. Chem.* 1985, **50**, 3091).

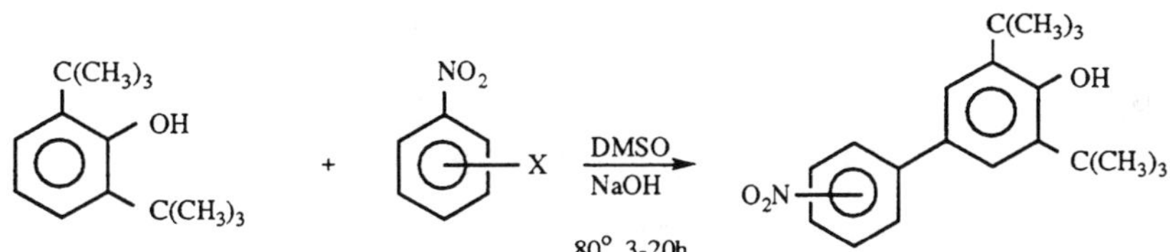

(3.81) **(3.82)**

X =	Yield
2-F	97%
2-Cl	71%
2-Br	79%
4-NO_2	77%

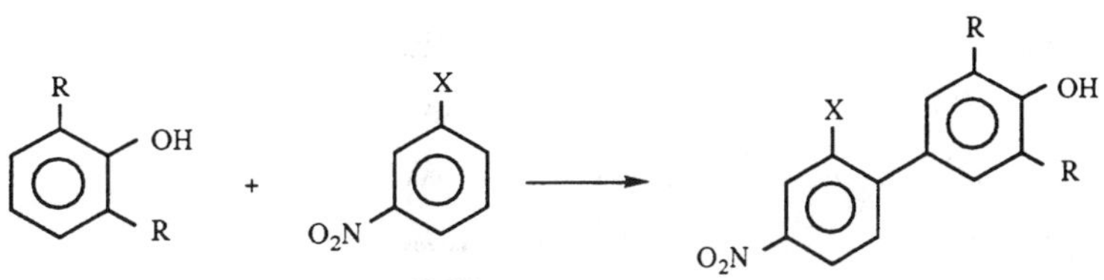

(3.83)

R	X	Yield
R = $C(CH_3)_3$,	X = NO_2	69%
"	X = CN	46%
R = CH_3,	X = NO_2	49%
R = H,	"	17%

Scheme 3.28

(i) *Dimerisation of Aryl Radicals; Termination Step of $S_{RN}1$ Reaction*

Direct evidence for the intermediacy of aryl radicals in radical nucleophilic aromatic substitution, the $S_{RN}1$ reaction, has been obtained by isolation of the radical dimer 4,4'-dicyanobiphenyl (39% yield) from electrolytically initiated reaction between 4-chlorobenzonitrile and pyridine-2-thiolate anion in liquid ammonia. The dimer was initially detected by cyclic voltammetry (R. Ellayeb, J.M. Savéant and A. Thiébaulte, *J. Amer. Chem. Soc.*, 1992, **114**, 10990).

(j) *Oxidative Coupling of Phenols and Phenol Ethers*

Recent synthetic studies of oxidative coupling have evidently concentrated on phenol ethers rather than free phenols; for a current synthetic application of the latter see Section 4b.

Syntheses of the lignans (±)-deoxyschizandrin (3.85, R = CH_3) and (±)-wuweizisu-C (3.85, R,R = $\cdot CH_2\cdot$) from their open-chain precursors (3.84) have been performed using vanadyl fluoride (VOF_3) (G.E. Schneiders and

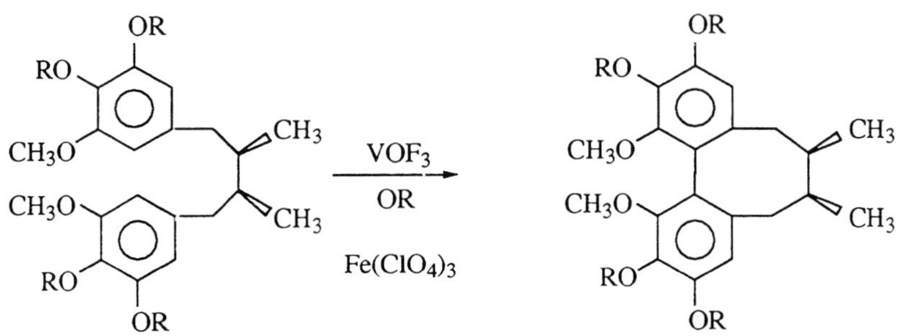

(3.84) **(3.85)**

R. Stevenson, *J. Org. Chem.* 1981, **46**, 2969, and references therein) and iron(3) perchlorate hexahydrate in acetonitrile (T. Takeya *et al.*, *Chem. Pharm. Bull.*, 1985, **33**, 3599).

The six-membered ring of the aporphine alkaloid (+)-ocoteine (3.87) is formed stereospecifically from the resolved precursor (3.86) using thallium(3) tristrifluoroacetate (D.A. Dickman and A.I. Mayers, *Tetrahedron Letters*, 1986, 1465). Ruthenium tetrakistrifluoroacetate is also effective in this cyclisation

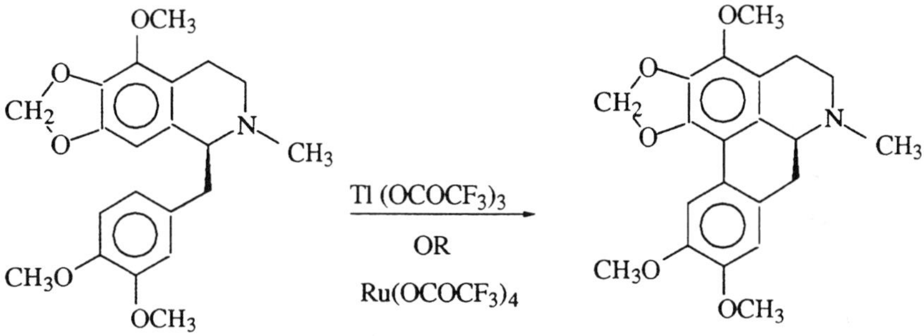

(3.86) **(3.87)**

(Y. Landais, D. Rambault and J.P. Robin, *Tetrahedron Letters*, 1987, 543).

In a procedure aimed at reducing hydrolysis of the reagent, thallium tristrifluoroacetate is generated *in situ* by reaction between thallium(3) oxide and trifluoroacetic acid in dichloromethane with trifluoroacetic anhydride to absorb liberated water. With boron trifluoride etherate this reagent gave the isostegane (3.89, Scheme 3.29) in 90% yield after a series of experiments, from dimethylmatairesinol (3.88). Isomerisation at 200°C gave a mixture from which the atropisomeric stegane (3.90) was isolated (R.C. Cambie, *et al.*, *Austral. J. Chem.*, 1984, **37**, 1775).

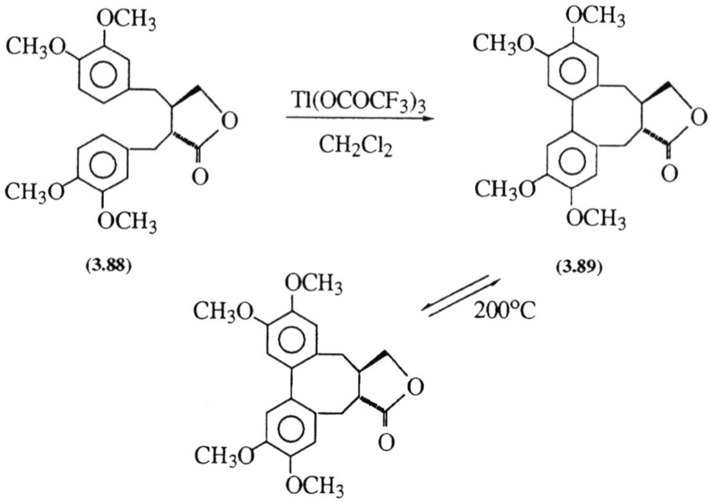

(3.88) (3.89) (3.90)

(Scheme 3.29)

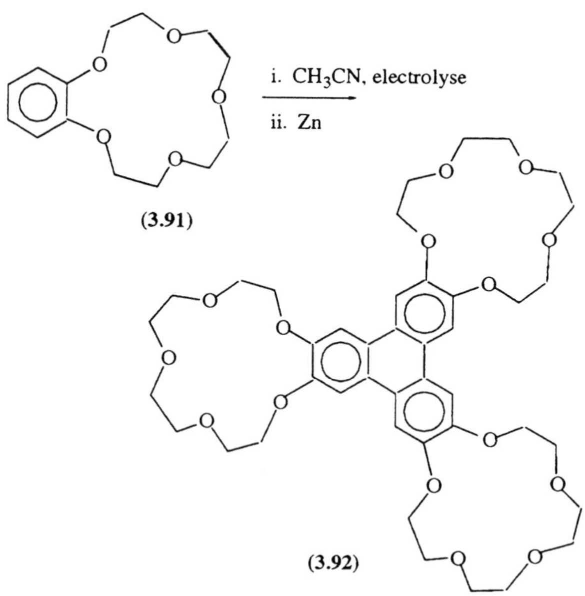

(3.91) (3.92)

Difficulties associated with nucleophilic interception of radical cations, and other side-reactions, in the electrolytic oxidation of the benzo-crown ether (3.91) to its triphenylene analogue (3.92), m.p. 285°C, are overcome by using a cell in which the precursor (3.91) solution in acetonitrile flows past the cathode and passes through a porous anode of carbon felt before falling into a flask containing a reducing agent, zinc powder. Quantities up to about one gram per hour can be processed with 30% conversion and recovery of the unreacted monomer (J. Simonet *et al., J.C.S. Chem. Commun.*, 1987, 984).

For a review of phenol oxidative coupling see D.W. Knight in *Comprehensive Organic Synthesis*, B.M. Trost Ed., Pergamon 1991, Chapter 2.3.

(k) Coupling by oxidation of Arylmetal Derivatives

(i) Oxidation of aryllithium reagents

A range of oxidants, usually transition metals in higher oxidation states have been used, generally with modest success, to couple aryllithium (or Grignard) reagents (for examples from older work see D.W. Knight, *loc. cit.* Section 1). More recently the preferred oxidant is iron(3) acetylacetonate which has found particular application in the construction of oligophenylene host molecules; for examples see Section 4(b).

(ii) Coupling of aryltributylstannanes with copper(2) nitrate

Symmetric biaryls are formed in 14-67% yield when substituted aryltributylstannanes are treated with hydrated copper(2) nitrate in THF at room temperature (S. Ghosal, G.D. Luke, and K.S. Kyler, *J. Org. Chem.*, 1987, **52**, 4296)

$$2ArSn(C_4H_9)_3 \xrightarrow[\text{THF, 23°C}]{Cu(NO_3)_2 3H_2O} Ar_2 + (C_4H_9)_3SnNO_3$$

Best yields are obtained with aryl groups substituted by methyl or methoxy at position 4, but a 14% yield of 2,2', 6,6'-tetramethoxy-4,4'-dimethylbiphenyl shows that the method is not fully suppressed by four *ortho*-substituents. A mechanism involving metal exchange leading to an arylcopper(2) cation $ArCu^+$ is suggested.

(iii) Oxidation of dilithium diarylcyanocuprates

This method is described in Section 3(a) (ii).

(l) Photochemical Cyclisation of Stilbenes with Oxidative Ring-opening of the Resulting Phenanthrenes

The biphenyl diol (3.94), isolated from the wood of *Taxus mairei,* was not obtained by attempted Ullmann coupling of the bromide (3.93) (presumably because of the unprotected hydroxyl groups). A mixture of three diastereomers of the required structure, arising by asymmetry of the two hydroxy-ethyl side chains and dissymmetry of the biphenyl unit, separable by chromatography on silica, was obtained by the sequence shown in Scheme 3.30 (Tong-Ing Ho *et al., Synthesis,* 1987, 795).

(m) Construction of an Aryl Group from the Side-chain of Phenylacetylenes

A short synthesis of the aglycone of gilvocarcin V (3.44, Scheme 3.31) uses the Dötz reaction of the chromium carbene complex (3.95), available from 2-bromoanisole with the phenylacetylene (3.96), prepared from vanillin in five steps. The resulting hydroxy-ester (3.97), obtained in 43% yield in a small scale experiment, is readily converted into the required lactone (3.44) using toluenesulphonic acid in benzene (K.A. Parker and C.A. Coburn, *J. Org. Chem.,* 1991, 56, 1666).

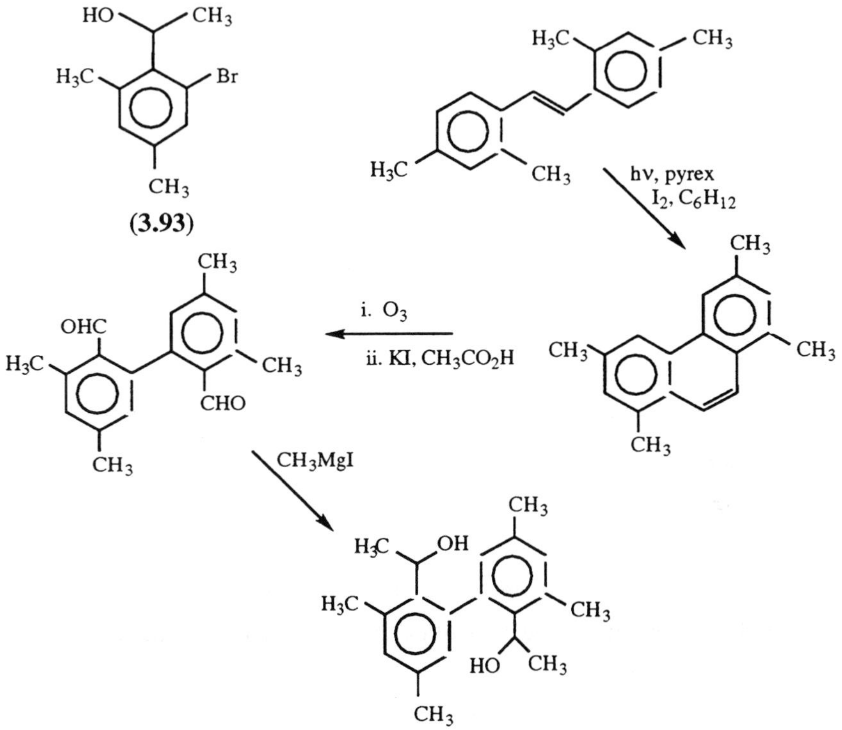

(3.93)

(3.94)

Scheme 3.30

While the Dötz reaction (K.H. Dötz, *Angew. Chem. Intn. Edn. Engl.* 1975, **14**, 644) is not a *general* arylbenzene synthesis since it leads necessarily to a 1-hydroxy-2-aryl-4-alkoxy-naphthalene, of which the gilvocarcin skeleton happens to be an example, the inclusion of this elegant synthesis at the end of a list of methods for aryl-aryl coupling provides a timely reminder that joining pre-formed aryl residues may not always be the best route to arylbenzenes.

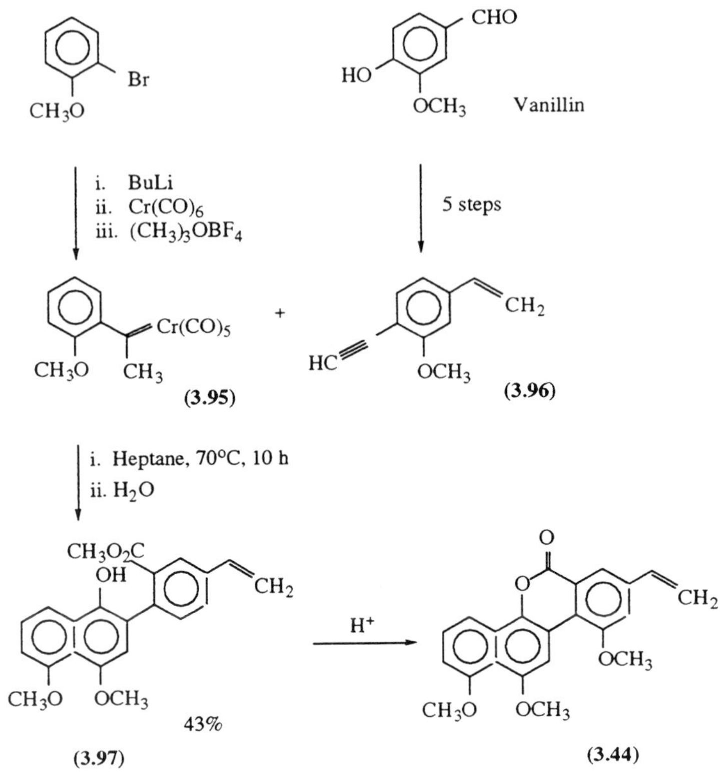

Scheme 3.31

4. Oligophenylene Host Molecules

(a) *Cationic Complexes*

The rigidity of the individual rings and the limited conformational freedom of oligo-phenylene systems has been applied in the design of specific host-molecules with cavities of predictable size, shape, and flexibility, particularly of the spherand (those whose cavities are lined with electron pairs and can accommodate only spherical entities and not parts of the host) and hemispherand types (those wherein at least half of the structural units are unable to fill their potential cavities by conformational change).

When such molecules complex metals, or other guest species, they need to adopt a particular conformation to bring their donor atoms into optimum positions to match the coordination geometry of the metal. This preliminary conformational change, termed pre-organisation, may contribute a major adverse entropy term to the free-energy of complexation. Thus the cyclosexiphenylene (4.1), which is effectively fixed in the correct conformation to coordinate lithium or sodium ions, has free-energies of complexation of these ions 17 and 13 Kcal. mole^{-1}, respectively, greater than its flexible open-chain analogue (4.2) (D.J. Cram *et al.*, *J. Amer. Chem. Soc.*, 1981, **103**, 6228).

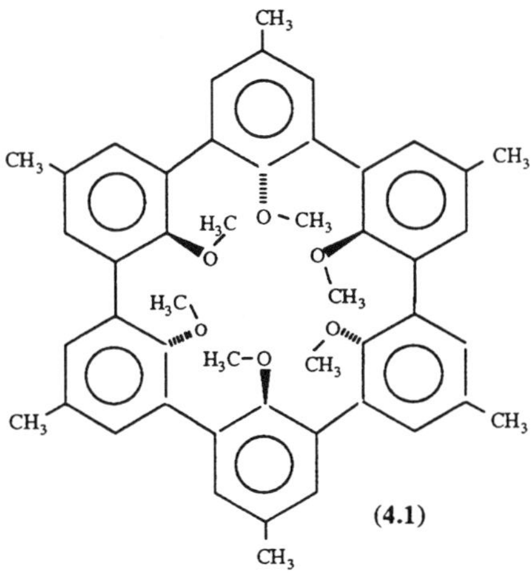

(4.1)

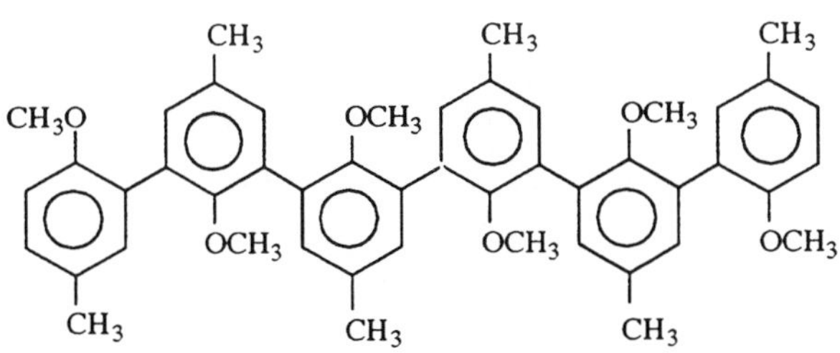

(4.2)

A large family of host molecules is based upon ether substituted terphenyl or quaterphenyl groups, analogous to (4.2), whose ends are bridged by more flexible chains incorporating various donor atoms. Examples are the quaterphenyls of type (4.3) (S.P. Artz and D.J. Cram, *J. Amer. Chem. Soc.*, 1984, **106**, 2160) and those in which the terphenyl unit forms a cap for a calixarene, which adopts a conical conformation (4.4) (W.I.I. Bakkev *et al.*, *J. Amer. Chem. Soc.*, 1994, **116**, 123 and references therein).

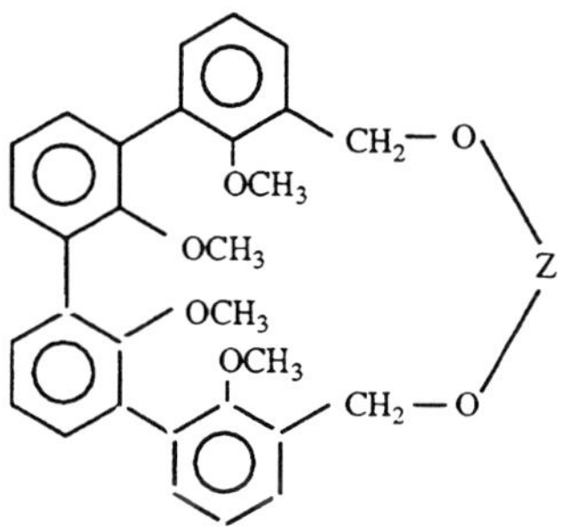

(4.3)

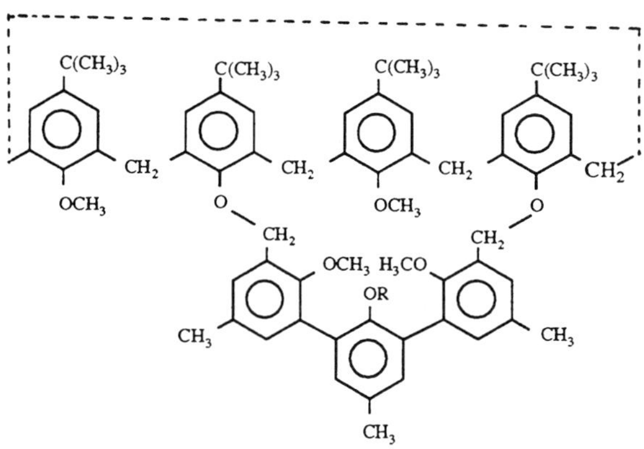

(4.4)

≡ 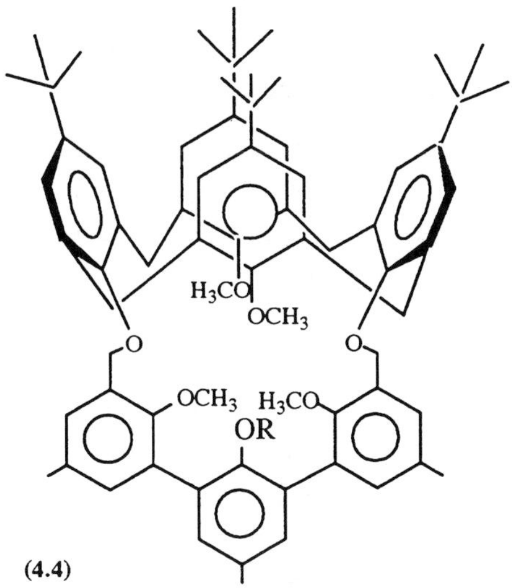

(4.4)

Assessment of the free energies of complexation, usually of the alkali metal and substituted ammonium ions, have shown how the complex stability and specificity for ions of particular size varies with the size and flexibility of the host cavity. In general, perhaps not surprisingly, greatest specificity is observed with smaller, more rigid cavities and between lithium and sodium ions.

Studies of the kinetic stability of these complexes have been made by measuring the rates at which one guest ion can replace another in the cavity, for example in relation to the size of various central alkoxy groups, R, in the terphenyl cap of calixarene (4.4).

When crown ether bridges are formed incorporating oxygen at the 2,2' and 6,6'-positions of biphenyl, as in (4.5), the biphenyl bond serves as a hinge so that conformational changes which occur when one crown ring complexes a metal ion are transmitted to the other which is thus optimised for complexation. The stability constants for complexes formed by the two rings are consequently coupled in an unusual way, which has been studied for sodium, potassium and ammonium ions. The singly bridged crowns (4.6) were prepared for comparison (G. Lindsten, O. Wennerström, and R. Isaksson, *J. Org. Chem.*, 1987, **52**, 547).

The optical isomers of the bis-crown ether (4.5) are described in Section 5. (d) ii.

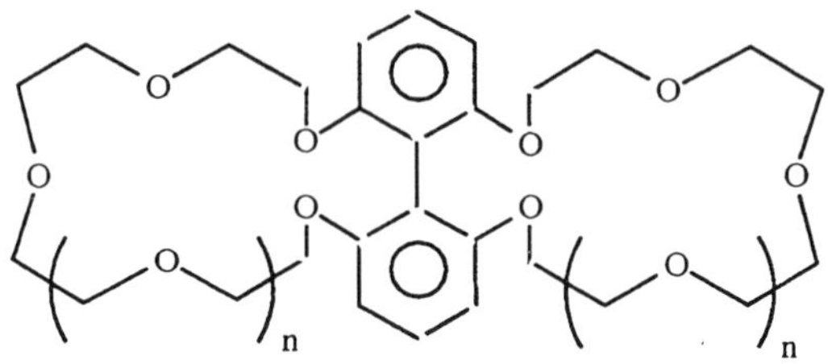

(**4.5**, n = 1 or 2)

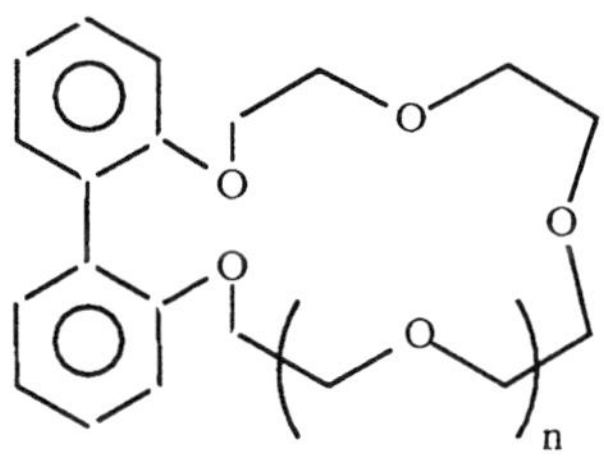

(**4.6**, n = 1 or 2)

(b) *Synthesis*

The polyether oligophenylenes required for these complexation studies are generally prepared by oxidative methods. Thus the first step in the synthesis of the cyclosexiphenylene (4.1) is oxidative phenol coupling of *p*-cresol to the trimer (4.7, Scheme 4.1). Bromination at the vacant positions *ortho* to hydroxyl, methylation of the latter functions and bromine-lithium exchange furnishes the corresponding dilithio-derivative whose oxidation with iron(3) acylacetonate gives the lithium complex (4.8) of the cyclic hexamer in 22% yield, probably with template assistance by the initially liberated lithium ion. The de-lithiated compound (4.1) crystallises when the complex (4.8), with tetrachloroferrate (3) as counter ion, is heated to 150°C in 80% aqueous methanol for five days, the insolubility of the netural species evidently displacing the complexation equilibrium (D.J. Cram *et al.*, *J. Amer. Chem. Soc.*, 1979, **101**, 6752).

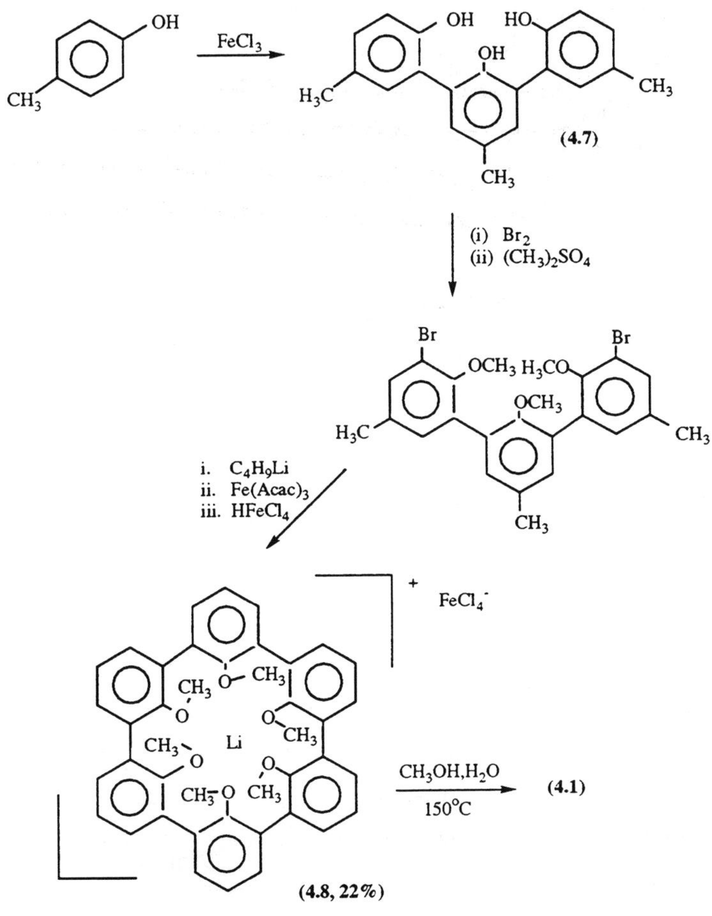

Scheme 4.1

The quaterphenylene hemispherands (4.3) are obtained starting from dibenzofuran (Scheme 4.2): lithiation (ortho to oxygen) and oxidative coupling with iron (3) acetylacetonate gives the dimer (4.9, R = H) which, after further lithiation and carboxylation, leads to the diacid (4.9, R = CO_2H). Furan ring opening by caustic alkali fusion and methylation of the resulting tetraphenol (4.10, R = CO_2H, R' = H), followed by reduction of the carboxyl groups with borane - THF complex, gives the diol (4.10, R = CH_2OH, R' = CH_3). The corresponding bis-bromomethyl intermediate (4.10, R = CH_2Br, R' = CH_3), obtained using phosphorus tribromide, can then be joined by ether links to a variety of diols (HO-Z-OH) using the Williamson synthesis with sodium hydride as base (S.P. Artz and D.J. Cram, *J. Amer. Chem. Soc.*, 1984, **106**, 2160).

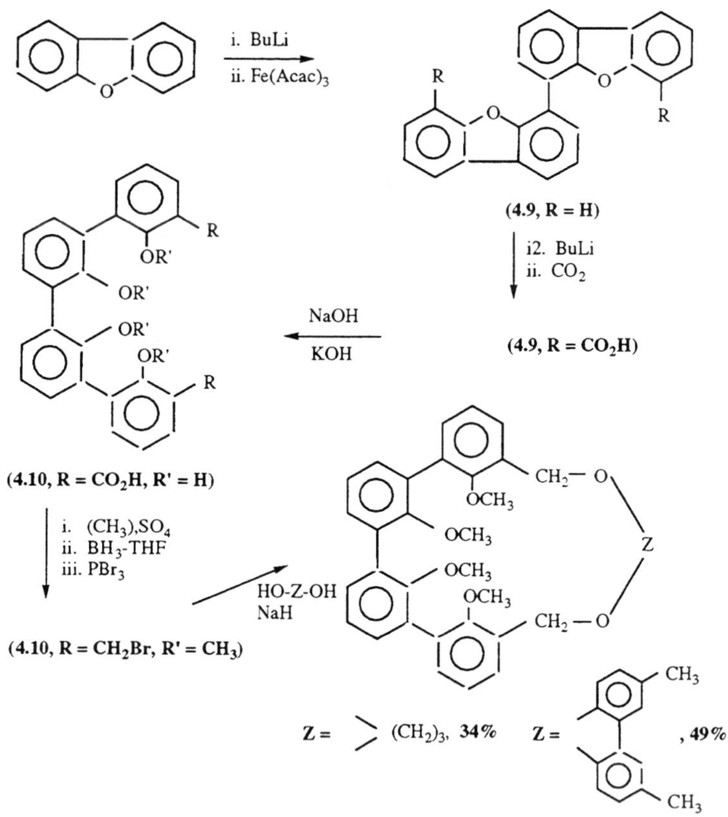

Scheme 4.2

A similar sequence starting from the terphenyl (4.7) gives compounds in the series represented by the calixarene (4.4).

The crown ether bridged biphenyls (4.5) and (4.6) are prepared from the corresponding di- or tetrahydroxybiphenyls (see Section 3b) by treatment with dichloro- or di-tosyloxy polyethers in the presence of base (G. Lindsten *et al.*, *loc. cit.*, Section 3b).

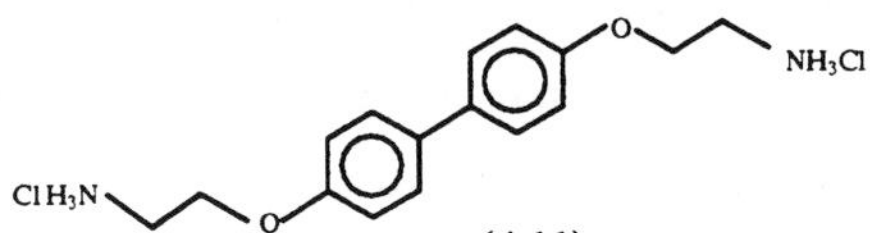

(4.11)

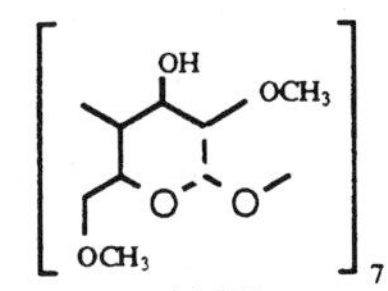

(4.12)

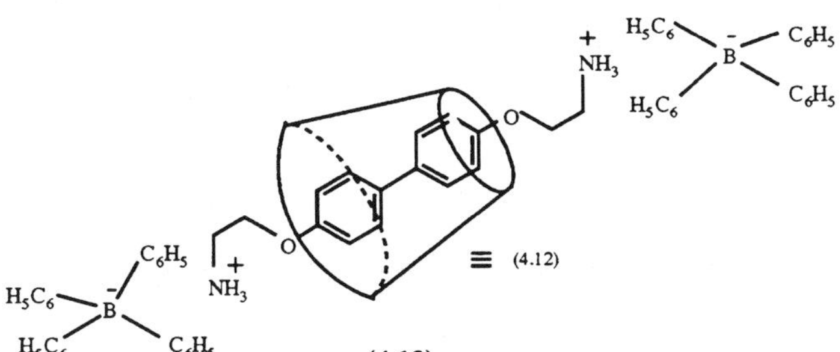

(4.13)

(c) *Biphenyl as the Central Spindle of a Rotaxane*

In contrast to the oligophenylene host molecules described above, a biphenyl group becomes the guest, prisoner perhaps, in the threaded-loop (rotaxane) inclusion compound (4.13). The sparingly soluble biphenyl diammonium salt (4.11) dissolves readily in water in the presence of heptakis (2,6-0-methyl)-β-cyclodextrin (4.12), indicating complex formation, and addition of sodium tetraphenylborate precipitates the ionically bound rotaxane (4.13) in 71% yield. The complex, characterised by NMR including nuclear Overhauser effects (NOE), does not dissociate in acetone solution, unless a trace of triethylamine is present, or on thin layer chromatography (TLC) on cellulose (T. Vankata, S. Rao and D.S. Lawrence, *J. Amer. Chem. Soc.*, 1990, **112**, 3614).

5. Chemical and Physico-chemical Properties of Biphenyls

(a) Substitution

(i) Selective para-chlorination

Chlorination of biphenyl under normal conditions, e.g. with chlorine and aluminium chloride, gives complex mixtures of isomers with varying degrees of chlorination (*i.e.* polychlorobiphenyls, PCBs) but reaction with chlorine in the presence of zeolites, in the melt or in solution, leads to selective *para*-chlorination. In the optimum experiment using lithium zeolite L and dichloromethane solvent at 40°C, 96.7% of the potential polymer intermediate 4,4'-dichlorobiphenyl is formed (see Section 6). Comparable success was achieved with zeolite L exchanged with potassium, sodium, strontium, silver or lead(2) ions, but other metals e.g. zinc and cerium(3) were much less selective (A. Botta, H.-J. Buysch, and L. Puppe, *Angew. Chem. Int. Ed. Engl.*, 1991, **30**, 1689).

Para, but not *ortho* or *meta*-terphenyl gave a similarly high yield of the 4,4"-dichloro-derivative. The hypothesis that reaction occurs within the cavities of the zeolite structure is supported by the observation that trimethylsilylation of the outer surface did not reduce either activity or selectivity.

(ii) Methylation studied by mass-spectrometric and radiolytic methods

Methylation of biphenyl in the gas phase at low pressure by dimethylhalonium ions ($(CH_3)_2F^+$ and $(CH_3)_2Cl^+$) generated from the halomethanes by gamma radiolysis in pyrex bulbs or by chemical ionisation with isobutane in the mass-spectrometer source has been examined. Analysis by gas chromatography (GC), in the former experiments, or by collision induced dissociation (CID) mass-spectrometry shows a preponderance of substitution *ortho* to the biphenyl bond which is attributed to initial formation of a complex between the electrophile and *both* benzene rings (F. Cacace *et al.*, J. Amer. Chem. Soc., 1986, **108**, 7495).

(iii) Dechlorination of polychlorobiphenyls (PCBs)

The detoxification of PCBs remains a subject for study from the distinct viewpoints of disposal of existing bulk stocks, cleaning of contaminated areas, and the mechanisms of natural degradation in the environment (see Section 2). Alkoxyborohydrides efficiently dechlorinate PCBs (S.M.H. Tabaei, C.V. Pittman and K.T. Mead, *J. Org. Chem.* 1992, **57**, 6669). While too expensive for the first application such reagents may be of value in the second.

A study of the photochemical hydrolysis of PCBs may have a bearing on their environmental degradation (J. Orvis, J. Weiss and R.M. Pagni, *J. Org. Chem.* 1991, **57**, 1851).

(b) Side-chain Cyclisation

(i) Formation of Heterocyclic compounds

Dehydration of 2,2'-dihydroxybiphenyls (5.1, X = 0) to dibenzofurans (5.2, X = 0) is catalysed by the solid insoluble perfluorinated sulphonic acid resin Nafion-H, which can be recovered for re-use, using solvents such as refluxing *o*-xylene (G.A. Olah *et al.*, *J. Org. Chem.* 1991, **56**, 3192).

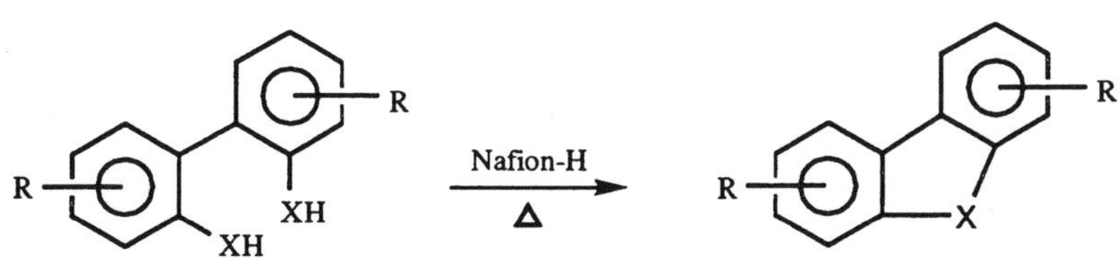

(5.1) **(5.2)**

The corresponding reaction of 2,2'-diaminobiphenyls (5.1, X = NH) to give dibenzopyrroles (carbazoles) (5.2, X = NH) needs higher temperatures but similarly high yields, 90-100% are attainable with alkyl or unsubstituted biphenyls using refluxing 4-*tert.*-butyltoluene or nitrobenzene. Bromo and alkoxy biphenyls gave *ca.* 60% of carbazoles (T. Yamamoto *et al.*, *J. Org. Chem*, 1991, **56**, 6248).

(ii) Cyclisation of alkenylbiphenyls

Direct ultraviolet irradiation of side-chain substituted 2-vinylbiphenyls (5.3) gives dihydrophenanthrene derivatives (5.4), in which the *cis* or *trans*-configuration of the starting alkene is retained, *via* a singlet excited state, but the triplet sensitised reaction is not stereospecific (R.Lapoyade *et al.*, *J. Chem. Soc. Chem. Commun.*, 1980, 664). A laser flash photolytic study of the life-times of the

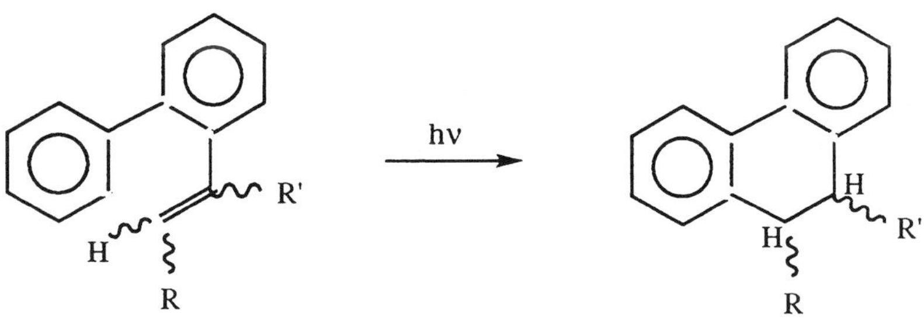

(5.3) **(5.4)**

intermediates in the triplet reaction shows that loss of stereospecificity is due to conformational changes in the excited state before cyclisation. Examples in which the alkene substituents (5.3, R and/or R') are phenyl do not cyclise under triplet sensitised conditions (S. Lazare, R. Lapoyade and R. Bonneau, *J. Amer. Chem. Soc.*, 1985, **107**, 6604).

Following up earlier thermal and photochemical studies of the valence isomerisation of 2,2'-distyrylbiphenyl (5.5), Mullen (K. Mullen *et al.*, *J. Amer. Chem. Soc.*, 1992, **114**, 688) has shown that electrochemical or alkali metal reduction leads either to the *trans*-dibenzyldihydrophenanthrene (5.6) or to the

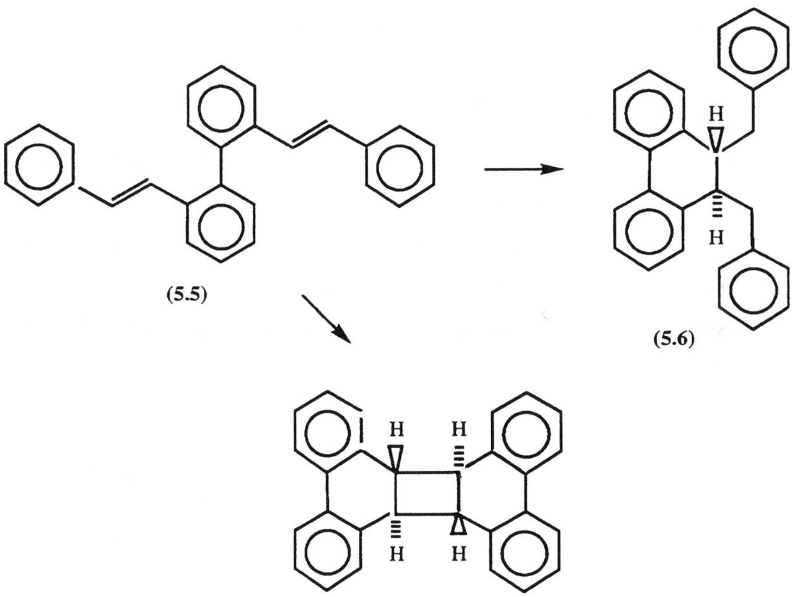

(5.5) (5.6) (5.7)

trans-cyclobutane (5.7) depending upon the conditions under which the di- and tri-anionic intermediates are quenched.

(c) Mechanisms and Electronic Effects

(i) Electronic interaction in biphenyl dinitrenes

Biphenyl-3,4'-, 3,3'- and 4,4'-dinitrenes (5.8, 5.9 and 5.10) were obtained in solid matrices by irradiation of frozen solutions of the corresponding diazides, prepared as in Scheme 5.1, and their electron spin resonance (ESR) spectra at 77K were used to compare exchange coupling effects of unpaired electrons (M. Minato, P.M. Lahti and H. van Willigen, *J. Amer. Chem. Soc.*, 1993, **115**, 4532).

The 3,3'-dinitrene (5.9) showed only the spectrum of a mononitrene, apparently owing to lack of any interaction between the two nitrene centres, in contrast to earlier work on the corresponding bis(phenylcarbene) (K. Itoah, *Pure Appl. Chem.*, 1978, **50**, 1251; see also K. Itoh *et al.*, *Chem. Phys. Lett.*, 1990, **173**, 265), but proof of the presence of the required dinitrene (5.9), although very probable, could not be established.

The 3,4'-dinitrene (5.8) shows a quintet resonance in spite of reduced conjugation due to non-coplanarity of the two rings, and this is considered to be the ground-state multiplicity as with the corresponding bis(phenylcarbene).

The ground-state of biphenyl-4,4'-dinitrene (5.10) is a singlet (no ESR signal) attributed to the quinonoid electron distribution (5.11), but raising the temperature revealed a low-lying triplet state at only 583 ± 120 cal. mole^{-1} higher energy.

(ii) Mechanism of the benzidine rearrangement

Further investigation of the acid catalysed conversion of 1,2-diphenylhydrazine (5.12, hydrazobenzene) into 4,4'-diaminobiphenyl (5.13, benzidine) by the kinetic isotope effect (KIE) using ^{13}C at the 1,1'- and 4,4'-positions, ^{14}C at the 1- and 4-positions, and ^{15}N,^{15}N labelled substrates confirmed earlier conclusions that the fomation of benzidine (5.13) is concerted but that of diphenylene (5.14) is stepwise, as expected for [5,5] and [3,5]-sigmatropic shifts. The proportion of these two products, 85% and 15% respectively, differs from earlier reports (70% and 30%) (W. Subotkowski, L. Kupczyk-Subotkowska, and H.J. Shine, *J. Amer. Chem. Soc.*, 1993, **115**, 5073).

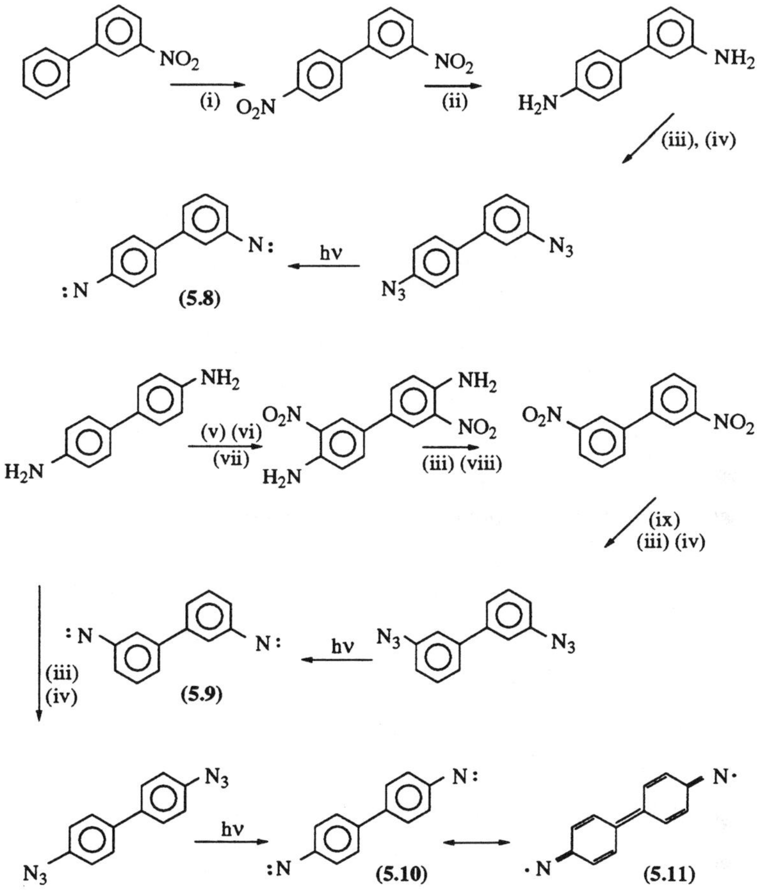

Scheme 5.10

Reagents:

(i)	HNO_3, H_2SO_4;	(ii)	H_2-Pt, 4 Atm;
(iii)	$NaNO_2$, acid;	(iv)	NaN_3, H_2O;
(v)	$(CH_3CO)_2O$, CH_3CO_2H;	(vi)	HNO_3;
(vii)	H_2SO_4, H_2O reflux;	(viii)	CuO, CH_3CO_2H, C_2H_5OH;
(ix)	$SnCl_2.2H_2O$, C_2H_5OH.		

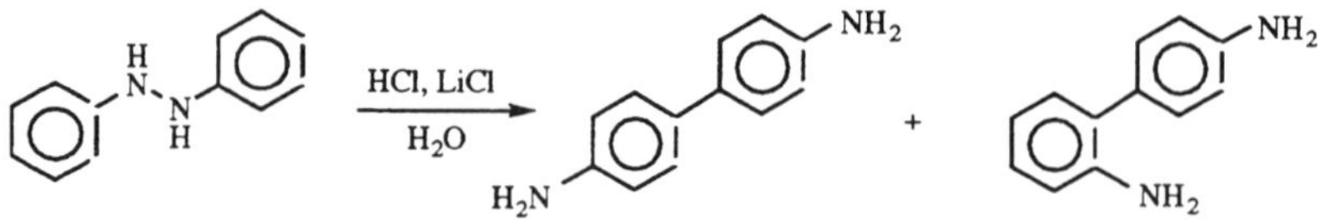

(5.12) (5.13) (5.14)

(iii) Mixed valence metal complexes

A highly symmetric cluster of ten manganese atoms in Mn(2) and Mn(3) oxidation states complexed by bridging 2,2'-dihydroxybiphenyl has the highest spin ground-state measured for a discrete molecular compound (D.A. Goldberg, A. Caneschi and S.J. Lippard, *J. Amer. Chem. Soc.*, 1993, **115**, 9299).

(d) Conformation and Optical Isomerism

(i) Twist angle of biphenyls

A function relating potential energy to the twist angle between the planes of the benzene rings of biphenyl has been derived by electron diffraction measurements in the gas phase and gives minimum energy at an angle of 44.4 ± 1.2° (O. Bastiansen *et al.*, *J. Mol. Struct.* 1985, **128**, 59; see also p. 115 for discussion of fluorinated derivatives).

The twist angle in the solid state of biphenyl derivatives with only hydrogen *ortho* to the biphenyl bond has been studied from the viewpoint of crystal packing effects in the crystal structures of 101 such compounds available from the Cambridge Structural Database. Boltzmann distribution functions for the twist angles were derived using the potential function of Bastiansen *et al.*, (*loc. cit*). Although the energy of the planar conformation is *ca.* 6 kJ. mole^{-1} above the minimum near 44° several examples showing nearly coplanar rings, including crystalline biphenyl itself above 80 K (see previous Supplement, p. 315 *et seq.*), are found to be correct and not, as previously suspected, the result of crystallographic disorder. Increased librational and torsional motion lowers the energy of these examples, by increasing the mean effective instantaneous distance between the *ortho* hydrogen atoms, which are systematically stabilised by crystal packing effects and consequently more common than was previously supposed (C.P. Brock and R.P. Minton, *J. Amer. Chem. Soc.*, 1989, **111**, 4586).

A theoretical study of non-bonded interactions and rotation barriers in biphenyl, its 2,6- and 2,2'-difluoro-derivatives and 2,2', 6,6'-tetrafluorobiphenyl in terms of the topological properties of their electron density distributions presents a more quantitative analysis than that based on van der Waals radii (J. Cioslowski and S.T. Nixon, *J. Amer. Chem. Soc.*, 1992, **114**, 4322).

The twist angle becomes 90° in biphenyls crowded by *ortho*-substituents, e.g. 2,2', 4,4', 6,6'-hexamethylbiphenyl (bimesityl) (R. Frohlich and H. Musso, *Chem. Ber.*, 1985, **118**, 4649).

A strained biphenyl (6.13) with a macrocyclic structure connecting the 4,4'-positions is described in Section 6(a) ii. The X-ray crystal structure of this compound shows that the biphenyl unit is bent so that the lines of the bonds to the 4- and 4'-positions meet at 65°; the twist angle is 31°. This remarkably high distortion is distributed between boat-like deformation of the benzene rings and deflection of the valence angles at the four *exo*-bonded carbon atoms (H.M. Colquhoun *et al.*, *J. Chem. Soc. Chem. Commun.*, 1990, 336).

(ii) Optical isomerism and substituent conformations

The temperature for interconversion of the optical isomers (atropisomers) of the doubly crown ether bridged biphenyls (4.5, n = 1 and 2) is shown to be above 440 K by broadening of the ABCD pattern of the nuclear magnetic resonance (NMR) signal of the methylene protons nearest to the biphenyl unit, which tends towards the AA'BB' type expected when interconversion is rapid. The corresponding activation energy for interconversion exceeds 100 kJ. $mole^{-1}$ leading to the expectation that the compound should be resolvable, and this was achieved by liquid chromatography on microcrystalline triacetyl cellulose eluted with ethanol (G. Lindsten *et al.*, *loc. cit.*, Section 4(a)). The dextro-rotatory isomer, eluted after the leavo, is shown by calculation and its circular dichroism (CD) spectrum to have the S absolute configuration. The observed optical rotations $[\alpha]^{D}_{24}$ (CH_2Cl_2) for the enantiomers of (4.5, n = 1) are **S** + 12.1°, **R** - 11.4° and for (4.5, n = 2) **S** + 9.4°, **R** - 9.0°.

The conformation of biphenyls bridged by a mono- or disubstituted four carbon atom chain (5.15, R,R' = H, Cl, Br, CH_3, CH_2OH, CO_2H or CO_2CH_3) have been studied by NMR and the results compared with those predicted by force-field calculations. Racemisation kinetics at 300 K and 314 K for the unsubstituted compound (5.15, R = R' = H) give $\Delta G^{\ddagger}$ 97 kJ. mole^{-1} for interconversion of the biphenyl twist angle from +60° to -60°. Substituents tend to adopt equatorial conformations with respect to the cyclo-octadiene ring and conformational inversions occur by reversing the biphenyl twist (P. Rashidi-Ranjbar and J. Sandström, *J. Chem. Soc. Perkin II*, 1990, 901).

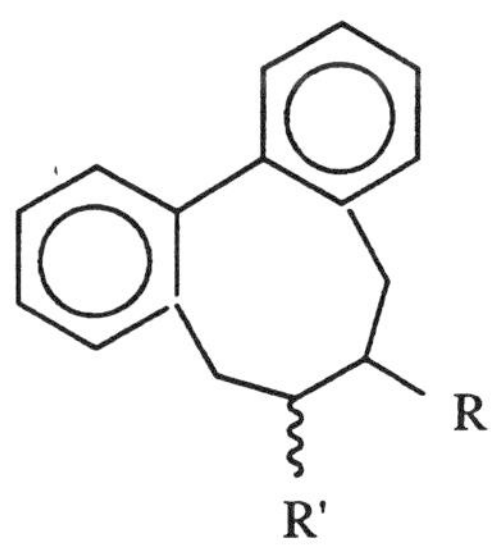

(5.15)

The rotational positions of the chromium tricarbonyl groups in the two diastereomers of its 1:1-complex with 2,2'-difluoro-6,6'-dimethylbiphenyl (5.16, m.p. 170-173°C) and (5.17, m.p. 126-129°C) have been studied by NMR measurement of the carbonyl-carbon to fluorine coupling constants. A carbonyl group predominantly eclipses fluorine rather than methyl in both isomers (P. Szczecinski and K. Wisniewski, *J. Organometal. Chem.* 1992, **423**, C13).

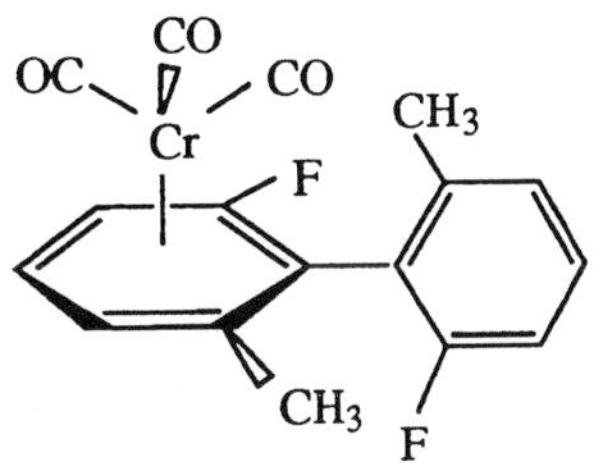

(5.16)

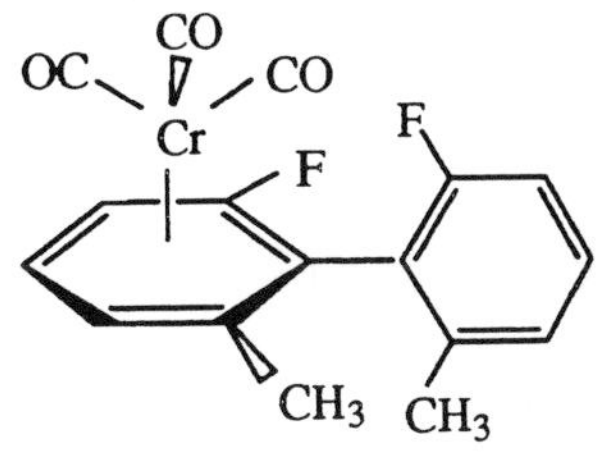

(5.17)

An unusual example with perpendicular benzene ring planes is decakis(dichloromethyl)biphenyl (5.18), prepared by chlorinating decamethylbiphenyl in carbon tetrachloride at reflux temperature using visible light (K. Mislow *et al.*, *J. Amer. Chem. Soc.*, 1988, **110**, 1917). The molecule is bent so that the angle defined by the *para* carbon atoms and the mid-point of the central bond is 165.2° while the latter is stretched to 1.534Å compared with 1.505Å in bimesityl (R. Frohlich *loc. cit.*, Section 5(d) i) and 1.493Å in biphenyl (G.P. Charboneau and Y. Delugeard, *Acta Crystallogr. Sect. B Struct. Crystallogr. Cryst. Chem.*, 1977, **B33**, 1586; for a summary of structural information on biphenyl see G. Häfelinger and C. Regelmann, *J. Comput. Chem.*, 1985, 368). X-ray diffraction by crystals of the 1:1 THF solvate shows that the conformations of the dichloromethyl groups are geared together so that the hydrogen atom of each group lies between the chlorine atoms of the adjacent one (5.19, shown schematically with chlorine omitted for clarity). With this conformation the molecule is chiral, the two enantiomers differing in the clockwise or anti-clockwise cycles of C-H bonds. Resolution was performed by chromatography on a column of cellulose tris(3,5-dimethylphenylcarbonate) eluted with hexane (hexane with 2% propanol gave no separation). Recemisation rates were measured at 20, 25 and 35°C monitoring by high pressure liquid chromatography (HPLC) on the same stationary phase, giving Arrhenius activation parameters E_a 24.9 Kcal. mole^{-1} and A $4.4.10^{13}$ S^{-1}, and $\Delta G^{\ddagger}$ 23.7 kcal.

mole^{-1}. Since this barrier is less than that reported for

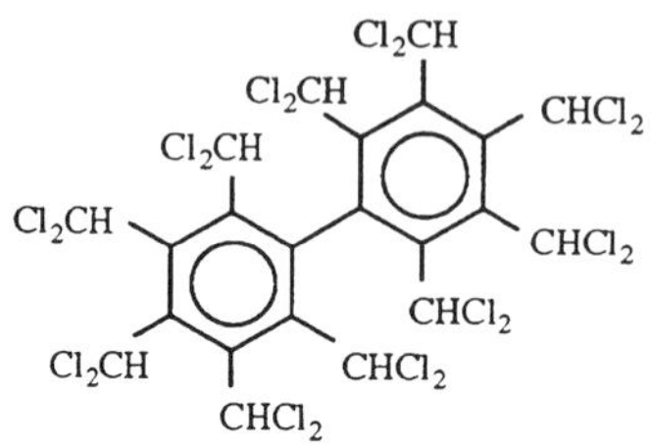

(5.18)

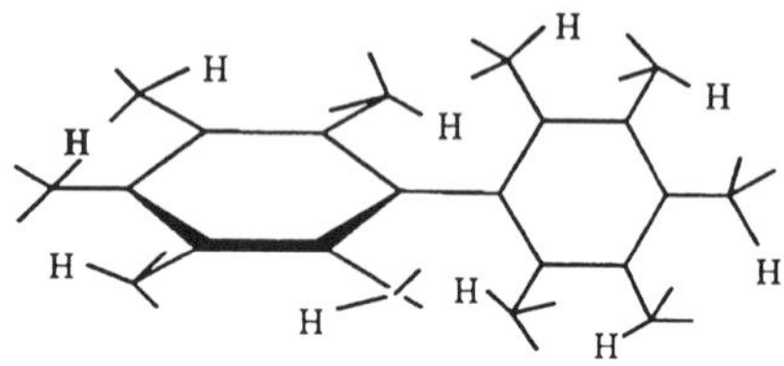

(5.19)

(+) -3,3'-diaminobimesityl, 29.8 kcal. mole^{-1} (W.W. Moyer and R. Adam, *J. Amer. Chem. Soc.*, 1929, **51**, 630), which has less bulky *ortho*-substituents and can only racemise by reversing the biphenyl twist, it is concluded that racemisation occurs by geared rotation of five $CHCl_2$ groups on one ring (ten such rotations regenerate the same enantiomer) and not rotation about the biphenyl bond.

(e) Liquid Crystalline and Related Ordered Phases of Biphenyls

(i) Liquid crystal effects

Polymer films for non-linear optical (NLO) applications, particularly second order harmonic generation (SHG) require initial 'poling', *i.e.* exposure to a strong electric field, in order to align the dipolar NLO active groups. The spontaneous ordering of molecules in ferro-electric liquid crystals has been shown to give the necessary alignment without poling by SHG measurements on liquid crystalline films of the alkoxybiphenylcarboxylate ester (5.20) of the NLO active *o*-nitroalkoxyphenol unit (D.M. Walba, *et al.*, *J. Amer. Chem. Soc.*, 1991, **113**, 5471).

A helically ordered phase has been observed in ferro-electric liquid crystals of biphenyl propiolate esters (e.g. 5.21), also based on

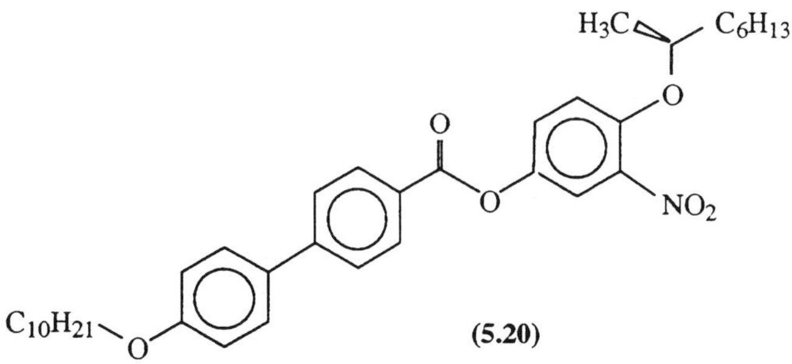

(5.20)

4-hydroxybiphenyl-4'-carboxylic acid (J.W. Goodby *et al.*, *J. Amer. Chem. Soc.*, 1989, **111**, 8119).

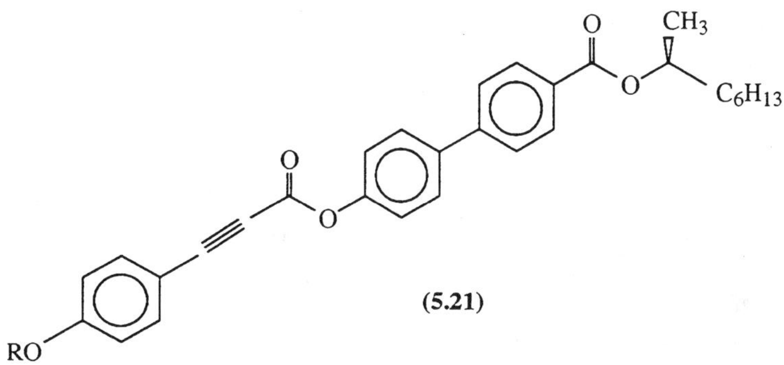

(5.21)

(ii) *Monolayers on metallic surfaces*

A study of self-assembling monolayers of 4-alkoxybiphenyl-4'-carboxylic acid on silver and copper surfaces, in which the carboxylic acid group bonds to the surface with the biphenyl unit at right angles to it, shows that the wettability and contact angle for water can be correlated with the tilt angle adopted by the alkoxy chains (Y.T. Tao, M.-T. Lee and S.-C. Cheng, *J. Amer. Chem. Soc.*, 1993, **115**, 9547).

(f) *Polyphenylenes*

(i) *Poly-para-phenylene (PpP)*

This material has attracted attention for half a century or more, firstly because it is potentially conjugated throughout the length of the chains, if adjacent rings can adopt an at least approximately coplanar conformation, leading to possible electrical conductivity and related properties; and secondly as a high temperature material.

A variety of preparations of unsubstituted PpP have been studied, Table 6.1, but comparison of the products is complicated by the insolubility of uniformly *para*-linked samples of more than even about 7 benzene rings, precluding $\overline{M}$ measurements except by physical methods applicable to solid samples such as end-group analysis by elemental composition, IR, or solid state (SS) NMR, and reflectance UV spectroscopy. It is inferred that soluble samples either have low $\overline{M}$ or contain *ortho, meta* or branching linkages. Most methods give material of relatively low DP; few exceed 20 benzene units and the majority reach only about 10, usually because of precipitation from the reaction medium.

Methods ref. 6, 7, 9 and 12 (Table 6.1) can be adapted to give thin supported films suitable for various physical measurements. The prepolymer (ref. 11) was obtained with some difficulty by Ziegler-Natta polymerisation of cyclohexa-1,3-diene. The acyloxy polymer (ref. 12) is formed by radical polymerisation of the corresponding cyclo-diene, available by esterifying the parent diol formed microbiologically from benzene by a genetically modified strain of *Pseudomonas putida* in which the enzyme responsible for the next step of oxidative degradation has been eliminated. The material formed by this promising route has high $\overline{M}$ but is soluble and therefore presumably of imperfect *para*-linked structure.

Table 6.1: Preparation of Unsubstituted PpP

Monomer or Prepolymer	Reagent	Reference
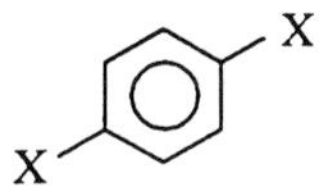 X = Cl or Br	Na or K	1
	Mg (Grignard), $NiCl_2$, bipy.	2
	Mg (Grignard), *cis* - $ClCH_2.CH{=}CH.CH_2Cl$	3
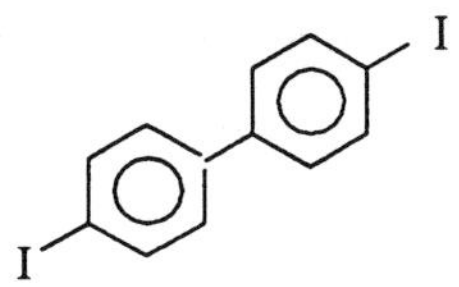	Cu	4
	$AlCl_3$, $CuCl_2$, O_2	5, 6
	HF, SbF_5 anodic oxidation	7
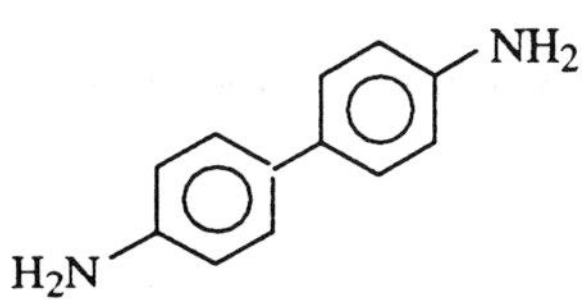	HCl, $NaNO_2$; CuCl	8
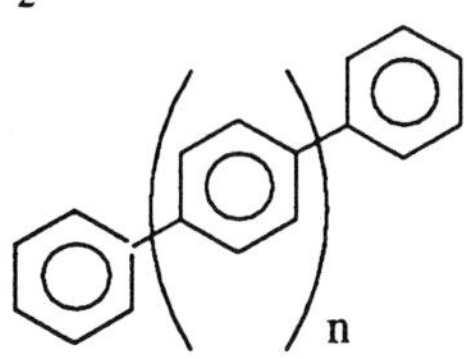 n = 0-4	AsF_5	9
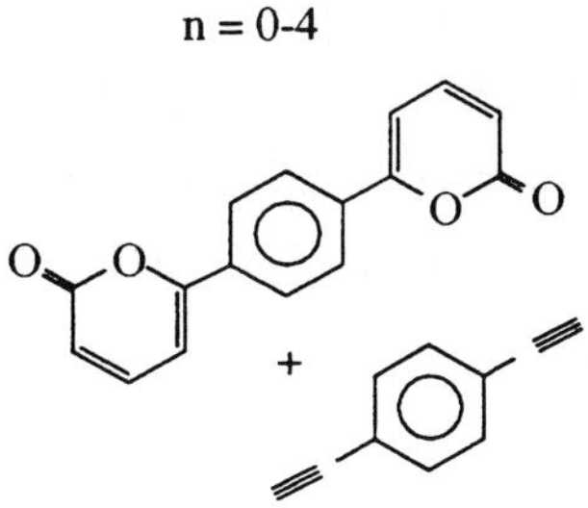	heat	10

Table 6.1 (continued)

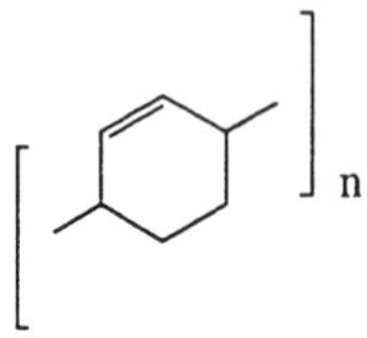	Chloranil	11
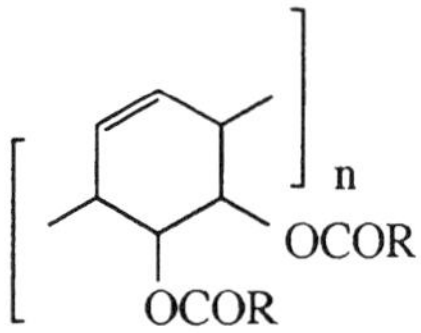	Heat	12

References: **1.** G. Edwards and G. Goldfinger, *J. Polym. Sci.* 1955, **16**, 589; **2.** T. Yamamoto, H. Hayashi and A. Yamamoto, *Bull. Chem. Soc. Japan*, 1978, **51**, 2091; **3.** S.K. Taylor *et al.*, *J. Polym. Sci. Polym. Lett. Edn.*, 1981, **19**, 85; **4.** G.K. Noren and J.K. Stille, *Macromol. Rev.* 1971, **5**, 385; **5.** P. Kovacic and A. Kyriakis, *J. Amer. Chem. Soc.*, 1963, **85**, 454, J.G. Speight, P. Kovacic and F.W. Koch, *J. Macromol. Sci. Rev. Macromol. Chem.*, 1971, **C5**, 295; **6.** H. Naarmann, *Angew. Makromol. Chem.*, 1982, **109/110**, 295; **7.** G. Brilmeyer and R. Jasinski, *J. Electrochem. Soc.*, 1982, **129**, 1950; **8.** A.A. Berlin, V.I. Liogonkin and V.P. Parnini, *J. Polym. Sci.*, 1961, **55**, 675; **9.** L.W. Shacklette *et al.*, in *Conductive Polymers*, R.B. Seymour, Ed., Plenum, New York, 1981, p. 115; **10.** H.F. Vankerckhoven, Y.H. Gilliams and J.K. Stille, *Macromolecules*, 1972, **5**, 541; **11.** C.S. Marvel and G.E. Hartzell, *J. Amer. Chem. Soc.*, 1959, **81**, 448, P.E. Cassidy, C.S. Marvel and S. Ray, *J. Polym. Sci.*, 1965, **A3**, 1553; **12.** G.D.H. Ballard *et al.*, *J. Chem. Soc. Chem. Commun.*, 1983, 954; Idem. *Macromolecules*, 1988, **21**, 294.

The properties of PpP depend on the method of preparation, *i.e.* on $\overline{M}$, inclusion of non-*para* links or chain branching and on the incidental, fortuitous, or intentional, presence of minor impurities and electronic imperfections including radical and ionic centres. Samples prepared *e.g.* by metal promoted dehalogenation methods, or otherwise after exposure to air, are light brown insoluble, infusible, crystalline solids which are electrical insulators, S = *ca.* 10^{-12} cm^{-1}, although PpP is properly classed as a semi-conductor since it has a vacant delocalised conductance electronic band separated energetically (by the *band-gap*, see below) from the filled π-electron levels.

Like other organic polymeric conductors (*e.g.* polyacetylene, polypyrrole, and polythiophen), PpP becomes conducting when 'doped' with electron accepting (e.g. AsF_5, $AlCl_3$) or donating (e.g. alkali metals) substances whose purpose is to form positive or negative charges in the chains, which are the charge carriers. Some methods of polymerisation give 'doped' air sensitive samples in which unpaired electrons can be detected by ESR. For example p-terphenyl can be deposited as an epitaxially grown crystalline film and polymerised, with simultaneous doping, by AsF_5 into a chain oriented film of conductivity 50 S cm^{-1} (ref. 9, Table 6.1). The $\overline{DP}$ of this material, composition $C_6H_{4 \cdot 42}(AsF_4)_{0.2}$, estimated by comparing the intensities of the IR bands at 800 cm^{-1} (C_6H_4 out-of-plane H-bend) and 765 cm^{-1} (C_6H_5 end-group o.o.p. bend), after evaporation of residual monomer, was 9 (*i.e.* 27 benzene rings); sexiphenylene similarly gave twice this length.

Shacklette *et al.*, (ref. 9) also prepared air sensitive potassium (0.5 mol.) doped pellets of sexiphenylene with golden metallic appearance and conductivity 0.5 S cm^{-1}, or 7.4 S cm^{-1} for the 16-unit oligomer, which are not further polymerised.

Films deposited by oxidation of benzene with $AlCl_3$-$CuCl_2$ (Kovacic method) *e.g.* under high shear conditions on a rotating (2,500 r.p.m.) disc are red, λmax 500 nm, and conducting until exposed to moisture (B. Tieke,

G. Bubeck and G. Lieser, *Makromol. Chem. Rap. Commun.* 1982, **3**, 261). AsF_5 doping of the best samples of Kovacic polymer gives conductivity 500 S cm^{-1} compared with 1200 S cm^{-1} for polyacetylene. Such measurements on compressed pellets depend on the degree of compaction and dopant penetration, however. For comparison of the properties of Kovacic PpP with that made in other ways see P. Kovacic and M.B. Jones (*Chem. Rev.* 1987, **87**, 357).

Although polymerisation by anodic oxidation is generally less successful for benzene than for pyrrole or thiophen, owing, probably, to its higher oxidation potential, films deposited in this way on gold or glassy carbon electrodes from HF or HF-AsF_5 electrolytes are electroactive, *i.e.* their conductivity is "switched on" only at a certain voltage (*ca.* 0.5 V) needed to form the cationic conducting state. The rate of switching, but not the final conductivity, depends on the rate of diffusion of the counter anion into the film (I. Rubinstein, *J. Electrochem. Soc.*, 1983, **130**, 1506). PpP prepared by anodic oxidation has $\overline{DP}$ less than 10, with some cross linking, and conductivity 10^{-2}-10^{-5} S cm^{-1}, although a value of 10^4 S cm^{-1} has been claimed.

A lithium battery with good power-to-weight and recharging performance has been constructed using electrochemical films of PpP for both electrodes (L.M. Goldenberg and P.C. Lacaze, *Synthetic Metals*, 1993, **58**, 271, and references therein).

Efficient emission of blue light has been demonstrated when a voltage is applied to a thin film of PpP between aluminium and indium-tin oxide layers on a glass support (G. Leising, *et al.*, Adv. Mater., 1992, **1**, 6). The film was formed by depositing the soluble pre-polymer, (ref. 12, Table 6.1) and aromatising it at 340°C *in vac.* The unusually short wave-length of the emitted light corresponds with the electronic band-gap, *ca.* 2.7 eV, of PpP (G. Leising, K. Pichler and F. Stelzer in *Electronic Properties of Conjugated Polymers III*, Eds. H. Kuzmany, M. Mehring and S. Roth, Springer, Heidelburg, 1989, p. 100). The air-stability and higher efficiency of PpP make it a more promising material for blue light emitting diodes (LEDs) than inorganic semiconductors like those commonly used for longer wavelengths.

Other applications of PpP described in patents include articles, moulded under pressure and sintered at 500-600°C, for use up to 400°C, composite materials, and lubricants (see Kovacic and Jones, *loc. cit.*).

A range of substituted examples of PpP have been prepard, but as any substituent is *ortho* to an adjacent ring, coplanarity and consequently conjugation, are reduced so that these are expected, and observed, to show lower electrical conductivity. Alternative interest in these systems stems from their unusual combination of rigid rod topology with solubility engendered by the substituents.

Pre-formed PpP has been substituted by reactions such as nitration, and Friedal-Crafts alkylation gives soluble derivatives of presumably random structure (M.B. Jones, P. Kovacic and D. Lanska, *J. Polym. Sci. Polym. Chem. Edn.*, 1981, **19**, 89). Schluter, Feast and coworkers obtained poly-2,5-dihexylphenylene (6.31, Scheme 6.10) from the dibromo-compound (6.29) *via* both the mono-Grignard reagent (6.30) with nickel(0) catalysis (Yamamoto method, Section 3(b) iv), and the bromophenyl boronic acid (6.33) with palladium(0) and aqueous alkali carbonate (Section 3(c) iii) (W.J. Feast, *et al.*, *Polymer*, 1989, 30, 1054 and 1060). $\overline{\mathrm{DP}}$ up to *ca.* 28 was achieved and the products showed the expected solubility, *e.g.* in chloroform or toluene.

This boronic acid approach has been adapted to incorporate acyl with alkyl side-chains by condensing the dibromide (6.34, Scheme 6.11) with the diboronic acid (6.35); reduction of the ketone functions of the resulting polymer (6.36) to secondary alcohol (6.37) and dehydration under Friedel-Crafts conditions gives regiospecific polycyclisation to the unusual long chain alkyl substituted ladder polymer (6.38) (V. Scheerf and K. Mullen, *Makromol. Chem. Rapid Commun.*, 1991, **12**, 484).

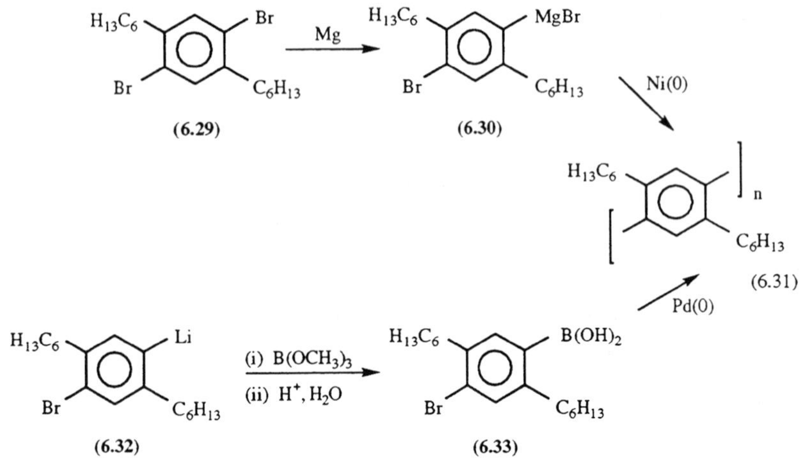

Scheme 6.10

PpPs substituted by phenyl or 4-substituted phenyl groups (6.39) have been prepared by nickel(0) catalysed polymerisation of the corresponding dichloroterphenyls under Colon-Kelsey conditions (Section 3(b) i) (J.K. Kallitsis and H. Naarmann, *Synthetic Metals*, 1991, **44**, 247). With $\overline{DP}$ in the range 20-80 the polymer (6.39, R = H) had low conductivity, 10^{-5} S cm^{-1} doped with iron(3) chloride, or 10^{-9} S cm^{-1} with iodine.

Phenyl-substituted PpP (6.41, Scheme 6.12) is also available by Diels-Alder polymerisation of the dicyclopentadienone (6.40) with p-diethynylbenzene (ref. 10, Table 6.1), although regiospecific addition of the alkyne to give only *para*-linkages is surprising in view of the relatively symmetric substitution of the diene.

Nickel(0) catalysed coupling of p-dihalobenzenes, and of the corresponding triflates (Section 3(b) v), monosubstituted by phenyl, *tert*-butyl, or carbomethoxy groups, to give polymers, $\overline{DP}$ 26-47, in which the substituents are presumably in random relative positions along the chains has been briefly described (C. Pugh *et al.*, *Makromol. Chem., Macromol. Symp.*, 1992, 54/55, 113).

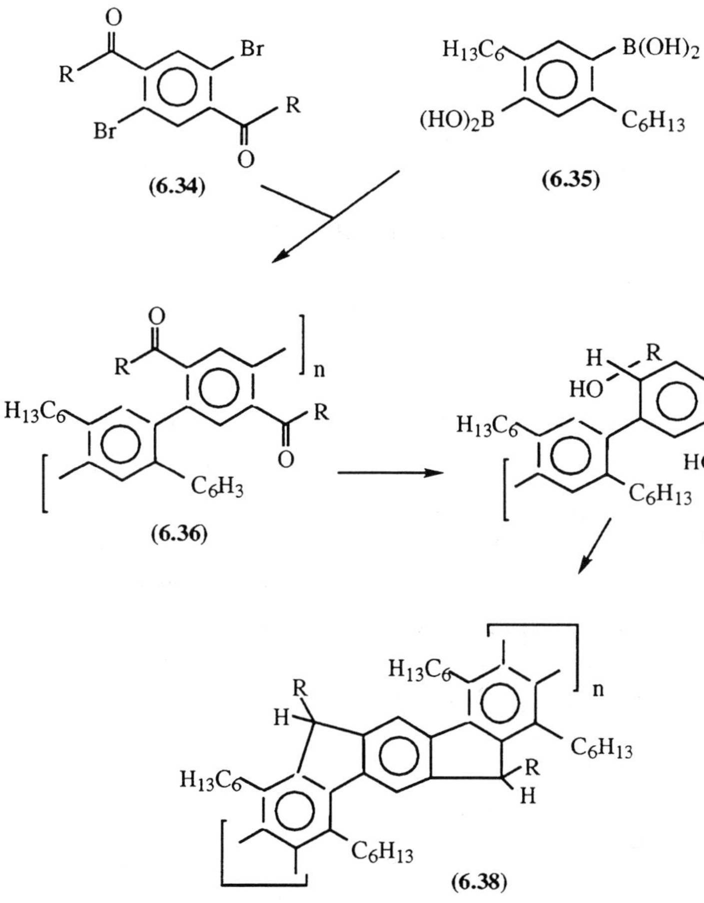

Scheme 6.11, R = p-$C_6H_4.C_{10}H_{21}$

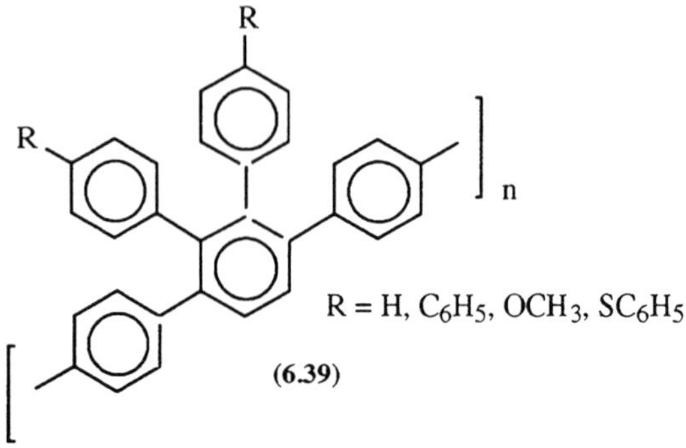

(6.39)

Soluble linear polymers (Scheme 6.13), whose rigidity is increased by steric crowding by fused terphenyl branches, having the polyarylene chain connected through nitrogen heterocyclic rings have been formed from the corresponding poly-diketone (6.42), available from the dichloromonomer using

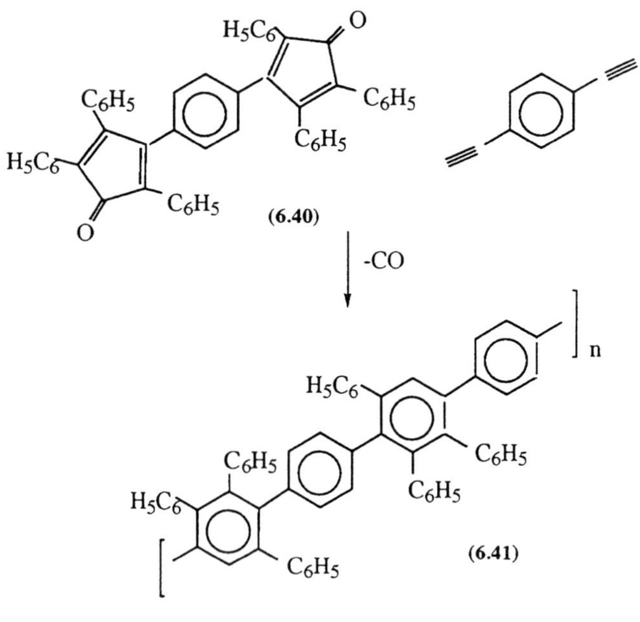

(6.40)

(6.41)

Scheme 6.12

nickel(0) (Colon-Kelsey), by condensation with nitrogen nucleophiles (R. Singh and A.S. Hay, *Makromol. Chem., Macromol. Symp.*, 1992, **54/55**, 357). With $\overline{DP}$ about 45, the polyketone (6.42) could be cast from dichloromethane into brittel films, Tg 283°C, while both the heterocyclic polymers (6.43) gave tough films, Tg near 330°C, which showed good thermal and oxidative stability, weight loss in air 5% at 520°C, and no crystallinity.

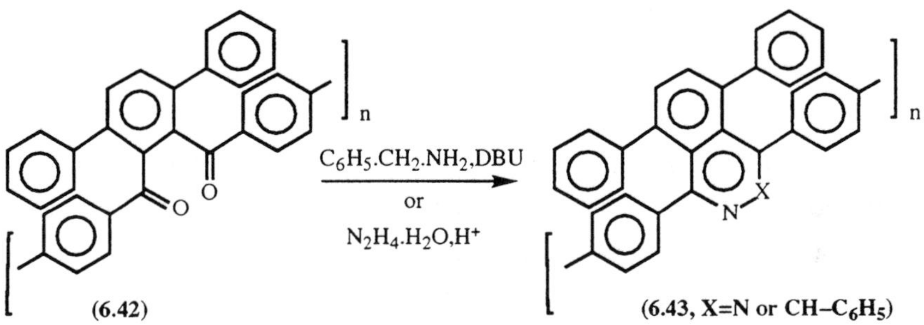

Scheme 6.13

DBU = 1,8-diazabicyclo[5.4.0]undec-7-ene

Electrochemical synthesis has been used to give PpP with pendant ether and polyether side chains (Y. Gache and J. Simonet, *J. Chim. Phys., Phys-Chim. Biol.*, 1992, **89**, 1027).

(ii) Polymetaphenylene (PmP)

This material is formed by dehalogenation of m-dihalobenzenes e.g. with a palladium catalyst (M. Busch, *loc. cit.*, Section 3 (c) v), or with nickel(0) by the Yamamoto method (T. Yamamoto *et al., loc. cit.,*), or under Colon-Kelsey conditions (Section 3(b) i) (M. Veda, Y. Seino and J. Sugiyama, *Polym. J.* 1993, **25**, 1319), but has been relatively little studied since the rings are not conjugated and electrical conductivity is not expected. In practice samples doped with

AsF_5 conduct relatively well, e.g. 10^{-3} S cm^{-1}, but this is attributed to cross-linking at *para*-positions by the dopant, an explanation supported by IR changes and insolubility of the doped material (L.W. Shacklette *et al.*, *J. Chem. Phys.* 1980, **75**, 1919; see also *Chem. Rev.* 1982, **82**, 209).

Meta-quinquephenylene has a U-shaped conformation in the crystal, leading to the suggestion that higher oligomers might form helices (P.W. Rabideau, *et al.*, *J. Chem. Soc., Chem. Commun.*, 1993, 1759). In a skillful study Colquhoun and coworkers isolated small quantities of m-deciphenyl from the mixture of oligomers formed by condensing m-dichlorobenzene with chlorobenzene ($\frac{1}{6}$ mol., Colon-Kelsey conditions), using *analytical* scale HPLC, and confirmed its helical structure by X-ray diffraction from a small crystal (D.J. Williams, H.M. Colquhoun and C.A. O'Mahoney, *J. Chem. Soc., Chem. Commun.*, 1994, 1643).

The helix has five benzene rings in each turn, the pitch is 11.22 Å, and the twist-angles between adjacent rings repeat in the sequence 53°, 34°, 44°, 34°.... The inter-ring bond lengths vary slightly in response to the twist angles, 1.49 Å (53°), 1.50 (34°) and 1.47 (44°), so that the shortest bond corresonds with the angle of minimum potential energy in biphenyl (Section 5 (d) i). Adjacent helices interleave in a tetragonal array. Remarkably, the helices continue from one molecule to the next without apparent interruption at the position of the two hydrogen end-groups and the crystal is disordered with respect to the chain end positions. Thus higher oligomers and PmP are expected to have the same crystal structure, and this is supported by the observations that the eleven membered oligomer crystallises in the same space-group with identical unit cell dimensions, and the powder diffraction pattern of crystalline PmP matches that calculated for this helix.

7. Biphenylene

(a) Molecular and electronic structure and spectra

Biphenylene (7.1), which forms very pale yellow relatively volatile crystals m.p. 110°C, is a special case in the arylbenzene group because distortion of the normally radial exocyclic bond angles of the two component benzene rings and destabilising interaction between their π-electron systems give it unique properties.

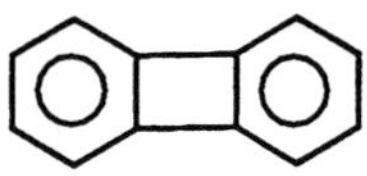

(7.1)

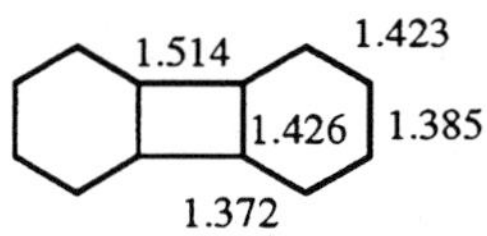

(7.2)

bond lengths in biphenylene, Å

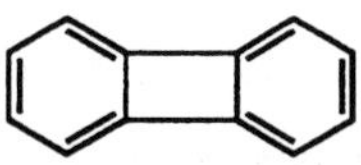

(7.3)

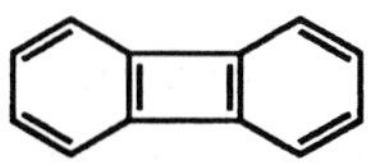

(7.4)

X-ray crystallography shows that the molecule is planar, with bond lengths shown in (7.2), and the four-membered ring is accurately rectangular (J.K. Fawcett and J. Trotter, *Acta. Crystallogr.*, 1966, **20**, 87). The inter-ring bond lengths 1.514 Å, or 1.524 Å in the vapour phase by electron diffraction (S.H. Bauer, C.F. Wilcox and A. Yokozeko, *J. Amer. Chem. Soc.*, 1974, **96**, 1026) are longer than the single bond between the benzene rings of biphenyl or bimesityl (Section 5 (d) ii). The molecule is symmetric but the benzene ring bonds show distinct alternation in length (see 7.2) so that the electron distribution represented by (7.3) is more important than those (e.g. 7.4) with alternative double bond positions and cyclobutadiene character in the central ring. This implies reduced delocalisation and aromatic stabilisation energy (resonance energy) for the benzene rings.

The observed heat of combustion of biphenylene leads to an overall stabilisation energy of 71.5 kJ mole^{-1} (*i.e.* compared to a hypothetical unstrained, localised poly-ene of the same structure); the corresponding figure for biphenyl, assumed unstrained and with unimpaired benzene resonance stabilisation, is 340 kJ mole^{-1} so that the strain energy of the σ-bond frame and reduced π-electron stabilisation together amount to 268.5 kJ mole^{-1} (H.D. Springall *et al.*, *J. Chem. Soc.*, 1962, 3895). There have been many estimates of the size and balance of the stabilising and destabilising effects in biphenylene *e.g.* strain energy 305 kJ mole^{-1} and resonance stabilisation 326 kJ mole^{-1} by a force-field computation which gives 444 kJ mole^{-1} for the standard enthalpy of formation (F. Vogtle, *Fascinating Molecules in Organic Chemistry*, John Wiley, Chichester 1992, p. 115, and references therein) compared with the experimental value 436.7 kJ mole^{-1} (J. Kao and N.L. Allinger, *J. Amer. Chem. Soc.*, 1977, 99, 975, and references therein). The compound behaves as a rather more than usually reactive aromatic hydrocarbon and evidently has considerable stability since it can be prepared using very high temperatures without specialised fast-cooling techniques (see Section 7 (b)); it is even found in the residues from detonation of high explosives (J.H. Johnson *et al.*, *J. Hazard Mater.* 1988, **18**, 161). The aromatic character is emphasised, if not exaggerated, by the calculated bond localisation energies, *i.e.* the reduction in π-electron stabilisation when one formal double bond is removed, as by an addition reaction, which are little different from those of benzene or naphthalene (D. Amic and M. Trinajstic, *J. Chem. Soc., Perkin* II, 1990, 1595).

The proton NMR spectrum of biphenylene is consistent with reduced diamagnetic ring currents in the benzene rings since the resonances are at relatively high field, 6.62 ppm (α-protons) and 6.73 ppm (β-protons), compared with benzene 7.27 ppm. These shifts, and in particular the higher field resonance of the α-protons, are rationalised on the basis that the four-membered ring sustains a paramagnetic ring-current characteristic of anti-aromatic (4n π-electron) cycles (H.P. Figeys *et al., Tetrahedron*, 1976, **32**, 2571). The ^{13}C

resonances 117.3 ppm (α-carbon), 128.1 (β-carbon) and 151.3 (bridgehead atoms) (A.J. Jones and D.M. Grant, *J. Chem. Soc. Chem. Commun.*, 1968, 1670) support this interpretation: ring-currents have little effect on the nuclei of *their own rings*, because their magnetic fields change sign between the inside and outside of the ring, so that the β-carbon resonance is close to that of benzene (127.5 ppm); the α-carbons, like the α-protons, experience an upfield shift from the paramagnetic current in the central ring and each pair of ring-junction nuclei are in the deshielding region of the other benzene ring. The large downfield shift of these atoms is, however, probably due in larger part to the effect of bond-angle distortion (see discussion of rehybridisation, Section 7 (c) i) which has been shown to have greater effect on ^{13}C resonances than that of ring-currents (H. Günther *et al.*, *Angew. Chem. Int. Edn. Engl.*, 1973, **12**, 243).

When more than one biphenylene system shares a common benzene ring, the angular isomer, (7.5) in which bond alternation in the middle ring can suit both, is expected to be more stable than the linear isomer (7.6) with more cyclobutadiene character, although calculations give conflicting results in the absence of experimental thermochemical data (J.W. Barton and R.B. Walker, *Tetrahedron Letters.*, 1978, 1005; K.P.C. Vollhardt *et al.*, *J. Amer. Chem. Soc.*, 1985, **107**, 5670). The tris-fused compound (7.7) is described as a derivative of cyclohexatriene (R. Dieraks and K.P.C. Vollhardt, *J. Amer. Chem. Soc.*, 1986, **108**, 3150).

The electronic spectrum of biphenylene shows a strong band near 250 nm and a weaker one near 350 nm tailing into the visible (R.M. Hochstrasser, *Can. J. Chem.* 1961, **39**, 765). The pattern is similar to that of, say, anthracene, but readily distinguished from it by the exceptional sharpness of the bands so that the short-wave doublet and long-wave multiplet structures are readily resolved. The general pattern and characteristic sharpness are retained in the spectra of most derivatives and some, but not all, of the nitrogen heterocyclic analogues of biphenylene (J.A.H. MacBride and S. Kanoktanaporn, *Tetrahedron Lett.*, 1977, 1817).

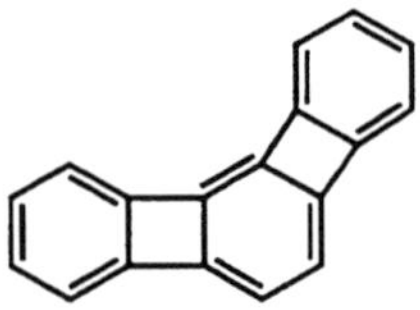

(7.5)

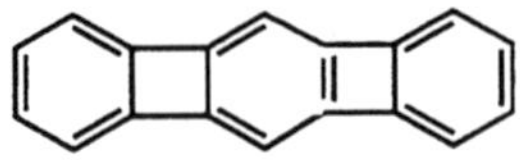

(7.6)

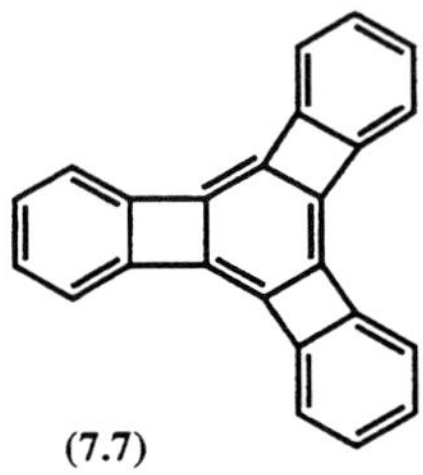

(7.7)

The infrared (IR) spectrum of biphenylene is not exceptional, but the CH out-of-plane bending vibration expected at *ca.* 740 cm^{-1} for an *ortho*-disubstituted benzene appears as a doublet at 750, 734 cm^{-1}; splitting of this band is apparently a characteristic of strained *ortho*-phenylenes, including indane and benzocyclobutene (R.F. Curtis and G. Viswanath, *J. Chem. Soc.*, 1959, 1670 and references therein; see also C. Pecile and B. Lunelli, *J. Chem. Phys.*, 1968, 1336).

(b) *Synthesis of Biphenylene*

(i) *Ring closure by the Ullmann reaction*

After a number of seemingly mistaken claims the authentic compound was first prepared by W.C. Lothrop (*J. Amer. Chem. Soc.*, 1941, 63, 1187) by cyclisation of 2,2'-di-iodobiphenyl (7.8, Scheme 7.1) by the Ullmann reaction (Section 3 (a) i) in which copper(1) oxide curiously gives better results than metallic copper. The synthesis has been repeated many times but yields vary unpredictably and seldom exceed 40% (see *e.g.* Springall *et al.*, *loc. cit.*).

Intermediate formation of the iodonium iodide (7.9), which can be used as starting material (e.g. J.F.W. McOmie *et al.*, *J. Chem. Soc. (C)*, 1966, 1767), does not affect the yield. This method has been used for several methyl and phenyl substituted examples.

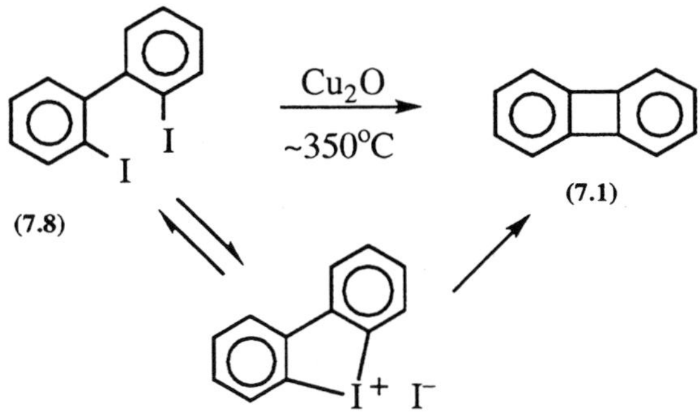

(7.8) (7.1) (7.9)

Scheme 7.1

Various *ortho*-dihalobenzenes give biphenylenes including the 1,5-dinitro-derivative, by two-step Ullmann coupling (see *e.g.* previous Supplement, p. 303).

(ii) *Benzyne dimerisation*

Biphenylene is readily formed by dimerisation of benzyne (7.10) provided electrophilic or nucleophilic species which intercept this intermediate are absent. These conditions are met (Scheme 7.2) by decomposition of benzene diazonium carboxylate, (7.10) which gives biphenylene in 30% yield from anthranilic acid on a five gram scale, but care (and nerve) is needed since this intermediate is very sensitively explosive if allowed to dry (F.M. Logullo, A.M. Sietz and L. Friedman, *Org. Synth.*, 1968, **48**, 12).

Various other benzyne precursors can be used (Scheme 7.2). The pyrolytic method from phthalic anhydride, and other similar compounds, (R.F.C. Brown, *loc. cit.*) minimises decomposition of the product by using low pressure so as to give short contact times in the hot zone.

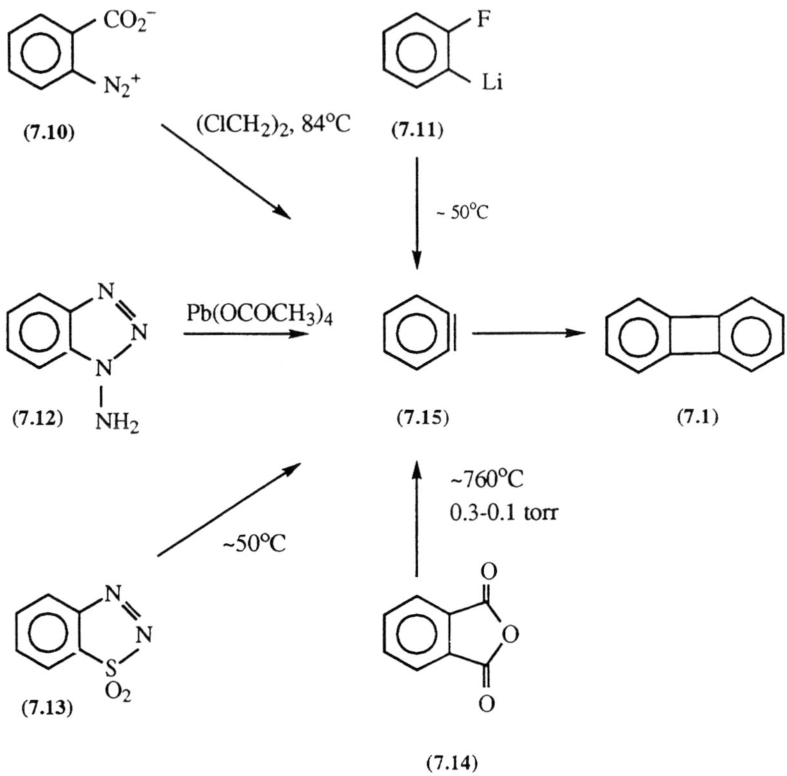

Scheme 7.2

References: (precursor 7.10) see text; (7.11) G. Wittig and L. Pohmer, Chem. Ber., 1956, **89**, 1334; (7.12) C.D. Campbell and C.W. Rees, *J. Chem. Soc.* (C), 1969, 742; (7.13) G. Wittig and R.W. Hoffmann, *Chem. Ber.*, 1962, **95**, 2718; (7.14) R.F.C. Brown *et al.*, Aust. *J. Chem.*, 1967, **20**, 139.

The range of derivatives preparable by this method is effectively limited to symmetrically tetra- or octa-substituted examples since dimerisation of unsymmetrical benzyne intermediates is not in general regiospecific so that a 4-substituted benzyne, for example, gives a mixture of 2,6- and 2,7-disubstituted biphenylenes whose separation is seldom practicable (contrast these observations with the mechanistic proposals of Section 3(f)). Polyhalo-, including octafluoro-biphenylenes have been made by anhydride pyrolysis (J.F.W. McOmie *et al.*, *J. Chem. Soc.*, 1969, 1994).

(iii) Pyrolysis of benzo[c]cinnolines

At 920°C dinitrogen can be extruded from benzocinnoline (7.16) to give biphenylene and at pressures of 10^{-2}–10^{-1} torr, to control the contact time (probably 20-50 msec) yields over 60% can be obtained using simple apparatus in which the product condenses in an extension to the furnace tube at room temperature (J.A.H. MacBride, *J. Chem. Soc. Chem. Commun.*, 1972, 1219; S. Kanoktanaporn and J.A.H. MacBride, *J. Chem. Research*: (S), 1980, 203; (M) 1980, 2901-2910). Application of the method to nitrogen heterocyclic analogues shows that cleavage to an aryne intermediate does not occur, since only one product is formed from unsymmetrical precursors; the nature of the, possibly biradical, intermediate has been considered (J.A.H. MacBride *et al.*, *J. Chem. Research*: (S) 1986, 236; (M) 1986, 2049-2063).

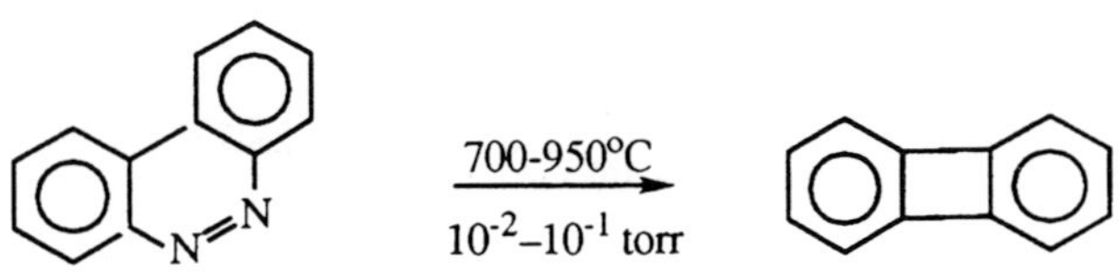

(7.16) (7.1)

Scheme 7.3

Application of this method to substituted derivatives is limited by the high temperature but stable substituents at the 4,5-positions (phenanthrene numbering) which strain the diazine ring, as in the octachloro-derivative (J.A.H. MacBride *et al.*, *J. Chem. Soc., Chem. Commun.*, 1983, 425), lower the pyrolysis temperature to near 700°C; thus 1,8-dimethylbiphenylene is formed in 53% yield, on the ten gram scale, using simple equipment but side-reactions reduce the yield of the diethyl homologue to 11% (C.F. Wilcox, K.R. Lassila, and S. Kang, *J. Org. Chem.*, 1988, **53**, 4333).

The fused biphenylenes (7.5) and (7.6) have been prepared by the nitrogen extrusion method (Barton and Walker, *loc. cit.*; J.W. Barton and D.J. Rowe, *Tetrahedron Lett.* 1983, **24**, 299) and, in higher yield, by the method of Vollhardt (*loc. cit.*) described below.

(iv) Addition of alkynes to o-arylenediynes catalysed by cyclopentadienylcobalt dicarbonyl

Following up an earlier observation (M.D. Rausch *et al.*, *J. Organometal. Chem.*, 1981, **212**, 247), Vollhardt and coworkers have developed a powerful and elegant biphenylene synthesis from three alkyne units, Scheme 7.4 (B.C. Berris, Y.-H. Lai, and K.P.C. Vollhardt, *J. Chem. Soc., Chem. Commun.*, 1982, 953 and subsequent papers *loc. cit.*). The substituents on the diyne

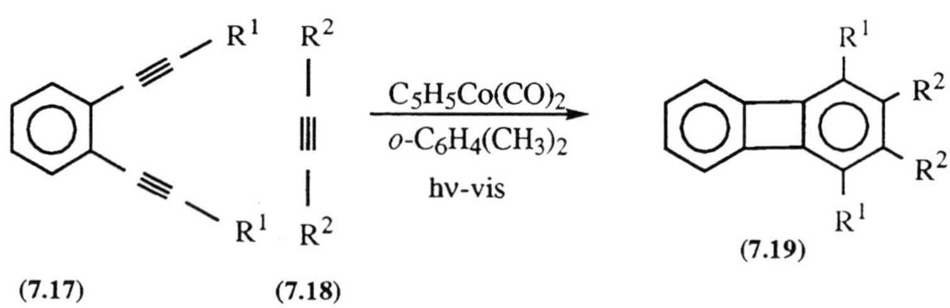

Scheme 7.4

(R^1, 7.17) are usually, but not necessarily, hydrogen and those on the mono-yne (R^2, 7.18) are usually trimethylsilyl, so that the initial product (7.19) can be converted into unsubstituted biphenylene by de-silylation with potassium hydroxide in methanol or into a di-iodobiphenylene (7.19, R^1 = H, R^2 =I) which can be coupled to give a diyne (7.19, R^1 = H, R^2 = -C≡C.Si(CH_3)$_3$) from which a second biphenylene group can be similarly assembled, as in synthesis of (7.5) (Vollhardt, *loc. cit.*).

This method can be adapted to form multiple biphenylene systems from poly-ynes, as in the alternative synthesis of (7.5) from 1,2,4,5-tetraethynylbenzene and the triply-fused example (7.7) from hexaethynylbenzene.

(c) Reactions of biphenylene

(i) Electrophilic substitution and addition

Biphenylene reacts readily with electrophiles and, within the limits of side-reactions of the type noted below, gives aromatic substitution products. Reaction is substantially faster at the β-positions, a result which is simply rationalised on the basis of greater destabilising cyclobutadiene character in the Wheland intermediate (7.20, Scheme 7.5) leading to α-substitution.

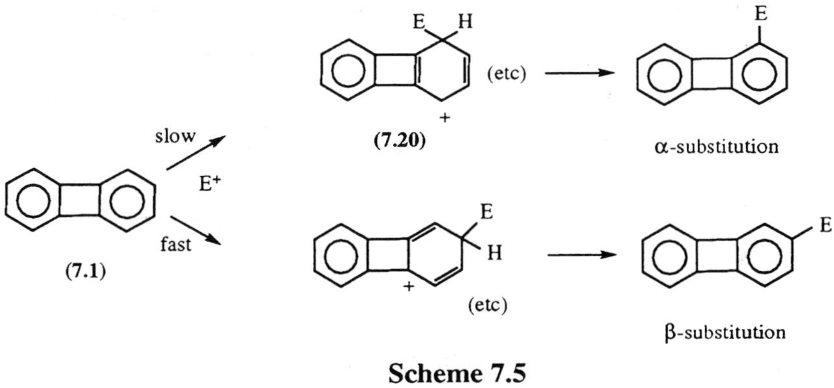

Scheme 7.5

If the initial substituent is positive charge stabilising (*i.e. ortho, para*-directing in benzene) a second substitution occurs at the remaining β-position of the same ring, giving a 2,3-disubstituted biphenylene; if it is electron withdrawing the second substituent enters a β-position in the other ring, the 2,6-isomer being favoured over the 2,7. Examples include nitration and Friedel-Crafts acylation (J.M. Blatchy, A.J. Boulton, and J.F.W. McOmie, *J. Chem. Soc.*, 1965, 4930 and references therein).

Monochlorination and bromination give the expected 2-halo-derivative but further halogenation leads to addition across the benzene bridgehead bond leading to the tetrabromo-derivative (7.21, Scheme 7.6) (J.F.W. McOmie *et al.*, *J. Chem. Soc.*, 1964, 1622; see also H. Kidokoro, M. Sato, and S. Ebine, *Chem. Lett.*, 1981, 9, 1269).

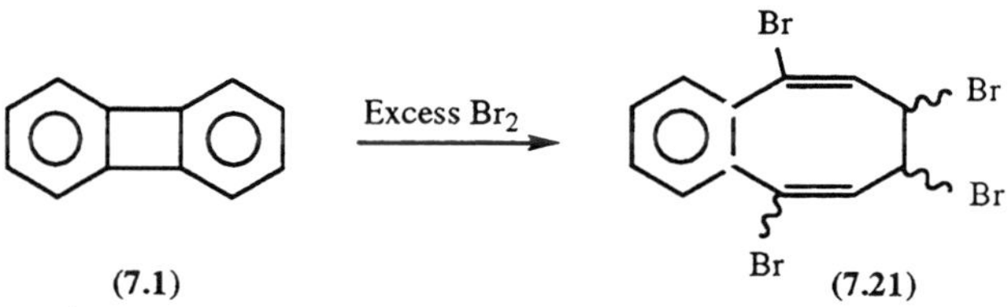

(7.1) (7.21)

Scheme 7.6

An interesting and synthetically useful reaction leading to 1-substituted derivatives (Scheme 7.7) proceeds by initial lithiation at this position, e.g. with butyllithium, and displacement of lithium by electrophiles such as halogen or ketones. 2-Lithiobiphenylene, available from the 2-bromo-derivative with

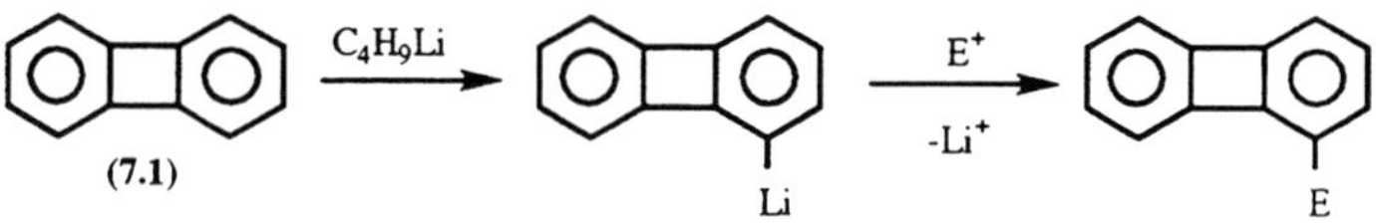

(7.1)

Scheme 7.7

butyllithium, reacts similarly (e.g. J.F.W. McOmie *et al.*, *J. Chem. Soc. (C)* 1966, 1767; see also F. Vögtle, *loc. cit.*).

Steitwieser has rationalised the acidity of the α-protons on the basis that the orthogonal σ-bonds formed by the bridgehead carbon atoms use orbitals of more p-character than the normal sp^2 hybrid, leaving the remaining σ-bond to these atoms, *i.e.* that to α-carbon, with higher s-character. Since s-orbitals interact more strongly with the nucleus than p-orbitals the effective electronegativity of the bridgehead carbon is increased, withdrawing electron density from the α-CH group and increasing its acidity (A. Streitwieser *et al.*, *J. Amer. Chem. Soc.*, 1968, **90**, 1357). The same effect probably underlies the reduced donor ability of the α-nitrogen atom compared with the β in 1,6-diazabiphenylene (7.22) (J.A.H. MacBride and P.M. Wright, *Tetrahedron Lett.*, 1981, **22**, 4545).

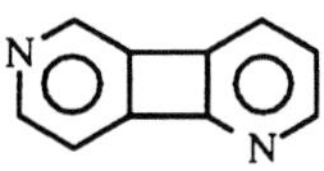

(7.22)

(ii) *Nucleophilic Substitution*

Reaction of octafluorobiphenylene (7.23, Scheme 7.8) with sodium hydroxide and isolation of the products after methylation with diazomethane gives 2-methoxyheptafluorobiphenylene (7.24) in 28% yield and smaller amounts, 7%, of the 1-methoxy-isomer (7.25) (J.F.W. McOmie *et. al.*, *J. Chem. Soc. (C)*, 1969, 1994).

Faster reaction at the β-position is consistent with formation of a Meisenheimer intermediate (7.26) with less cyclobutadiene character than that

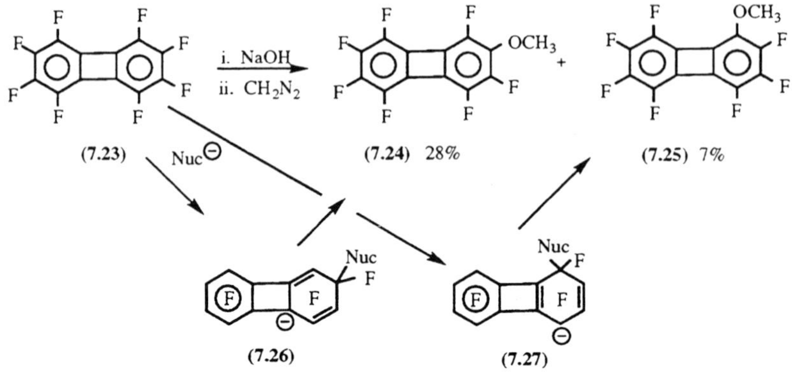

Scheme 7.8

(7.27) leading to α-substitution. Nucleophilic substitution in perfluoroaromatic compounds is, however, favoured at positions which are not *para* to fluorine since in intermediates such as (7.27), of which the *para*-quinonoid forms are generally most important, the energy is raised by repulsion between the carbanion and fluorine electron pairs so that β-substitution is to be expected irrespective of any effect of the ring-system itself. Since the ratio of the isolated products was only 4:1, noting that only about one third of the material was accounted for, the directing effect of the ring system may be small. Hexafluoro-1,8-diazabiphenylene reacts with nucleophiles almost specifically at the 4-position so that the *para*-directing effect of ring nitrogen over-rides the effect of the biphenylene system; frontier molecular orbital treatments of nucleophilic substitution of fluorinated biphenylenes have been discussed (J.A.H. MacBride *et al.*, *J. Chem. Research (S)*, 1990, 172; *(M)*, 1990, 1301-1333 and references therein).

Nucleophilic displacement of hydrogen by amide ion occurs when 2-nitrobiphenylene (7.28, Scheme 7.9) reacts with sodamide giving the 3-amino-derivative (7.29) but the 1-nitro-compound does not react (J.W. Barton, A.R. Grinham and K.E. Whitaker, *J. Chem. Soc. (C)*, 1971, 1384). It is likely that the intermediate is a Meisenheimer adduct (7.30); the 1-nitro-analogue (7.31) would have unfavourable cyclobutadiene character.

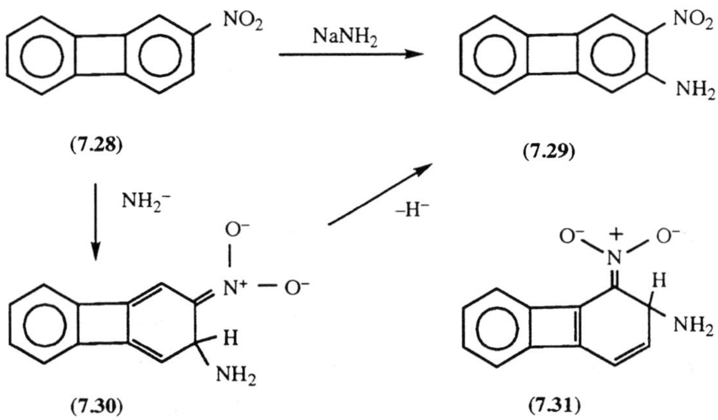

Scheme 7.9

(iii) *Four-membered ring opening reactions*

Catalytic hydrogenation of biphenylene over Raney nickel gives biphenyl (W. Baker, M.P.V. Boarland and J.F.W. McOmie, *J. Chem. Soc.*, 1954, 1476) and heating in the liquid phase to *ca.* 400°C forms tetraphenylene (7.33, Scheme 7.10). This reaction, which is also catalysed by transition metals, involves a biradical intermediate (7.32) since deuterium exchange with C_6D_6 is observed. In a flow experiment with nitrogen carrier gas at 735°C traces of undeuterated triphenylene (7.34) show that benzyne is present (L. Friedman and D.F. Lindlow, *J. Amer. Chem. Soc.*, 1968, 90, 2324).

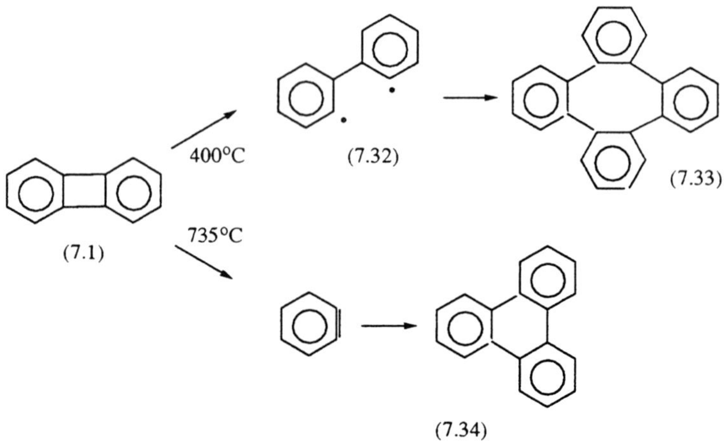

Scheme 7.10

Biphenylene end-caps on acetylenic polymer chains serve as thermal cross-linking groups, with formation of diarylphenanthrene units, by addition of biradicals of type (7.32) to triple bonds, for high performance graphite re-inforced composite materials (S. Sloessel *et al.*, *Polym. Sci.*, 1988, **36**, 184).

Chapter 21

DI-, TRI- AND TETRA-PHENYLMETHANES, THEIR DERIVATIVES AND OXIDATION PRODUCTS

JOHN D. HEPWORTH AND STEPHANE G. R. GUINOT

1. Diphenylmethane, its derivatives and oxidation products

(a) Diphenylmethane and its nuclear-substituted derivatives

(i) General methods of preparation

The synthesis of diphenylmethanes by the Friedel Crafts reaction is well established such that new work tends towards improving yields, seeking milder reaction conditions and reducing both costs and environmental problems associated with the usual Lewis acid catalysts. Clays, such as montmorillonite, show both Brønsted and Lewis acidity which depend on the water content and the exposed Al^{3+} and Fe^{3+} ions, respectively. Understandably, therefore, this type of material has been advocated as a new generation of catalysts for Friedel Crafts alkylations.

The benzylation of benzene occurs in 57% yield in the presence of montmorillonite over 24 hours at room temperature. When the clay is pretreated with $ZnCl_2$, the yield is increased to 80% after only 15 minutes. Such supported reagents exhibit good regioselectivity with the *para*-isomer predominating when substituted benzene derivatives are alkylated (J. H. Clark *et al.*, Chem. Commun., 1989, 1353; *idem*, J. Chem. Res.(S), 1991, 74).

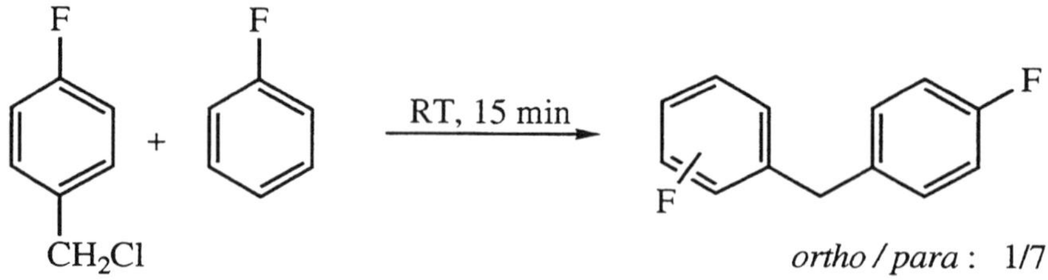

Paraformaldehyde reacts at 80 °C with benzene to give *ca.* 70% of diphenylmethane, whilst phenol gives a mixture of the 4,4'-, 2,2'- and 4,2'-dihydroxydiphenylmethane in which the last predominates. In a similar manner, toluene and bromine react in the presence of bentonite to give a mixture of 2- and 4-methyldiphenylmethane in a 1:2 ratio in an 85% overall yield (M. Salmon, E. Angeles and R. Miranda, Chem. Commun., 1990, 1188). The use of an Al-exchanged form of the clay facilitates the reaction (R. Miranda *et al.*, Synth. Commun., 1994, 24, 727).

An alternative heterogeneous catalyst to clays is the perfluorinated sulfonic acid resin, Nafion-H ™. This catalyst effects the benzylation of a range of aromatic hydrocarbons with benzyl alcohols with very little dibenzyl ether $PhCH_2OCH_2Ph$ being produced through dehydration (T. Yamato *et al.*, J. Org. Chem., 1991, 56, 2089).

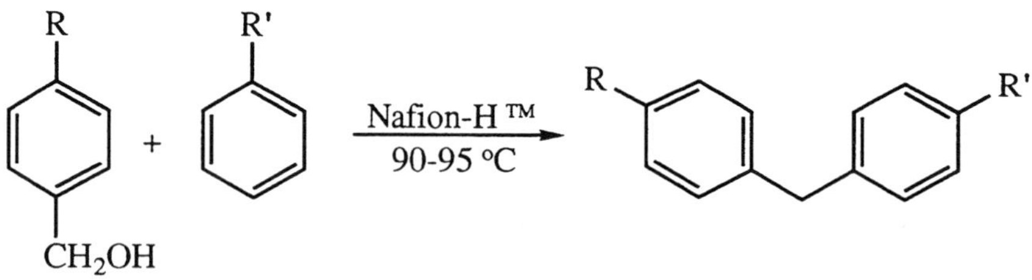

R'= H, R=	H	4-Me	4-tBu	4-Br	$4\text{-}NO_2$
% yield	54	81	85	70	52

R= H, R' =	Me	Et	$1,2\text{-}Me_2$	$1,3\text{-}Me_2$	$1,4\text{-}Me_2$	$1,3,5\text{-}Me_3$
% yield	70	73	70	71	73	79

The reaction also brings about intramolecular benzylation.

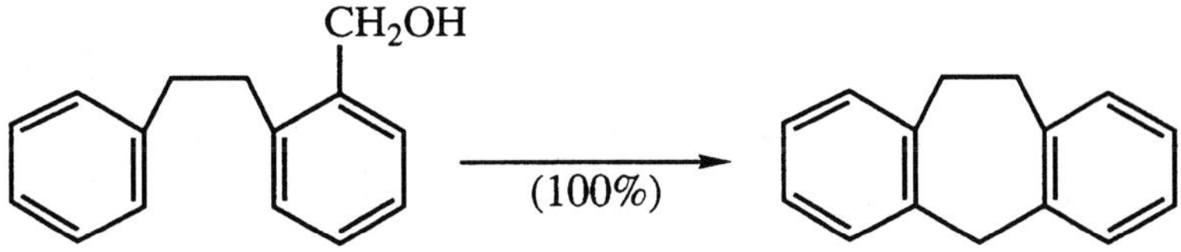

Bis(hydroxymethyl)benzenes afford dibenzylbenzenes in *ca.* 75% yield, whilst combination of the two approaches leads to cyclobenzylation (T. Yamato *et al.*, J. Chem. Res. (S), 1991, 242; *idem*, Chem. Express, 1990, 5, 941).

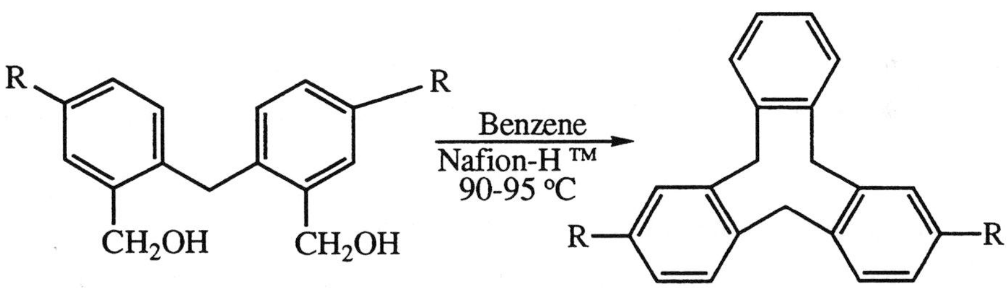

The protonation and subsequent "tri- and tetra-merisation" of dimethoxybenzyl alcohols occurs in CH_2Cl_2 in high yield, leading to cycloveratrylenes (1) and (2). The synthesis, conformations and applications of cyclotriveratrylenes and cryptophanes have been reviewed (A. Collet, Tetrahedron, 1987, 43, 5725).

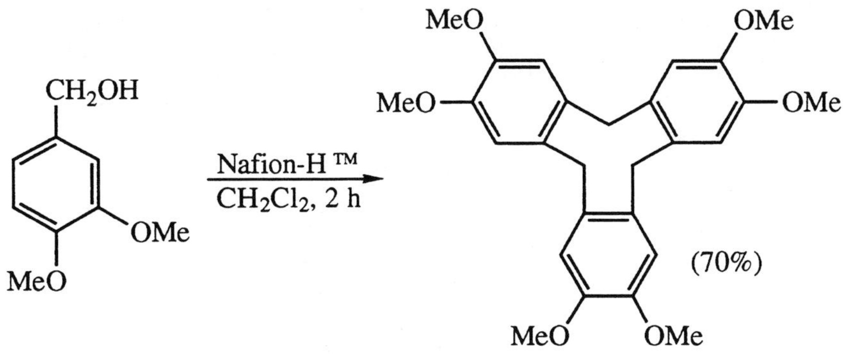

(1)

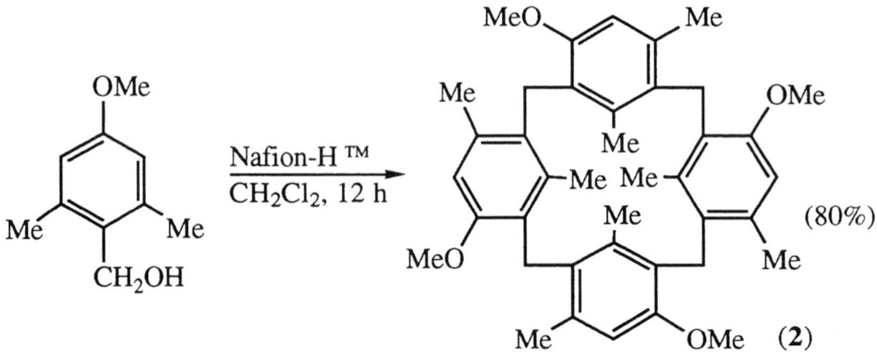

(**2**)

The diphenylmethane unit has been incorporated into a range of cyclophanes. [1.1.1.1]-Paracyclophane, which is rigid with a square cavity, has been obtained from the acyloin derived from (3) *via* the *trans, trans*-diene (4) and a subsequent Diels Alder reaction with vinyl phenyl sulfoxide (Y. Miyahara, T. Inazu and T. Yoshino, Tetrahedron Letters, 1983, 5277). Both [1.1.1.1.1]- and [1.1.1.1.1.1]- paracyclophanes can be obtained from 4-bromobenzophenone by a repeated sequence of reduction to the diphenylmethane, formylation and reaction with 4-lithiobromobenzene and culminating in a Friedel Crafts cycloalkylation in trifluoroacetic acid (G. W. Gribble and C. F. Nutaitis, Tetrahedron Letters, 1985, 6023).

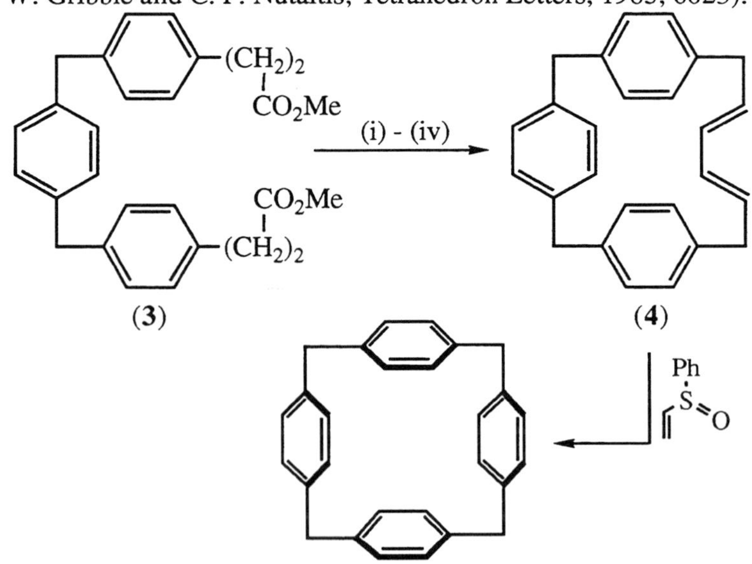

(i) Na, Me_3SiCl, Δ then H^+; (ii) $NaBH_4$; (iii) Ac_2O, NaOAc; (iv) 550 °C/ 0.1 mm

The $[1_4]$- and $[1_5]$- orthocyclophanes have been obtained from 2,2'-dibromodiphenylmethane. Reaction of the derived lithio compound with 2,2'-diformyldiphenylmethane and subsequent oxidation to the conformationally rigid diketone and deoxygenation gave the $[1_4]$ macrocycle. The $[1_5]$- cyclophane resulted when similar methodology was applied to 1,2-bis(2-formylbenzyl)benzene. NMR studies indicate flexible conformations for these macrocycles, with the larger molecule exhibiting the greater flexibility. It is noteworthy that bridged macrocyclic structures result when the cyclic diketones are reduced under Clemmensen and McMurry conditions (W. Y. Lee, C. H. Park and Y. D. Kim, J. Org. Chem., 1992, 57, 4074).

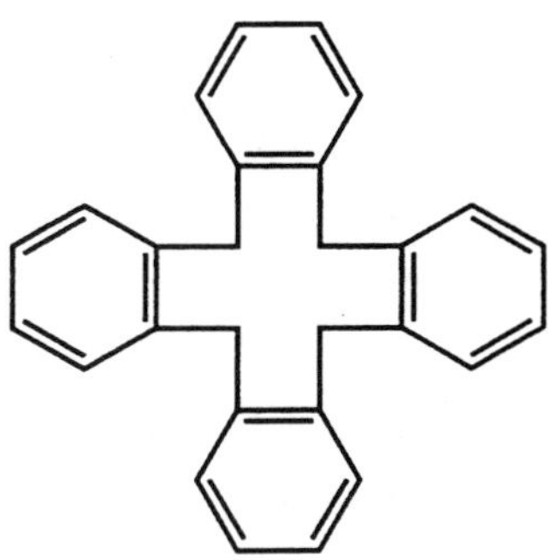

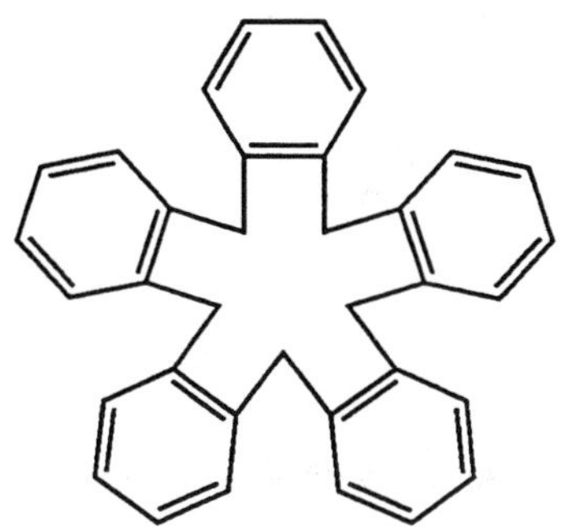

Orthocyclophanes result from an intramolecular Friedel Crafts cycloalkylation of terminal benzylbenzilic alcohols with repeating benzyl units. When an α,ω–benzylbenzilic diol (5) is used, a (1,2)(4,5)-biscyclisation occurs leading to a bisorthocyclophane with two 9-membered rings (6) (W. Y. Lee, W. Sim and K. D. Choi, J. Chem. Soc., Perkin Trans. 1, 1992, 881). The chemistry of $[1_n]$-orthocyclophanes has been reviewed (W. Y. Lee, Synlett, 1994, 765).

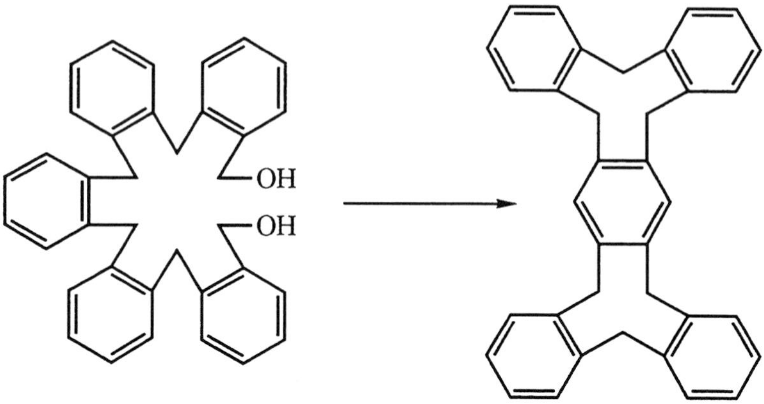

(5) (6)

The silane reduction of diaryl ketones to diarylmethanes can be improved by activation of the cationic site of the carbonyl group. Thus, hydride transfer from triethylsilane to benzophenones is facilitated by trifluoromethanesulfonic acid and the reduction to diarylmethanes is fast, high yielding and effective in the presence of either electron withdrawing or electron donating substituents. However, the reaction is somewhat retarded by sterically bulky groups (G. Olah, M. Arvanagli and L. Ohannesian, Synthesis, 1986, 770). A particular effective ionic hydrogenation involves pyridinium poly(hydrogen fluoride) as the co-reagent with Et_3SiH (G. Olah, Q. Wang and G. K. S. Prakash, Synlett, 1992, 647). In a similar manner, boron trifluoride etherate allows successful silane reduction of benzophenones bearing electron withdrawing substituents. Used in conjunction with $(C_2H_5)_3SiD$, a useful route to deuterated diarylmethanes is available (I. Smonov, Synth. Commun., 1994, 24, 1999).

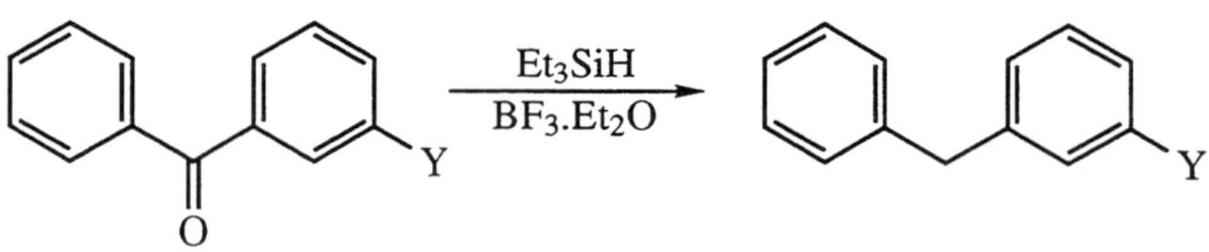

Y:	H	NO_2	CF_3	CO_2Me
%:	92	95	85	97

Diphenylsilane in the presence of aluminium chloride reduces only those benzophenones with an electron withdrawing substituent to the hydrocarbon (Y. Inoue and S. Tanimito, Bull. Inst. Chem. Res., Kyoto Univ., 1991, 69, 489).

Catalytic-transfer reduction of benzophenone using ammonium formate in acetic acid in the presence of Pd-C is temperature dependent such that diphenylmethanol is formed at 70 °C, but diphenylmethane at 110 - 120 °C (S. Ram and L. D. Spicer, Tetrahedron Letters, 1988, 3741; *idem*, Synth. Commun., 1992, 22, 2673). 9,10-Dihydroanthracene (DHA) effects the hydrogenation of benzhydryl fluorides at high temperature in the absence of a catalyst. The reaction, which appears to proceed by a molecule-induced homolysis, may well be applicable to other related systems (F. Schaffer *et al.*, Chem. Ber., 1994, 127, 557).

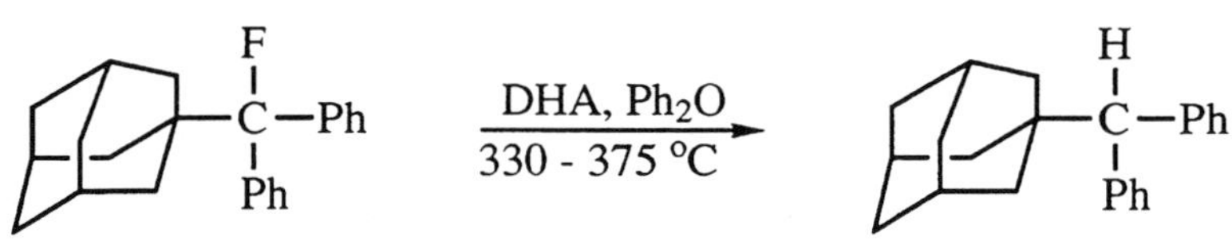

Benzophenones can be electrochemically reduced to diarylmethanes in a one-pot procedure which is based on the cleavage of the C-O bond during cathodic reduction of oxalate esters (N. Islam, D. W. Sopher and J. H. P. Utley, Tetrahedron, 1987, 43, 2741).

$$\mathrm{ArCOAr} \xrightarrow[\text{MeCN}]{\text{Hg cathode}} \mathrm{ArCH(OH)Ar} \xrightarrow[\text{Hg cathode}]{(\mathrm{CO_2Et})_2} \mathrm{ArCH_2Ar}$$

Substituted diarylmethanes are formed when the anions derived from the cleavage of diarylmethyl methyl ethers with lithium metal are treated with electrophiles (U. Azzena *et al.*, Synth. Commun., 1994, 24, 591).

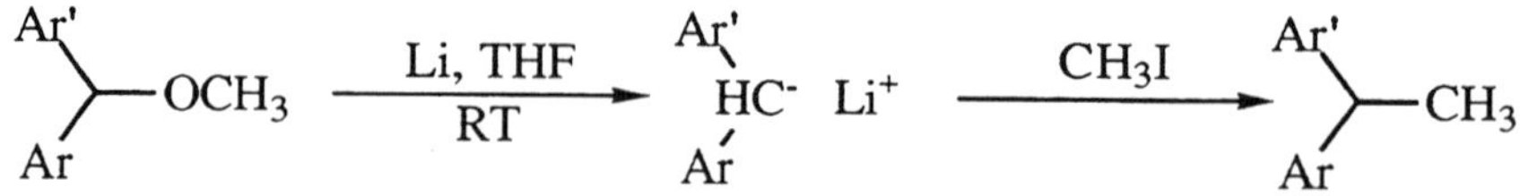

Phenylcopper reacts with benzyl bromide in THF to give diphenylmethane (G. W. Ebert and R. D. Rieke, J. Org. Chem., 1988, 53, 4482) and the copper-catalysed cross coupling of an aryl Grignard reagent with dibromomethane provides good yields of a range of diarylmethanes, some of which are difficult to obtain by other methods (T. Yamamoto *et al.*, Org. Prep. Proced. Int., 1991, 23, 617).

$$2\,ArMgBr + CH_2Br_2 \xrightarrow[\text{THF - HMPA}]{\text{CuBr, }\Delta} ArCH_2Ar$$

The oxidation of mesitylene by oxygen and H_2PtCl_6 in aqueous trifluoroacetic acid leads to the substituted diphenylmethane (7) in good yield (G. V. Nizova and G. B. Shul'pin, Izv. Akad. Nauk. SSSR, Ser. Khim., 1987, 935).

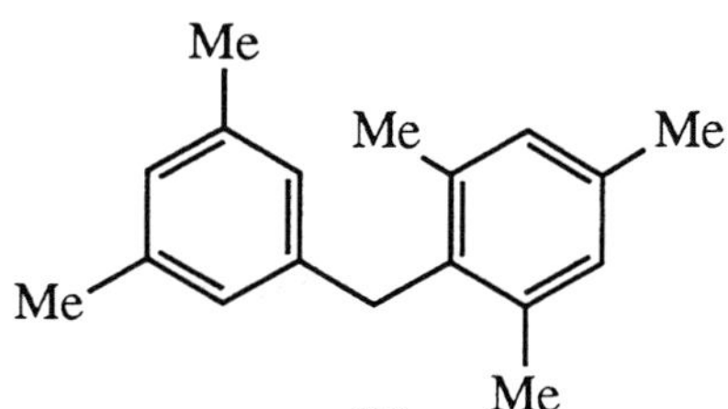

(7)

(ii) Properties and reactions

The acidities of a range of substituted diphenylmethanes have been measured in cyclohexylamine (A. Streitwieser, Jr., E. R. Vorpagel and C.-C. Chen, J. Amer. Chem. Soc., 1985, 107, 6970) and in dimethyl sulfoxide (F. G. Bordwell, X. Zhang and J.-P. Cheng, J. Org. Chem., 1991, 56, 3216). The former work includes data on examples of acidity too low to be measured in

DMSO and shows that pK_a values for *meta*-substituted compounds lie on a normal Hammett plot. However, the *para*-substituted series deviates from the *meta* correlation even when σ^- parameters or various dual substituent parameter approaches are applied. Electron withdrawing substituents in the α-position exert an acidifying effect, but have variable effects on the homolytic bond dissociation energies. Thus, Ph and CN appear to stabilise the incipient radicals, whilst SPh, CO_2Et and SO_2Ph are destabilising.

It appears that the carbanions derived from diphenylmethanes are sp^2 hybridised at the ionising centre (S. E. Browne *et al.*, J. Amer. Chem. Soc., 1984, 106, 1432; S. Top *et al., ibid.*, 1988, 110, 6426). Studies of the influence of *p*-nitro groups and the tricarbonylchromium(0) function on the acidity of diphenylmethane indicate that these functions enhance the coplanarity of the aromatic rings in the carbanion (F. Terrier *et al.*, Organometallics, 1985, 4, 1291). However, introduction of *o*-nitro substituents prevents mutual coplanarity which is reflected in the acidities of polynitrodiphenylmethanes (F. Terrier *et al.*, J. Org. Chem., 1990, 55, 2610). Theoretical and Raman / infrared studies on molecules with a geminal diphenyl group indicate that the equilibrium conformation for diphenylmethane is a gable form, whereas benzophenone and other examples prefer a helical conformation (I. Baraldi, E. Gallinella and M. Scopon, J. Chim. Physique, 1986, 83, 653; *idem*, Spectrochim. Acta, 1987, 43A, 1045).

A variety of reagents has been used to achieve the oxidation of diphenylmethane to benzophenone. Transition metal oxidants which are successful include barium permanganate in refluxing CH_3CN (H. Firouzabadi *et al.*, Tetrahedron, 1990, 46, 6869), pyridinium chlorochromate in benzene (R. Rathore, N. Saxena and S. Chandrasekaran, Synth. Commun., 1986, 16, 1493), benzeneseleninic anhydride in chlorobenzene at 120 °C (D. H. R. Barton, R. A. H. F. Hui and S. V. Ley,

J. Chem. Soc., Perkin Trans. 1, 1982, 2179) and cerium (IV) trifluoromethanesulfonate at room temperature in CH_3CN (T. Imamoto, Y. Koide and S. Hiyama, Chem. Letters, 1990, 1445). *tert*-Butyl hydroperoxide (TBHP) brings about benzylic oxidations in the presence of various catalysts, such as CrO_3 (J. Muzart, Tetrahedron Letters, 1987, 2131), bis(tributyltin oxide)dioxochromium (VI) (J. Muzart, Synth. Commun., 1989, 19, 2061) and pyridinium dichromate (N. Chidambaram and S. Chandrasekaran, J. Org. Chem., 1987, 52, 5048). These methods utilise the oxidising power of the chromium without requiring stoichiometric quantities. TBHP is used as the reoxidant in the room temperature benzylic oxidation of Ph_2CH_2 by a chromium pillared montmorillonite. Yields are higher than those achieved with other Cr oxidising systems (B. M. Choudary *et al.*, J. Org. Chem., 1992, 57, 5841). The aerial oxidation of diphenylmethane and some derivatives occurs at 150 ^oC in the presence of alumina-supported potassium or sodium fluoride. The unsupported fluorides exhibit no catalytic effects (J. H. Clark *et al.*, J. Chem. Research (S), 1994, 102). Diphenylmethane can be oxidised to benzophenone by oxygen after initiation of the reaction by alkali metal hydrides in a polar aprotic solvent. In the presence of 18-crown-6 ether and in DMSO, $NaBH_4$ causes partial oxidation to the benzhydrol (G. S. Grigoryan *et al.*, Dokl. Akad. Nauk. SSR, 1987, 295, 1124; *idem*, Zh. Org. Khim., 1989, 25, 1036). 2-Nitrobenzenesulfonyl chloride and potassium superoxide, a source of 2-nitrobenzene peroxysulfonyl radicals, oxidise diphenylmethane at -35 ^oC in CH_3CN (Y. H. Kim, K. S. Kim and H. K. Lee, Tetrahedron Letters, 1989, 6357). Oxidation by the Gif IV system also yields benzophenone (D. H. R. Barton *et al.*, Tetrahedron Letters, 1989, 6615). The photochemical bromination of diphenylmethanes leads to benzophenones as a consequence of hydrolysis of the initially formed dibromo- or diphenyl-methanes during work-up (S. Mataka *et al.*, Bull. Chem. Soc. Jpn., 1994, 67, 2336).

Hydride transfer from 4,4'-bis(dimethylamino)diphenylmethane to the triphenylmethyl carbocation yields the corresponding diarylmethyl cation in CH_3CN, but involves abstraction from the 4-substituent in SO_2 (N. N. Bychkov *et al.*, Zh. Org. Khim., 1987, 23, 1516).

The anion of diphenylmethane is efficiently generated by treatment with complex bases such as $NaNH_2$-NaOtBu doped with a potassium halide and is readily alkylated by benzyl chloride (B. Jamart-Gregoire *et al.*, Chem. Commun., 1992, 1459). Addition of Ph_2CH^- to dibromoalkanes leads to 1-bromo-ω,ω-diphenylalkanes (R. A. Bunce and J. P. Sullivan, Synth. Commun., 1990, 20, 865). Reaction with esters and anhydrides affords diphenylmethyl ketones, Ph_2CHCOR (R. A. Bunce and E. D. Dowdy, Synth. Commun., 1990, 20, 3007).

The Gattermann-Koch formylation of diphenylmethane in HF-SbF_5 under 20 atmospheres pressure of carbon monoxide affords the 4,4'-diformyl derivative with only traces of the 2,4'- and 2,2'-isomers. The major product is benzaldehyde, which arises by cleavage of the methylene unit (M. Tanaka and Y. Souma, Chem. Commun., 1991, 1551).

The diphenylmethyl cation is sufficiently electrophilic to react with benzene in trifluoromethanesulfonic acid (TFSA), forming triphenylmethane. However, the α-methyldiphenylmethyl cation, which can be observed by NMR spectroscopy, gives 1,1-diphenylethylene on quenching with water, indicating localisation of the charge at the α-position. When such cations have an α-carbonyl substituent, an intramolecular cyclisation leads to a fluorene derivative. An intermediate carbodication is proposed, and indeed its ^1H NMR spectrum has been recorded. This dication undergoes a 4π electrocyclisation to the fluorene (T. Ohwada and K. Shudo, J. Amer. Chem. Soc., 1988, 110, 1862).

Diphenylmethane reacts with hexafluoropropene to give a mixture of the 1:1 adduct, $Ph_2CHCF_2CHFCF_3$, and a rearranged major product, $PhCH_2CF_2CFPhCF_3$. The latter compound arises by a 1,3-phenyl shift in an intermediate radical, caused by steric hindrance of the usual chain transfer process (R. N. Haszeldine, C. M. Raynor and A. E. Tipping, J. Chem. Soc., Perkin Trans. 2, 1983, 2801).

(iii) Benzylbenzoic acids

Benzyl benzoate undergoes an intramolecular rearrangement on vapour phase reaction over alumina. The initial product, 2-benzylbenzoic acid, either decarboxylates to diphenylmethane or cyclodehydrates to anthrone and ultimately anthracene (K. Ganesan and C. N. Pillai, Tetrahedron, 1989, 45, 877).

(b) Benzhydryl halides, benzhydrylidene dihalides, benzhydrylamines and their nuclear substituted derivatives

(i) General methods of preparation and properties

Benzyltrimethylammonium tetrachloroiodate, prepared from the quaternary ammonium chloride and iodine trichloride in dichloromethane, behaves as "solid chlorine". In the presence of radical initiators such as 2,2'-azobisisobutyronitrile (AIBN), it converts diphenylmethane into benzhydryl chloride and triphenylmethane into triphenylmethyl chloride in high yields (S. Kajigaeshi *et al.*, Tetrahedron Letters, 1988, 5783).

Benzhydryl alcohols are converted into their bromides and chlorides on treatment with BBr_3 and $SnCl_4$ in the absence of solvents, respectively. Both benzhydryl halides undergo transhalogenation with these reagents (A. Amrollah-Madjdabadi, T. N. Pham and E. C. Ashby, Synthesis, 1989, 614).

Benzhydrylamine is formed in 94% yield by the reaction of phenyllithium with *N*-(trimethylsilyl)phenylmethanimine, $Ph_2C{=}NSiMe_3$ (A. Hirao *et al.*, Synthesis, 1982, 461). Benzhydrylamine resins, which are useful as polymeric supports for solid phase peptide synthesis, can be obtained from a divinylbenzene-styrene copolymer by treatment with the butyllithium-tetramethylethylenediamine complex and subsequent reaction with the imine (A. Hirao *et al.*, Chem. Commun., 1983, 25).

A number of new benzhydrylamine and benzhydryl ether and thioether derivatives have been synthesised by alkylation using benzhydryl chloride or benzhydrylamine (S. Corsano *et al.*, Arch. Pharm. (Weinheim), 1989, 322, 873).

A range of novel benzhydryl ethers have been prepared by the reaction of benzhydryl 2-chloroethyl ether with charge delocalised organolithium reagents. The products undergo a facile acid catalysed hydrolysis and an easy phenylthiolysis affording the corresponding sulfide, Ph_2CHSPh. On treatment with lithium metal or lithium naphthenide, cleavage of the ether leads to the formation of diphenylmethyllithium, from which diphenylmethane and diphenylacetic acid result on hydrolysis and carbonation, respectively (C. G. Screttas and M. Micha-Screttas, J. Org. Chem., 1982, 47, 3008).

$$Ph_2CHOCH_2CH_2Cl + Ph_3CLi \longrightarrow Ph_2CHOCH_2CH_2CPh_3$$

Various substituted benzophenones can be converted into benzhydrylidene difluorides by way of their dithiolane derivatives through reaction with sulfuryl chloride and pyridinium poly(hydrogen fluoride) (G. K. S. Prakash *et al.*, Synlett, 1993, 691).

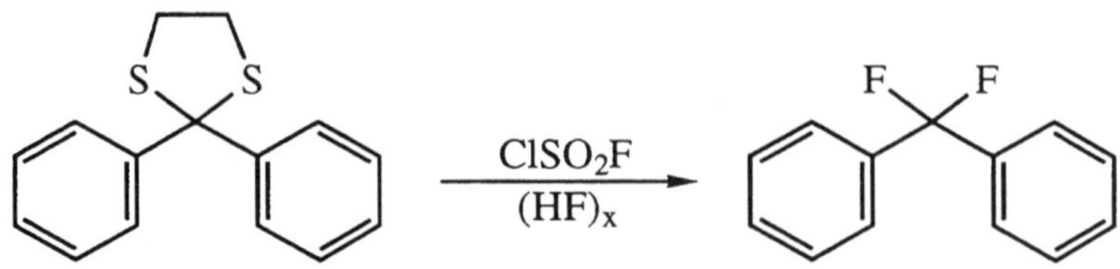

The sterically hindered *N*-(diphenylmethyl)bis(pentamethylphenyl)ketene imine (8), prepared from bis(pentamethylphenyl)acetic acid by the route shown below, is much more stable than the corresponding diphenylketene imine. On heating this compound in CCl_4 or on melting, a red colour develops which is associated with formation of the relatively stable radical, cyanobis(pentamethylphenyl)methyl. On reduction, this radical is converted into bis(pentamethylphenyl)acetonitrile (L. F. Clarke, A. F. Hegarty and P. O'Neill, J. Org. Chem., 1992, 57, 362).

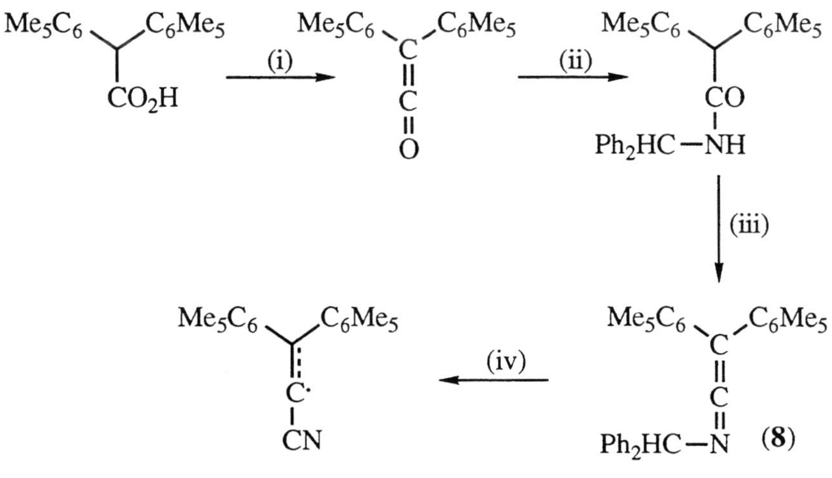

Reagents: (i) $SOCl_2$; (ii) Ph_2CHNH_2; (iii) $SOCl_2$, pyridine; (iv) CCl_4, Δ.

Benzophenone arylhydrazones are converted into 1-(arylazo diphenylmethyl)pyridinium salts on treatment with a solution of bromine in pyridine. Further reaction with a variety of *N* and *O* nucleophiles leads to *N*- and *O*- functionalised arylazodiphenylmethyl derivatives (H. Gstach and J. G. Schantl, Synth. Commun., 1986, 16, 741; *idem*, Monatsh., 1987, 118, 851).

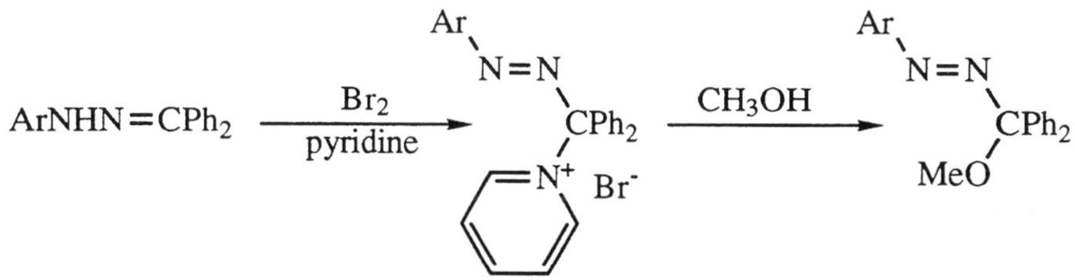

Oxidation of benzhydryl halides to benzophenone has been accomplished using sodium hypochlorite (S. Cacchi and F. La Torre, Chem. Ind. (London), 1986, 286) and benzhydryl cyanides are converted to the benzophenone by stirring them in air in DMSO in the presence of potassium carbonate (S. S. Kulp and M. J. McGee, J. Org. Chem., 1983, 48, 4097).

Diphenylmethyl has been advocated as a protecting group for sulfonamides. Reaction of benzhydrylamine with methanesulfonyl chloride forms the sulfonamide which can be used in further reactions. The protecting group is stable to strongly acidic and basic conditions and can be removed by catalytic hydrogenation (M. A. Poss and J. A. Reid, Tetrahedron Letters, 1992, 7291).

A series of papers discusses the mechanism of the reaction of benzhydryl cations with alkenes (H. Mayr *et al.*, J. Amer. Chem. Soc., 1990, 112, 4446, 4454, 4460) and the ionisation of benzhydryl bromide has been studied (E. A. Ponomareva, T. L. Pervishko and C. F. Dvorko, Organic Reactivity, 1982, 19, 20). *Ab initio* calculations of the stability of the benzhydryl cation are in good agreement with data derived from solvolysis studies. The cation exhibits C_2 symmetry with both rings twisted some 18° about the Ph-C^+ bonds (T. L. Amyes, J. P. Richard and M. Novak, J. Amer. Chem. Soc., 1992, 114, 8032). The diphenylmethyl radical shows a complex chemistry (A. Bromberg, K. H. Schmidt and D. Meisel, J. Amer. Chem. Soc., 1985, 107, 83; R. W. Redmond *et al.*, J. Phys. Chem., 1989, 93, 6397; W. Adam and R. S. Oestrich, J. Amer. Chem. Soc., 1992, 114, 6031).

(c) Benzhydrol and its nuclear substituted derivatives

(i) General methods of preparation, properties and functional derivatives

The well established routes to benzhydrols continue to be used. Thus, 4-dialkylaminophenyl lithium and 4-dialkylaminobenzaldehydes react to give Michler's hydrol and its derivatives in high yield (G. Hallas and M. M. Mitchell, J. Soc. Dyers Colourists, 1986, 102, 15).

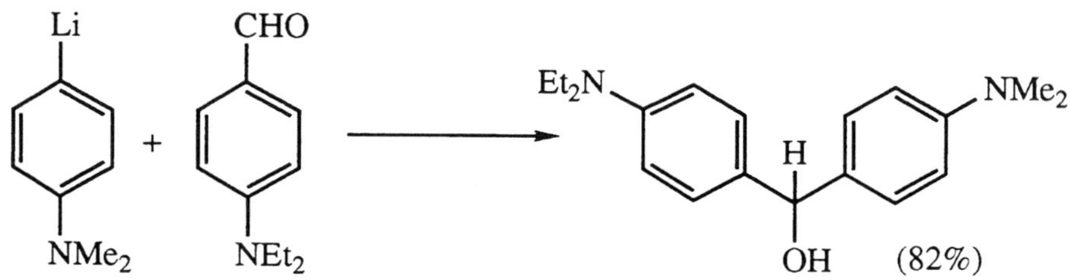

Examples involving the complementary route by reduction of benzophenones include the formation of the highly substituted derivative (9) using $LiAlH_4$ (S. Antus *et al.*, Tetrahedron, 1982, 38, 133) and of a range of cycloheteroalkyl analogues of Michler's hydrol (10) by sodium borohydride in propan-2-ol (S. F. Beach *et al.*, J. Chem. Soc., Perkin Trans. 2, 1989, 1087).

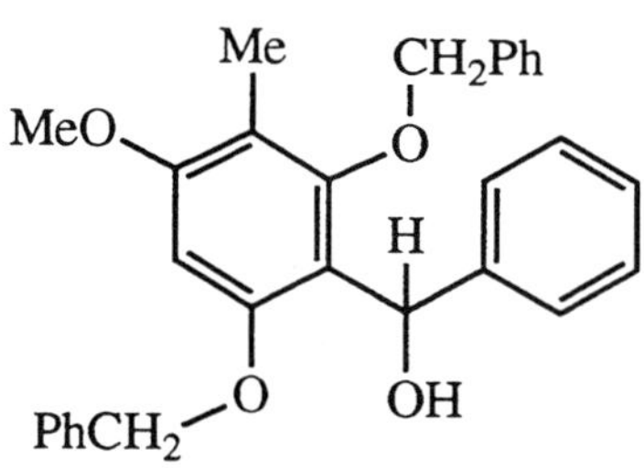

(9)

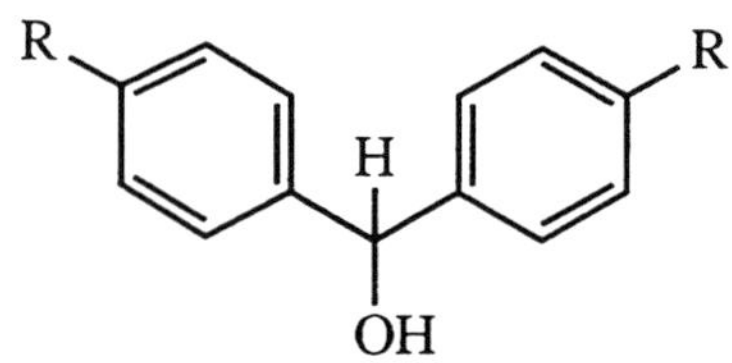

(10)

R= pyrrolidino, piperidino, morpholino

There is particular interest in the asymmetric reduction of prochiral benzophenones. The use of $LiAlH_4$ pre-treated with an excess of (R)-(-)-2-(2-isoindolinyl)butan1-ol or the (S)-(+)-enantiomer to reduce a range of

2-substituted benzophenones resulted in high yields of the corresponding optically active benzhydrols with high (>88%) enantiomeric excess. This reagent is very much less effective when used to reduce *meta*-substituted benzophenones (E. Brown, A. Lézé and J. Touet, Tetrahedron Asymm., 1992, 3, 841). 4-Chlorobenzophenone has been reduced to the S-benzhydrol in low yield, but in 100% e.e. by *Debaromyces marama* micro-organisms (G. Spassov *et al.*, Tetrahedron Asymm., 1993, 4, 301). The nature of a *para*-substituent in *S*-benzhydrols seems to influence their configuration. Electron releasing groups favour the *S*-(-) benzhydrols, but electron withdrawing substituents the *S*-(+) enantiomers (B. Wu and H. S. Mosher, J. Org. Chem., 1986, 51, 1904).

Benzhydrol-2,3,4,5,6-d_5, which is chiral by virtue of deuterium substitution, of 85% enantiomeric purity has been obtained from enantiomerically pure 4'-nitrobenzophenone-2,3,4,5,6-d_5 by reduction and deamination of the diazonium salt (T. Makino *et al.*, J. Org. Chem., 1985, 50, 5357).

Further work on the ^{13}C NMR spectra of the cations derived from substituted benzhydrols by ioniation in superacids indicates that for donor substituents Hammett plots of the chemical shifts of the cationic carbon against the σ^{C+} values are linear. However, electron withdrawing groups exert a shielding effect resulting in an upward deviation from the established correlation line. The deviations are attributed to competition between resonance and inductive π-polarisation effects (D. P. Kelly and M. J. Jenkins, J. Org. Chem., 1984, 49, 409).

Benzhydrol undergoes both reduction and reductive alkylation *via* the carbocation on treatment with trialkylboron reagents in trifluoromethanesulfonic acid. The bulkier the alkyl group, the greater the tendency towards simple reduction, since boron-carbon insertion becomes increasingly more difficult than C-H bond insertion (G. Olah, A. Wu and O. Farooq, J. Org. Chem., 1991, 56, 2759).

Reduction of benzhydrols to diphenylmethanes can be achieved by tributyl tin hydride by way of the di(benzothiazol-2-yl)disulfide (11) (Y. Watanabe *et al.*, Tetrahedron Letters, 1986, 5385) and triethylborane together with tributylstannane reduces the dithiocarbonate to the hydrocarbon (K. Nozaki, K. Oshima and K. Utimoto, Tetrahedron Letters, 1988, 6125). The reaction of benzhydrol with Raney nickel in refluxing toluene leads to a 2:1 mixture of benzophenone and diphenylmethane in 95% yield (M. E. Krafft *et al.*, J. Org. Chem., 1988, 53, 3158).

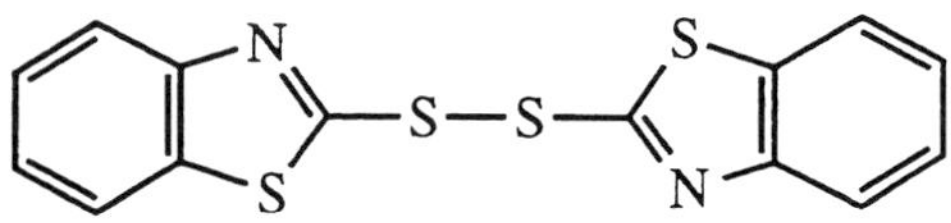

(11)

(d) Benzophenone and its nuclear substituted derivatives

(i) General methods of preparation

The synthesis of benzophenones is so well established and documented that few developments have occurred over the past twelve years, although the advent of new oxidants together with changes in reaction conditions have resulted in new oxidative routes.

The regioselective lithiation of unprotected 4-(methoxymethyl)phenols, followed by reaction with benzaldehydes and rapid oxidation of the resulting benzhydrol leads to some highly substituted benzophenone derivatives (J. M. Saa *et al.*, J. Org. Chem., 1988, 53, 4263).

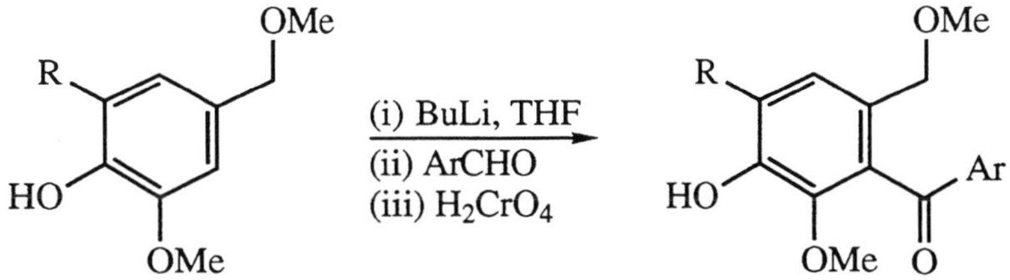

Balanol, a fungal metabolite, is a highly substituted ester of 4-benzoylbenzoic acid. Two syntheses of the benzophenone moiety are based on directed metallation methodology and involve oxidation of a benzhydrol (S. P. Hollinshead, J. B. Nichols and J. W. Wilson, J. Org. Chem., 1994, 59, 6703).

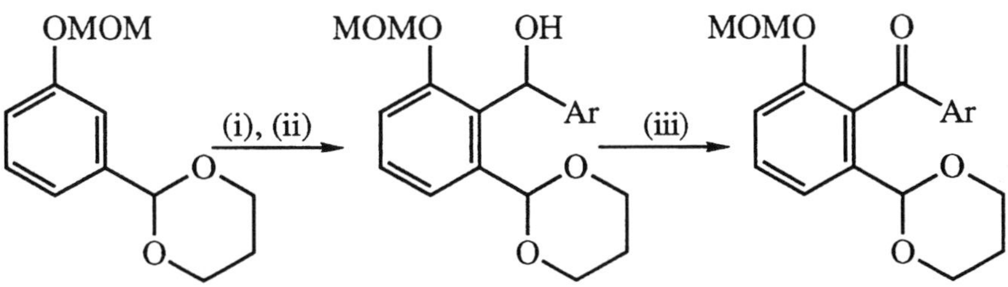

(i) n-BuLi, C_6H_{12}, RT, DMF; (ii) ArLi, THF, 0 °C; (iii) MnO_2, CH_2Cl_2.

Peroxyacetic acid oxidises benzhydrol to benzophenone in the presence of sodium bromide, the active species being acetyl hypobromite (T. Morimoto *et al.*, Bull. Chem. Soc. Jpn., 1992, 65, 703). Benzhydrol is slowly oxidised quantitatively to Ph_2CO by trichloromelamine (S. Kondo *et al.*, J. Org. Chem., 1993, 58, 5003).

Mechanistic studies on the oxidation of benzhydrols by bromamine T (S. Gunasekaran and N. Venkatasubramanian, J. Chem. Soc., Perkin Trans. 2, 1983, 949), pyridinium bromochromate (N. Narayanan and T. R. Balasubramanian, J. Chem. Res.(S), 1991, 336) and pyridinium chlorochromate (K. K. Banerji, Indian J. Chem., 1983, 22B, 413) have been reported.

Both aromatic and heteroaromatic hydrocarbons are benzoylated at 0 - 25 °C by trichloromethylbenzene in the presence of $AlCl_3$, giving access in the former case to a range of substituted benzophenones (N. R. Ayyangar *et al.*, Synthesis, 1991, 322). Under alkaline phase-transfer conditions, phenols are *C*-benzoylated *para* to the hydroxy group, but at 100 °C *ortho* acylation occurs, providing an easy route to hydroxybenzophenones (C. Sarangi and Y. R. Rao, J. Chem. Res. (S), 1994, 392).

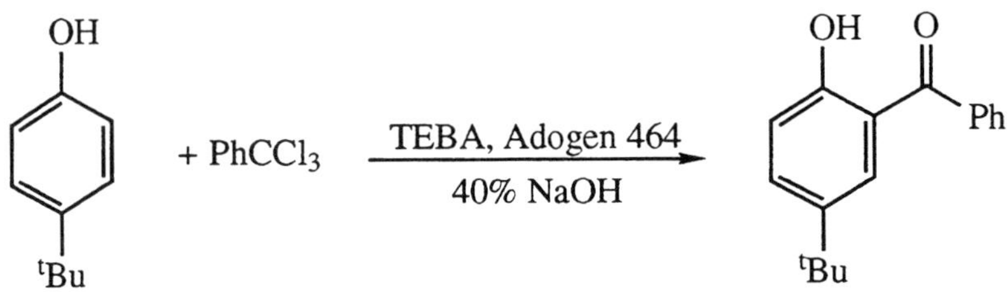

Arylcopper reagents react with benzoyl chloride in dimethoxyethane to give benzophenone derivatives, the process being especially successful when the aryl unit bears an electron withdrawing group (G. W. Ebert and R. D. Rieke, J. Org. Chem., 1988, 53, 4482).

Zeolites are alternative catalysts to the usual Lewis acids for aromatic benzoylation and show good regioselectivity. For example, 2-nitroaniline gives 2-amino-3-nitrobenzophenone on reaction with either benzoyl chloride (84%) or benzotrichloride (97%) on refluxing in dichloromethane (V. Paul *et al.*, Tetrahedron Letters, 1994, 2601).

The nucleophilic displacement of halide from 4,4'-dihalogenobenzophenones by cyclic secondary amines is efficient for the difluoro compound, affording a variety of analogues of Michler's ketone. However, the dichloro derivative undergoes monosubstitution, except under forcing conditions (S. F. Beach *et al.*, J. Chem. Soc., Perkin Trans. 2, 1985, 107 and 1989, 1087).

Some methoxy and methylenedioxy derivatives of benzophenone have been obtained from gallic acid, 3,4,5-trihydroxybenzoic acid, involving a benzilic acid rearrangement and subsequent oxidation (S.-H. Zee and S.-Y. Chou, J. Chinese Chem. Soc., 1990, 37, 517).

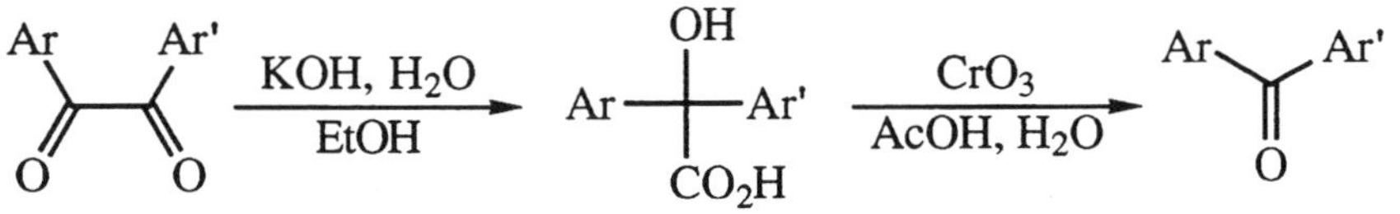

A novel route to substituted derivatives of 2-aminobenzophenones is based on the facile ring opening of pyrylium salts by nucleophiles. Reaction of 2,4,6-triarylpyrylium salts with variously substituted benzyl cyanides, pyridyl acetonitriles and aldehydes under basic conditions leads to 2-amino-3,4,6-triarylbenzophenones (T. Zimmermann and P. G. Jones, J. Prakt. Chem., 1993, 335, 351; T. Zimmermann, *ibid.*, 1993, 335, 644; *idem*, J. Heterocyclic Chem., 1994, 31, 277). In a like manner, the corresponding thiopyrylium salts react with arylacetaldehydes to give 2,4,5-triarylthiobenzophenones (T. Zimmermann, Synthesis, 1994, 252).

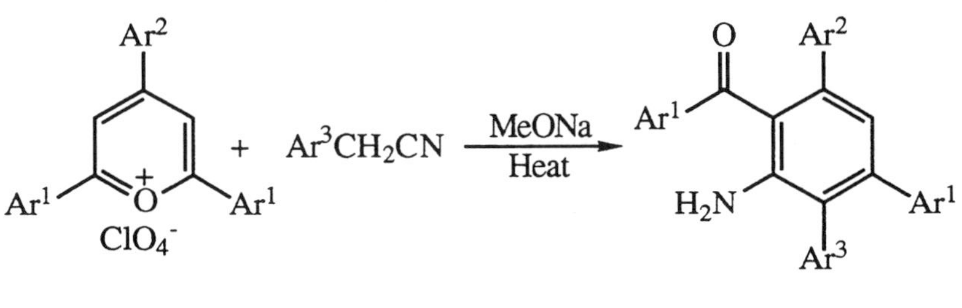

The aromatisation of cyclohexenones with mercury (II) acetate has been used to prepare 2,6-dihydroxybenzophenone starting from benzoyl chloride and cyclohexan-1,3-dione (J. E. Oliver, K. R. Wilzer and R. M. Waters, Synthesis, 1990, 1117).

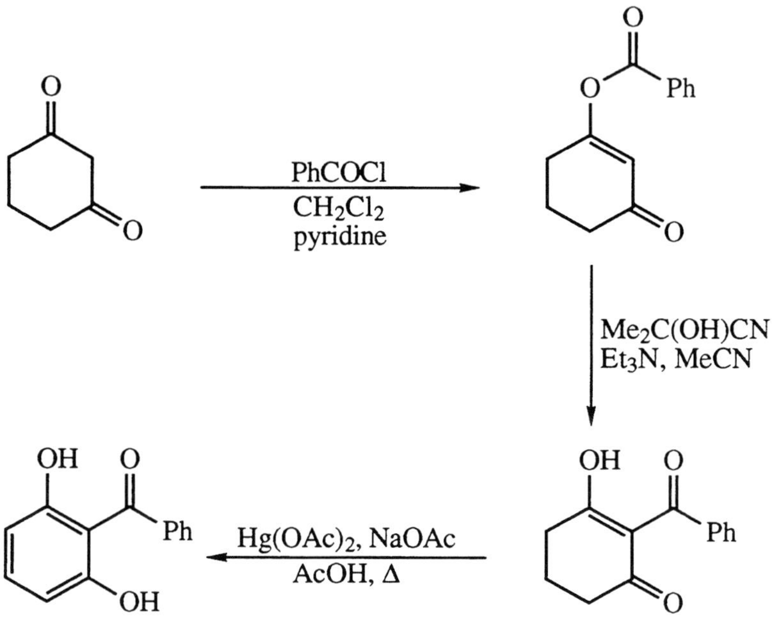

(ii) Properties and reactions

The preferred conformation of benzophenone is helical with the rings twisted out of the C=O plane in opposite directions (12), rather than the arrangement with the two rings twisted in the same direction, *i.e.* the gable form (13) (I. Baraldi, E. Gallinella and F. Momicchioli, J. Chem. Phys., 1986, 83, 655). Various estimates of the torsion angle have been proposed, though X-ray diffraction indicates an angle of 30° (G. M. Lobanova, Kristallografiya, 1984, 13, 1968), a value supported by a figure of 29° determined for 4,4'-dichlorobenzophenone (M. M. Granger and M. F. Coillot, Acta Crystallogr. Sect. C, 1985, 41, 542). Theoretical values based on NDO methods give an angle of 90° (K. M. Weitzel and H. Bassler, J. Chem. Phys.,

1986, 84, 3). However, MNDO fails to account correctly for conjugation energy, whereas Gaussian methods (STO-3G) predict an energy minimum conformation when the angle of twist is 36°. These calculations also suggest that rotation of the rings occurs in a disrotatory manner and proceeds through a perpendicular intermediate (R. J. Abraham and I. S. Haworth, J. Chem. Soc., Perkin Trans. 2, 1988, 1429).

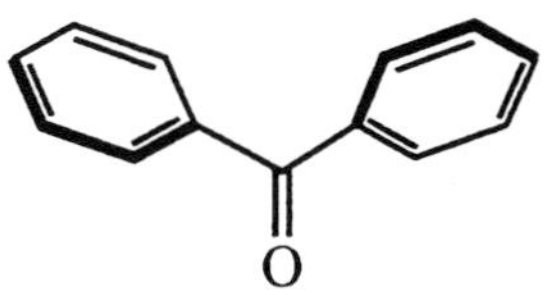

(**12**)

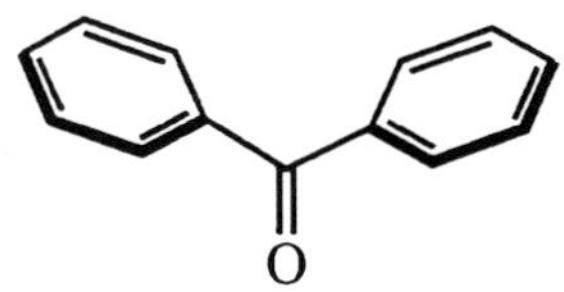

(**13**)

4,4'-Bis(dimethylamino)benzophenone, Michler's ketone, exhibits pronounced solvatochromism, which in the case of phenolic solvents is attributed to charge-transfer complexation (C. Spange and D. Keutel, Liebigs Ann. Chem., 1992, 423). X-Ray data for the 1:1 complex of the ketone with pentafluorophenol indicate coordination to the carbonyl group (T. Gramstad *et al.*, Acta Chem. Scand. Ser. B, 1987, 41, 555).

The vibrational spectra of some *p*-disubstituted benzophenones have been determined (V. Volovsek *et al.*, J. Raman Spectros., 1991, 22, 35 and 43).

Benzophenone *O*-oxide has been isolated in an argon matrix through the reaction of diphenylcarbene with oxygen. Its infrared and ultraviolet spectra have been recorded (W. Sander, Angew. Chem. Int. Ed. Engl., 1986, 25, 255).

$$Ph_2C{=}N_2 \xrightarrow{h\nu} Ph_2C: \xrightarrow[35\ K]{O_2} Ph_2C{=}O^+{-}O^-$$

The photochemical addition of benzophenone to alkenes (the Paterno-Büchi reaction) to produce oxetanes has been extended to methyl vinyl sulfides. The reaction is both regiospecific and stereoselective, leading to the *trans*-4-alkyl-3-methylthio-oxetanes (T. H. Morris, E. H. Smith and R. Walsh, Chem Commun., 1987, 964). With 2-methylselenophene, the selenooxetane (14) is produced (C. Rivas, D. Pacheco and F. Vargas, Acta Sud Am. Quim., 1982, 2, 1). Under similar conditions, benzophenone adds to both double bonds of allenic esters affording a mixture of mono- and di-oxetanes together with benzopinacol (M. P. S. Ishar and R. P. Gandhi, Tetrahedron, 1991, 47, 2211).

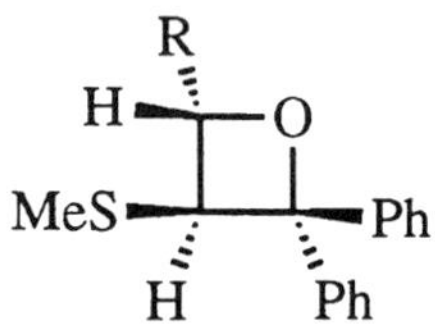

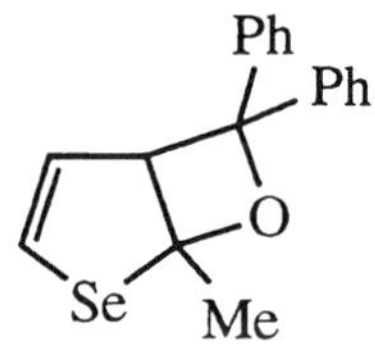

(**14**)

The photocyclisation of highly substituted 2-methoxymethylbenzophenones is stereoselective and leads to 1-aryl-1-hydroxy-2-methoxy benzocyclobutenes in high yield (G. Coll *et al.*, J. Org. Chem., 1992, 57, 6222).

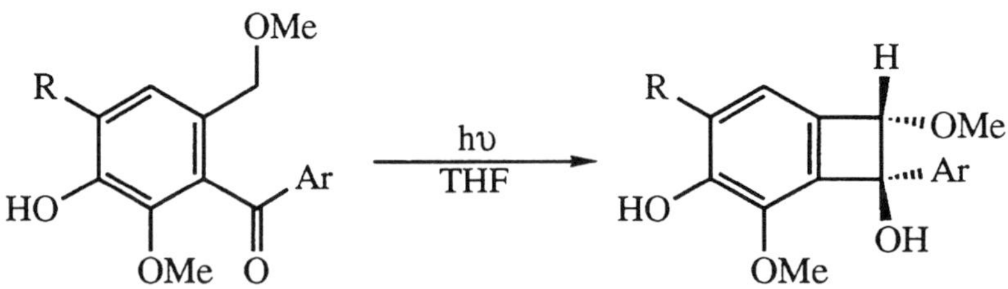

Benzophenone can be *tert*-butylated by reaction with *tert*-butyllithium followed by oxidation of the 1,6-adduct by thionyl chloride. A vinylsulfuryl chloride intermediate is postulated, which undergoes loss of sulfur monoxide and HCl. Some 1,4-addition also occurs, yielding the 2-substituted benzophenone. The 4-*tert*-butylbenzophenone can be further butylated (G. A. Olah, A. Wu and O. Farooq, Synthesis, 1991, 1179).

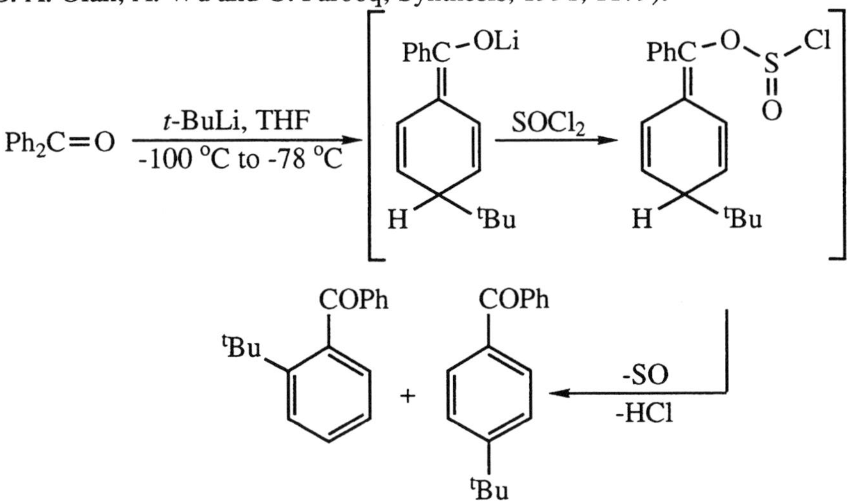

There has been much debate on the mechanism of the reaction of benzophenone with Grignard reagents. Based largely on the determination of carbon kinetic isotope effects, it now seems that the rate determining step involves an inner-sphere electron transfer. An electron is donated by the Grignard reagent to the ketone with a concerted transfer of magnesium to oxygen, producing a magnesium benzophenone ketyl (T. Holm, J. Amer. Chem. Soc., 1993, 115, 916).

The reaction with organolithium reagents follows a similar pathway, although there seems to be some differences in the fine detail of the mechanisms (H. Yamataka, N. Miyano and T. Hanafusa, J. Org. Chem., 1991, 56, 2573).

A one electron transfer is also involved in the Meerwein-Pondorf-Verley reduction of benzophenone, generating a benzophenone ketyl. Reduction to benzhydrol proceeds by abstraction of a β-hydrogen atom from the alkoxide (E. C. Ashby and J. N. Argyropoulos, J. Org. Chem., 1986, 51, 3593).

The reduction of benzophenone by the action of various hydrides proceeds by a polar mechanism, although reaction with diisobutylaluminium hydride (DIBAL) involves a rate determining electron transfer from DIBAL to the ketone (H. Yamataka and T. Hanafusa, J. Org. Chem., 1988, 53, 772). In the case of reduction by lithium dialkylamides containing a β-hydrogen atom, an electron transfer process is discounted, the reaction proceeding by β-hydride transfer (M. Newcomb and M. T. Burchill, J. Amer. Chem. Soc., 1984, 106, 2450).

The mechanism of the Clemmensen reduction of benzophenone is complex. Two main pathways have been recognised, their relative importance depending on the reaction conditions. Both proceed through benzhydryl chloride and both involve ionic and radical reactions. In one, reduction follows a single electron transfer from zinc to the ketone with concomitant bonding of zinc to oxygen. The other proceeds *via* formation of benzhydrylidene dichloride. Dimeric products arise not only from the initially formed radical, but also from other radical species (M. L. Di Vona and V. Rosnati, J. Org. Chem., 1991, 56, 4269). When amalgamated zinc is used, a new coupling product, 1,1,2,2-tetraphenylethane, has been isolated (S. K. Talapatra, Tetrahedron, 1990, 46, 6047). The reductive deoxygenation of benzophenone by organoaluminium Lewis acids provides an alternative to the Clemmensen reaction. Whereas diisobutylaluminium hydride alone converts the ketone solely into benzhydrol, when used in conjunction with isobutylaluminium dichloride or aluminium bromide, diphenylmethane is formed exclusively. The ease of the reaction, which appears to be driven by the formation of a dialuminoxane species, is controlled by the stability and ease of formation of the carbocation intermediate (J. J. Eisch, Z.-R. Liu and M. P. Boleslawski, J. Org. Chem., 1992, 57, 2143).

Carbon dioxide is incorporated electrochemically into benzophenone to yield benzilic acid in the presence of an iron-sulfur catalyst (T. Tomohiro, M. Kodaka and H. Okuno, Chem. Express, 1991, 6, 755).

Reaction of *o*-fluorophenyl lithium with dicarbonyl (η^5-cyclopentadienyl)-iodoiron, (η^5-C_5H_5)Fe$(CO)_2$I, forms an aryliron complex which on treatment with $(MeCN)_3Cr(CO)_3$ yields the metallated benzophenone (15) (I. R. Butler *et al.*, Organometallics, 1990, 9, 1964).

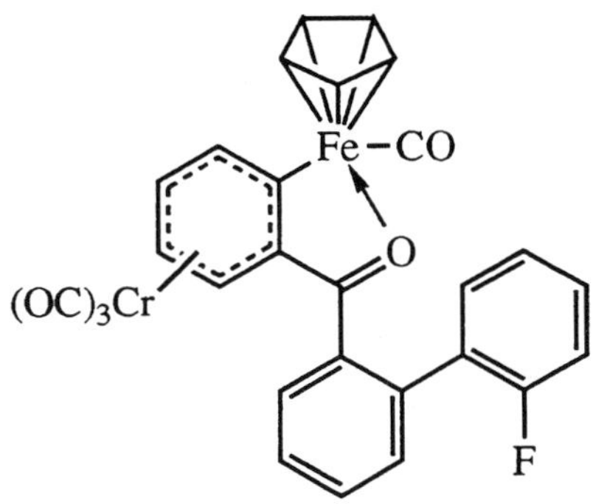

(**15**)

(iii) Functional derivatives of benzophenone

Benzophenone *N*-arylimines cannot usually be prepared directly by reaction of the diaryl ketone with 2-substituted aniline derivatives. An indirect route to such sterically crowded anils involves formation of the anil from a benzaldehyde, followed by reaction with an aryl lithium and oxidation with DDQ. NMR studies indicate that with an unsubstituted diphenylmethyl unit, one phenyl ring is coplanar with the C=N bond, whilst the other is twisted out of the C=N plane. The *N*-aryl moiety is twisted such that it approaches perpendicularity with the C=N plane. Anils with an unsymmetrically substituted diphenylmethyl system exist as mixtures of the *E*- and *Z*-diastereomers. The latter configuration is preferred in which the more sterically demanding *C*-aryl group is perpendicular to the C=N bond and *syn* to the *N*-aryl ring (L. Streckowski *et al.*, J. Org. Chem., 1989, 54, 2464).

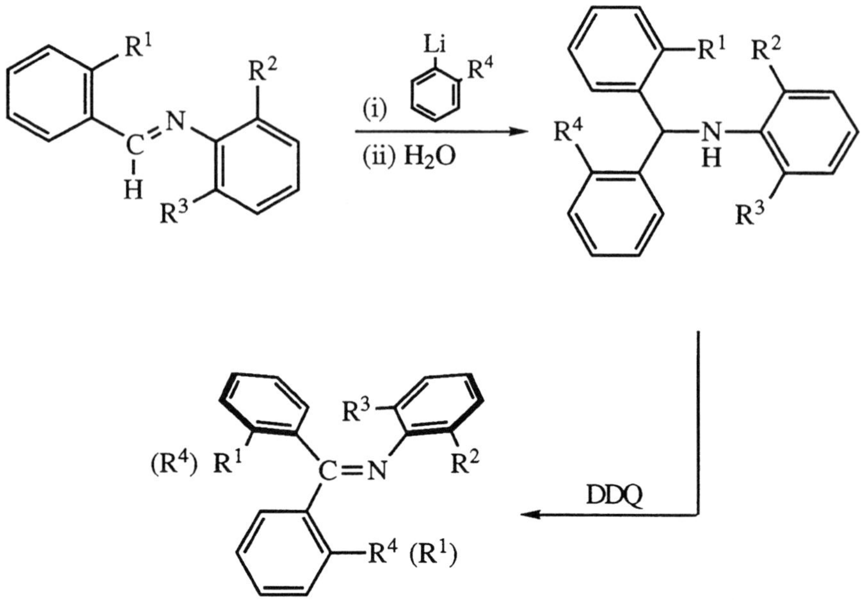

Decarboxylative chlorination of carboxylic acids can be achieved by photolysis of their benzophenone oxime esters providing a route to both aryl and alkyl chlorides (M. Hasebe and T. Tsuchiya, Tetrahedron Letters, 1988, 6287).

$$RCOON{=}CPh_2 \xrightarrow[CCl_4]{h\nu} R{-}Cl$$

Potassium bromate effects the regeneration of carbonyl compounds from their semicarbazones, whilst the oxidative hydrolysis of thiosemicarbazones by $KBrO_3$ is catalysed by mercury (II) acetate (K. Ramakrishnan and V. S. Srinvasan, J. Chem. Soc., Perkin Trans. 2, 1986, 1557; *idem, ibid.*, 1991, 1523). Benzophenone is also regenerated from its semicarbazone on treatment with $PhI(OAc)_2$ in aqueous acetonitrile (D. W. Chen and Z. C. Chen, Synthesis, 1994, 773).

The deprotection of the dithiane, dithiolane and diphenyldithioacetal derivatives of some benzophenones occurs photochemically on irradiation in the presence of molecular oxygen using 2,4,6-tri(4-chlorophenyl)pyrylium perchlorate as a photosensitiser (M. Kamata *et al.*, Tetrahedron, 1994, 50, 12821).

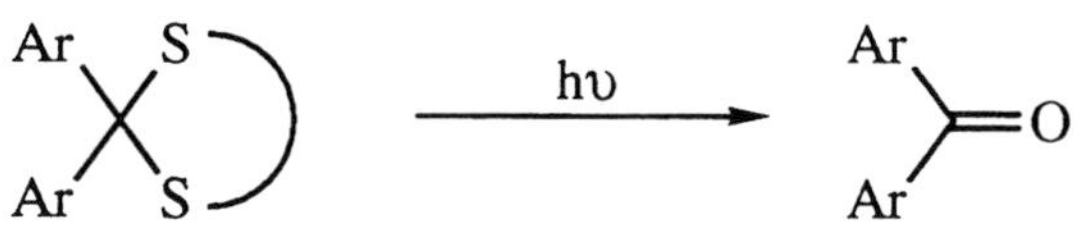

The anion derived from benzophenone benzoylhydrazone (16) is converted into an oxadiazolidine (17) in the presence of acid chlorides. Electron transfer from the anion to the acid chloride to give a radical anion - radical pair is suggested as the critical step.

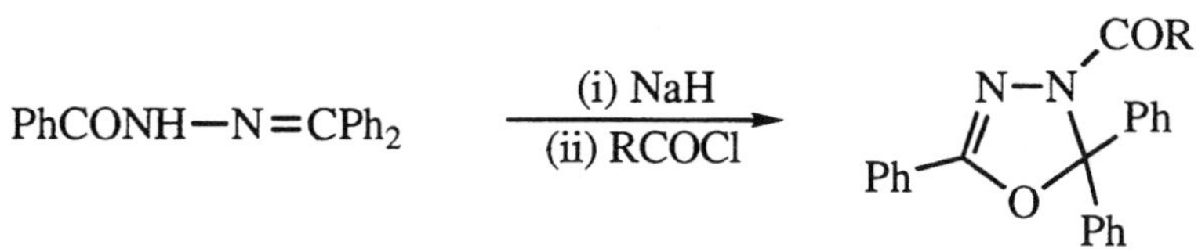

(16) **(17)**

In the absence of an acid chloride, products such as the enol ether (18) and the *N*-methylamide (19) are obtained through attack of an electrophile on the open chain anion (20) (D. Armesto *et al.*, Tetrahedron Letters, 1988, 3581).

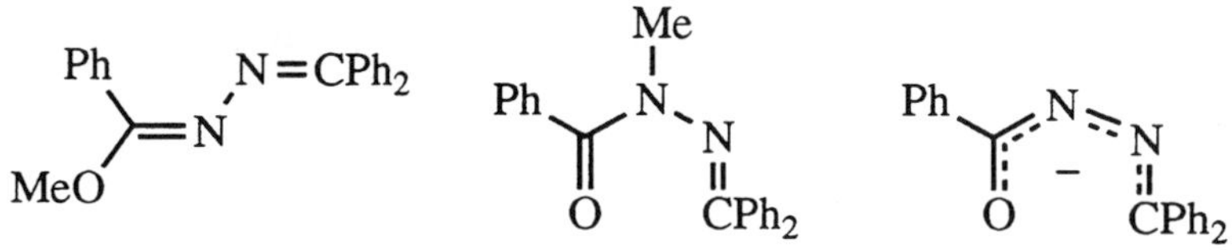

(18) **(19)** **(20)**

The dianion obtained by sequential treatment of 2-bromobenzophenone tosylhydrazone with methyllithium and butyllithium undergoes selective attack at carbon by various electrophiles. Reaction with acid chlorides is particularly useful, providing access to a range of heterocyclic compounds (J. T. Sharp and C. E. D. Skinner, Tetrahedron Letters, 1986, 869).

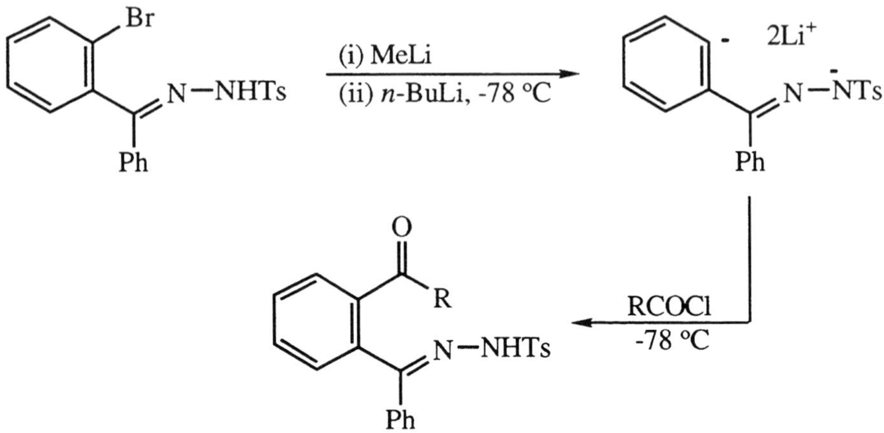

(iv) Benzoylbenzoic acids

Aroyl chlorides are converted into 2-benzoylbenzoic acids by reaction with 2-lithioaryloxazolines (K. J. Edgar and C. K. Bradsher, J. Org. Chem., 1982, 47, 1585) and the reaction of arylcoppermagnesium reagents with phthalic anhydride leads to 2-aroylbenzoic acids together with varying amounts of 3,3-diphenylphthalide. In the presence of dimethyl sulfide, less of the phthalide is produced. Phenyl lithium however yields the phthalide exclusively (M. T. Rahman and S. K. Nahar, Organometallics, 1992, 425, 201).

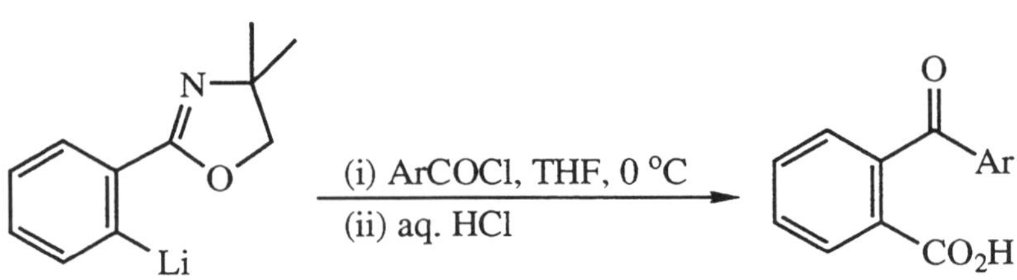

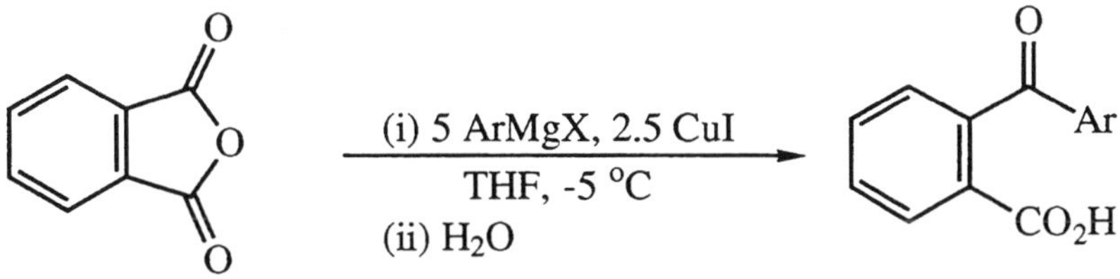

A Grignard reagent also features in a synthesis of 4-benzoylbenzoic acid isotopically labelled at the carboxyl function. Protection of the carbonyl group in 4-bromobenzophenone precedes Grignard formation and carboxylation with $^{14}CO_2$ (K. L. Nakamaye and R. G. Yount, J. Labelled Comp. Radiopharm., 1985, 22, 607).

Simultaneous oxidation of the methyl and methylene functions in some 5-*tert*-butyl-2-methyldiphenylmethanes occurs with $KMnO_4$ in aqueous pyridine resulting in the formation of 2-aroyl-4-*tert*-butylbenzoic acids. Cyclodehydration of these acids with concentrated sulfuric acid leads to the formation of 2,6- and 2,7- disubstituted anthraquinones in a ratio which depends on the nature of R. The isomers arise by direct intramolecular acylation and through a Hayashi rearrangement involving the acyl cation (Yu. V. Pozdnyakovich, R. P. Savyak and S. M. Shein, Zh. Org. Khim., 1986, 22, 595).

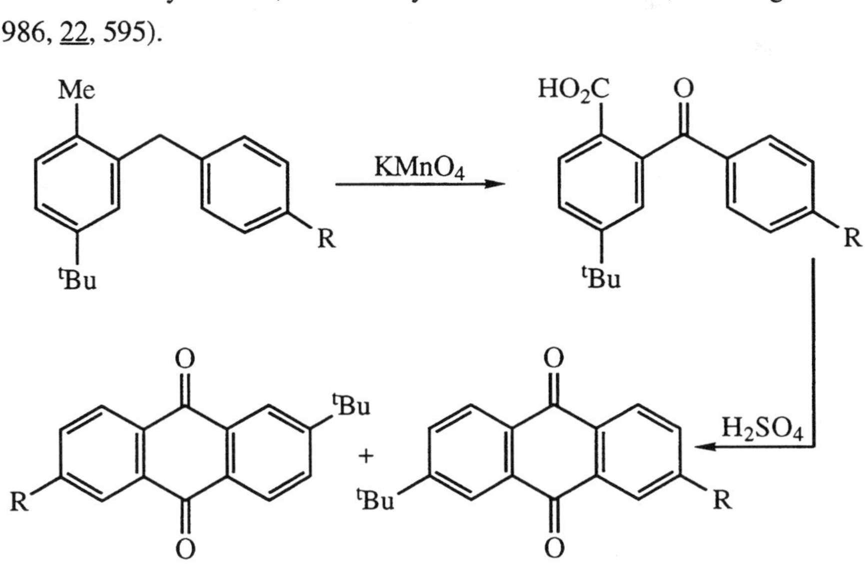

The hydrazone derivatives of 2-aroylbenzoic acids cyclise to phthalazinones (S. V. Yakovlev and L. A. Pavlova, Zh. Org. Khim., 1988, 24, 2433; F. A. Yassin *et al.*, Rev. Roum. Chim., 1991, 36, 201).

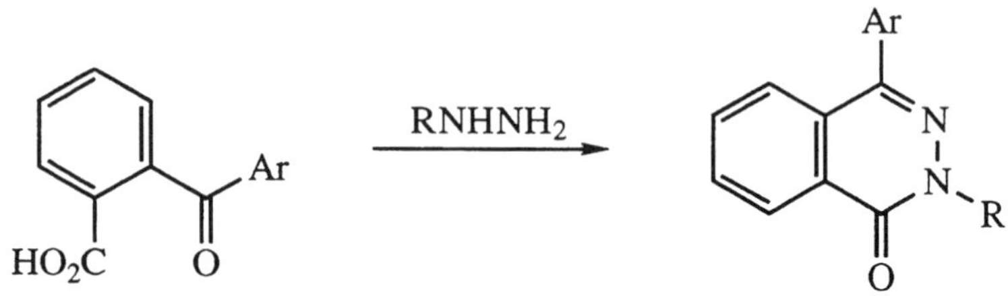

The crystal structures of the anhydrous and hydrated forms of 2-benzoylbenzoic acid have been described (R. A. Lalancette, P. A. Vanderhoff and H. W. Thompson, Acta Cryst., 1990, C46, 1682).

2 Triphenylmethane, its derivatives and oxidation products

(a) Triphenylmethane and its nuclear substituted derivatives

(i) General methods of preparation and properties

Further examples of the synthesis of triphenylmethane derivatives by the condensation of a substituted benzaldehyde or benzhydrol with an activated aromatic compound have been reported as illustrated below.

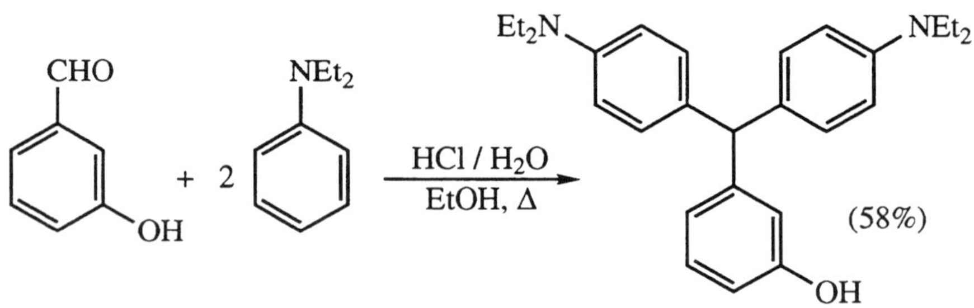

(B. M. Fox *et al.*, J. Chem. Soc., Perkin Trans. 2, 1982, 987)

The recently developed and rapidly expanding synthetic potential of benzotriazoles has been applied to the synthesis of asymmetrically substituted triphenylmethanes. For example, mesitylene is benzotriazolylalkylated on treatment with an aromatic aldehyde and 1-(benzenesulfonyl)-benzotriazole. Subsequent reaction with *N,N*-dimethylaniline in the presence of zinc chloride affords the triarylmethane *via* the carbocation (A. R. Katritzky *et al.*, Heterocycles, 1994, 38, 345).

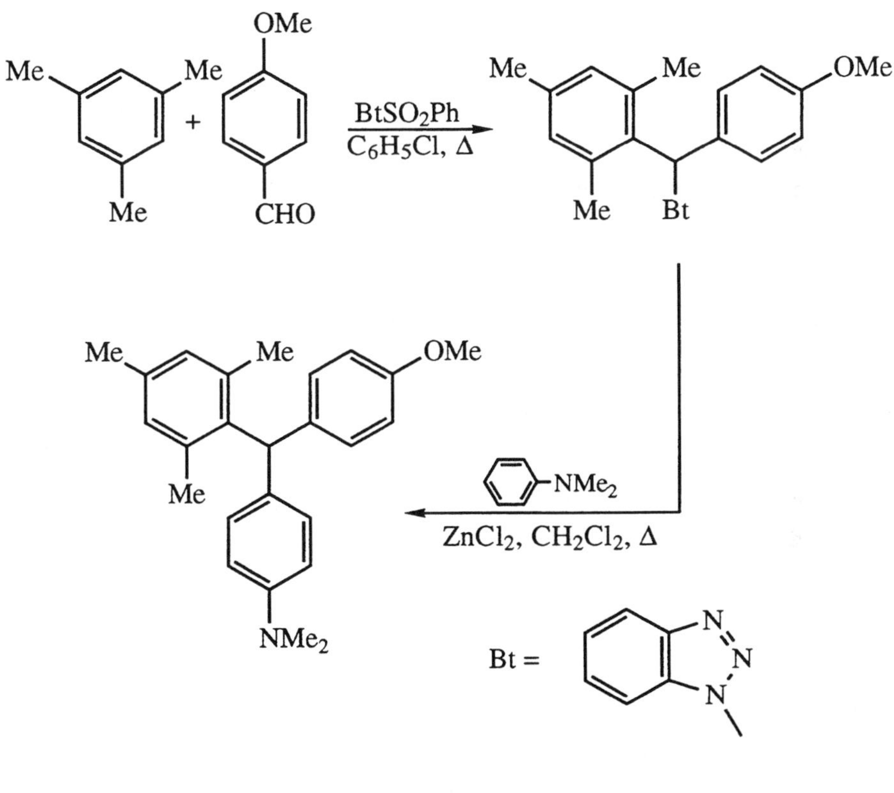

The cleavage of tetraarylpinacolones leads to triphenylmethanes in high yields under reducing or basic conditions (T. Tanwari, J. Chem. Soc. Pak., 1990, 12, 351; M. Stranberg and J.-P. Anselme, J. Chem. Educ., 1990, 67, 616).

Triphenylmethanes result from the reduction of triphenylmethanols using for example the reagents listed below. This reduction has also been effected by an NADH model compound, 9,10-dihydro-10-methylacridine, in the presence of perchloric acid, which allows hydride transfer from the dihydroacridine to the triphenylmethyl cation (M. Ishikawa *et al.*, Bull. Chem. Soc. Jpn., 1989, 62, 3754).

$$Ar_3COH \longrightarrow Ar_3CH$$

Reagent	Conditions	Yield %	Reference
R_3B	CF_3SO_3H, -30 °C - RT	100	1
$NaBH_4$	$AlCl_3$, THF, reflux, 2h.	95	2
Me_2SiCl_2	NaI, MeCN, 10 min.	95	3
NaSEt	DMF, reflux, 9h.	100	4
RCH_2OH	propan-2-ol, $HClO_4$, 50 °C	85	5
Li	DBB, THF, ultrasound, 48h.	41	6

References:

1. G. A. Olah *et al.*, J. Org. Chem., 1991, 56, 2759.
2. A. Ono *et al.*, Synthesis, 1987, 736.
3. J. M. Wiggins, Synth. Commun., 1988, 18, 741.
4. R. Berscheid *et al.*, Chem. Ber., 1992, 125, 1687.
5. N. Wada *et al.*, Chem. Commun., 1993, 1462.
6. R. Karaman *et al.*, Tetrahedron Letters, 1990, 6155.

4,4'-Dihydroxytriphenylmethane is formed by the base catalysed reaction of 2,4,6-tribromophenol with tropylium perchlorate followed by catalytic dehydrogenation. A norcaradiene intermediate is postulated which rearranges to a quinone methide which is subsequently attacked by a second molecule of the phenol. A reductive dehydrohalogenation completes the sequence (T. Morita *et al.*, Bull. Chem. Soc. Jpn., 1993, 66, 337).

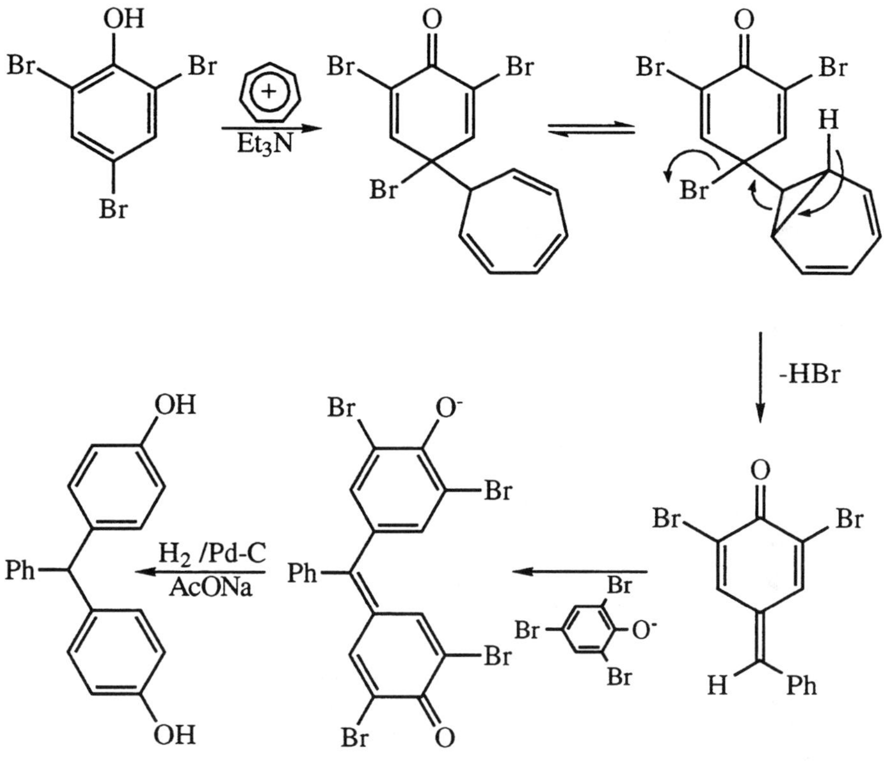

The isolation of the first naturally occurring triarylmethane derivative, melanervin (21) has prompted syntheses of this and related materials. Direct alkylation of a flavanone with diphenylmethanol and its derivatives occurs at the 6- and 8- positions when these are unsubstituted, with preferential attack at the former site (S. Antus *et al.*, Tetrahedron, 1982, 38, 133).

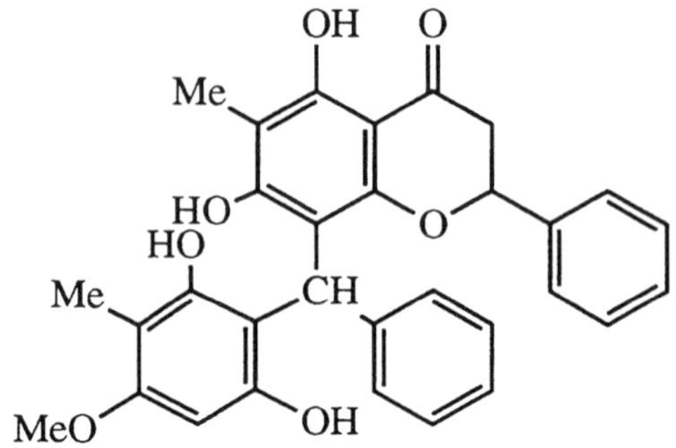

(**21**)

In an alternative approach, benzhydrylation of variously substituted acetophenones is followed by cyclisation to the chromone (A. C. Jain *et al.*, Indian J. Chem., 1985, 24B, 393 and 1023; *idem, ibid.*,1986, 25B, 755).

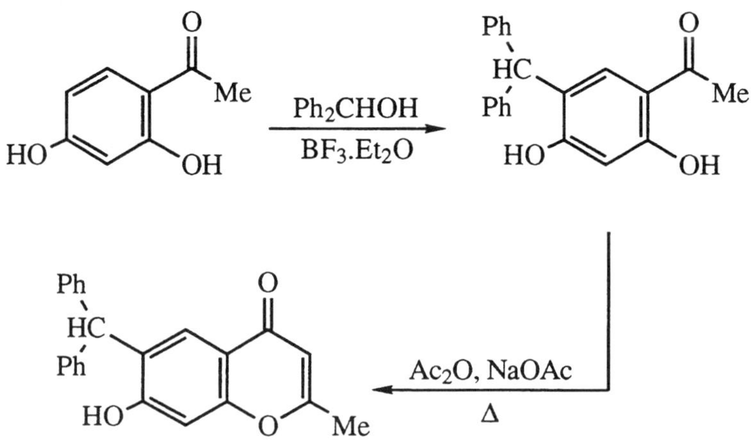

An excellent and comprehensive review of the photochemistry and photophysics of triphenylmethane dyes in both solid and liquid media which covers the literature until 1991 has been published (D. F. Duxbury, Chem. Rev., 1993, 93, 381). Studies of the rates of hydroxylation of a variety of triphenylmethane dyes have shown that the process is under entropy control (S. F. Beach *et al.*, J. Chem. Soc., Perkin Trans. 2, 1985, 107).

Interest in infrared absorbing dyes has focussed attention on ethynologues of triphenylmethane dyes (22). Reduced crowding about the central carbon atom arising from the spacer effect of the ethynyl unit results in an increased planarity of the cation which together with substituent effects leads to significant red shifts of the absorption band. Thus, whilst Malachite Green and Crystal Violet absorb at 621 and 589 nm, respectively, their ethynologues absorb at 688 and 663 nm (S. Nakatsuji *et al.*, Chem. Letters, 1986, 329). The effects of bridging the triphenylmethane cation across the 2' and 2" positions with a heteroatom (23) also brings about a pronounced bathochromic shift associated with increased planarity about the xanthylium system (S. Akiyama *et al.*, J. Chem. Soc., Perkin Trans. 1, 1988, 3157).

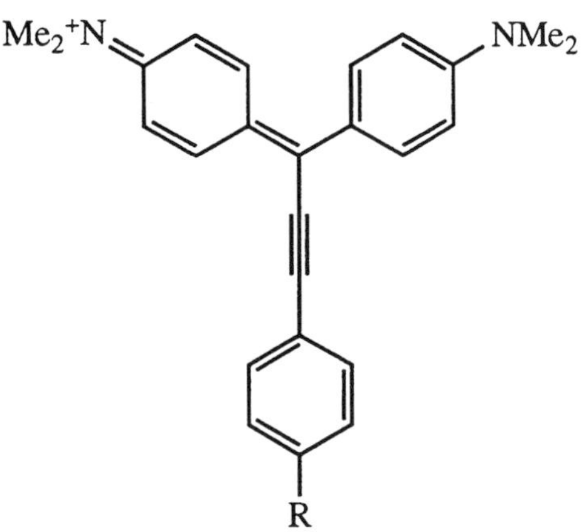

(22)

R= H; λ_{max}= 688 (493) nm
R= NMe$_2$; λ_{max}= 663 nm

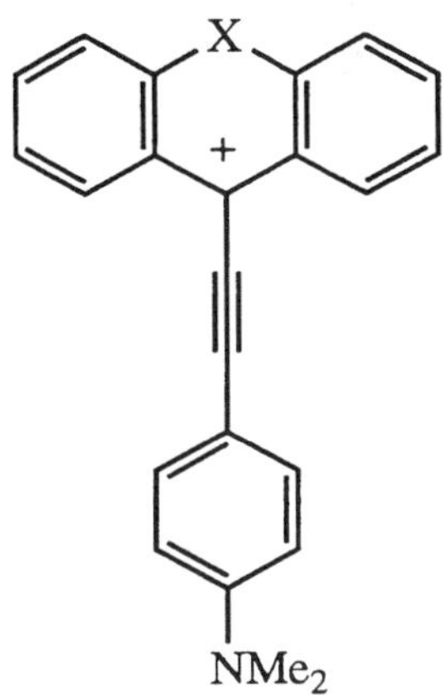

(23)

X= O; λ_{max}= 737 nm
X= S; λ_{max}= 775 nm

The reaction of ozone with the dye Malachite Green (24a) occurs by two distinct routes. Attack at the terminal dimethylamino unit leads to the *N*-formyl-*N*-methylamino compound (24b), whilst the products arising from attack at the central carbon atom are primarily 4-(*N,N*-dimethyl amino)benzophenone and 4-(*N,N*-dimethylamino)phenol (M. Matsui *et al.*, Bull. Chem. Soc. Jpn., 1984, 57, 3312).

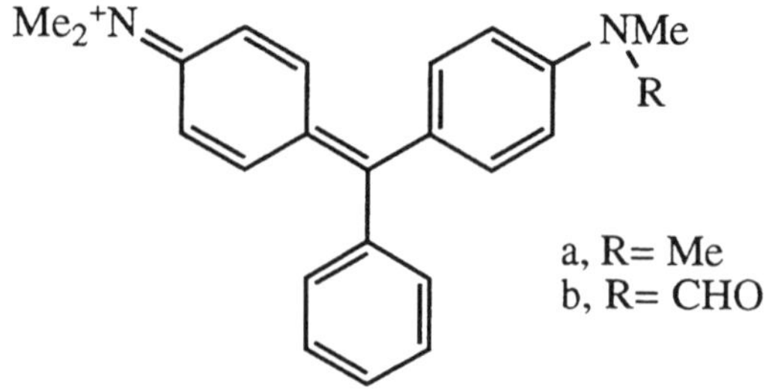

(24)

On the other hand, lacking both the cationic site and amino substituents, triphenylmethane itself is unreactive towards ozone, enabling a relative sequence of reactive sites in both the gas and liquid phase to be formulated:

$$C{=}C \;>\; NR_2 \;>\; C{=}N \;>\; \text{aromatic ring}$$

(D. Grosjean, P. M. Whitmore and G. R. Cass, Environ. Sci. Technol., 1989, 23, 1164; R. Atkinson and W. L. P. Carter, Chem. Rev., 1984, 84, 437).

On treatment with a Cs - K - Na alloy in THF at -75 °C, triphenylmethane forms a red dianion involving a 1,2-phenyl shift. On protonation, the cyclohexadiene derivative (25) is produced, in which the two allylic hydrogen atoms are non-equivalent in the NMR spectrum. The corresponding dianion from 1,1,1-triphenylethane cyclises in a stereo-specific manner to form 9-methyl-9-phenyl-*cis*-2,4a,4b,7-tetrahydrofluorene (E. Grovenstein *et al.*, Organometallics, 1990, 9, 2587).

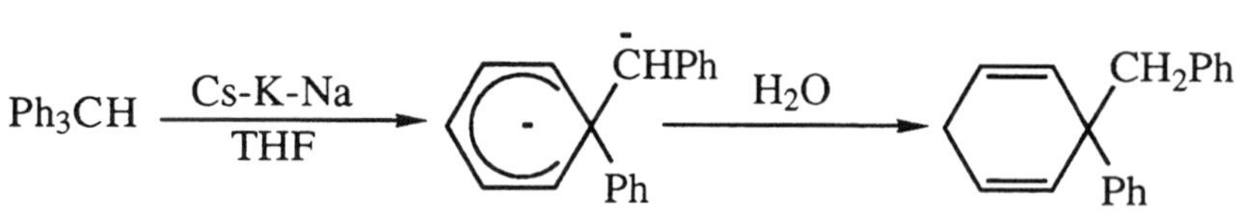

(25)

Thermolysis of triphenyl(phenylazo)methane, ($Ph_3C-N=NPh$), and its deuterated derivative in 2,2-dimethoxypropane and sodium methoxide confirms that trityl radicals exhibit ambident reactivity in hydrogen abstraction reactions (P. Huszthy *et al.*, J. Chem Soc., Perkin Trans. 2, 1985, 491).

(b) Triphenylmethyl halides and their nuclear substituted derivatives

Fluorination of triphenylmethane to give triphenylmethyl fluoride occurs on reaction with caesium fluoroxysulfate, $CsSO_4F$, at room temperature (S. Stauber and H. Zupan, Tetrahedron, 1992, 48, 5875) and by photolysis in CH_3CN using AgF as the source of fluorine and TiO_2 as a catalyst (C. M. Wang and T. E. Mallouk, J. Amer. Chem. Soc., 1990, 112, 2016).

(c) Triphenylmethanol and its nuclear substituted derivatives

(i) General methods of preparation

The reaction of organometallic compounds with benzophenone and its derivatives continues to provide new examples of triphenylmethanol derivatives. In particular, this method lends itself to the synthesis of unsymmetrically substituted alcohols:

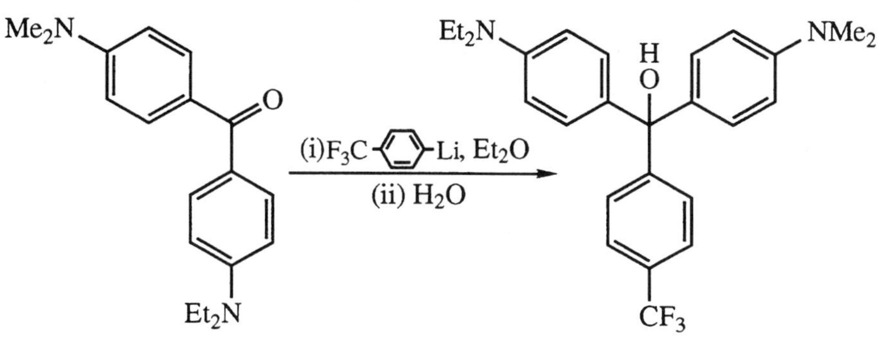

(G. Hallas and M. M. Mitchell, J. Soc. Dyers Colourists, 1986, 102, 15)

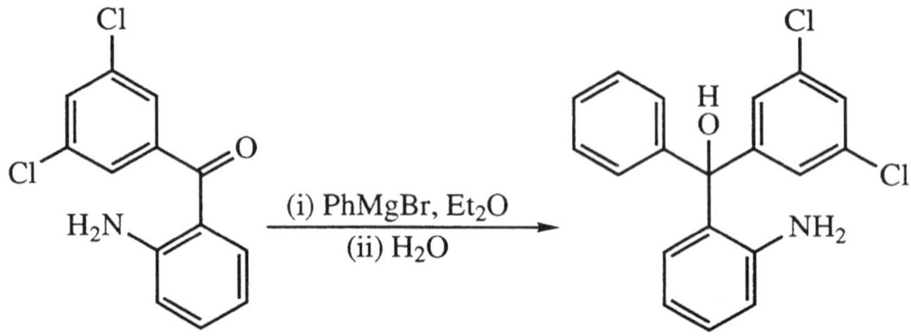

(R. Bolton *et al.*, J. Chem. Soc. Perkin Trans. 2, 1986, 405)

A comprehensive series of triphenylmethanols bearing different tertiary amino functions in the *para*-positions has been prepared (S. F. Beach *et al.*, J. Chem. Soc., Perkin Trans 2, 1989, 1087).

The crowned derivative of Malachite Green carbinol (26) has been obtained by the reaction of 4'-lithiated benzo-15-crown-5 with Michler's ketone. Reaction of the derived cation with KCN yields the corresponding leucocyanide (K. Kimura, M. Kaneshige and M. Yokoyama, Chem. Commun., 1994, 1103).

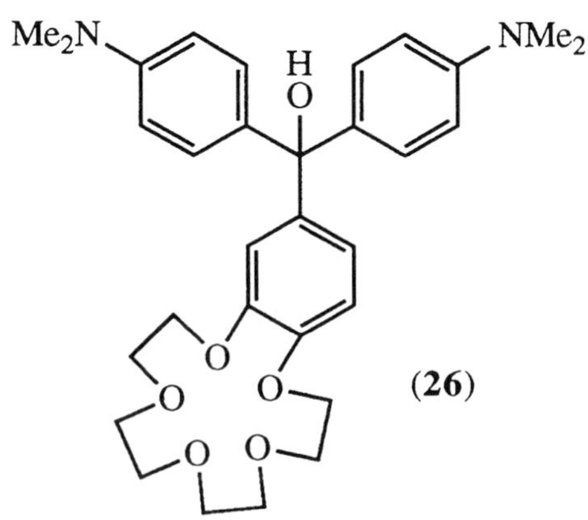

(26)

A number of reagents can effect the oxidation of triphenylmethane into triphenylmethanol, some of which are tabulated below. Oxidation by $K_3Fe(CN)_6$ proceeds *via* the triphenylmethyl radical and the carbocation (A. K. Bhattacharjee and M. K. Mahanti, Gazz. Chim. Ital., 1983, 113, 1).

$$Ph_3CH \longrightarrow Ph_3COH$$

Reagent	Conditions	Yield %	Reference
$KMnO_4$	$Bu_4N^+HSO_4^-$, KOH, CH_2Cl_2	80	1
$KMnO_4$	Et_3N, H_2SO_4	46	2
NaOCl	$Bu_4N^+HSO_4^-$, NaBr, C_6H_6	69	3
CrO_2Cl_2	CCl_4, ultrasound	99	4

References

1. S. M. Gannon and J. G. Krause, Synthesis, 1987, 915.
2. W.-S. Li and L. K. Liu, Synthesis, 1989, 293.
3. J. Correia, J. Org. Chem., 1992, 57, 4555.
4. F. A. Luzzio and W. J. Moore, J. Org. Chem., 1993, 58, 512.

1,6-Bis(4-methoxycarbonylphenoxy)hexa-2,2-diyne (27), obtained from ethyl 4-hydroxybenzoate by reaction with 3-bromopropyne followed by oxidative dimerisation, yields the bistriphenylmethanol (28) on reaction with a four-fold excess of 4-methoxyphenylmagnesium bromide (R. Berscheid and F. Vogtle, Synthesis, 1992, 58).

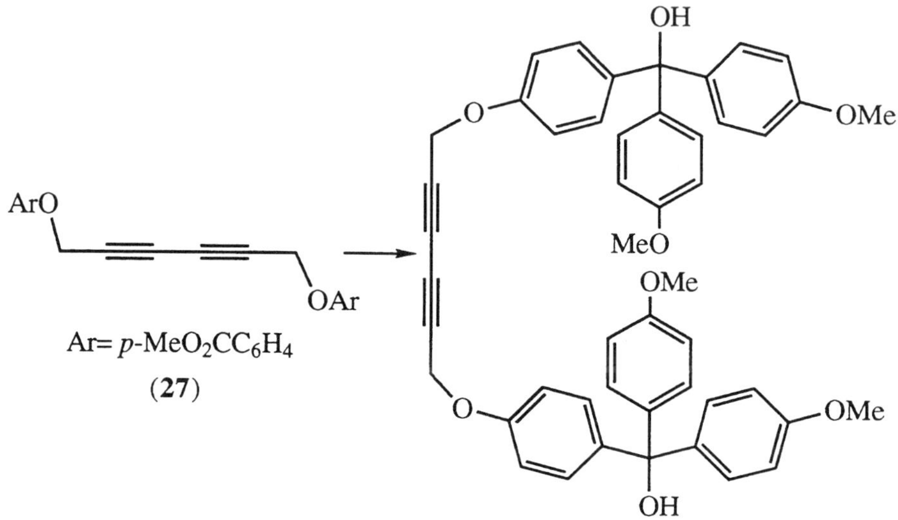

Ar= *p*-$MeO_2CC_6H_4$

(**27**)

(**28**)

A series of multiple bridged triphenylmethane analogues has been obtained utilising the above methodology. Thus, use of 4,4'-dipropynyloxy benzophenone leads to the doubly bridged system, whilst rosolic acid yields the triply bridged example (29) and 1,4-bis[bis(4-hydroxy phenyl)methyl]benzene the four-fold bridged compound. Further elaboration is possible using dihalogenoaromatic compounds to link two 4-hydroxyphenyldiphenylmethane units together. The orange to red cations derived from these macrocyclic species are concave and have a large rigid cavity which can contain a range of guest molecules (R. Berscheid, M. Nieger and F. Vogtle, Chem. Ber., 1992, 125, 1687 and 2539).

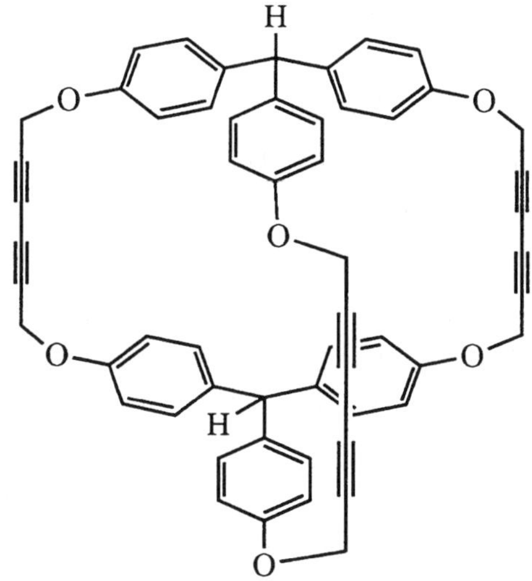

(**29**)

The cation derived from 2-chloro-1,3-dimethylbenzene by dissolution in trifluoromethanesulfonic acid reacts with 1,4-dinitrobenzene to give tris(4-chloro-3,5-dimethylphenyl)methanol. The reaction, which is also shown by chlorotoluenes, is initiated by hydride transfer from the chloro compound to the protonated nitro compound, followed by a series of benzylation and debenzylation reactions (R. P. Austin and J. H. Ridd, J. Chem. Soc., Perkin Trans. 2, 1994, 1205).

Triphenylmethanol exists in the solid state as hydrogen bonded pyramidal tetramers which are oriented along two interpenetrating sites (G. Ferguson *et al.*, Acta Cryst., 1992, C48, 1272). The molecule forms crystalline inclusion compounds with methanol, $Ph_3COH.MeOH$ (1:1), and dimethyl sulfoxide, $Ph_3COH.DMSO$ (2:1), a property which offers potential for the purification of both solvents (E. Weber, K. Skobridis and I. Goldberg, Chem. Commun., 1989, 1195).

On treatment with Lawesson's reagent, 2,4-bis(4-methoxyphenyl)-1,3-dithiaphosphetane-2,4-disulfide, triphenylmethanol is converted quantitatively into triphenylmethanethiol (T. Nishio, Chem. Commun., 1989, 205), the crystal structure of which has been determined (G. Bernadelli, M. Geoffroy and R. Franzi, Z. Krystogr., 1991, 195, 147). Sulfur extrusion is observed when triphenylmethanesulfenyl chloride is treated with 4,4'-dimethoxythiobenzophenone. A dithiirane or a thiosulfine is postulated as an intermediate (R. C. Williams and D. N. Harpp, Tetrahedron Letters, 1991, 7633). Uses of the thiol and triphenylmethanesulfenyl chloride have been reviewed (A. Senning, Sulfur Letters, 1991, 12, 183). The *S*-trityl function can be introduced into carbohydrates through its tetrabutylammonium salt. Removal of the trityl group by sequential reaction with phenylmercury (II) acetate and hydrogen sulfide results in the overall introduction of a thiol group at the anomeric position of the sugar (M. Blanc-Muesser, L. Vigne and H. Driguez, Tetrahedron Letters, 1990, 3869). The use of the S-trityl protecting group for the uracil residue in oligoribonucleotide synthesis has been described (H. Takaku, K. Imai and M. Nagai, Chem. Letters, 1988, 85).

The dediazonisation of unsymmetrically substituted derivatives of 2-amino-triphenylmethanol leads to the formation of 9-arylfluoren-9-ols. However, the direction of ring closure depends on the method of decomposition of the diazonium salt, radical intermediates showing greater selectivity than carbocation species (R. Bolton, R. S. Mguni and G. H. Williams, J. Chem. Soc., Perkin Trans. 2, 1986, 405).

3. Tetraphenylmethane and its derivatives

Tetraphenylmethane has been prepared in good yield by a Friedel Crafts alkylation of aniline with triphenylmethanol, followed by diazotisation and deamination (M. Grimm, B. Kirste and H. Kurreck, Angew. Chem. Int. Ed. Engl., 1986, 25, 1097) and an excellent yield of 4-hydroxytetraphenyl methane results when phenol is alkylated by Ph_3COH in the presence of aluminium phenoxide (V. A. Koshchii, Ya. B. Kozlikovskii and A. A. Matyusha, Zh. Org. Khim., 1988, 24, 1508).

$$Ph_3COH \xrightarrow[(60\%)]{(i)\ (ii)\ (iii)} Ph_4C$$

(i) aniline hydrochloride, (ii) isopentyl nitrite, (iii) H_3PO_2

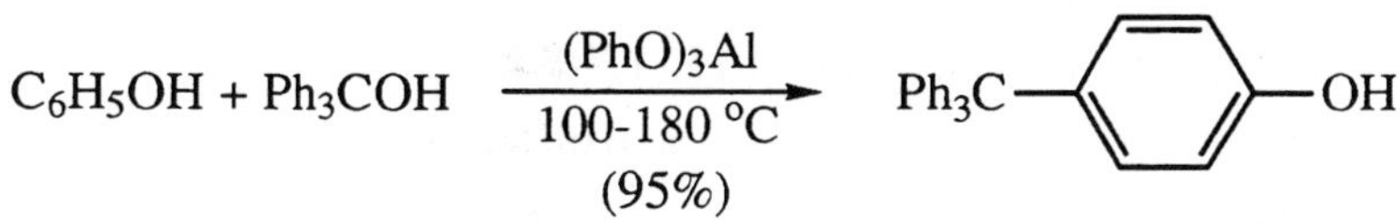

Bromination of tetraphenylmethane yields tetra(4-bromophenyl)methane, m.p. 312-313 °C, which on treatment with KCN in the presence of palladium (II) acetate gives the corresponding nitrile. Hydrolysis and reaction with diazomethane afford the tetracarboxylic acid and its methyl ester. The ester is the starting material for the preparation of tetrakisgalvinol (30) from which a fourfold tetraradical with tetrahedral symmetry can be derived by oxidation with potassium hexacyanoferrate (III) (M. Grimm *et al., loc. cit.*).

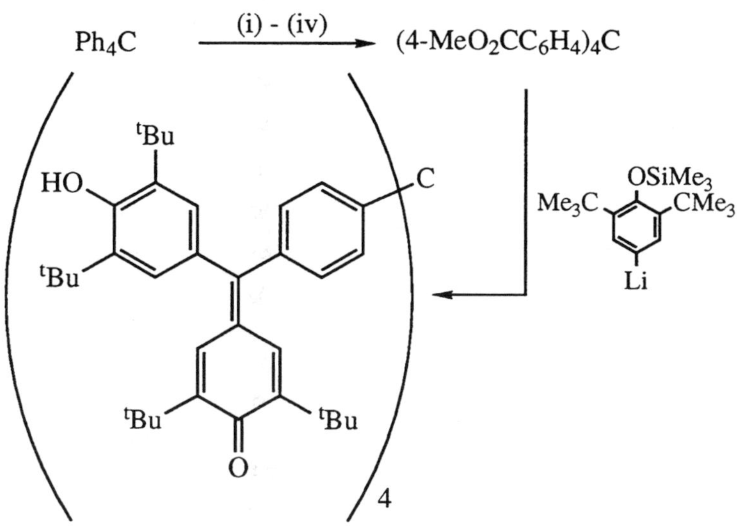

(30)

Reagents: (i) Br_2; (ii) KCN, Pd(II); (iii) KOH; (iv) CH_2N_2

Infrared and Raman studies have permitted the assignment of the whole of the vibrational spectrum of tetraphenylmethane (N. E. Schlotter and B. Hudson, J. Chem. Phys., 1982, 76, 4844; B. Zelei, Spectrochim. Acta, 1986, 42A, 1119).

X-Ray crystallographic studies combined with ^{1}H NMR measurements show that the cyclohexa-1,4-diene ring in *cis*-1,4-dihydro-4-tritylbiphenyl (31) is almost planar. A consequence is that the *ortho* protons of the aromatic ring of the partially reduced biphenyl unit lie in the π-cloud of one of the trityl rings resulting in a high field signal (6.3 ppm) in the ^{1}H NMR spectrum (M. C. Grossel *et al.*, J. Org. Chem., 1993, 58, 6654).

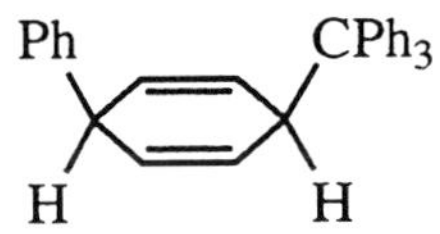

(31)

Second Supplements to the 2nd Edition of Rodd's Chemistry of Carbon Compounds, Vol.III D/E/F (Partial), edited by M. Sainsbury

Chapter 22

DIPHENYL-ALKANES, -ALKENES, AND -ALKYNES, THEIR DERIVATIVES AND OXIDATION PRODUCTS

A. ARMSTRONG and R.A. STOCKMAN

1. *Introduction*

In the twelve years since this area was last surveyed, synthetic organic chemistry has continued to advance, particularly in terms of chemo-, regio- and stereoselectivity. This Chapter, necessarily a subjective selection from the recent literature, will attempt to emphasise developments in these aspects and to illustrate the use of new and more efficient reagents for accomplishing well-known transformations. On the whole, the text will follow the order of the corresponding Chapter in the supplement to the second edition, by Heaney. In general, however, the emphasis of the present account will be on the *synthesis* of diphenyl-alkanes, alkenes, and alkynes and their derivatives; many of the transformations included here represent interconversion of one or more of the title compounds, and such reactions are classified according to the compound synthesised, rather than by that being transformed.

2. *Diphenyl-alkanes and -alkenes, in which the phenyl groups are attached to the same carbon atom*

(a) 1,1-Diphenylalkanes and their derivatives

The venerable Friedel-Crafts reaction is prevalent amongst methods for the preparation of 1,1-diphenylalkanes and their derivatives. The development of new reaction conditions is still of interest and the advantages of perfluorooctane sulfonic acid as a catalyst have been described (X. Fu, M. He, Q. Lei, and B. Luo, *Synth. Commun.*, 1991, **21**, 1273). Laszlo has recently studied the use of K10 montmorillonite resin in the presence of various first row transition metal ions, although the yield improvement over conventional reagents is modest (P. Laszlo and A. Mathy, *Helv. Chim. Acta.*, 1987, **70**, 577). A more interesting example of the use of heterogeneous reaction conditions is seen in a new method for the preparation of methylated diphenylmethane derivatives, useful as insulating oils in high voltage electrical devices. Salmón describes a simple synthesis catalysed by commercially available betonitic earth (M. Salmón, E. Angeles, and R. Miranda, *J. Chem. Soc., Chem. Commun.* 1990, 1188). The reaction was carried out in carbon disulfide at reflux in the presence of bromine and a catalytic amount of suspended bentonite, affording the *ortho*- and *para* - isomers (85%), in a ratio of 1:2, respectively. The process is believed to involve initial free-radical benzylic bromination, followed by the Friedel-Crafts step.

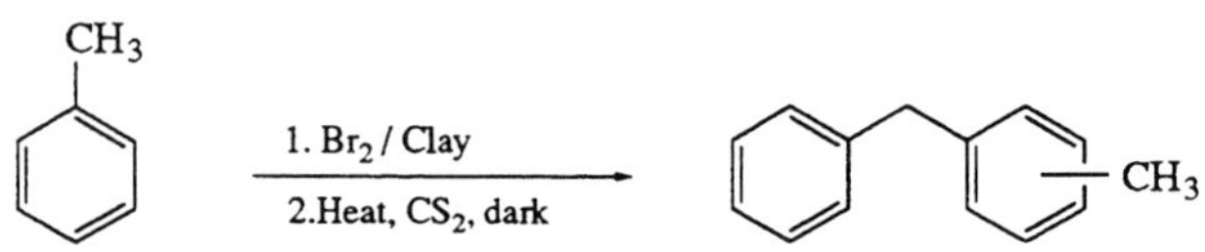

Tandem, double Friedel-Crafts reactions are particularly useful since they allow introduction of both phenyl groups of the target, often in one reaction vessel. An example of this process is the conversion of the morpholine derivative (**1**) to (**2**) upon reaction with an excess of benzene in the presence of a Lewis acid (L. Eberson, M. Malmberg, and K. Nyberg, *Act. Chem. Scand., Ser. B*, 1984, **38**, 345).

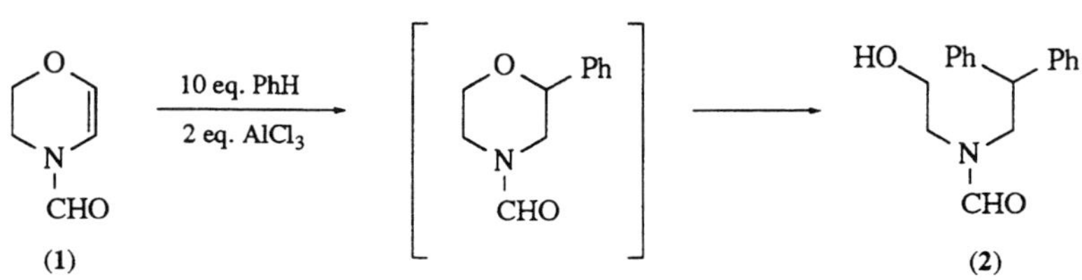

(1) (2) (68%)

Ample demonstration of the utility of the Friedel-Crafts reaction, including the tandem process, is provided by a synthesis of the anti-depressant Sertraline (G. J. Quallich, M. T. Williams, and R. C. Friedmann, *J. Org. Chem.*, 1990, **55**, 4971). Initial regioselective acylation of 1,2-dichlorobenzene was followed by $NaBH_4$ reduction to afford the lactone (**3**). Formation of the key tetralone (**4**) was then achieved by reaction with benzene and one of several suitable acid catalysts.

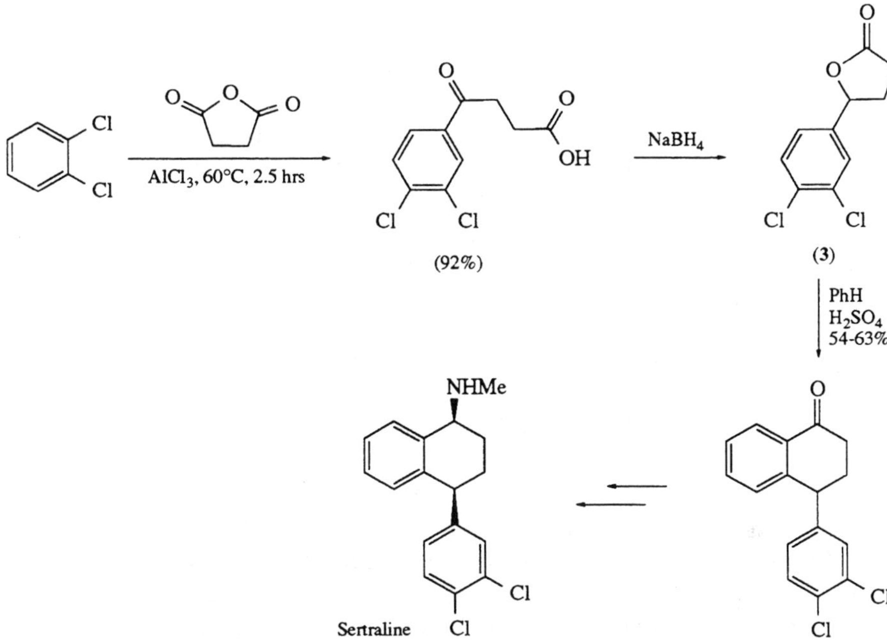

(4)

A facile preparation of 3-aryl-3-(3,4-dimethoxyphenyl)propionic acids relies on the Lewis acid-mediated alkylation of veratrol with ethyl cinnamates. These products are useful intermediates in the synthesis of various isoquinoline derivatives (S. Kano, Y. Yuasa, and S. Shibuya, *Heterocycles*, 1984, **22**, 2327).

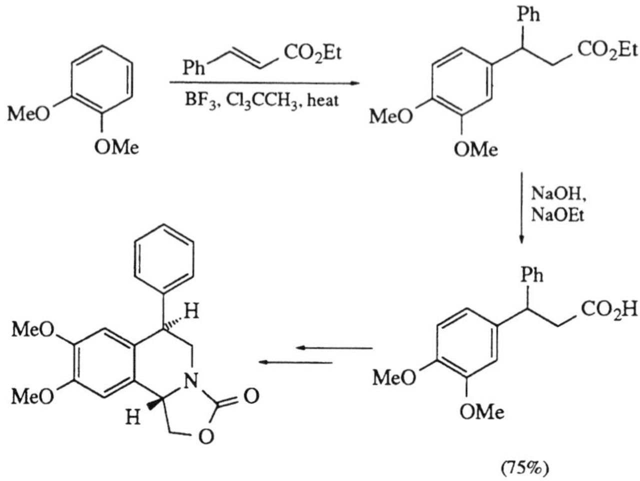

1,1-Diphenylacetic acids can be obtained in a Friedel-Crafts process when 3-(bromomethyl)-5,6-dihydro-1,4,2-dioxazines are used as masked α-acyl cation equivalents (S. Shatzmiller and S. Bercovici, *J. Chem. Soc., Chem. Commun.* 1990, 327).

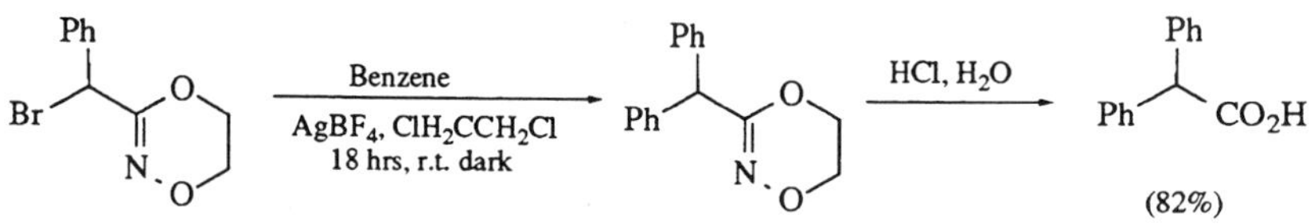

Another method for the synthesis of 1,1-diphenylacetic acid derivatives is provided by the work of Watanabe and co-workers (T. Mitsudo, M. Kadokura, and Y. Watanabe, *J. Org. Chem.*, 1987, **52**, 1695). An interesting palladium-catalysed coupling of allylic carbonates with diphenyl ketene proceeds rapidly under mild conditions. If an allylic acetate is used in place of the allylic carbonate, the product is a 1,3-diene (*vide infra*).

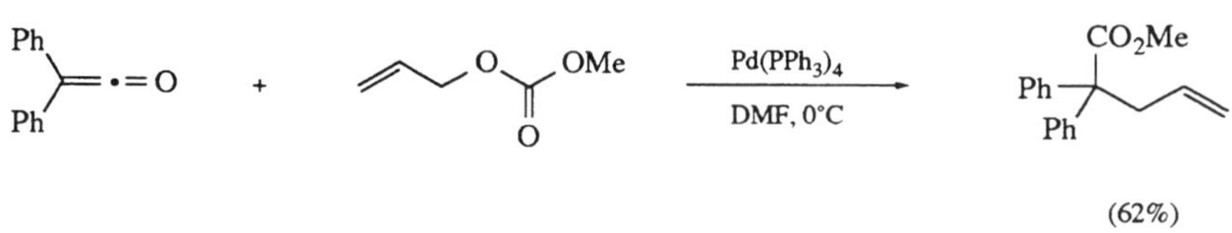

Addition of aryl lithiums to aromatic aldehydes and ketones is a common strategy for the synthesis of 1,1-diphenylalkanols. A report of a practical, high yielding preparation of 2,6-dichlorophenyllithium is of particular interest given the fact that many biologically active compounds contain this aromatic unit. The anion is stable at temperatures below –50 °C, but decomposes on warming (T. Kress and M. Leanna, *Synthesis,* 1985, 804).

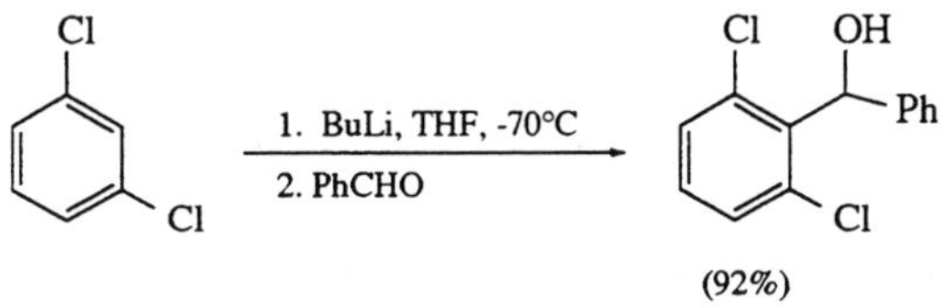

A 1,1-diarylamine is the product of the reaction between an aryl lithium and an imine derived from benzaldehyde. Treatment with DDQ effects oxidation to the corresponding imine (L. Strekowski, M.T. Cegala, D.B. Harden, and S.B. Kong, *J. Org. Chem.*, 1989, **54**, 2464). Since imines of this type are often difficult to prepare directly when the aldehyde or amine components are sterically hindered, for example by *ortho*-substitution, this two step procedure offers a useful alternative.

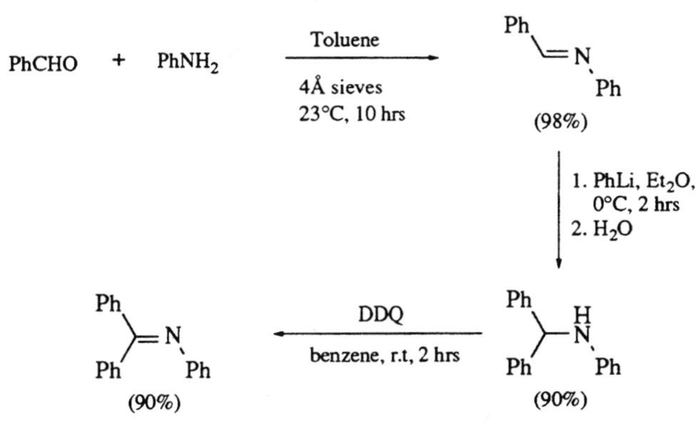

Traditionally, it was thought that benzonitriles are not suitable for Birch reduction, due to competing reduction of the nitrile group. Schultz has reported the successful reduction of benzonitrile to the dihydro derivative using lithium in ammonia-THF solution (A. G. Schultz and M. Macielag, *J. Org. Chem.*, 1986, **51**, 4983). Following quenching with benzyl bromide and suitable work-up, elimination of HCN occurs to give the 1,1-diphenylalkane.

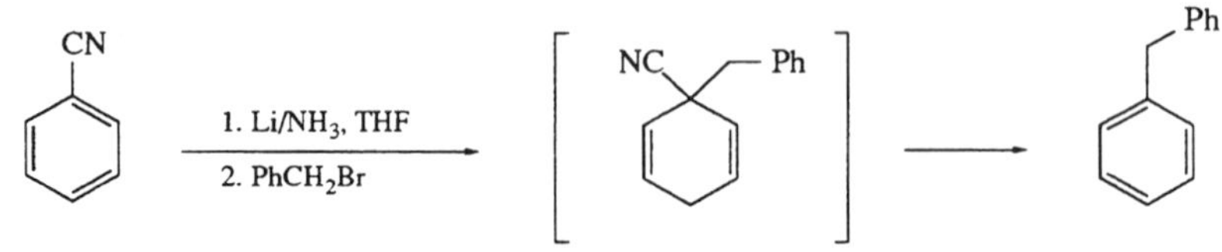

An interesting tandem addition – rearrangement occurs when arynes, generated by treatment of 2-bromoanisoles with LDA, are treated with carbanions derived from benzylic nitriles (S.P. Khanapure, L. Crenshaw, R.T. Reddy, and E.R. Biehl, *J. Org. Chem.*, 1988, **53**, 4915). The intermediate α-lithio-2-cyanodiarylmethane (**5**) can also be trapped with aldehydes (L. Crenshaw, S.P. Khanapure, U. Siriwardane and E.R. Biehl, *Tetrahedron Lett.* 1988, **29**, 3777).

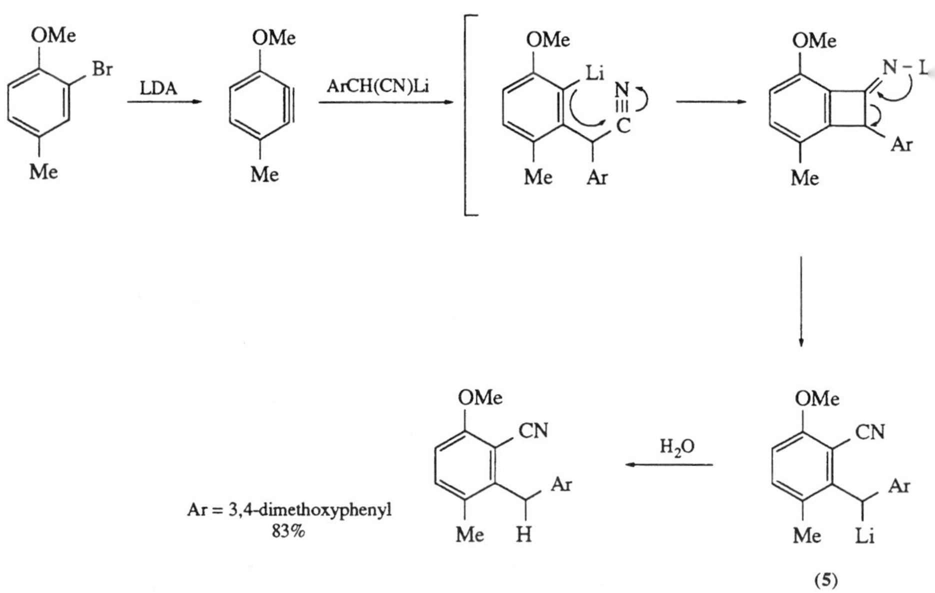

The Fujiwara group has demonstrated the ytterbium metal-promoted cross-coupling reaction between diaryl ketones and a variety of electrophiles, including ketones, nitriles, epoxides, and carbon dioxide (Z. Hou, K.

Takamine, O. Aoki, H. Shiraishi, Y. Fujiwara, and H. Taniguchi, *J. Chem. Soc., Chem. Commun.*, 1988, 668). Several useful 1,1-diphenyl methanol derivatives are accessible using this methodology, including unsymmetrical pinacols.

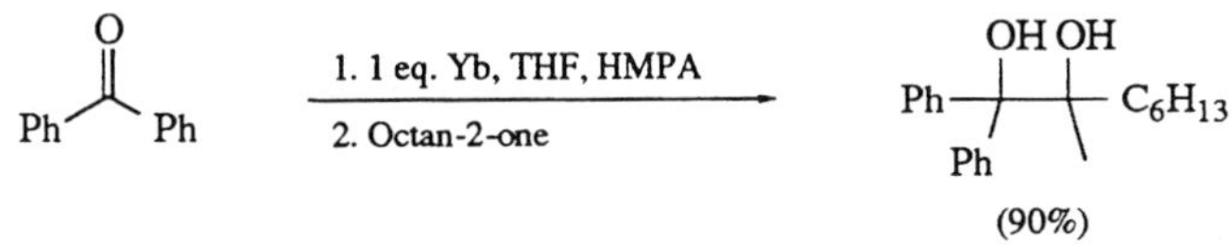

Transformation of a benzylic carbonyl group to the corresponding benzylic halide can conveniently be achieved by reduction with lithium aluminium hydride at low temperature, followed by quenching with a hydrogen halide (C. Bilger, R. Royer, and P. Demerseman, *Synthesis*, 1988, 905). This useful transformation would usually be carried out in two or more steps.

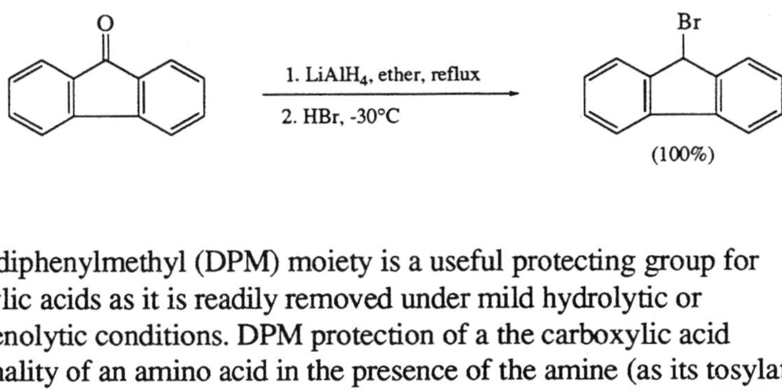

The diphenylmethyl (DPM) moiety is a useful protecting group for carboxylic acids as it is readily removed under mild hydrolytic or hydrogenolytic conditions. DPM protection of a the carboxylic acid functionality of an amino acid in the presence of the amine (as its tosylate salt) can be achieved (C. Froussios and M. Kolovos, *Synthesis*, 1987, 1106).

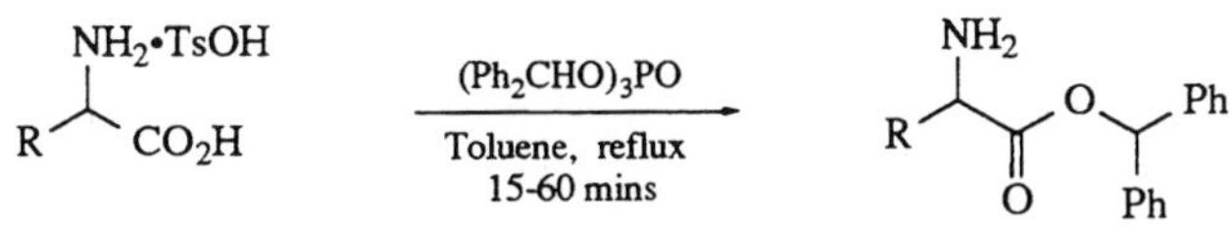

(b) 1,1-Diphenyl-alkenes, -allenes, and -ketenes and their derivatives

(i) Preparative methods for simple 1,1-diphenyl alkenes

The use of silver triflate to promote intermolecular Friedel-Crafts vinylation has been reported (T. Kitamura, S. Kobayashi, H. Taniguchi, and Z. Rappoport, *J. Org. Chem.*, 1982, **47**, 5003). The reaction is postulated to proceed *via* a stabilised, α-aryl vinyl cation, and produces the *para* product with high selectivity.

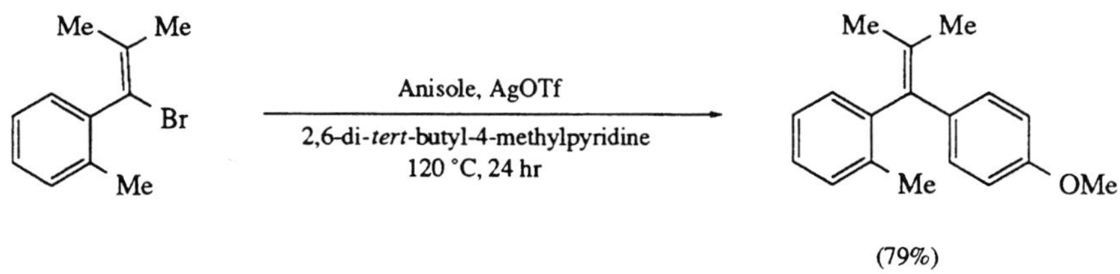

Addition of aryl palladiums, formed *in situ*, to propargylic alcohols occurs in good yields; the presence of cuprous iodide improves yields and shortens reaction times. High regioselectivity is obtained, presumably for steric reasons, when the alcohol-bearing carbon is hindered, providing a useful route to 1,1-diphenylalkenes (A. Arcadi, S. Cacchi, and F. Marinelli, *Tetrahedron*, 1985, **41**, 5121).

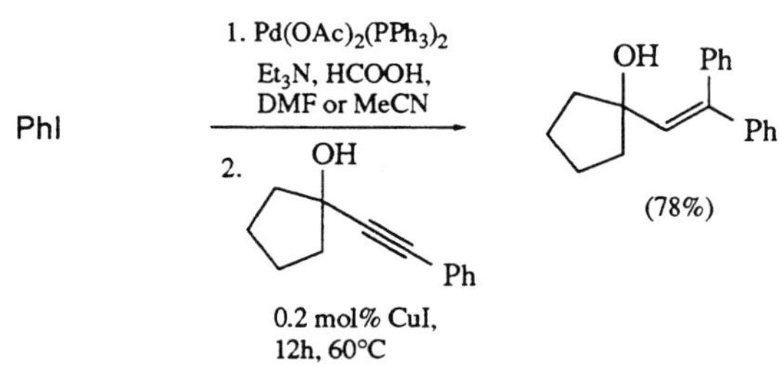

The reaction of Grignard reagents with dithioacetals is catalysed by $NiCl_2(PPh_3)_2$, and can be used to provide alkenes in good yields; a mechanism has been proposed (Z.-J. Ni and T.-Y. Luh, *J. Chem. Soc., Chem. Commun.*, 1987, 1515).

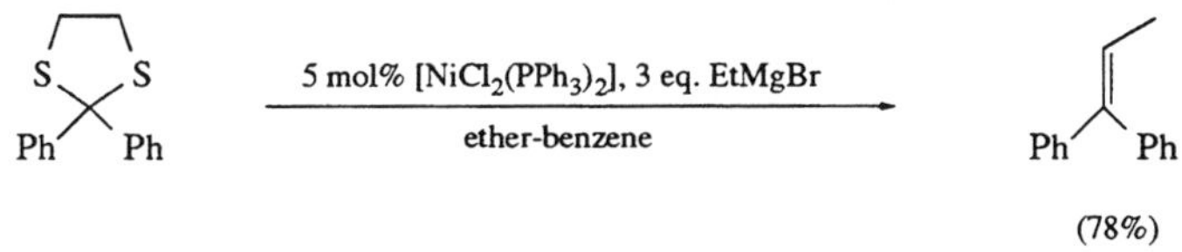

Binger and co-workers have described various reactions of 1,1-diphenyl cyclopropene; the [2+1] cycloaddition with electron deficient olefins in the presence of a nickel (0) catalyst leads to 1,1-diphenylethenylcyclopropane derivatives (P. Binger, J. McMeeking, and H. Schäfer, *Chem. Ber*., 1984, **117**, 1551). The configuration of the starting unsaturated ester is largely retained in the new cyclopropane ring.

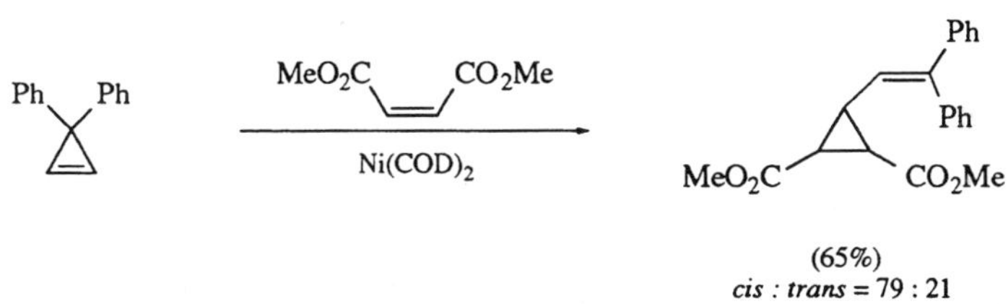

Rhodium catalysed coupling of a 1,1-diphenylcarbene and an allylic alcohol provides a 1,1 diphenyl-homoallylic alcohol in a novel and interesting process (J. Wolf, L. Brandt, A. Fries and H. Werner, *Angew. Chem., Int. Ed. Engl*., 1990, **29**, 510).

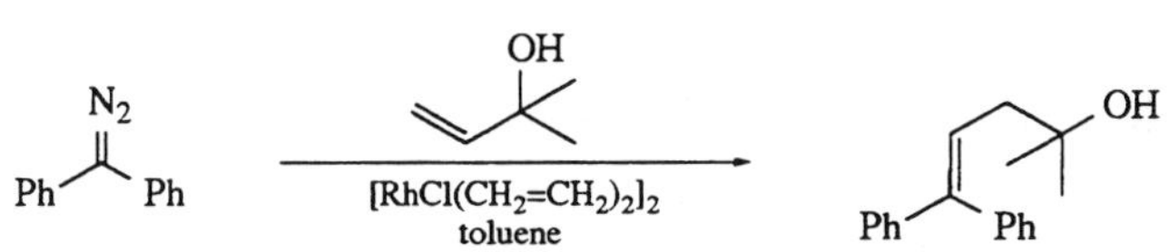

(ii) Preparative methods for 1,1-diphenylalkenes bearing a conjugated or heteroatomic group

One of the most widespread methods for alkene formation involves nucleophilic addition to a carbonyl group followed by elimination of water. Given the high degree of conjugation in the product, this approach becomes particularly useful for the synthesis of 1,1-diphenylalkenes, especially those bearing substituents affording still further mesomeric stabilisation. As a recent example, addition of bis(cyclopentadienyl) yttrium chloride to benzophenone and subsequent dehydration has been reported (C. Qian and A. Qiu, *Tetrahedron Lett.*, 1988, **29**, 6931).

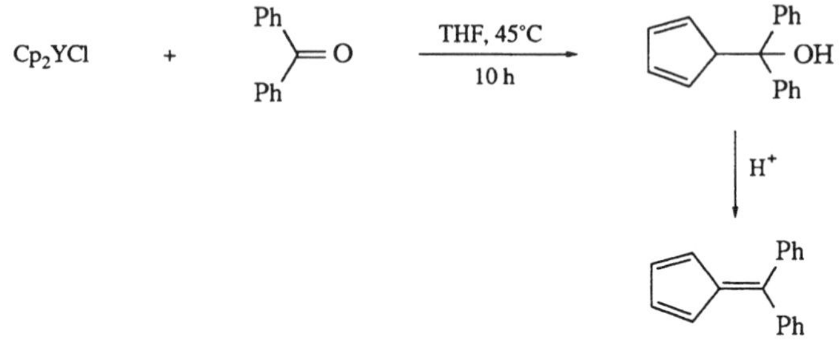

(57%)

As mentioned in Section 2a, the reaction of diphenylketene with allylic acetates is a useful route to the corresponding 1,1-diphenyl-1,3-dienes (T. Mitsudo, M. Kadokura, and Y. Watanabe, *J. Org. Chem.*, 1987, **52**, 1695).

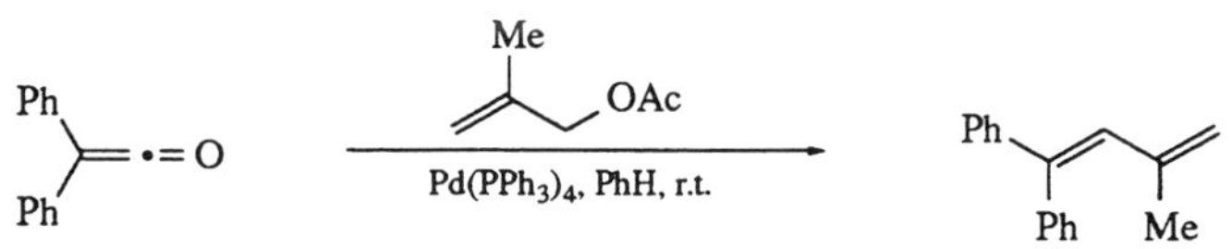

(51%)

Attempted [2+2]-cycloaddition of a thioketone to the bromoketeneacetal (**6**) resulted in formation of β, β-diphenyl-α,β-unsaturated esters (**7**) (P. Ooms and W. Hartmann, *Angew. Chem., Int. Ed. Engl.*, 1985, **24**, 976). This reaction is formally a condensation of an ester enolate equivalent with a thioketone.

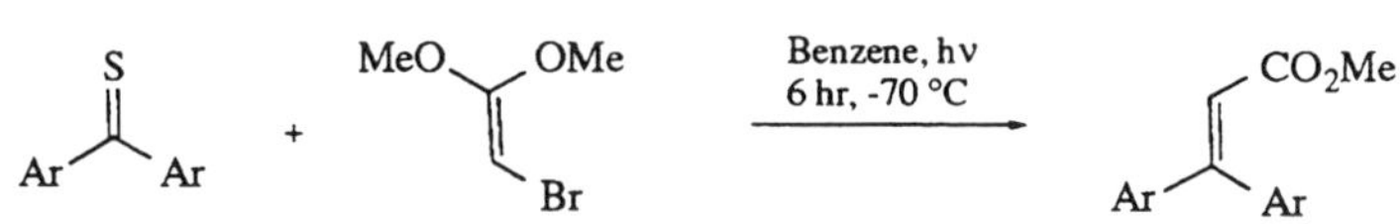

(**6**) (**7**)

Ar = 4-methoxyphenyl

Another route to β, β-diphenyl-α,β-unsaturated esters utilises the Meyer-Schuster rearrangement. The starting diphenyl propargyl alcohol can be readily prepared by addition of the methoxyacetylide anion to benzophenone (G.A. Olah, A. Wu, O. Farooq, and G.K.S. Prakesh, *Synthesis*, 1988, 537).

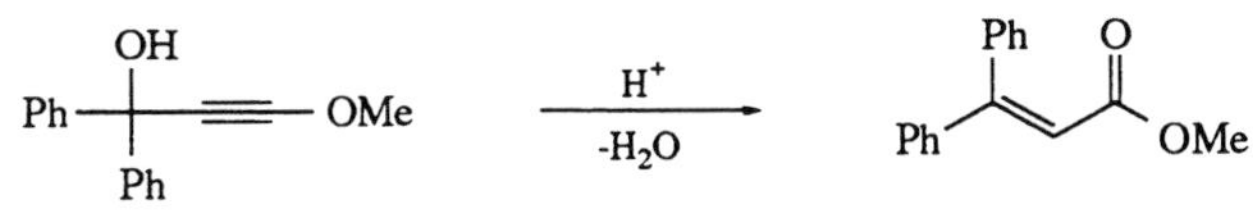

(70%)

Several routes have been developed for the preparation of 1,1-diphenyl alkenes bearing a halogen atom in the 2-position. Vinyl fluorides are important targets in view of their potential use as irreversible enzyme inhibitors, but traditional methods have been lengthy and have involved reagents that are difficult to handle. Several new approaches have been developed, mostly involving olefination of ketones, including benzophenone. McCarthy and Peet and co-workers demonstrated the utility of the α-fluoro lithiosulfone (**8**) (M. Inbasekaran, N.P. Peet, J.R. McCarthy, and M.E. LeTourneau, *J. Chem. Soc., Chem. Commun.*, 1985, 678). Following dehydration, reductive removal of the sulfone moiety was achieved in high yield using aluminium amalgam.

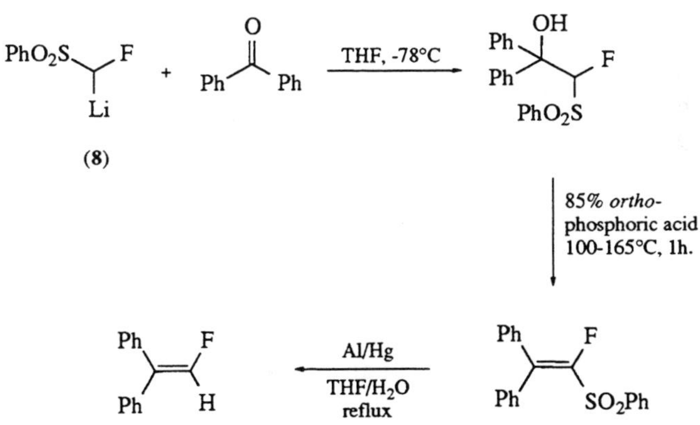

A similar method, but avoiding the acidic dehydration step, was later reported by the same group (J.R. McCarthy, D.P. Matthews, M.L. Edwards, D.M. Stemerick, and E.T. Jarvi, *Tetrahedron Lett.*, 1990, **31**, 5449). α-Fluoromethyl phenyl sulfone was converted into its 1,1-dianion before quenching with diethylchlorophosphonate to prepare a Wadsworth-Emmons reagent *in situ.*

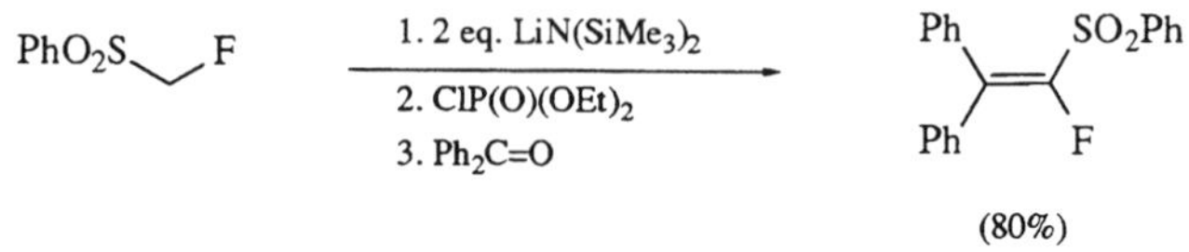

A closely-related variation on a phosphorus-based olefination, this time the Horner-Emmons reaction of phosphine oxides, has been shown to afford 1,1-difluoro-2,2-diphenylethene from benzophenone (M.L. Edwards, D.M. Stemerick, E.T. Jarvi, D.P. Matthews, and J.R. McCarthy, *Tetrahedron Lett.*, 1990, **31**, 5571).

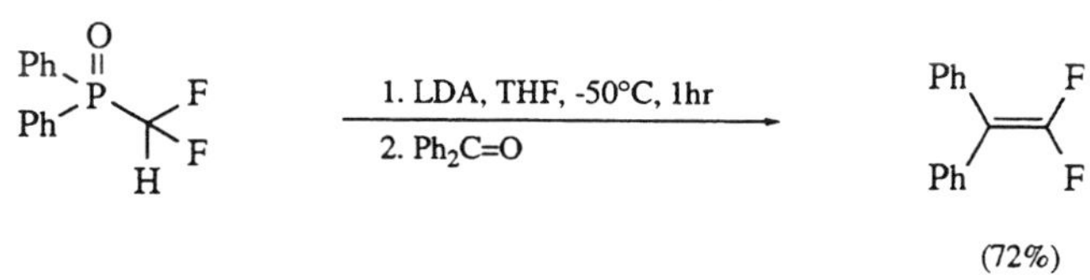

Synthetic methods for vinyl iodides have also attracted attention. One interesting reaction, involving formation of a new carbon-carbon bond to a phenyl ring, involves reaction between phenylacetylene and benzene in the presence of bis(pyridine)iodine (I) tetrafluoroborate (J. Barluenga, M. Rodriguez, J. González, and P. Campos, *Tetrahedron Lett.*, 1990, **29**, 4207). Toluene, xylene, anisaldehyde, and naphthalene were all used with success in place of benzene itself. The intramolecular variant of the reaction provides a useful route to substituted dihydronaphthalene derivatives (J. Barluenga, J.M. González, P.J. Campos and G. Asensio, *Angew. Chem., Int. Ed. Engl.*, 1988, **27**, 1546) and an alternative to Birch reduction or Diels-Alder strategems.

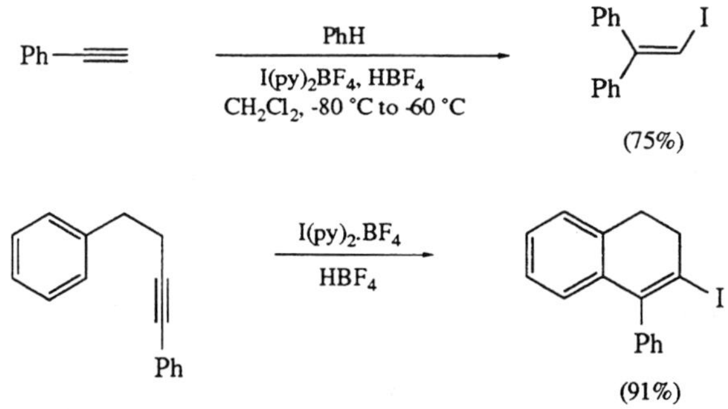

A new reagent system for alkene synthesis from 1,2-dibromoalkanes has been reported, and shown to be applicable to the synthesis of 1,1-diphenylalkanes (A.P. Suárez, M.R. Mazzieri, and A.G. Suárez, *J. Am. Chem. Soc.*, 1989, **111**, 763). The reaction conditions, involving use of an iron (0) catalyst in carbon tetrachloride at reflux, are mild in comparison to conventional methods. Mechanistic studies have not proved conclusive.

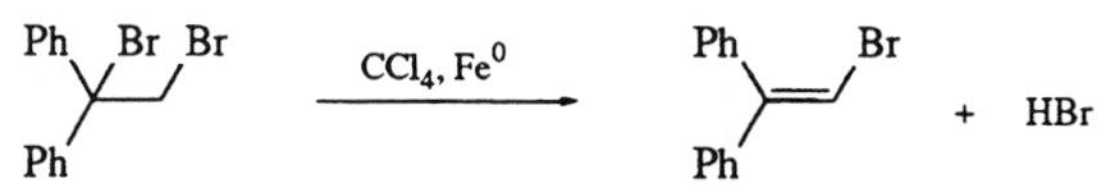

The synthesis of 1-chloro-2,2-diphenylalkenes has also received attention, and a mild, selective method for the chlorination of 1-1-diphenylethene has been described (N. Kamigata, T. Satoh, and M. Yoshida, *Bull. Chem. Soc. Jap.*, 1988, **61**, 449). The process is mechanistically interesting; it is believed that interaction between the benzeneseleninyl chloride and aluminium trichloride leads to formation of salt (**9**). Electrophilic addition of chlorine to the alkene is followed by attack on the resulting carbocation by a selenium species, which is then lost by elimination to regenerate the alkene.

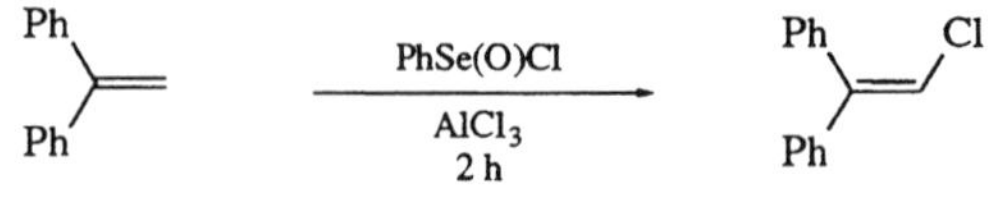

(94%)

Possible mechanism:

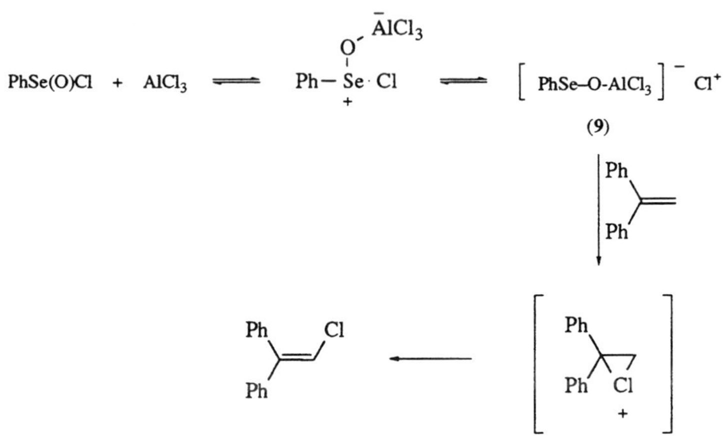

Finally, given the immense importance of vinyl silanes in synthesis, it is not surprising that new methods for their preparation are to be found in the literature. The nickel-catalysed reaction of Grignard reagents with thioacetals, described earlier for the synthesis of simple 1,1-diphenylalkenes, has been extended to the use of α-trimethylsilyl methylmagnesium chloride (Z.-J. Ni and T.-Y. Luh, *J. Org. Chem.*, 1988, **53**, 2129).

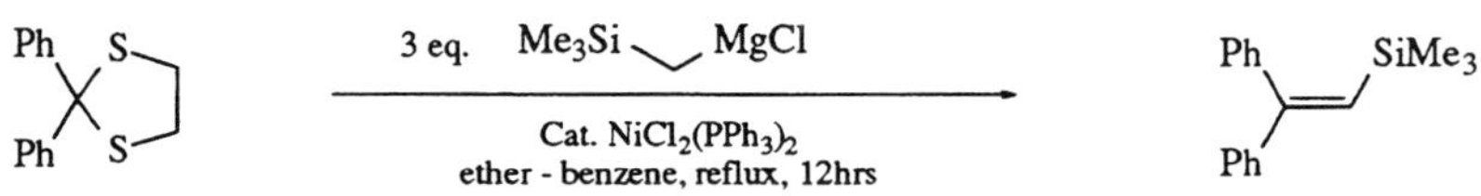

(79%)

(iii) 1,1-Diphenyl allenes, -ketenes and their derivatives

Olefination of 1,1-diphenylketenes provides an obvious entry to the corresponding allenes. Use of a suitably substituted Horner-Wittig reagent allows preparation of allenyl sulfones, which are of potential biological interest (H. Fillion, A. Hseine, M. Pera, V. Defaud, and B. Refouvelet, *Synthesis*, 1987, 708). The amount of base used in this reaction must be kept to a minimum as the products are base-sensitive.

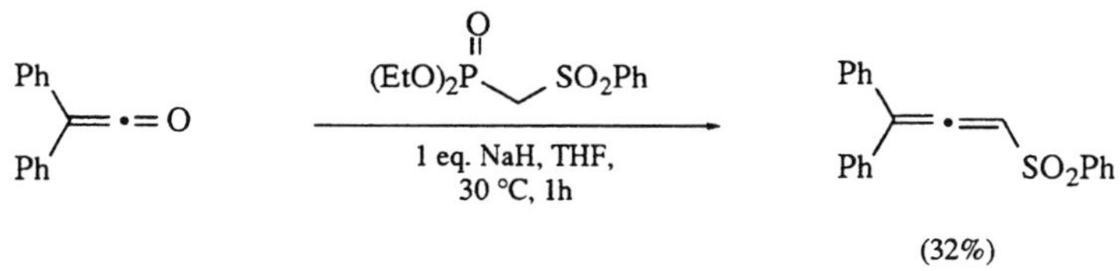

Allenic carbenes can be prepared by the reaction of 1,1-dihalocyclopropanes with aqueous sodium hydroxide under phase transfer conditions. The reaction of this species with 1,3-dienes has been studied (H. Sugita, K. Mizuno, T. Mori, K. Isagawa, and Y. Otsuji, *Angew. Chem. Int. Ed. Engl.*, 1991, **30**, 984). The major products are usually the cyclopropanes, derived from 1,2-addition to the diene, but appreciable amounts of cyclopentenes (from 1,4-addition) are often produced as well.

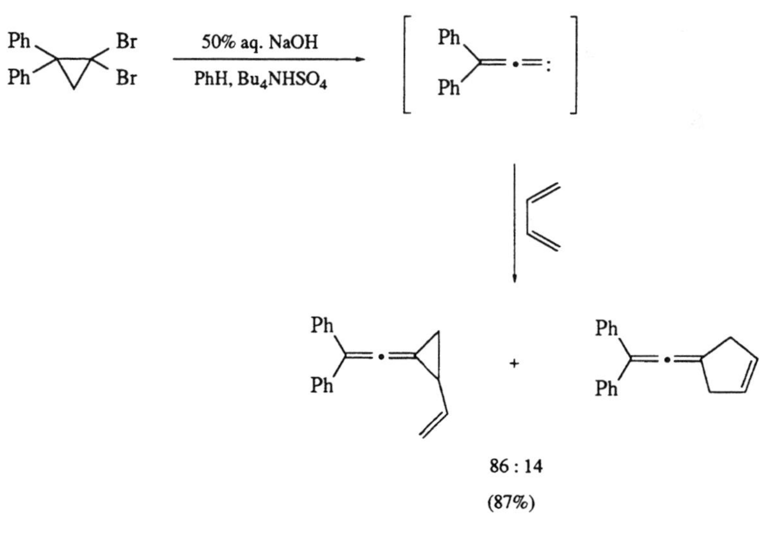

An interesting transformation of diphenylketenes is represented by their reaction with epoxides, catalysed by tributyltin iodide (I. Shibata, A. Baba, and H. Matsuda, *Bull. Chem. Soc. Japan*, 1986, **59**, 4000). The products are diphenylketene acetals. The authors propose that the epoxide is initially opened by the tributyltin iodide, giving the organotin β-iodoalkoxide. Reaction of this species with diphenylketene results in formation of the cyclic acetal, with regeneration of the tributyltin iodide.

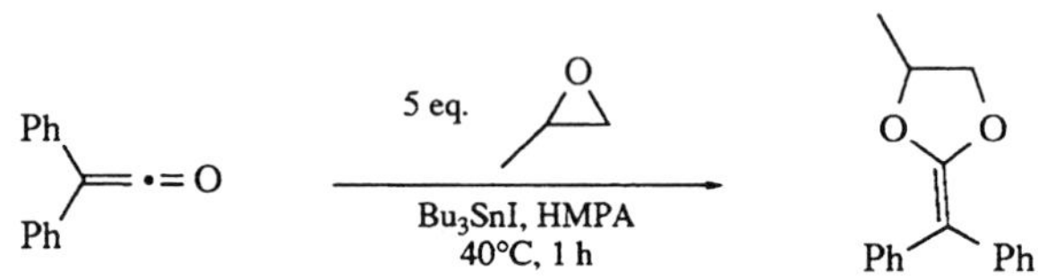

Proposed mechanism:

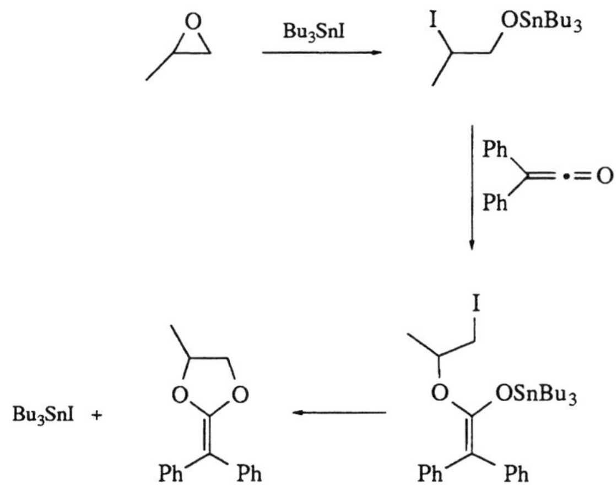

3. *Diphenyl-alkanes and -alkenes, in which the phenyl groups are attached to different carbon atoms*

The emphasis in this Section will be on the chemistry of 1,2-diphenyl-alkanes and -alkenes since, as *n* increases, the two phenyl groups in 1,*n*-diphenyl compounds become increasingly independent in terms of the chemistry involved.

(a) 1,2-Diphenylalkanes and their derivatives

Included in this Section are 1,2-diphenyl epoxides and aziridines; while these could equally have been considered as derivatives of stilbene, the phenyl-bearing carbons are formally alkanic in nature.

(i) Preparative methods for simple 1,2-diphenylalkanes

Heck-type addition of aryl halides to styrene provides a useful route to 1,2-diphenylalkanes. New reaction conditions have been reported, using a tetraethylammonium triflate / dimethylformamide electrolytic solution with a platinum anode and a lead cathode (S. Torii, H. Tanaka, and K. Morisaki *Chem. Lett.* 1985, 1353).

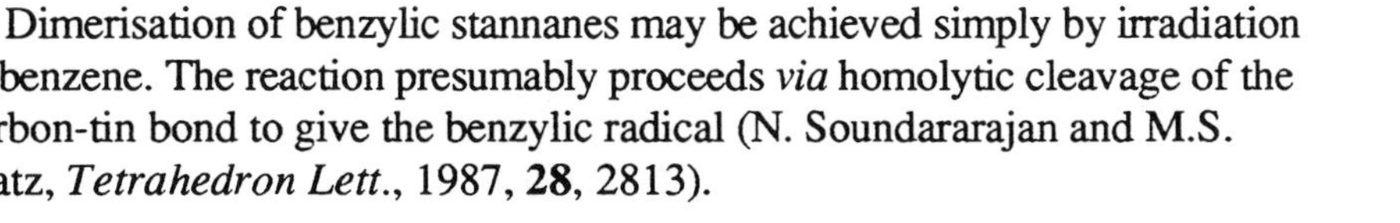

Dimerisation of benzylic stannanes may be achieved simply by irradiation in benzene. The reaction presumably proceeds *via* homolytic cleavage of the carbon-tin bond to give the benzylic radical (N. Soundararajan and M.S. Platz, *Tetrahedron Lett.*, 1987, **28**, 2813).

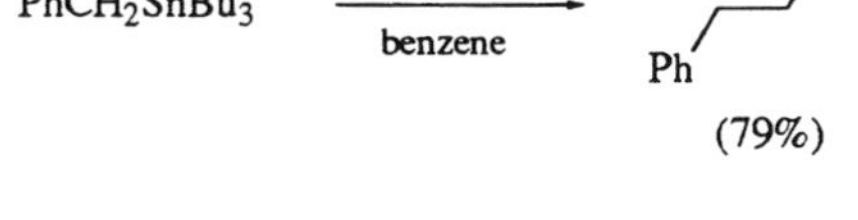

Krief has developed a synthesis of 1,1-diphenylcyclopropanes starting from diselenoacetals. Initial lithium-selenium exchange is followed by reaction with styrene to produce the cyclopropane in 77% yield, and with high diastereoselectivity (*trans* : *cis* = 97:3) (A. Krief, P. Barbeaux, and E. Guittet, *Synlett*, 1990, 509).

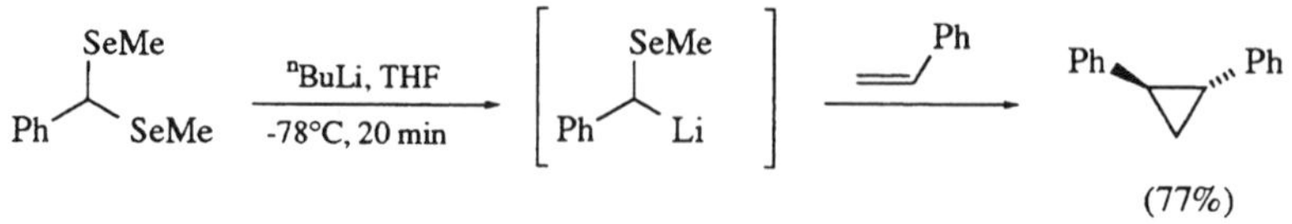

(ii) Preparative methods for 1,2-diphenylalkanes bearing conjugated or heteroatomic groups

Lewis acid-mediated reaction of a benzylic ether with a phenyl-substituted allyl silane has been demonstrated by Katzellenbogen (R. Mohan and J. A. Katzellenbogen, *J. Org. Chem.*, 1984, **49**, 1238). The product was used in the synthesis of derivatives of the non-steroidal oestrogen, hexestrol.

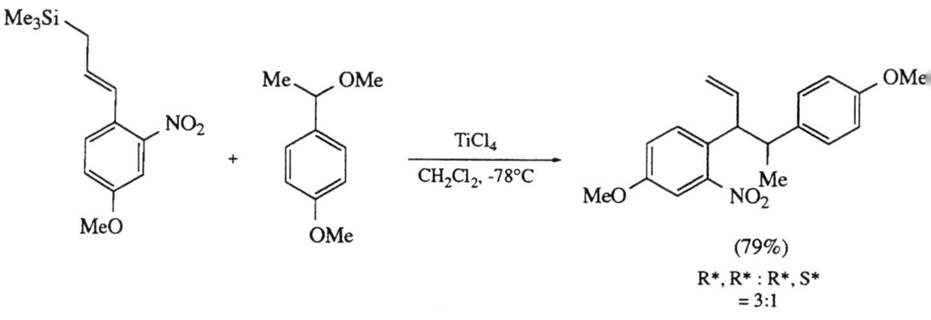

The benzoin condensation is one of the oldest methods for the preparation of α-hydroxy ketones. Lappert has described the use of bi(1,3-dialkylimidazolidin-2-ylidenes) as effective catalysts for this reaction (M. F. Lappert and R. K. Maskell, *J. Chem. Soc., Chem. Commun.*, 1982, 580). Best results are obtained in the absence of a co-solvent.

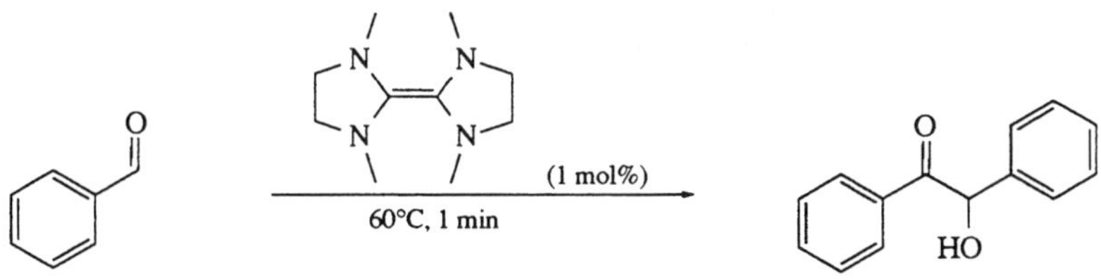

(84%)

The addition of benzylic anions to benzaldehydes provides the most direct entry to α-hydroxy-1,2-diphenylalkanes. Unfortunately, the synthesis of benzyl lithiums from the corresponding benzyl halides is a very poor-yielding process due to the reactivity of the product towards the starting halide. Similarly, attempted direct deprotonation at a benzylic position is often thwarted by competing ring-metallation. The development of new preparative methods is therefore of great interest. Krief has exploited the facile cleavage of the carbon-selenium bond by butyl lithium at low temperature; halogen-substituents in the starting benzylic selenides are tolerated (M. Clarembeau, A. Krief, *Tetrahedron Lett.*, 1985, **26**, 1093).

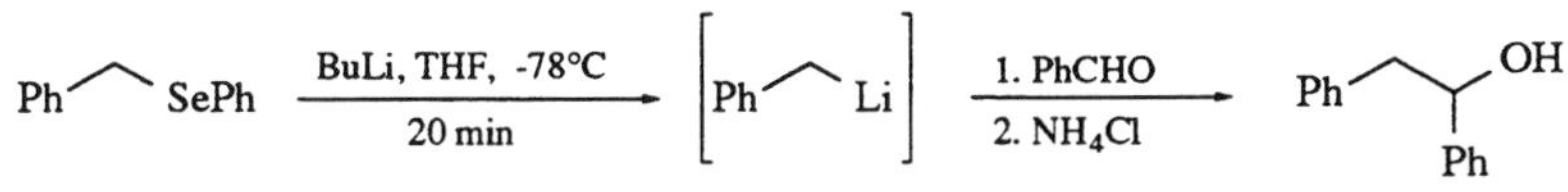

(86%)

Fluoride ion-catalysed cleavage of the carbon-silicon bond provides another useful method for the generation of a benzyl anion equivalent for addition to aldehydes (B. Bennetau and J. Dunogues, *Tetrahedron Lett.*, 1983, **24**, 4217). The initially-formed trimethylsilyl ether is hydrolysed to the alcohol on work-up.

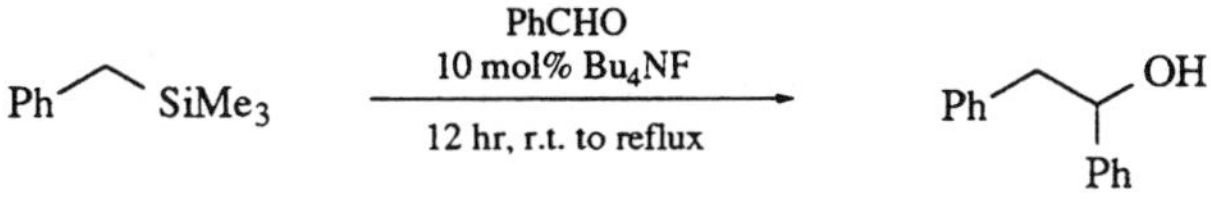

(92%)

Direct deprotonation at a benzylic position is, of course, facilitated by the presence of activating groups in the *ortho*-position. Following reaction of the anion with benzaldehyde, cyclisation onto an amide activating group can be used to effect a synthesis of isocoumarins (M. Watanabe, M. Sahara, M. Kubo, S. Furukawa, R. J. Billedeau, and V. Snieckus *J. Org. Chem.* 1983, **49**, 742).

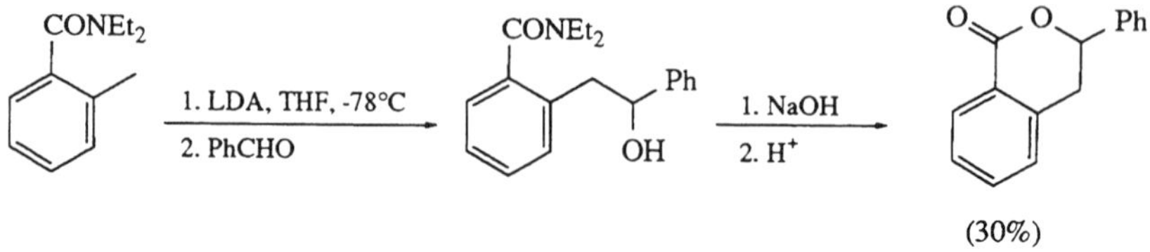

(30%)

Allylic acetates can be converted into the corresponding allyl metals in the presence of palladium-tin (II) chloride (Y. Masuyama, R. Hayashi, K. Otake, and Y. Kurusu, *J. Chem. Soc., Chem. Commun.*, 1988, 44). The presence of a phenyl substituent on the allyl unit offers further anion stabilisation. Reaction with benzaldehyde provides a 1,2-diphenyl-homoallylic alcohol with modest diastereoselectivity.

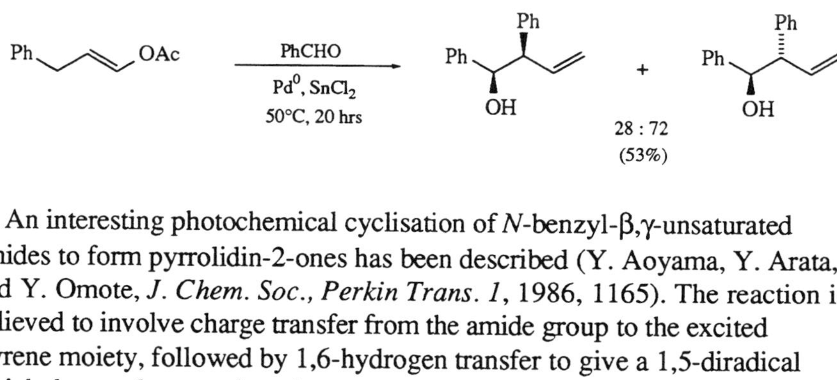

An interesting photochemical cyclisation of *N*-benzyl-β,γ-unsaturated amides to form pyrrolidin-2-ones has been described (Y. Aoyama, Y. Arata, and Y. Omote, *J. Chem. Soc., Perkin Trans. 1*, 1986, 1165). The reaction is believed to involve charge transfer from the amide group to the excited styrene moiety, followed by 1,6-hydrogen transfer to give a 1,5-diradical which then undergoes ring closure.

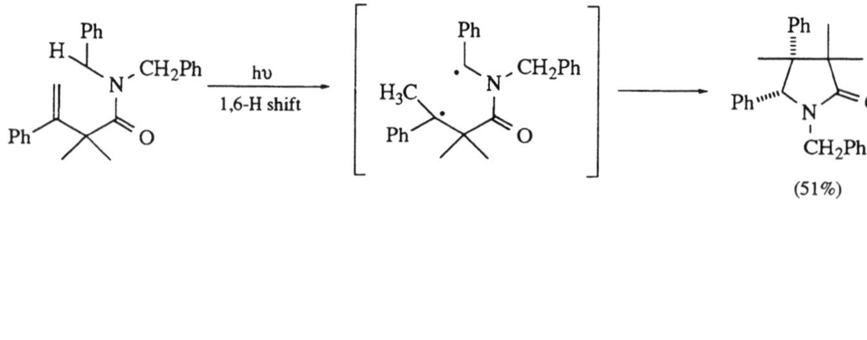

(51%)

Reaction of titanium tetraisopropoxide and a lithium amide results in formation of a complex that can undergo addition to benzaldehydes. The resulting intermediate can be transformed in the same pot to an α-amino-1,2-diphenylalkane by addition of benzyl magnesium chloride (H. Takahashi, T. Tsubuki, and K. Higashiyama, *Synthesis*, 1988, 238).

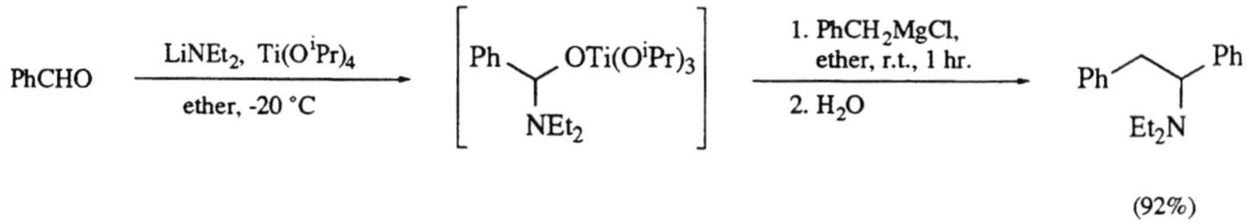

Several procedures have been reported that allow the formation of 1,2-diamines by reductive dimerisation of benzaldehyde-derived imines. Use of a lead-aluminium redox system was developed by Torii and co-workers (H. Tanaka, H. Dhimane, H. Fujita, Y. Ikemoto, and S. Torii, *Tetrahedron Lett.*, 1988, **29**, 3811), while Roskamp and Pedersen advocate the use of catalytic niobium tetrachloride (E.J. Roskamp and S.F. Pedersen, *J. Am. Chem. Soc.*, 1987, **109**, 3152). The latter method was highly diastereoselective (19:1) for the *dl*-isomer over the *meso*.

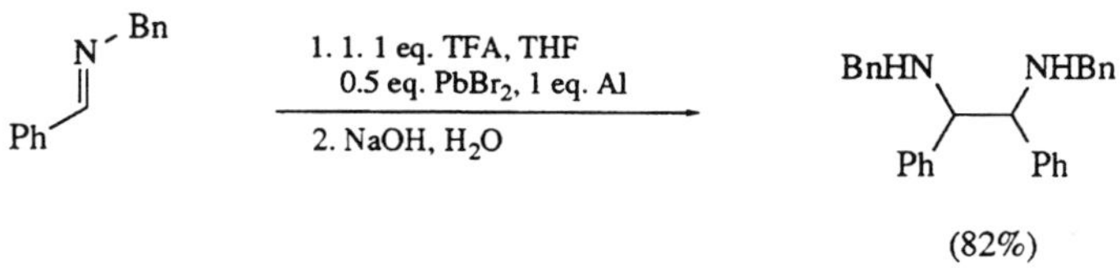

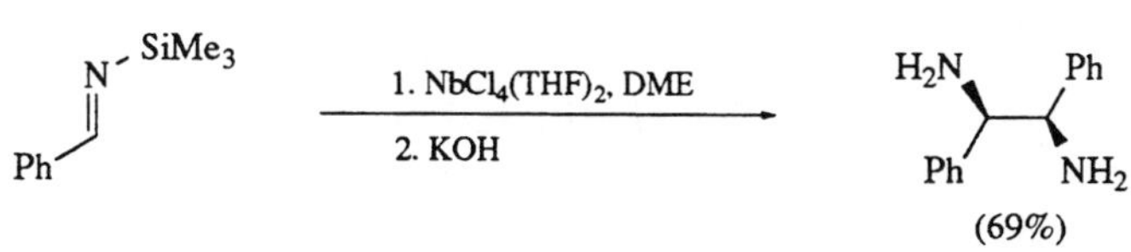

In a variation on this process, Seebach has demonstrated the direct conversion of benzaldehydes into 1,2-diamines (R. Imwinkelried and D. Seebach, *Helv. Chim. Acta.*, 1984, **67**, 1496).

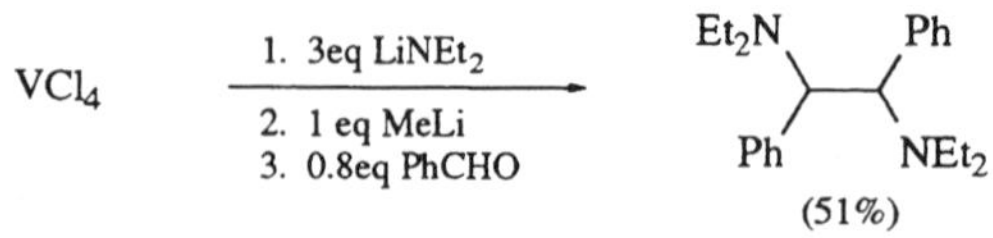

Treatment of aldoximes with sodium bromite and a catalytic amount of tributyltin chloride affords the nitrile oxide, which can be trapped with styrene to form a 3,4-diphenylisoxazoline (O. Moriya, H. Nakamura, T. Kageyama and Y. Urata, *Tetrahedron Lett.*, 1989, **30**, 3987).

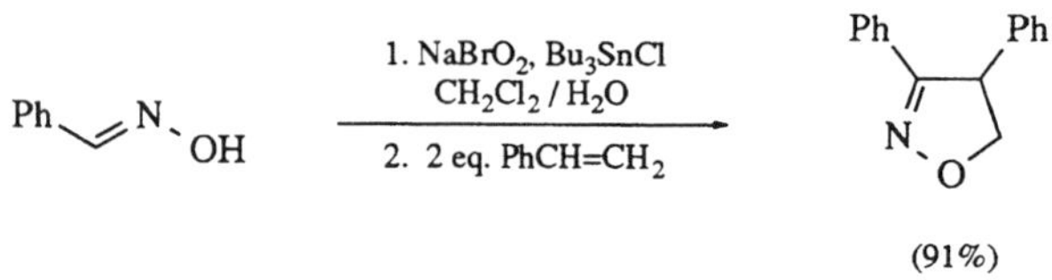

In the presence of trifluoromethane sulfonic acid, nitroethene undergoes diprotonation to give a dicationic species which is capable of participating in Friedel-Crafts processes. Reaction in this way with two equivalents of benzene provides the oxime of deoxybenzoin (T. Ohwada, A. Itai, T. Ohta, and K. Shudo, *J. Am. Chem. Soc.*, 1987, **109**, 7036).

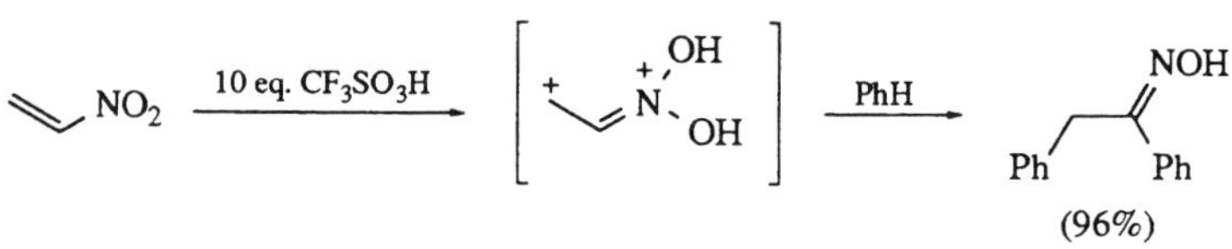

Addition of phenyl-substituted sulfur ylids to benzaldehydes provides a connective approach to stilbene oxides. The Padwa group has reported that the fluoride-induced desilylation of (**10**), followed by proton transfer, affords the requisite sulfur ylids (A. Padwa and J. R. Gasdaska, *J. Org. Chem.*, 1986, **51**, 2857). Reaction with benzaldehyde provides *trans*-stilbene oxide.

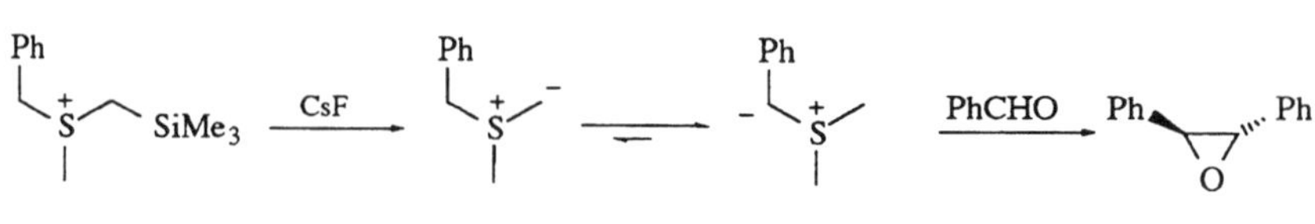

(**10**)

The asymmetric synthesis of stilbene oxides by this approach has also received attention. The chiral sulfur ylid derived from (**11**) affords reasonable enantioselectivity (L. Breau, W.W. Ogilvie, and T. Durst, *Tetrahedron Lett.*, 1990, **31**, 35).

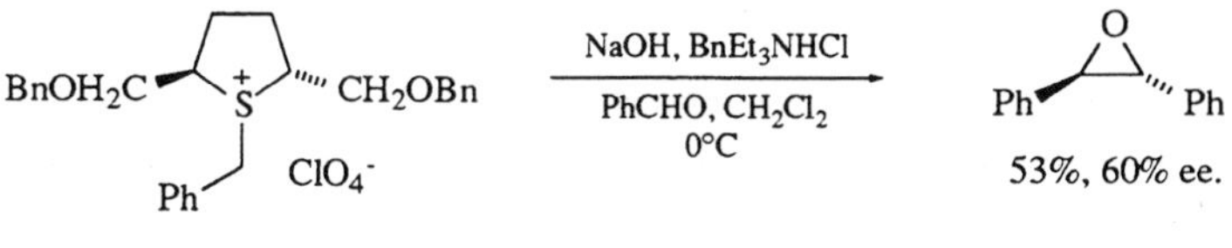

(**11**)

The ylid generated *in situ* by deprotonation of amine-*N*-oxide (**12**) reacts with alkenes in an intermolecular [3+2]-cycloaddition process (G. Roussi and J. Zhang, *Tetrahedron Lett.*, 1988, **29**, 3481). Even non-activated alkenes such as stilbene may be used. Removal of the *N*-substituent of the product pyrrolidine is possible, although three steps are required.

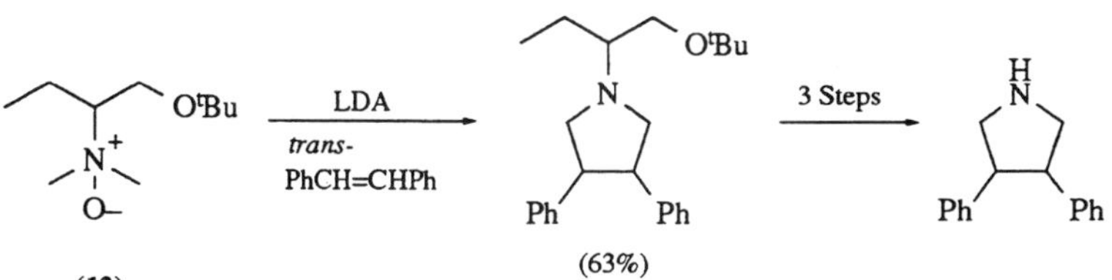

(**12**)

(b) Preparative methods for simple stilbenes or 1,2-diphenylethenes

(i) Carbonyl olefination reactions

The strength of the silicon-fluorine bond provides the driving force in a new method for the formation of phosphonate anions (T. Kawashima, T. Ishii, and N. Inamoto, *Tetrahedron Lett.*, 1983, **24**, 739). Subsequent Wadsworth-Emmons reaction with benzaldehyde affords stilbene, as the *E*-isomer exclusively. Caesium fluoride proved to be the best fluoride source of several examined. The procedure can also be applied to acetophenone, providing α-methylstilbene in moderate yield (67%) as a 25:75 *E*:*Z* mixture.

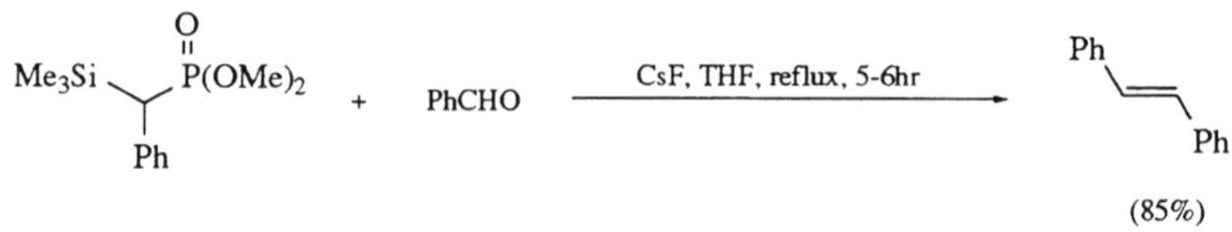

Dawson and co-workers have studied the effect of the base in determining the *E*:*Z* ratio in olefination reactions of the type shown below. Labelling studies demonstrated that product equilibration was possible *via* deprotonation of the allylic methyl group. High selectivity for the thermodynamically more stable *E*-product was therefore accomplished by treating the product mixture with base. The *E*-isomers represent a new class of aromatic retinoids that are of interest in the treatment of many proliferative diseases of the skin such as cancer, psoriasis, and acne (M.I. Dawson, K. Derdzinski, P.D. Hobbs, R.L.-S. Chan, S.W. Rhee, and D. Yasuda, *J. Org. Chem.*, 1984, **49**, 5265).

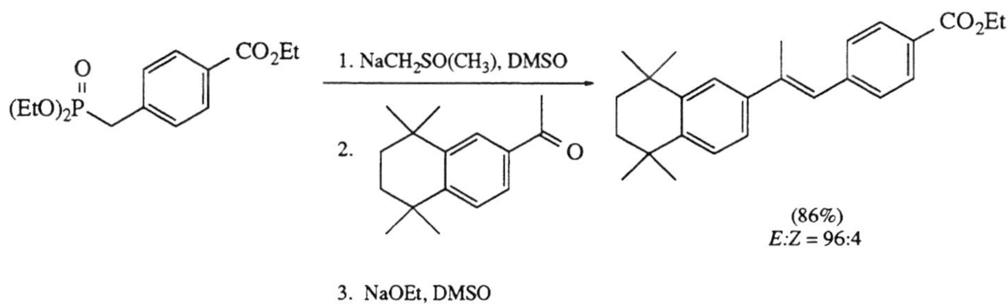

An interesting olefination based on α-(fluorosulfonyl) carbanions has been reported (S. Kagabu, K. Hara, and J. Takahashi, *J. Chem. Soc., Chem. Commun.*, 1991, 408). Contrary to the behaviour of the corresponding chloride, deprotonation of phenylmethanesulfonyl fluoride is not followed by elimination of halogen, and so it is possible to trap the anion with an aldehyde. Formation of a cyclic intermediate, ring opening and loss of sulfur dioxide provide a stereospecific route to stilbene. Since the diastereoselectivity of the anion addition to the aldehyde is low, however, the stilbenes are usually formed as a mixture of geometrical isomers.

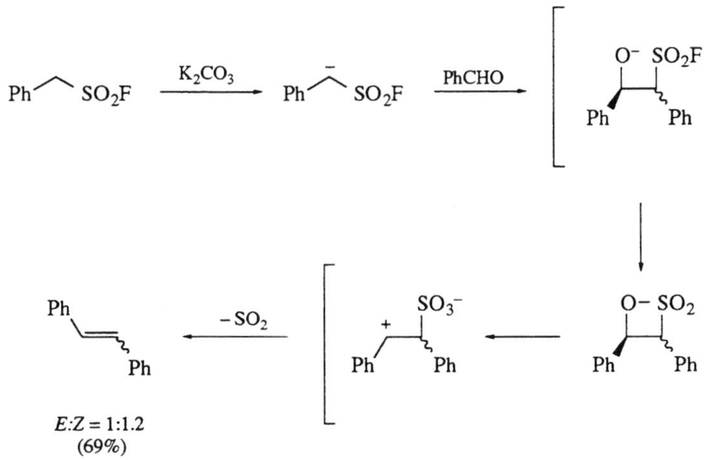

(ii) Arylation of styrene derivatives

Heck-type reaction between a styrene and an aryl palladium offers a powerful method for stilbene synthesis. The aryl palladiums are usually derived from the corresponding halides, but a novel route from *N*-nitroso-*N*-arylacetamides has been described (K. Kikukawa, M. Naritomi, G.-X. He, F. Wada, and T. Matsuda, *J. Org. Chem*., 1985, **50**, 299). These nitroso species can be prepared *in situ* by nitrosation of the acetamides. The overall yield of the process is generally excellent (90-97%).

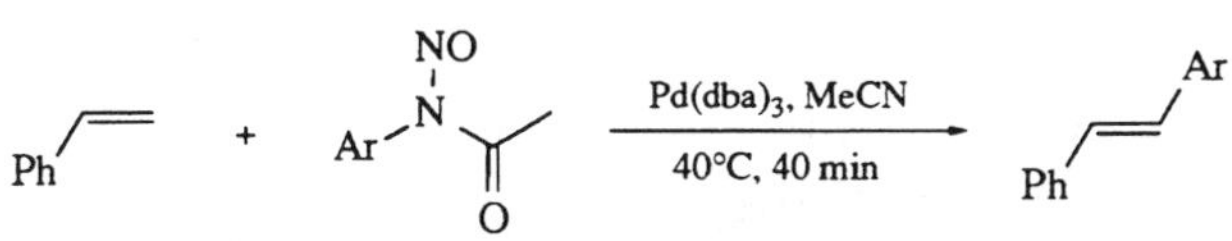

Another new approach to aryl palladiums begins with aryl chlorides, which undergo decarbonylation in the reaction pot (H.-U. Blaser and A. Spencer, *J. Organomet. Chem*., 1982, **233**, 267).

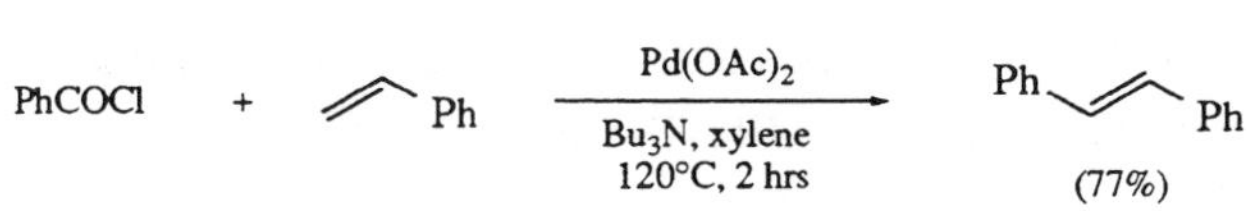

Palladium chemistry has been used in a new route to unsymmetrical 2-formyl stilbenes (I.R. Girling and D.A. Widdowson, *Tetrahedron Lett.*, 1982, **23**, 1957). Cyclopalladation of aryl t-butyl imines ($PdCl_2$, AcOH, NaOAc) proceeds in high yield in the *ortho* position, with additional ring substituents exerting the expected directing effects. Following coupling with styrene, hydrolysis of the imine yields the target compounds.

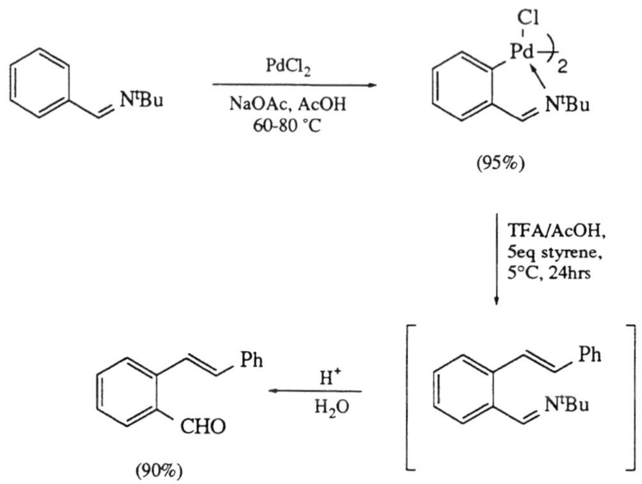

One-pot, double arylation of vinyl acetate is an attractive strategy for the synthesis of symmetrical stilbenes, but poor yields have been achieved with conventional palladium catalysts. Choudary and Sarma have reported an improved process using a polymer supported palladium reagent (B.M. Choudary and M.R. Sarma, *Tetrahedron Lett.*, 1990, **31**, 1495).

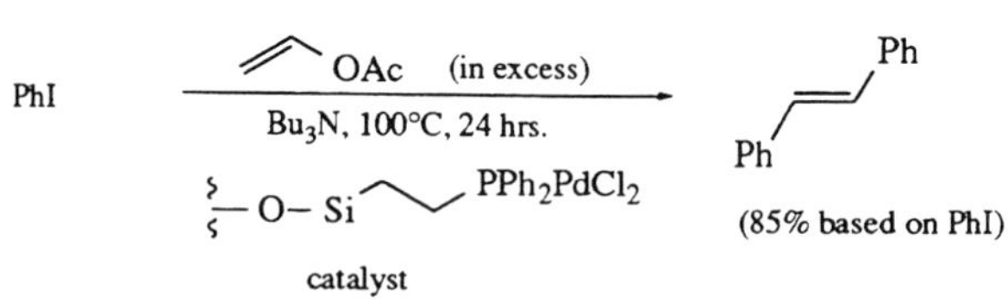

Arylation of alkenyl bromides has been accomplished under mild conditions with biphosphine platinum complexes (J.M. Brown, N.A. Cooley, and D.W. Price, *J. Chem. Soc., Chem. Commun.*, 1989, 458).

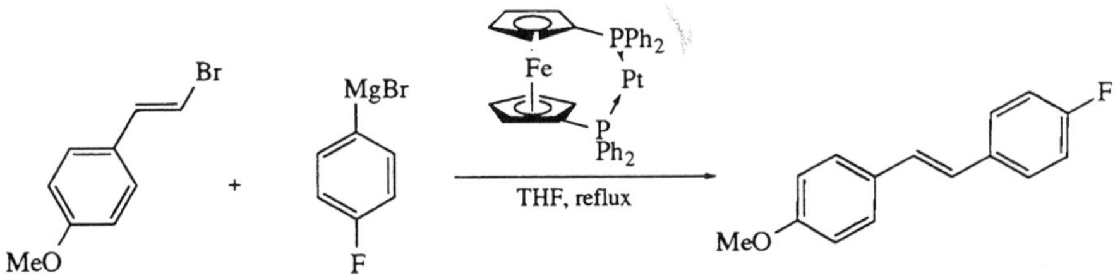

(95%)

(iii) Reduction of alkynes

The discovery of new reagents for the reduction of alkynes to alkenes continues, with chemoselectivity an important goal. Stryker's copper (I) hydride reagent, $[(Ph_3P)CuH]_6$, reduces 1,2-diphenylacetylene to *cis*-stilbene in good yield (91%) (J.F. Daeuble, C. McGettigan, and J.M. Stryker, *Tetrahedron Lett.*, 1990, **31**, 2397). This transformation may also be accomplished using sodium borohydride / ferrous chloride in the presence of a dithiol ligand (M. Kijima, Y. Nambu, and T. Endo, *Chem. Lett.*, 1985, 1851), or with zinc / methanol / *N, N'*-dimethyl paraquat in the presence of palladium on carbon (H. Shosenji, Y. Nakano, and K. Yamada, *Chem. Lett.*, 1988, 1033). An impressive example is the reduction of the acetylene moiety in the presence of an aldehyde using palladium on charcoal with ammonium formate as the hydrogen source. The *Z*- isomer is the major product (S. Ram and R.E. Ehrenkaufer, *Synthesis*, 1988, 91; J.R. Weir, B.A. Patel, and R.F. Heck, *J. Org. Chem.*, 1980, **45**, 4926).

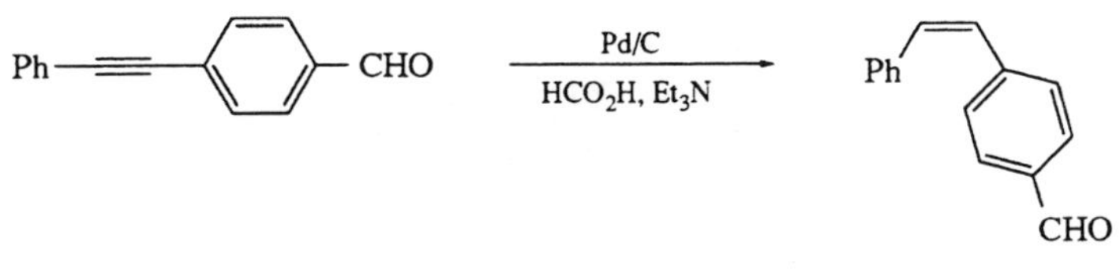

(iv) Reduction of dihalides, diols and epoxides

Deoxygenation procedures proceeding *via* cyclic intermediates derived from 1,2-diols have the considerable advantage that they are stereospecific. Modified conditions for the Corey-Winter reduction of thionocarbonates have been reported (E. J. Corey and P.B Hopkins, *Tetrahedron Lett.*, 1982, **23**, 1979) which are especially suitable for complex or sensitive molecules.

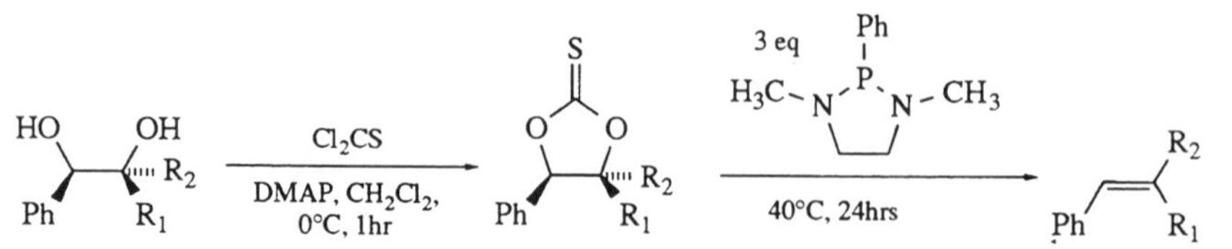

R_1=H, R_2=Ph 94%
R_1=Ph, R_2=H 85%

Another mild and stereospecific procedure for deoxygenation of 1,2-diols described recently involves formation of the dimethylformamide-mixed acetal derivative, followed by treatment with triflic anhydride. This reaction is believed to proceed *via* quaternisation at nitrogen followed by deprotonation at the acetal centre to give a nitrogen ylid, and subsequent formation of the alkene with loss of CO_2 (J.L. King, B.A. Posner, K.T. Mak, and N.C. Yang, *Tetrahedron Lett.*, 1987, **28**, 3919).

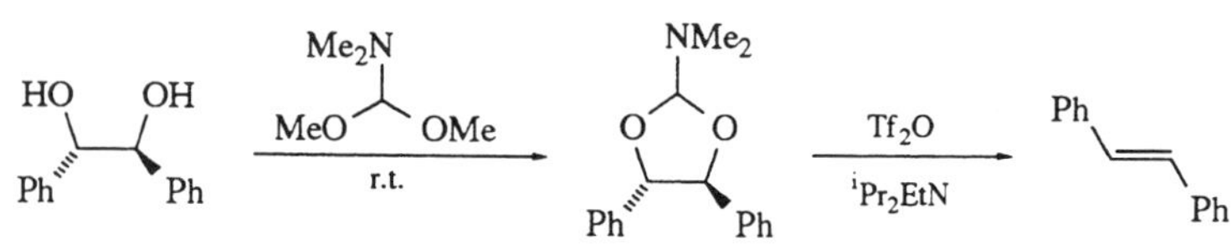

A convenient and practical procedure for the conversion of 1,2-dibromides to alkenes involves use of sodium sulfide in a water / benzene system in the presence of a phase-transfer catalyst (J. Nakayama, H. Machida, and M. Hoshino, *Tetrahedron Lett.*, 1983, **24**, 3001). The reaction exhibits good *anti*-specificity; hence, *E*-stilbene is produced when the *meso*-1,2-dibromide precursor is employed, whereas the *d,l*-dibromide affords predominantly (87:13) the *Z*-alkene. Simple treatment of 1,2-dibromides with base usually results in elimination of HBr, and so this is not generally a feasible method for reductive debromination; however, a triaminomethane, $HC(N(CH_3)Ph)_3$, has been reported to give good results for this transformation (A. Schönberg, E. Singer, and W. Stephan, *Chem. Ber.*, 1987, **120**, 1581).

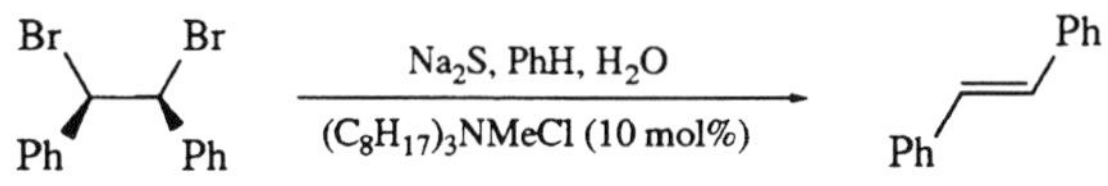

(93%)

Sodium sulfide has also been employed for the conversion of 1,2-dinitroalkanes to alkenes; either diastereomer provides almost exclusively the *E*-stilbene product and a two fold electron transfer mechanism has been proposed (L. Crombie and B.S. Roughley, *Tetrahedron Lett.*, 1986, **42**, 3147).

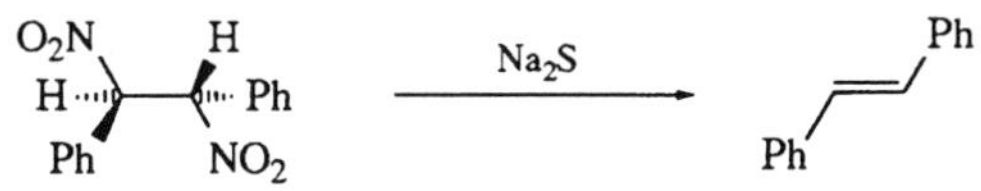

The use of manganese-lithium ate complexes for the deoxygenation of stilbene oxides has been described. High selectivity for *E*-stilbene is observed, independent of the configuration of the starting epoxide, suggesting that a freely rotating intermediate is involved (T. Kauffmann and M. Bisling, *Tetrahedron Lett.*, 1984, **25**, 293).

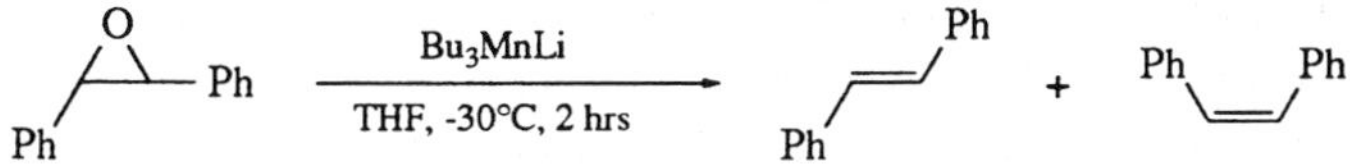

trans	100 : 0	(66%)
cis	9:1	(59%)

Reaction of epoxides with lithium tetralkylcerates results in simultaneous alkylation and deoxygenation, thus providing a synthesis of substituted alkenes (Y. Ukaji and T. Fujisawa, *Tetrahedron Lett.*, 1988, **29**, 5165). For the example shown with *trans*-stilbene oxide, high stereoselectivity is obtained for the *E*-product; however, *cis*-stilbene oxide reacts far less stereoselectively. A disadvantage of the process is that regioselectivity, which is not an issue in the case of stilbene oxide, can be poor if the epoxide is unsymmetrical. The authors propose a mechanism which involves deprotonation of the epoxide followed by ring opening to produce a carbenoid, which inserts into the carbon-cerium bond followed by β-elimination of cerium oxide.

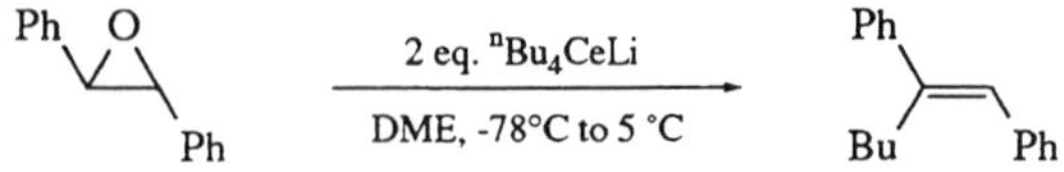

E:*Z* = 84:16 (98%)

(v) Miscellaneous methods

Pyrolysis of aryl cinnamates, examined many years ago by Skraup, has been re-investigated as a route to unsymmetrical stilbenes (W. Chamchaang, N. Chantarasiri, S. Chaona, and C. Thebtaranonth, *Tetrahedron*, 1984, **40**, 1727). Yields are reasonable, and there is no scrambling of aromatic substitution patterns.

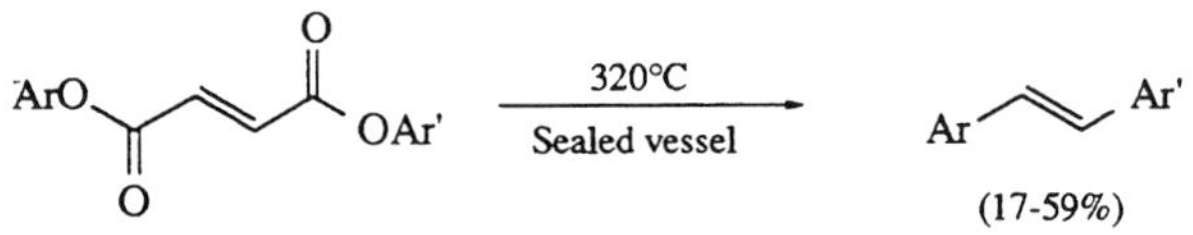

(17-59%)

The dimerisation of several aryldiazomethanes using rhodium (II) acetate or iodorhodium (III) tetraphenylporphyrin has been examined, and has been found to afford predominantly the *Z*-stilbene products (B.K.R. Shankar and H. Shechter, *Tetrahedron Lett.*, 1982, **23**, 2277). Attempted extension of the method to tetrasubstituted alkenes, using secondary aryldiazomethanes, was not successful. A mechanistic study has appeared, and the *Z*-selectivity in the reaction has been explained in terms of secondary orbital interactions (K. Ishiguro, Y. Sawaki, and H. Iwamura, *Chem. Lett.*, 1989, 1853).

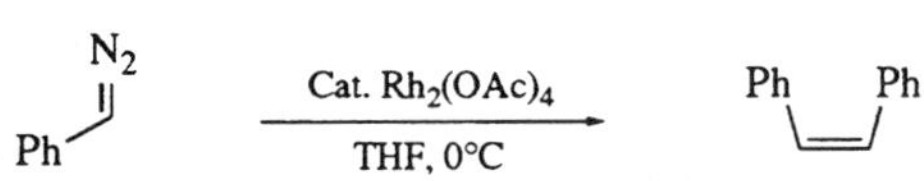

Z:*E* = 4:1 (84%)

Dithioacetals have been shown to undergo desulfurisation-dimerisation when treated with $W(CO)_6$ in boiling chlorobenzene, but no stereoselectivity is observed (L.L. Yeung, Y.C. Yip, and T.-Y. Luh, *J. Chem. Soc. Chem. Commun.*, 1987, 981). In another dimerisation process, *E*-stilbenes can be prepared by reaction of primary benzylic bromides with LDA in THF (R.S. Mali and P.G. Jagtap, *Synth. Commun.* 1991, **21**, 841).

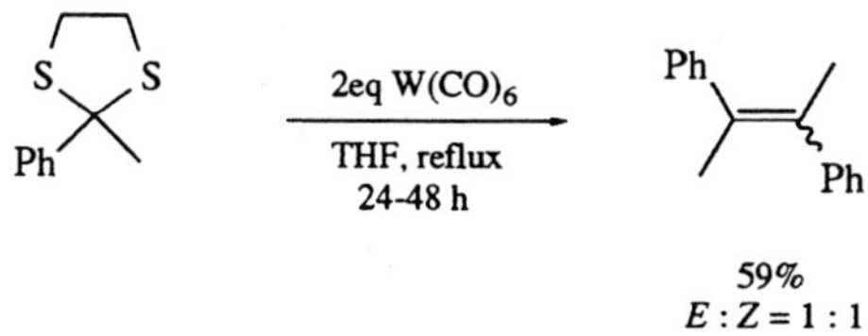

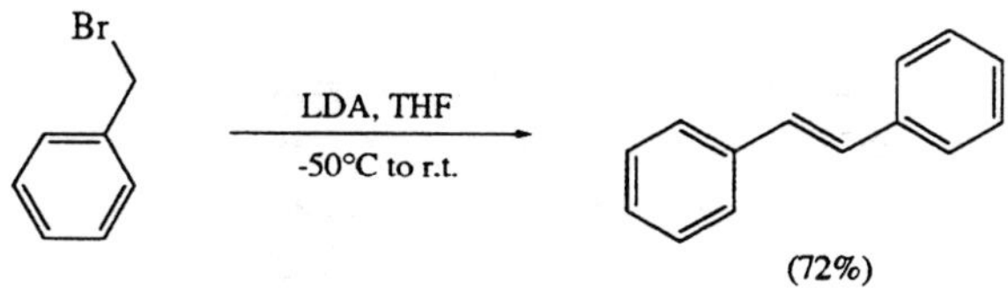

ortho-Manganated aryl carbonyl compounds are readily prepared and react with alkynes to provide a new route to indenols and indenones (N.P. Robinson, L. Main, and B.K. Nicholson, *J. Organomet. Chem.*, **364**, C37).

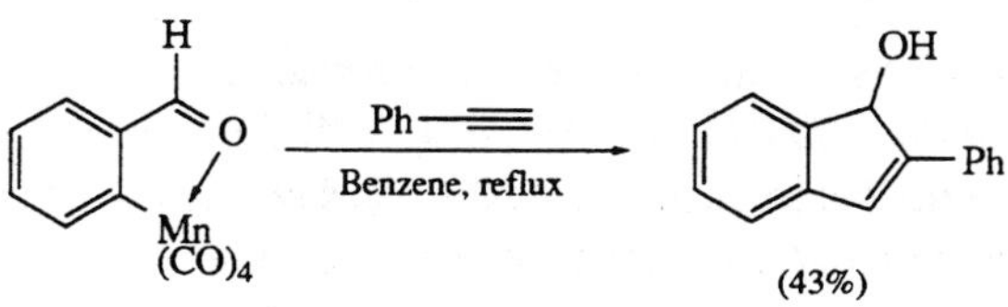

A novel route to stilbenes has been described which is unusual in that it is the aromatic ring that is formed in the transformation. An α–keto dithioketeneacetal is treated with an allyl Grignard reagent to produce a tertiary alcohol, which upon treatment with boron trifluoride etherate in benzene at reflux undergoes cationic cycloaromatisation (C.V. Askokan, H.Ila, and H. Junjappa, *Synthesis*, 1987, 284). Removal of sulfur can be achieved using nickel (II) chloride / sodium borohydride. The procedure is not applicable, however, to the synthesis of 4-methoxystilbene, the α–keto dithioketeneacetal precursor of which afforded intractable tars under Lewis-acidic conditions.

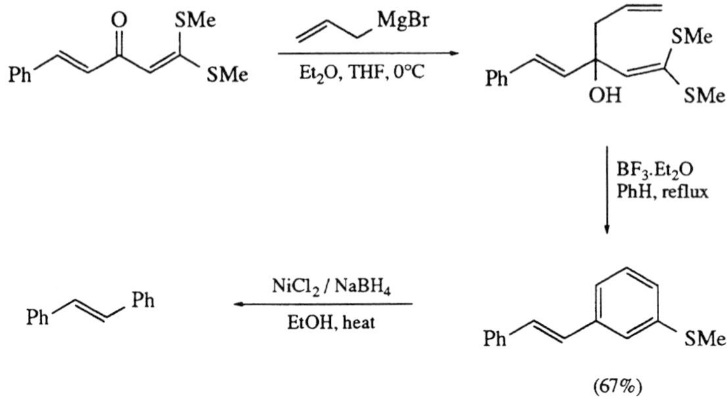

(c) Preparative methods for stilbenes with conjugated or heteroatomic substituents

A new approach to stilbene *bis*-aldehydes begins with intramolecular McMurry-type coupling of a 1,4-diketone. Double allylic bromination is followed by the key step, in which silver perchlorate is used to effect transformation to the dialdehyde (F. Toda, Y. Takehira, Y. Kataoka, K. Mori, T. Sato, and M. Segawa, *J. Chem. Soc., Chem. Commun.*, 1984, 1234). The mechanism of this reaction is not clear.

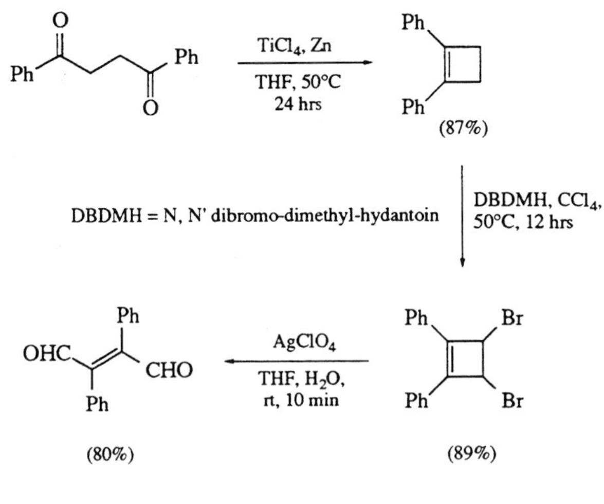

An interesting synthesis of stilbene carboxylic acid is provided by the discovery of the carbon analogue of the Hofmann reaction (C.J. Kowalski and K.W. Fields, *J. Am. Chem. Soc.*, 1982, **104**, 321); the proposed mechanism is outlined below.

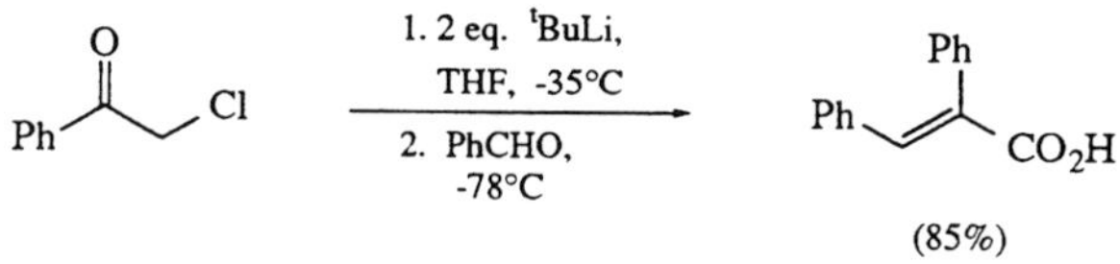

Proposed mechanism:

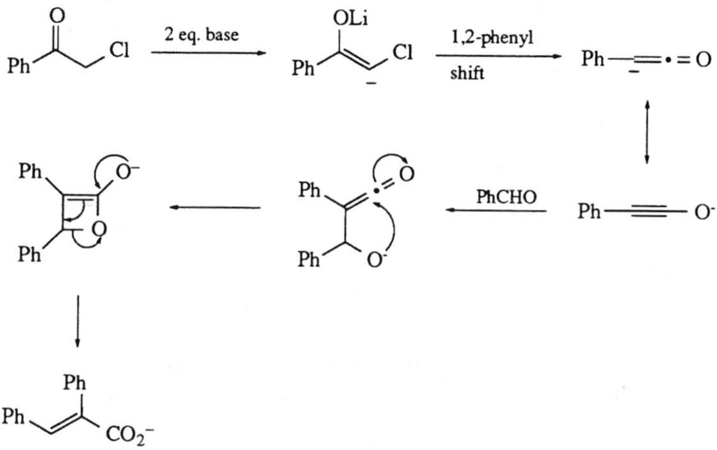

Another approach to stilbene carboxylic acid involves carbonylation of vinyl halides with octacarbonylidcobalt (M. Miura, K. Okuro, A. Hattori, and M. Nomura, *J. Chem. Soc., Perkin Trans. 1*, 1989, 73). For the example shown, the thermodynamically more stable *E*-product is obtained irrespective of the configuration of the starting vinyl bromide, suggesting ready stereochemical interconversion of the presumed vinyl cobalt intermediate.

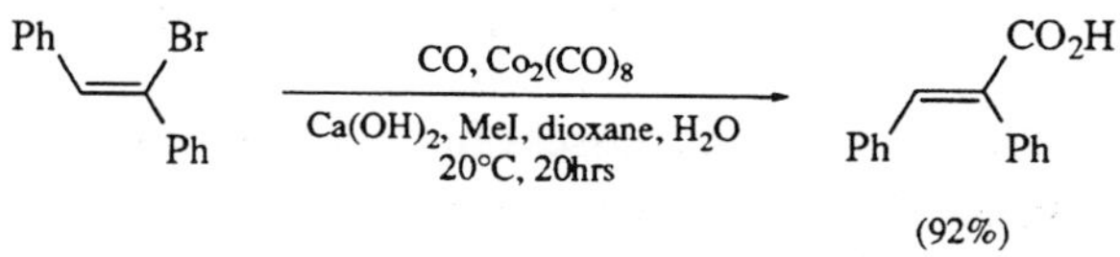

The ring opening of diphenyl-substituted cyclopropenimines with various heteroatomic nucleophiles, including alcohols and amines, has been studied (T. Eicher, R. Graf, H. Konzmann, and R. Pick, *Synthesis*, 1987, 887). The products are substituted stilbenes.

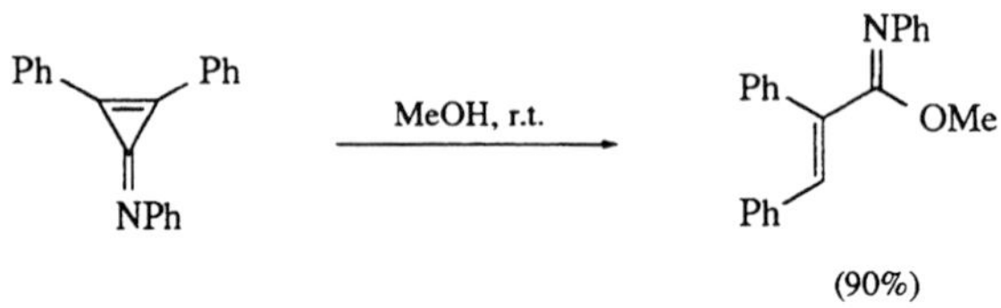

Thermal [2,1,3] elimination of methanethiol from a 1,5-benzothiazepine produces a benzothiazole-substituted 1,2 diphenyl alkene (G. Kaupp, *Chem. Ber.*, 1984, **117**, 1643); it is not known whether this process is general.

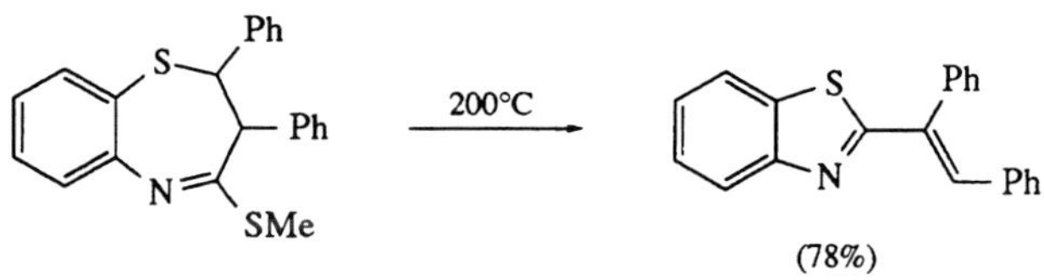

A novel diphenyl-substituted spiro-thiete is the product of [2+2] photochemical cycloaddition between diphenylacetylene and *N*-methylthiophthalimide (J.D. Coyle, P.A. Rapley, J. Kamphuis, and H.J.T. Bos *Tetrahedron Lett.* 1985, **26**, 2249). ^{13}C NMR demonstrated that the product was not in equilibrium with the isomeric ring-opened unsaturated thioketone.

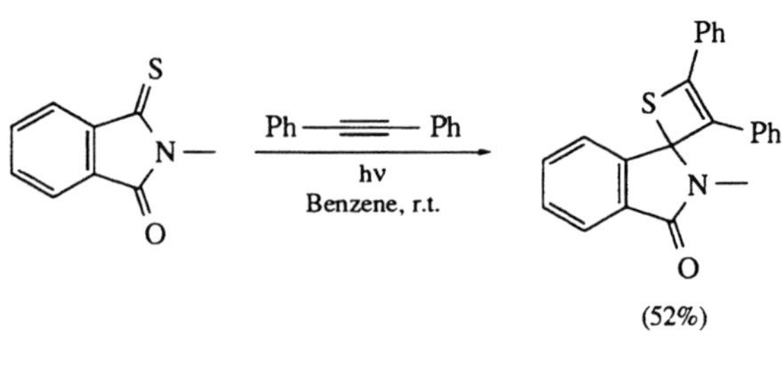

Some 1,4-oxathiin derivatives possess fungicidal properties and methods for their synthesis are therefore of interest. Hemithioacetals derived from αbromoketones undergo base-mediated rearrangement with loss of bromine to form this ring system (J. Mattay and C. Dittmer, *J. Org. Chem.*, 1986, **51**, 1894).

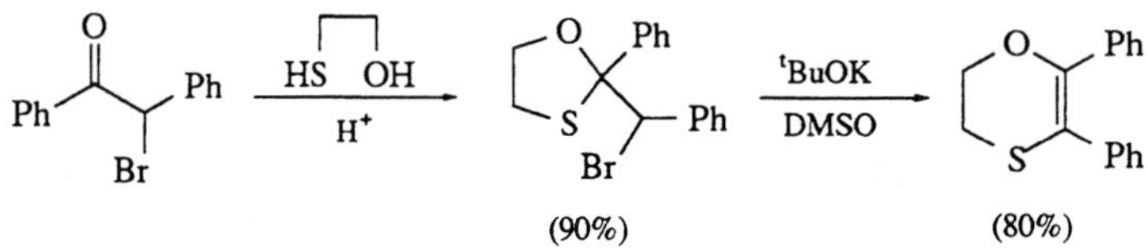

Addition of iodine in HIO_3 / MeOH to a 1-phenyl propargyl alcohol results in an α-iodo vinyl cation which undergoes a 1,2-phenyl shift to give a keto-substituted stilbene (J.J. Janas, E.T. Asirvatham, and E. McNelis, *Tetrahedron Lett.*, 1985, **26**, 1967). In the absence of iodine, the only product seen is that of simple substitution of the hydroxyl group for methoxy.

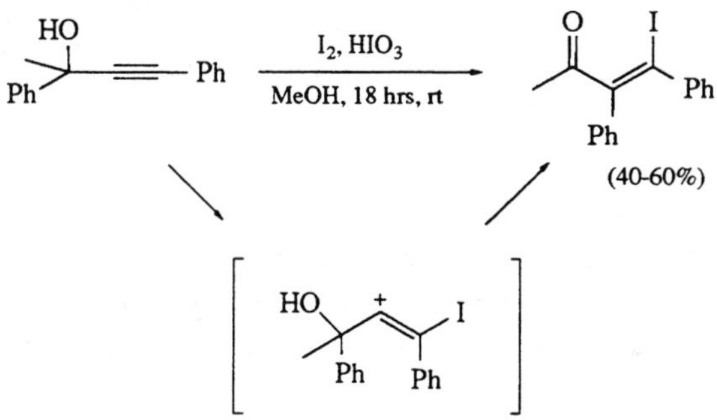

4. *Diphenyl-acetylenes and -allenes*

Palladium-catalysed coupling of aryl bromides to terminal acetylenes is nicely exemplified by a step in Vollhardt's terphenylene synthesis (R. Diercks and K.P.C. Vollhardt, *Angew. Chem. Int. Ed. Engl.*, 1986, **25**, 266).

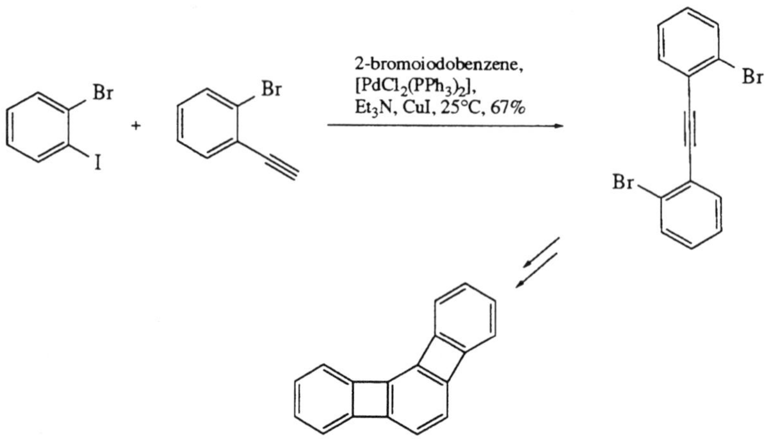

It has been shown that palladium can also insert into the carbon-oxygen bond of aryl fluoroalkane sulfonates, prepared *in situ* from phenols. Again, coupling to terminal alkynes is possible (Q.-Y. Chen and Y.-B. He, *Synthesis*, 1988, 896). The method was applied to the alkynylation of oestrogen.

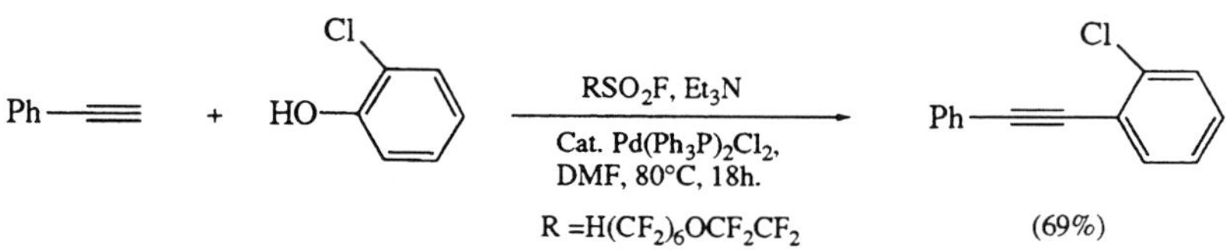

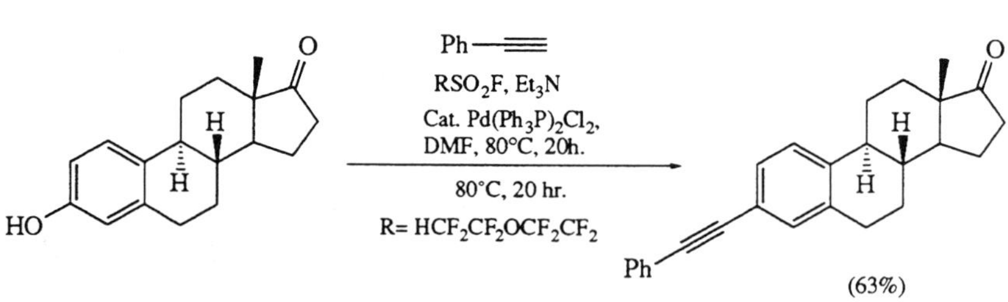

Ando and co-workers have reported the use of potassium fluoride on alumina for several base-mediated transformations. *meso*-Stilbene dibromide can be converted to phenylacetylene in good yield, although side products may be obtained at lower temperatures and in different solvents (J.

Yamawaki, T. Kawate, T. Ando, and T. Hanafusa, *Bull. Chem. Soc. Japan*, 1983, **56**, 1885).

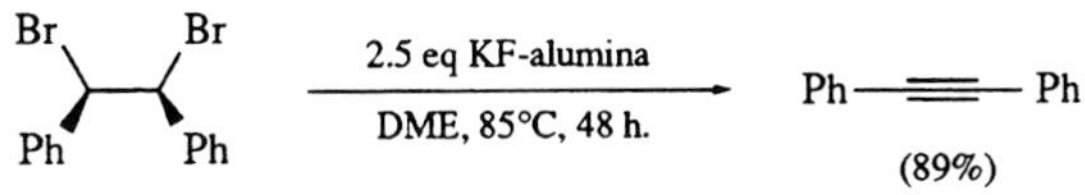

Heating tolane (perchlorodiphenylethene) with copper powder in a sealed vessel has been shown to allow the preparation of perchlorodiphenylacetylene (M. Ballester, J. Castañer, J. Riera, and O. Armet, *J. Org. Chem.*, 1986, **51**, 1100). Isolation of the product in good yield, however, required continuous extraction with chloroform for 4 days.

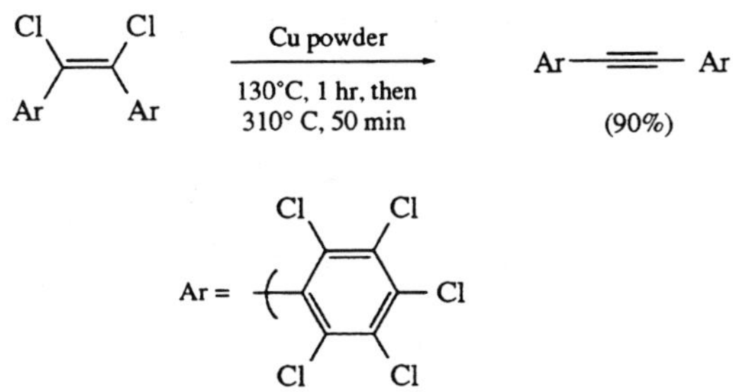

In attempting to develop conditions for halogen exchange at vinylic carbon, Suzuki and co-workers discovered that 1,2-dibromostilbene could be debrominated cleanly to diphenylacetylene upon reaction with potassium iodide / copper (I) iodide in HMPT (H. Suzuki, M. Aihara, H. Yamamoto, Y. Takamoto, and T. Ogawa, *Synthesis*, 1988, 236). However, reaction under the same conditions of alkenes bearing a hydrogen substituent, such as β,β-dibromostyrene, resulted in elimination of HBr followed by dimerisation, affording 1,4-diphenylbutadiyne.

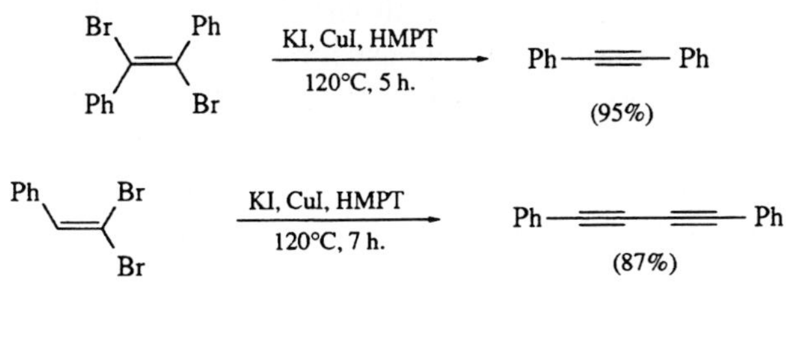

Various copper (II) salts have been shown to effect the decomposition of vicinal hydrazones to the corresponding acetylenes (J. Tsuji, H. Kezuka, Y. Toshida, H. Takayanagi, and K. Yamamoto, *Tetrahedron*, 1983, **39**, 3279). The process can be made catalytic in copper (II) in the presence of oxygen to re-oxidise the copper (I) salts produced.

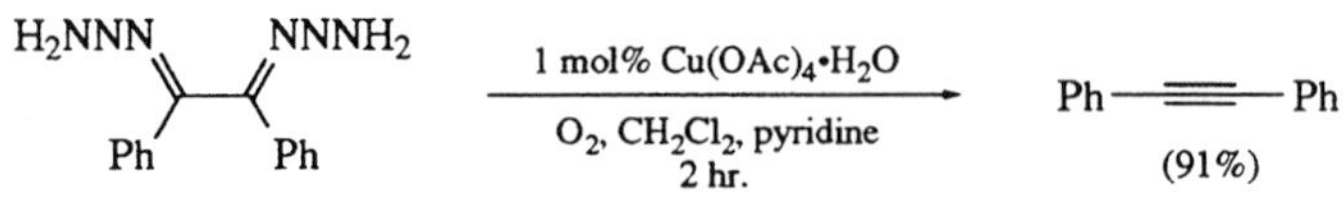

Reductive dimerisation coupled with acylation consumes four equivalents of benzoyl chloride to provide the enediol ester (**13**). Further reduction effects conversion to diphenylacetylene (L. Horner and K. Dickerhof, *Chem. Ber.*, 1983, **116**, 1615). The process was shown to be successful for several other substituted benzoyl chlorides.

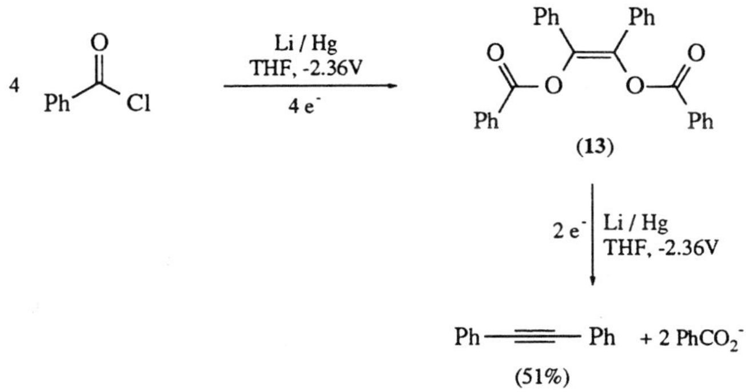

Hendrickson has developed a useful method for the dehydration of benzyl ketones. Reaction between triphenylphosphine oxide and trifluoromethanesulfonic anhydride produces a salt *in situ* which reacts with the ketone to give an enol phosphonium derivative; subsequent loss of triphenylphosphine oxide yields the alkyne (J.B. Hendrickson and M.S. Hussoin, *Synthesis*, 1989, 217). The usual reagent system of choice, phosphorus pentachloride / pyridine, often affords vinyl halide by products.

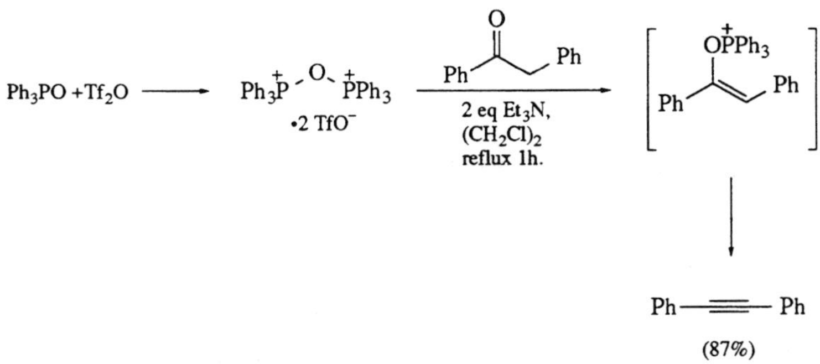

A useful and general method for the synthesis of allenes has been described, utilising the boron-analogue of the Wittig reaction (A. Pelter, K. Smith, and K.D. Jones, *J. Chem. Soc., Perkin Trans. 1*, 1992, 747). The requisite boron-stabilised alkenyl carbanions are obtained from stannyl acetylenes by hydroboration followed by transmetallation.

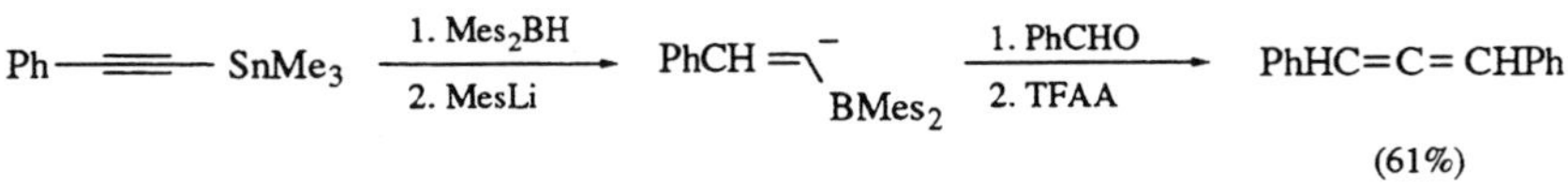

Mes = mesityl (2,4,6-trimethylphenyl)

The stereochemistry of the palladium (0)-catalysed phenylation of optically active 1-haloallenes has been studied in detail by Elsevier and Vermeer (C.J. Elsevier and P. Vermeer, *J. Org. Chem.*, 1985, **50**, 3042). For chloro- and bromo-allenes, a moderate to high degree of inversion was observed, whereas the corresponding iodides reacted with retention. The authors rationalised these results in terms of two competing processes – insertion of palladium into the carbon-halogen bond with retention, and a palladium (0)-induced *anti*-S_N2'-like reaction followed by suprafacial [1,3]-shift of palladium leading to overall inversion.

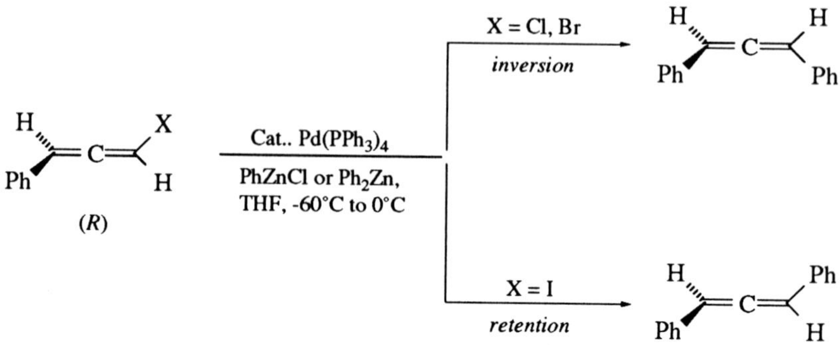
X = Cl, Br
inversion
Cat.. Pd(PPh3)4
PhZnCl or Ph2Zn,
THF, -60°C to 0°C
(R)
X = I
retention

Guide to the Index

This index is constructed in a similar manner to the volume indexes of the first edition of the Chemistry of Carbon Compounds. However, to make the index easier to use, more descriptive entries have been made for the commonly occurring individual, and groups of chemicals.

The indexes cover primarily the chemical compounds mentioned in the text, and also include reactions and techniques, where named, and some sources of chemical compounds such as plant and animal species, oils, etc.

Chemical compounds have been indexed alphabetically under the names used by authors, editing being restricted to ensuring uniformity of entries under the same heading. In view of the alternative nomenclature that can often be used, a limited amount of cross-referencing has been done where it is considered to be helpful, but attention is particularly drawn to Convention 2 below.

For this and the succeeding volumes, the indexing conventions listed below have been adopted.

1. Alphabetisation

(a) A letter by letter alphabetical sequence is followed for entries, firstly for the main entry, followed by the descriptive entry.

(b) The following prefixes have not been counted for alphabetising:

n-	*o-*	*as-*	*meso-*	*C-*	*E-*
	m-	*sym-*	*cis-*	*O-*	*Z-*
	p-	*gem-*	*trans-*	*N-*	
	vic-			*S-*	
		lin-		*Bz-*	
				Py-	

Some prefixes and numbering have been omitted in the index, where they do not usefully contribute to the reference.

(c) The following prefixes have been alphabetised:

Allo	Epi	Neo
Anti	Hetero	Nor
Bis	Homo	Pseudo
Cyclo	Iso	

2. Cross references

In view of the many alternative trivial and systematic names for chemical compounds, the indexes should be searched under any alternative names which may be indicated in the main body of the text. Only a limited amount of cross-referencing has been carried out, where it is considered that it would be helpful to the user.

3. Derivatives

Simple derivatives are not normally indexed if they follow in the same short section of the text.

4. Collective and plural entries

In place of "– derivatives" the plural entry has normally been used. Plural entries have occasionally been used where compounds of the same name but differing numbering appear in the same section of the text.

5. Main entries

The main entry of the more common individual compounds is indicated by heavy type. Multiple entries, such as headings and sub-headings over several pages are shown by "–", e.g., 67–74, 137–139, etc.

Index